small arms survey 2012

moving targets

CAMBRIDGE UNIVERSITY PRESS

Cambridge, New York, Melbourne, Madrid, Cape Town,
Singapore, São Paulo, Delhi, Mexico City

Cambridge University Press
The Edinburgh Building, Cambridge CB2 8RU, UK

Published in the United States of America by Cambridge University Press, New York

www.cambridge.org
Information on this title: www.cambridge.org/9780521146876

© Small Arms Survey, Graduate Institute of International
and Development Studies, Geneva 2012

This publication is in copyright. Subject to statutory exception
and to the provisions of relevant collective licensing agreements,
no reproduction of any part may take place without the written
permission of Cambridge University Press.

First published 2012

Printed in the United Kingdom by MPG Books Group

A catalogue record for this publication is available from the British Library

Library of Congress Cataloguing in Publication data

ISBN 978-0-521-19714-4 Hardback
ISBN 978-0-521-14687-6 Paperback

Cambridge University Press has no responsibility for the persistence or
accuracy of URLs for external or third-party Internet websites referred to in
this publication, and does not guarantee that any content on such websites is,
or will remain, accurate or appropriate.

FOREWORD

Illicit trafficking in small arms remains a deadly challenge for international peace and security. Across the world, violence carried out with small arms and light weapons undermines our efforts to promote sustainable development, protect human rights, build safer cities, improve public health, and help countries emerge from conflict. The casualties include children, the stability of entire societies, and public confidence in institutions. The opportunity costs—people whose lives have been cut short; countries made fragile and unattractive for investment—are equally profound.

The publication of this edition of the *Small Arms Survey* is timely. This year, the United Nations will convene the Second Review Conference of the decade-old UN Programme of Action on Small Arms and Light Weapons, giving Member States an opportunity to review progress and ensure that this framework continues to guide international action.

Like previous editions, the *Small Arms Survey 2012: Moving Targets* provides original research and analysis that can improve policy-making. It can also contribute to the development of measurable goals for small arms control—an objective I articulated most recently in my 2011 report on small arms to the UN Security Council.

I commend the *Small Arms Survey 2012* as an authoritative volume to Member States and all stakeholders committed to reducing the devastating toll that small arms inflict on individuals, communities, and entire countries and regions. Let us work together to solve the big problems caused by small arms.

—**Ban Ki-moon**
Secretary-General of the United Nations

CONTENTS

About the Small Arms Survey vi
Notes to readers vi
Acknowledgements vii
Introduction 1

Chapter 1. A Fatal Relationship: Guns and Deaths in Latin America and the Caribbean
Introduction 9
Firearm homicides: an overview 10
Unpacking patterns and trends 15
Types and origins of firearms used in homicides 25
Mapping an agenda for future research 32
Conclusion 34

Chapter 2. When Business Gets Bloody: State Policy and Drug Violence
Introduction 41
Drug violence in the Americas: actors, violence, and state responses 42
Mexico 51
Central America's Northern Triangle 60
Rio de Janeiro 64
Conclusion 72

Chapter 3. A Matter of Survival: Non-lethal Firearm Violence
Introduction 79
Non-lethal gun violence in perspective 80
Assessing the scale and scope of non-lethal firearm violence 84
Challenges to non-fatal injury surveillance 94
Conclusion 101

Chapter 4. Blue Skies and Dark Clouds: Kazakhstan and Small Arms
Introduction 107
A safe haven under threat? 109
Civilian small arms: under control, in demand 116
Still secret: small arms and the state 128
Conclusion 139

Chapter 5. Between State and Non-state: Somaliland's Emerging Security Order
Introduction 147
Setting the stage: small arms and (in)security in Somaliland 148
Political violence 156
Criminal violence and local (in)security 159
Communal violence 166
Conclusion 168

Photo Essay. Troubled Waters: Somali Piracy

Chapter 6. Escalation at Sea: Somali Piracy and Private Security Companies
Introduction 191
Somali piracy in context 192
Armed guards at sea 204
Conclusion 212

Chapter 7. Precedent in the Making: The UN Meeting of Governmental Experts
Introduction 219
The MGE: a short history 220
The MGE 223
Marking 223
Record-keeping 227
Cooperation in tracing 229
National frameworks 231
Regional cooperation 233
International assistance and capacity-building 234
Conclusion 236

Chapter 8. Piece by Piece: Authorized Transfers of Parts and Accessories
Introduction 241
Terms and definitions 243
Estimating the international small arms trade 247
Transfers of parts for small arms and light weapons 251
Accessories for small arms and light weapons 261
Conclusion 275

Chapter 9. Point by Point: Trends in Transparency
Introduction 283
The Transparency Barometer 284
Unpacking the seven parameters: national practices 294
Conclusion 307

Chapter 10. Surveying the Battlefield: Illicit Arms in Afghanistan, Iraq, and Somalia
Introduction 313
Use of terms 314
War weapons: unpacking the data 315
Illicit small arms and light weapons in Iraq 317
Illicit small arms and light weapons in Afghanistan 330
Illicit small arms and Somalia 336
Conclusion 348

Index

ABOUT THE SMALL ARMS SURVEY

The Small Arms Survey is an independent research project located at the Graduate Institute of International and Development Studies in Geneva, Switzerland. Established in 1999, the project is supported by the Swiss Federal Department of Foreign Affairs and current contributions from the Governments of Australia, Belgium, Canada, Denmark, Finland, Germany, the Netherlands, Norway, Sweden, the United Kingdom, and the United States. The Survey is grateful for past support received from the Governments of France, New Zealand, and Spain. The Survey also wishes to acknowledge the financial assistance it has received over the years from foundations and many bodies within the UN System.

The objectives of the Small Arms Survey are: to be the principal source of public information on all aspects of small arms and armed violence; to serve as a resource centre for governments, policy-makers, researchers, and activists; to monitor national and international initiatives (governmental and non-governmental) on small arms; to support efforts to address the effects of small arms proliferation and misuse; and to act as a clearinghouse for the sharing of information and the dissemination of best practices. The Survey also sponsors field research and information-gathering efforts, especially in affected states and regions. The project has an international staff with expertise in security studies, political science, law, economics, development studies, sociology, and criminology, and collaborates with a network of researchers, partner institutions, non-governmental organizations, and governments in more than 50 countries.

NOTES TO READERS

Abbreviations: Lists of abbreviations can be found at the end of each chapter.

Chapter cross-referencing: Chapter cross-references are fully capitalized in brackets throughout the book. One example appears in Chapter 2, which reviews recent empirical trends and theoretical explanations of drug violence in Latin America: 'Trends in homicide rates show a similar pattern, with the Northern Triangle countries suffering from rates two to three times that of Panama, Nicaragua, and Costa Rica—and Belize occupying a middle position (LATIN AMERICA AND THE CARIBBEAN).'

Exchange rates: All monetary values are expressed in current US dollars (USD). When other currencies are also cited, unless otherwise indicated, they are converted to USD using the 365-day average exchange rate for the period 1 September 2010 to 31 August 2011.

Small Arms Survey: The plain text—Small Arms Survey—is used to indicate the overall project and its activities, while the italicized version—*Small Arms Survey*—refers to the publication. The *Survey*, appearing italicized, relates generally to past and future editions.

Small Arms Survey
Graduate Institute of International and Development Studies
47 Avenue Blanc, 1202 Geneva, Switzerland

t +41 22 908 5777 **f** +41 22 732 2738

e sas@smallarmssurvey.org **w** www.smallarmssurvey.org

ACKNOWLEDGEMENTS

This is the 12th edition of the *Small Arms Survey*. Like previous editions, it is a collective product of the staff of the Small Arms Survey project, based at the Graduate Institute of International and Development Studies in Geneva, Switzerland, with support from partners. Numerous researchers in Geneva and around the world have contributed to this volume, and it has benefited from the input and advice of government officials, advocates, experts, and colleagues from the small arms research community and beyond.

The principal chapter authors were assisted by in-house and external contributors, who are acknowledged in the relevant chapters. In addition, chapter reviews were provided by: Marcelo Aebi, James Bevan, Catrien Bijleveld, Jurgen Brauer, Rustam Burnashev, Alex Butchart, Irina Chernykh, Helen Close, Neil Corney, Angelica Duran-Martinez, Diego Fleitas, Stephanie Gimenez Stahlberg, William Godnick, Cornelius Graubner, Laura Hammond, Tracy Hite, Bruce Hoffman, Paul Holtom, Richard Jones, Michael Knights, Murat Laumulin, Jonah Leff, Stuart Maslen, Arturo Matute Rodríguez, Ken Menkhaus, Malgorzata Polanska, Daniël Prins, Jorge Restrepo, Margaret Shaw, Anthony Simpson, Murray Smith, Tim Stear, and Joanna Wright.

Small Arms Survey 2012

Editors	Glenn McDonald, Emile LeBrun, Eric G. Berman, and Keith Krause
Coordinator	Glenn McDonald
Publications Manager	Alessandra Allen
Designer	Richard Jones, Exile: Design & Editorial Services
Cartographer	Jillian Luff, MAP*grafix*
Copy-editor	Tania Inowlocki
Proofreader	Donald Strachan

Principal chapter authors

Introduction	Glenn McDonald and Emile LeBrun
Chapter 1	Elisabeth Gilgen
Chapter 2	Benjamin Lessing
Chapter 3	Anna Alvazzi del Frate
Chapter 4	Nicolas Florquin, Dauren Aben, and Takhmina Karimova
Chapter 5	Dominik Balthasar and Janis Grzybowski
Photo essay	Alessandra Allen and Nicolas Florquin
Chapter 6	Nicolas Florquin
Chapter 7	Glenn McDonald
Chapter 8	Janis Grzybowski, Nicolas Marsh, and Matt Schroeder
Chapter 9	Jasna Lazarevic
Chapter 10	Matt Schroeder and Benjamin King

Eric G. Berman, Keith Krause, Emile LeBrun, and Glenn McDonald were responsible for the overall planning and organization of this edition. Alessandra Allen managed the editing and production of the *Survey*. Tania Inowlocki copy-edited the book; Jillian Luff produced the maps; Richard Jones provided the design and the layout; Frank Benno Junghanns laid out the photo essay; Donald Strachan proofread the *Survey*; and Margaret Binns compiled the index. Daly Design created the illustrations in Chapter 8. John Haslam, Carrie Parkinson, and Daniel Dunlavey of Cambridge University Press provided support throughout the production of the *Survey*. Natacha Cornaz, Sarah Hoban, Chelsea Kelly, Natalia Micevic, Mihaela Racovita, and Jordan Shepherd fact-checked the chapters. Olivia Denonville and Martin Field helped with photo research. Benjamin Pougnier, Cristina Tavares de Bastos, and Carole Touraine provided administrative support.

The project also benefited from the support of the Graduate Institute of International and Development Studies, in particular Philippe Burrin and Monique Nendaz.

We are extremely grateful to the Swiss government—especially to the Department for Foreign Affairs and to the Swiss Development Cooperation—for its generous financial and overall support of the Small Arms Survey project, in particular Serge Bavaud, Siro Beltrametti, Erwin Bollinger, Prasenjit Chaudhuri, Alexandre Fasel, Thomas Greminger, Jürg Lauber, Armin Rieser, Paul Seger, Julien Thöni, Claude Wild, and Reto Wollenmann. Financial support for the project was also provided by the Governments of Australia, Belgium, Canada, Denmark, Finland, Germany, the Netherlands, Norway, Sweden, the United Kingdom, and the United States.

The project further benefits from the support of international agencies, including the International Committee of the Red Cross, the UN Development Programme, the UN High Commissioner for Refugees, the UN Office for the Coordination of Humanitarian Affairs, the UN Office for Disarmament Affairs, the UN Institute for Disarmament Research, the UN Office on Drugs and Crime, and the World Health Organization.

In Geneva, the project has benefited from the expertise of: Aino Askgaard, Silvia Cattaneo, Pieter van Donkersgoed, Philip Kimpton, Patrick Mc Carthy, Suneeta Millington, and Tarja Pesämaa.

Beyond Geneva, we also received support from a number of colleagues. In addition to those mentioned above, and in specific chapters, we would like to thank: Philip Alpers, Steven Baxter, Wolfgang Bindseil, Andrew Cooper, Steven Costner, Hanne B. Elmelund Gam, Gillian Goh, Nicholas Iaiennaro, Roy Isbister, Kimberly-Lin Joslin, Guy Lamb, Chris Loughran, Jim McLay, Daniël Prins, Robert Muggah, Joy Ogwu, Steve Priestley, Anthony Simpson, Alison August Treppel, and Dawn Wood-Memic.

Our sincere thanks go out to many other individuals (who remain unnamed) for their continuing support of the project. Our apologies to anyone we have failed to mention.

—**Keith Krause**, Programme Director
Eric G. Berman, Managing Director
Anna Alvazzi del Frate, Research Director

A suspected drug gang member takes position during an operation at Grota favela, Rio de Janeiro, November 2010. © Sergio Moraes/Reuters

Introduction

INTRODUCTION

In July 2001, UN member states met in New York to adopt the Programme of Action on Small Arms (PoA) and, more fundamentally, accelerate national, regional, and international efforts to tackle the small arms problem. The PoA focuses on arms control; it seeks to curb small arms proliferation by reinforcing controls over manufacture, international transfer, storage, and final disposal. A spin-off measure, the International Tracing Instrument, adopted in December 2005, is designed to strengthen weapons marking, record-keeping, and tracing. The UN small arms process also laid the groundwork for a parallel effort, centred on the Geneva Declaration on Armed Violence and Development (adopted in June 2006), aimed at enhancing our understanding of, and ability to respond to, the dynamics of armed violence.

Over the past decade, our knowledge of a range of small arms issues—including the scope of the arms trade and the negative impact of small arms in conflict and non-conflict settings—has grown considerably. But what of international efforts to combat small arms proliferation and armed violence?

In August–September 2012, UN member states will meet in New York for the PoA's Second Review Conference. States are asked to 'review progress made' in the PoA's implementation, but the reality is that the tools that would allow the UN membership to make such an assessment have yet to be developed. There is little doubt that the PoA and other similar measures, including many at the regional level, have spurred a wide range of activity, from improved marking practices to the destruction of surplus stocks. The Geneva Declaration's focus on measurability and programming has similarly catalysed a series of practical measures aimed at preventing and reducing armed violence.

But, in both cases, the big picture remains hazy. The broad outlines of action can be discerned, but not many of the details of that activity. Moreover, we do not yet know what impact small arms measures, when implemented, have on weapons proliferation or on individual or community security.

The first edition of the *Small Arms Survey*, launched in 2001 at the time of the PoA's adoption, reviews what was then known about small arms supply, control efforts, and effects. Eleven years later, the *Small Arms Survey 2012: Moving Targets* attempts something similar. Drawing on existing sources of information—but also new and previously untapped sources—it presents several critical trends in small arms supply and misuse. In essence, rather than presenting an updated snapshot of the small arms situation, the 2012 *Survey* focuses on the question of change.

CHAPTER HIGHLIGHTS

Reports of increasing rates of violence in Latin America and the Caribbean regularly make the headlines, but what are the trends exactly? Two chapters in this edition examine this region, where the proportion of homicides committed with firearms is higher than the global average in 21 of 23 reviewed countries. As Chapter 1 illustrates, a strong correlation appears to exist between homicide and firearm homicide rates; in Latin American countries that have a high overall homicide rate, firearms are responsible for a higher ratio of the deaths. While this relationship has also been observed in other regions, this chapter presents the first in-depth review of its manifestations in Latin America and the Caribbean.

In Chapter 2, an analysis of time-series data demonstrates how drug-related violence can push countries that are ostensibly at peace into something resembling civil war. In Mexico, major drug-trafficking organizations have responded in kind to President Felipe Calderón's 'declaration of war' against them. While the roles and motivations of the cartels are clearly important factors in drug violence, they do not, in and of themselves, explain the recent explosion of violence in Mexico. Instead, it appears that the government's strategy of blanket suppression has triggered the escalation in cartel-led violence.

Perhaps because it is so obviously disruptive to society, homicide is usually captured in state statistics; policy-makers and many interest groups measure gun violence in terms of lives lost. Non-lethal violence is seldom routinely monitored and is less well understood. But, as Chapter 3 indicates, it outpaces lethal violence at the global level by a wide margin, leaving huge physical, psychological, economic, and social costs in its wake. A plausible estimate of three non-fatal gun assaults for every fatal one would mean that at least half a million non-fatal injuries are sustained as a result of intentional firearm violence each year worldwide. For now, however, such estimates have a large margin of error; without improved, more systematic surveillance of non-lethal violence, our picture of the global burden of armed violence will remain incomplete.

As our understanding of armed violence, including associated trends, advances, so too does our knowledge of the global trade in small arms and light weapons, their parts, accessories, and ammunition. In 2009, the Small Arms Survey began a four-year project to develop a more precise estimate of the value of the global authorized trade in these weapons, previously calculated at around USD 4 billion per year. The fourth and final phase of the project, presented in Chapter 8, indicates that the current figure is at least twice that—USD 8.5 billion or more. Part of this increase stems from better information and the use of rigorous modelling techniques tailored to specific components of the trade (small arms, light weapons, parts, accessories, and ammunition). Another part, however, is the result of absolute increases in the value of small arms transfers, including, it appears, the importation of firearms by US civilians and the acquisition of a range of weaponry by armed forces fighting in Afghanistan and Iraq.

Chapter 8 notes that many gaps remain in our knowledge of the authorized trade. Estimates, however robust, are no substitute for transparent reporting by states on arms exports. Chapter 9, on transparency, takes stock of ten years of arms export reporting, highlighting the significant improvements some states have made in their reporting, but also the fact that the average level of transparency among the world's 52 major exporters of small arms remains poor—averaging less than half of all points available in the 2012 Transparency Barometer.

These gains in our understanding of authorized small arms transfers, however partial, appear almost monumental compared to the thicket of uncertainty that the illicit trade represents. As the Survey wraps up its multi-year authorized transfers project, it begins a new one designed to increase our knowledge of illicit small arms, a topic of particular relevance to the PoA. The first phase of the project, presented in Chapter 10, focuses on three war zones: Afghanistan, Iraq, and Somalia. Although a lack of data precludes a definitive mapping of insurgent arms in these countries, multiple sources of information nevertheless tend to the same conclusion in all three cases: non-state armed groups are using older-generation weapons. To a great extent, the legacy of state collapse and plundered stockpiles, rather than new supply, appears to determine the arsenals of today's insurgents. In fact, their relative lack of advanced weaponry suggests that arms control efforts of the past decade are having some effect. Yet it is equally clear that once governments lose control of their stockpiles, the consequences can be felt for decades.

The looting of government arsenals during the Somali civil war is a case in point. In fact, it provides the backdrop, not only for a review of insurgent weapons in Somalia in Chapter 10, but also for two stand-alone chapters relating to the country. Chapter 5, on the autonomous region of Somaliland, examines the consolidation of state authority in the territory and other factors that have led to far better levels of security than are found elsewhere in present-day Somalia. Chapter 6 discusses one of the spillover effects of persistent insecurity and underdevelopment in north-eastern Somalia, namely Somali pirate attacks in the Gulf of Aden, the Indian Ocean, and the Red Sea. This chapter reviews the protective measures undertaken by shipping companies and governments, most notably the deployment of international naval forces and the increased use of private security guards. As discussed in both Chapter 6 and the accompanying photo essay, the greater presence of armed protectors, coupled with their actual use of force in some cases, appears to have secured a drop in the number of successful pirate attacks, yet also seems to be inducing pirates to adopt more aggressive tactics.

Many of the situations portrayed in the *Small Arms Survey 2012* are in flux. So, too, is the UN Programme of Action. Chapter 7 reviews the most recent event in the PoA calendar, the Open-ended Meeting of Governmental Experts, held in May 2011. Although, as the chapter notes, the meeting does not tell us much about the state of national implementation of the International Tracing Instrument, the information states shared on obstacles to implementation and means of overcoming them does represent a crucial first step towards strengthened implementation over the longer term. No final answers, then, but a process that, fundamentally, remains a work in progress.

The *Small Arms Survey 2012* features three sections: a thematic section highlighting trends in armed violence; two country-specific studies; and a final section that focuses on weapons and markets. More information on the chapters in each of these sections follows.

Trends section

Chapter 1 (Latin America and the Caribbean): Many countries in Latin America and the Caribbean not only experience significantly higher homicide rates than other parts of the world, but also have higher proportions of firearm homicides than the global average of 42 per cent. Firearms were used in an average of 70 per cent of homicides in Central America, in 61 per cent in the Caribbean, and in 60 per cent in South America. At the same time, there are significant differences among countries in the region.

This chapter sheds light on patterns and trends of homicides and firearm homicides in Latin America and the Caribbean. It shows that higher overall homicide rates are frequently linked to higher proportions of firearm homicides. In addition, there appears to be a link between rising homicide rates over time and an increase in the proportion of firearm homicides. While it is unclear whether firearm homicides are driving overall homicide rates or vice versa, there is clearly a relationship between the two. This chapter discusses numerous factors that may explain why the relationship is especially pronounced in many countries in Latin America and the Caribbean.

Chapter 2 (Drug violence): In December 2006, the Mexican president called in the army to wage an all-out war on the country's drug cartels. While the action has led to the capture and death of numerous cartel leaders, it has not succeeded in destroying the cartels themselves. Worse, it is has precipitated major counter-offensives against the army as well as inter-cartel violence that have left some 47,000 Mexicans dead. State pressure also appears to be leading some cartels to migrate south into Central America, where they threaten to alter the dynamics of the drug trade in El Salvador, Guatemala, and Honduras.

The Mexican response differs significantly from the one recently taken by officials in Brazil. There, in an effort to wrest control of Rio de Janeiro's *favelas* from prison-based drug syndicates, state security forces began a new programme in 2008 to retake and then occupy favelas with long-term community-oriented police forces. The programme is unusual in that it moves against only the most violent syndicates and aims not to eradicate the illicit drug trade itself, but instead to reduce the worst drug-related violence and re-establish state authority.

This chapter reviews various facets of drug violence between organized actors in Mexico and Brazil. It finds that the onset and intensity of systemic drug violence on the scale seen in Mexico and Rio de Janeiro are highly variable and sensitive, not only to drug trafficking and market structure, but also to state anti-narcotics policy and enforcement. In contrast to the Mexican army's attempt to suppress and even destroy its domestic cartels, Rio's multi-faceted programme appears to be successfully 'pacifying' its syndicates, although it is too early to tell whether the reduction in levels of armed violence can be sustained.

Chapter 3 (Non-lethal violence): While most state and local authorities keep records on the number and characteristics of homicides, including firearm homicides, very few do so for violence that does not result in death. Yet non-fatal intentional firearm injuries far outnumber gun homicides. This chapter provides an overview of the available data on the incidence of non-lethal firearm violence in non-conflict settings, including estimates for countries where data collection is relatively robust. It notes that, while survivors of gun violence often face long-term medical, economic, and social difficulties, such consequences are seldom counted among the costs of gun violence because they are not captured in official statistics. The chapter reviews the challenges to conducting non-fatal firearm injury surveillance and describes interim systems and entry points for interested donors, arguing that capturing non-lethal violence is crucial for a full understanding of the burden of armed violence on societies.

Country studies

Chapter 4 (Kazakhstan): This analysis of the security situation in Kazakhstan highlights paradoxical trends. On the one hand, the country has been spared the civil war and ethnic strife that has affected some of its Central Asian neighbours. Crime has decreased significantly since the mid-1990s, while the government keeps rather tight controls over civilian ownership of firearms. On the other hand, concerns with Kazakhstan's internal stability are growing and include a domestic homicide rate that still exceeds the regional and global average, an increase in the use of firearms in crime, and a recent surge in terrorist violence, as well as cases of social, ethnic, and political violence. The six large-scale, unplanned explosions at munitions sites that have occurred in Kazakhstan since 2001 also highlight problems in the management of state stockpiles of arms and ammunition. While it would be alarmist to speak of an approaching storm, Kazakh skies are not entirely clear.

Chapter 5 (Somaliland): Although Somaliland has its origins in a violent struggle and experienced episodes of large-scale violence through the mid-1990s, it has not only largely freed itself of armed conflict, but has also established a relatively high level of security for its population. This chapter looks at the trajectory of violence during the ongoing state formation process in Somaliland and examines patterns of cooperation between state and non-state security providers at the local level in urban areas. It suggests that the rising capacity of the state has curtailed armed opposition to a large extent, inducing various local non-state groups to collaborate with the state, thereby allowing for a relatively high degree of security.

> **Definition of small arms and light weapons**
>
> The Small Arms Survey uses the term 'small arms and light weapons' to cover both military-style small arms and light weapons as well as commercial firearms (handguns and long guns). Except where noted otherwise, it follows the definition used in the *Report of the UN Panel of Governmental Experts on Small Arms* (UN document A/52/298):
>
> **Small arms:** revolvers and self-loading pistols, rifles and carbines, sub-machine guns, assault rifles, and light machine guns.
>
> **Light weapons:** heavy machine guns, grenade launchers, portable anti-tank and anti-aircraft guns, recoilless rifles, portable anti-tank missile and rocket launchers, portable anti-aircraft missile launchers, and mortars of less than 100 mm calibre.
>
> The term 'small arms' is used in this volume to refer to small arms, light weapons, and their ammunition (as in 'the small arms industry') unless the context indicates otherwise, whereas the terms 'light weapons' and 'ammunition' refer specifically to those items.

Weapons and markets

Photo essay: This essay illustrates the stark reality of Somali piracy as seen through the lenses of more than a dozen professional and amateur photographers. The photo essay considers the root causes and enablers of Somali piracy, such as criminal networks, official corruption, and a lack of alternative economic opportunities. It also portrays some of the many measures undertaken to respond to pirate activities—from local resistance and incarceration of pirates to deployments of NATO vessels and private security guards.

Chapter 6 (Somali piracy): The frequency of pirate attacks worldwide reached record levels in 2010, largely attributable to Somali groups. While international naval forces and private security companies have increased their presence in affected waters in response, pirates have reacted by using captured vessels as 'mother ships' to allow them to strike at ever-greater distances from the coast. Pirate groups have also become more inclined to resort to lethal violence during attacks and to abuse their hostages.

The persistent threat of Somali piracy has prompted the shipping industry to turn to private security companies to protect vessels, with encouraging results as of late 2011. A number of states have sought to facilitate the provision of private armed security by, for instance, allowing security firms to rent out government-owned firearms. But embarking private armed guards on ships poses complex legal challenges, with the lack of clear rules for the use of force and firearms further complicating matters.

Chapter 7 (UN update): In 2011, the UN convened a new type of PoA meeting, an Open-ended Meeting of Governmental Experts (MGE). Chapter 7 distils key elements of these discussions, which focused on the current sticking points in the implementation of the International Tracing Instrument, along with means of overcoming the same. While the chapter highlights the range of 'implementation challenges and opportunities' that MGE participants discussed in relation to weapons marking, record-keeping, and tracing, it also notes the uncertainty concerning the meeting's future legacy. UN member states have yet to develop specific means of following up on the ideas, proposals, and lessons learned that were shared at the MGE.

Chapter 8 (Authorized transfers): This year's transfers chapter is the final part of a four-year project to estimate the annual value of authorized transfers of small arms and light weapons, their parts, accessories, and ammunition. It examines the trade in parts and accessories and summarizes and re-evaluates some of the findings produced between 2009 and 2011. The chapter concludes that the average annual value of authorized international transfers is at least USD 8.5 billion. This is a substantial revision of the long-standing USD 4 billion estimate. The annual value

Residents look out of their windows as soldiers occupy Rocinha favela, Rio de Janiero, November 2011. © Mauricio Lima/The New York Times/laif

of transfers of parts of small arms and light weapons is estimated to be worth approximately USD 1.4 billion. Annual transfers of sights are estimated to be worth more than USD 350 million.

Chapter 9 (Transparency): This chapter reviews ten years of reporting on the small arms trade by 52 countries exporting at least USD 10 million worth of small arms, light weapons, their parts, accessories, and ammunition. It highlights which countries have achieved top scores, noting in which reporting categories states can improve their level of transparency. The chapter concludes that over the past ten years, major exporting states have been increasingly transparent in reporting their small arms and light weapons transfers; however, the average score of all states remains below 50 per cent of the maximum possible score in the Small Arms Survey's Transparency Barometer, leaving most states with considerable room for improvement.

Chapter 10 (Illicit small arms): This chapter inaugurates a multi-year project aimed at improving public understanding of illicit small arms and light weapons by acquiring and analysing new and hitherto under-utilized data on illicit weapons. The project consists of three phases, each focusing on illicit weapons in countries that share a defining characteristic. Chapter 10 summarizes findings from the first phase, which looks at illicit weapons in conflict zones.

The Small Arms Survey gathered data on more than 80,000 illicit small arms and light weapons in Afghanistan, Iraq, and Somalia. One striking finding is the absence of the newest small arms and light weapons. Most appear to be older Soviet- and Chinese-designed models that have been circulating in these countries for many years. Also noteworthy is the comparatively small number of portable missiles, most of which are older Soviet-designed systems that are much less capable than modern missiles.

CONCLUSION

The 2012 edition of the *Small Arms Survey* seeks to increase our scrutiny of what is changing, and not changing, in relation to small arms proliferation and armed violence. It remains to be seen whether UN member states will develop the tools that allow for a better assessment of progress made in PoA implementation. Without such an assessment, it is impossible to answer a second, arguably more important question, namely what impact PoA implementation is having on the illicit small arms trade. In general terms, we lack the tools needed to determine where progress has been made, which issues and regions require the most attention, and where the challenges are greatest.

In any case, it appears clear from the findings presented in this volume that small arms proliferation remains a critical problem in many countries and regions, and that armed violence, both lethal and non-lethal, continues to undermine the security and well-being of people and societies around the world. Over the past decade, knowledge of small arms issues has expanded, with many aspects of the authorized trade, weapons diversion, and lethal armed violence now better understood. Yet crucial gaps remain in these—more extensively surveyed—fields, while other areas, including the illicit trade, are largely uncharted. Much has been learned and achieved in the decade since the adoption of the PoA, but the remaining gaps loom large. Knowledge, like security itself, continues to fall short of key needs.

—**Glenn McDonald and Emile LeBrun**

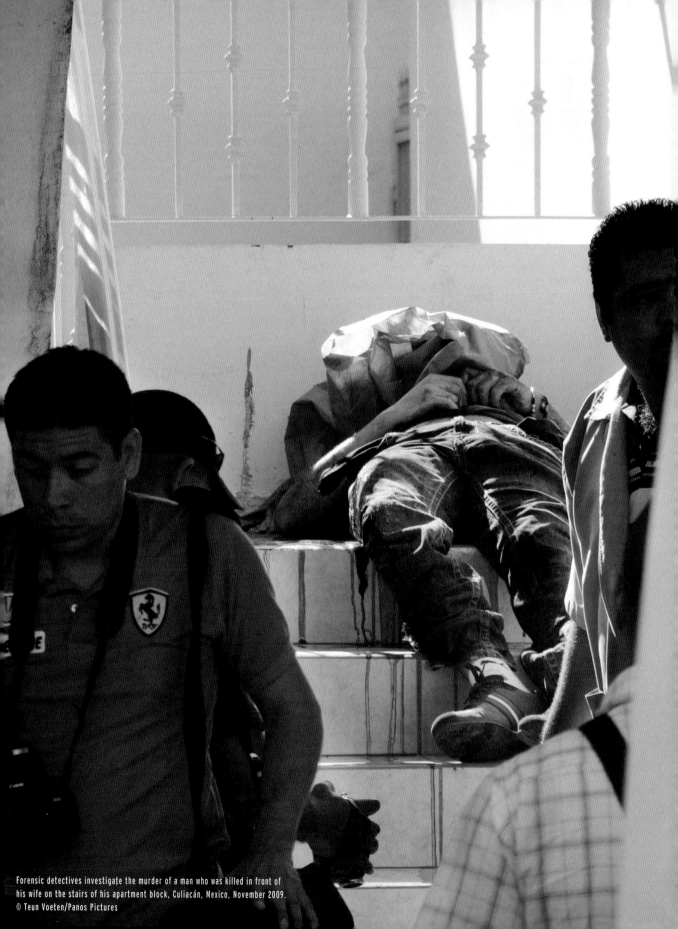

Forensic detectives investigate the murder of a man who was killed in front of his wife on the stairs of his apartment block, Culiacán, Mexico, November 2009.
© Teun Voeten/Panos Pictures

A Fatal Relationship
GUNS AND DEATHS IN LATIN AMERICA AND THE CARIBBEAN

INTRODUCTION

Armed violence is a defining problem for contemporary Latin America and the Caribbean (Davis, 2006, p. 178).[1] Many countries in the region suffer from rates of armed violence as high or higher than those in countries affected by war (Geneva Declaration Secretariat, 2011, pp. 51–65).

Some countries in the region not only show significantly higher homicide rates than countries elsewhere, but their security situation is also deteriorating. Homicide rates are generally increasing, and the brutality of the violence in some countries can be extraordinary. For example, in May 2011, neighbours of a remote ranch in northern Guatemala found the bodies of 27 farm workers, including two women and three teenagers. All but two had been decapitated (ICG, 2011, p. 2). While the killings were gruesome, even by the standards of the country's long history of violence, attacks of this kind are not unheard of in several countries in Latin America and the Caribbean.

Firearms are a defining feature of armed violence in the region. Specifically, the proportions of homicides committed with firearms in the majority of countries in Latin America and the Caribbean are higher than elsewhere in the world, although they vary considerably within the region.[2] To date, firearm homicide levels and trends have not been subject to any comprehensive reviews; nor have studies been carried out on the types of firearms most commonly used in the region. This chapter aims to address this knowledge gap.

The main findings are:

- In 21 of 23 countries in Latin America and the Caribbean for which data has been reviewed, proportions of homicides committed with firearms were higher than the global average (42 per cent). The exceptions were Cuba and Suriname.
- Having experienced increases in homicide rates between 1995 and 2010, El Salvador, Guatemala, Honduras, Jamaica, and Venezuela all suffer from very high homicide rates (>30 per 100,000). Together with Brazil, Colombia, Panama, and Puerto Rico, these countries also exhibit very high proportions of homicides committed with firearms (>70 per cent).
- In contrast, Argentina, Chile, Cuba, Peru, Suriname, and Uruguay report low homicide rates (<10 per 100,000), improving or stable trends between 1995 and 2010, and a proportion of firearm homicides below 60 per cent
- Like the rest of the world, countries in Latin America and the Caribbean appear to show a positive relationship between the national homicide rate and the percentage of firearms used in homicides. That is, higher homicide rates are usually accompanied by higher percentages of firearms used in homicides.
- While evidence is scarce, it does suggest that pistols and revolvers predominate in firearm homicides in Latin America and the Caribbean; however, further research is needed on the different types of firearms and perpetrators involved in different categories of armed violence.

The chapter is divided into four parts. The first section provides an overview of firearm homicides in Latin America and the Caribbean and discusses the methodology used to establish a database on homicides and firearm homicides. It then analyses the relationship between national homicide rates and the proportion of firearm homicides, presents trends in homicide rates between 1995 and 2010, and provides a cross-country comparison of firearm homicides on the basis of 2010 data. The second section unpacks some patterns and characteristics of firearm homicides in the different countries and sub-regions. It looks at Mexico, the Northern Triangle (El Salvador, Guatemala, and Honduras), southern Central America (Costa Rica, Nicaragua, and Panama), the Caribbean, Colombia and Venezuela, Brazil, and the Southern Cone sub-region (Argentina, Chile, Paraguay, and Uruguay). The third section presents the results of an extensive literature review on the type and origins of firearms most commonly involved in homicides in the region. The final section outlines topics for further research, including the availability of firearms, the presence of youth gangs and drug-trafficking organizations, and impunity in relation to firearm homicides.

FIREARM HOMICIDES: AN OVERVIEW

Armed violence has a range of impacts—from loss of property to feelings of insecurity and fear, emotional suffering, physical injury, and death. Homicide, commonly defined as an 'unlawful death purposefully inflicted on a person by another person', is a useful, but imperfect, indicator for assessing levels of armed violence (UNODC, 2011, p. 10). While homicide data is typically more accessible than other types of armed violence measures, it does not reflect non-fatal types of violence and crime, such as armed robberies, kidnappings, assaults, sexual violence, or non-fatal firearm injuries (NON-LETHAL VIOLENCE). Nor does it include suicides or unintentional firearm deaths. Furthermore, homicides are rarely recorded as such if the body is not found; consequently, in unresolved cases of enforced or involuntary disappearances that result in the killing of victims, the deaths are not necessarily recorded (Geneva Declaration Secretariat, 2011, p. 50). Thus, although homicide rates are useful proxies, they frequently under-count the actual numbers of deaths and only provide a partial picture of armed violence victimization.

> **Box 1.1 Compiling statistics on firearm homicides**
>
> In most countries, all natural and non-natural deaths are certified and registered. Information on a death generally comes from death certificates—which are typically filled out at hospitals, health clinics, emergency rooms, mortuaries, or forensic institutes. These records are usually integrated into national health statistics. Ideally, deaths are coded according to the International Classification of Diseases, currently in its tenth revision. A homicide is recorded as a fatal 'assault',[3] which covers 'injuries inflicted by another person with intent to injure or kill' (WHO, n.d.). Sub-categories can capture the type of weapon used to commit the assault.
>
> The criminal justice system is the second major source of homicide data and its statistics are among the most comprehensive. Since this information typically concerns illegal killings, the police and the criminal justice system investigate the intent of the killing, creating statistics on intentional homicides (Geneva Declaration Secretariat, 2011, p. 50). While the most important sources of criminal justice data are the statistics of the national police, the records from forensic institutes and legal medicine bureaus are also key.
>
> How is relevant data recorded following a homicide? The Instituto de Medicina Lega (Institute of Forensic Medicine, IML) in San Salvador, for one, indicates that the first step concerns the scientific investigation of the crime scene. Once IML staff have secured and registered the elements at the crime scene, they transfer the body of the victim to the IML, where they identify the cause of death and secure any further evidence in and on the body of the victim (such as bullets). This evidence is processed by the Laboratorio de Investigación Científica del Delito (Laboratory for the Scientific Investigation of Crime), which undertakes a ballistic examination of the bullets and cartridges recovered at a crime scene, among other relevant evidence. In the Salvadoran case, homicide statistics are produced by the IML and include information on whether a firearm or other instrument was used to commit the homicide.[4]

Establishing a methodology

For purposes of this chapter, the Small Arms Survey created a database on homicides and firearm homicides for the period 1995 to 2010. The data is drawn from public health and criminal justice statistics (see Box 1.1); source material used for the *Global Burden of Armed Violence 2011* report (Geneva Declaration Secretariat, 2011); and the United Nations Office on Drugs and Crime (UNODC) homicide statistics (UNODC, n.d.). Additional information and clarifications were directly requested from national statistical offices of all countries in Latin America and the Caribbean.

Once all sources had been compiled, several time series from varying sources were available for most countries. The most reliable source for each country was then selected based on careful consideration of the characteristics of the available data and communication with the statistical offices. Preference was given to sources that included the longest time series of data on homicides and proportions of homicides committed with firearms. Whenever several comprehensive data sources were available for a single country, a selection was made based on the following criteria: the accessibility of the sources (publicly available data); the clarity of the sources (such as provided definitions); consistency in the elaborated time series (regular reporting); and up-to-date reporting.[5] The resulting Small Arms Survey Database covers the years between 1995 and 2010 and includes a total of 34 countries in the three sub-regions of Central America, the Caribbean, and South America for which data on homicide rates is available for at least five consecutive years.[6]

At the global level, 42 per cent of firearms are committed with firearms.

Patterns of firearm homicides

At the global level, 42 per cent of homicides are committed with firearms (UNODC, 2011, p. 10). Yet the majority of countries in Latin America and the Caribbean have significantly higher proportions of firearm homicides than this global average. For the year 2010, or the latest year for which data is available, the Small Arms Survey Database suggests that firearms were used in an average of 70 per cent of homicides in Central America; that average drops to 61 per cent in the Caribbean and 60 per cent in South America.[7] Overall, the figures stand in stark contrast to those of Asia and Europe, where only 22 and 24 per cent of homicides are carried out with firearms, respectively (Geneva Declaration Secretariat, 2011, p. 99).

In trying to explain the more frequent use of firearms in homicides in Latin America and the Caribbean, it is important to note that overall homicide rates in the region are much higher than elsewhere. Indeed, a statistical analysis of data on the instrument of homicides confirms that there is a positive relationship between national homicide rates and the proportion of homicides committed by firearms.[8] Globally, in countries with high or increasing levels of violence, the use of firearms—effective instruments for committing homicides—is growing. This is true for countries in Latin America and the Caribbean as well. Figure 1.1 presents data on national homicide rates and the percentage of homicides committed with firearms in all countries in Latin America and the Caribbean for which data was available between 1995 and 2010.

Figure 1.1 includes both cross-sectional (different countries) and longitudinal (same country, different years) data. What can be observed is that higher overall homicide rates seem to be linked to higher overall proportions of firearm homicides. It is unclear whether firearm homicides are driving overall homicide rates or vice versa. Whatever the causality, there is clearly an important relationship between the two.

A breakdown of national homicide trends and firearm homicides allows for a second observation.[9] There appears to be a link between increasing homicide rates over time and an increase in the proportion of firearm homicides in

Figure 1.1 **Relationship between homicide rates per 100,000 and proportion of firearm homicides in 30 countries in Latin America and the Caribbean, 1995–2010**[10]

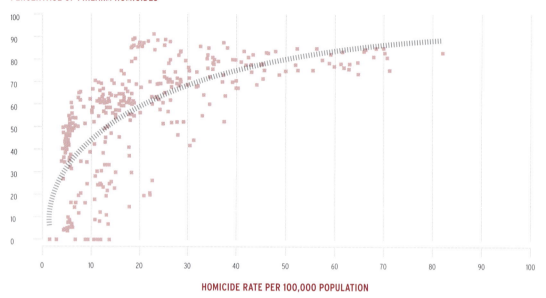

Source: Small Arms Survey Database

many countries in Latin America and the Caribbean. Similarly, a decrease in homicide rates is reflected by a decrease in the proportion of firearm homicides.[11]

Figure 1.2 presents national homicide rates and the proportion of firearm homicides on the basis of 2010 data (or data from the latest available year) in 23 countries in Latin America and the Caribbean. The figure confirms that, in general, the countries with higher homicide rates (top of the figure) exhibit higher proportions of firearm homicides, while the countries with lower homicide rates show lower proportions. There are a number of outliers, however, including Puerto Rico and Cuba.

Puerto Rico has the highest proportion of homicides committed with firearms across the entire region—91.5 per cent. While the country experiences homicide rates comparable to those of other Latin American and Caribbean countries, the percentage of homicides committed with a gun is disproportionately high. As a commonwealth of the United States, the island is significantly affected by access to the US firearms market. In 2010, the US Bureau of Alcohol, Tobacco, Firearms and Explosives traced 444 firearms that were confiscated by the police in Puerto Rico. In 322 cases, the source was a seller in the United States; 166 firearms were sold in Florida alone (ATF, 2010). The majority of firearms were confiscated due to illicit ownership or in the context of investigations. Out of all the confiscated firearms, only nine could be directly traced back to a homicide (ATF, 2010). While these figures do not provide a full picture of the origins of guns involved in homicides in Puerto Rico, they point to arms flows from the United States into Puerto Rico.

According to one report, lax airline regulations on travelling from the US mainland to Puerto Rico contribute to the islanders' ease of access to US firearms. Airlines seldom require passengers to show proof that their firearms are registered in Puerto Rico. Indeed:

To travel with five Glock handguns, three AK-47 assault rifles and 11 pounds of ammunition, an American Airlines passenger simply needs to inform an employee and make sure the weapons are packed securely and safely before they're placed onto the luggage conveyor belt (Rivera-Lyles, 2007).

According to a former agent of the Federal Bureau of Investigation, it is not uncommon for passengers to buy weapons on the US mainland and resell them on the black market or directly to members of drug-trafficking organizations in Puerto Rico, earning USD 50–100 per gun (Rivera-Lyles, 2007).

In contrast, **Cuba**, geographically very close to Puerto Rico, has one of the lowest homicide rates in Latin America and the Caribbean (4.5 per 100,000). Of special note is that less than five per cent of these homicides are committed with firearms (4.8 per cent). Cuba's low proportion of gun homicide is exceptional, not just in comparison to its neighbours, but also to other parts of the world. The Ministry of Public Health in Cuba indicates that in the last decade, the vast majority of homicides were committed with knives—71.8 per cent in 2010. Almost one-quarter (23.4 per cent) of homicides included 'other instruments'.[13] But even if the 23.4 per cent were counted as firearm homicides, the proportion of firearm homicides would remain far lower than in most countries in the region.

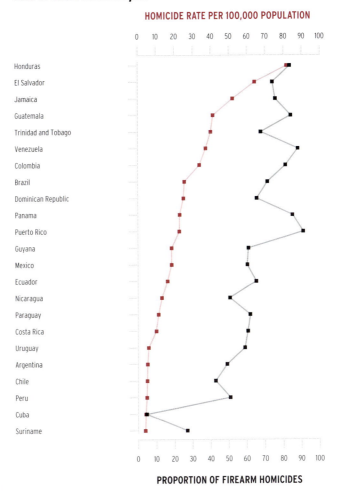

Figure 1.2 **National homicide rates and proportion of firearm homicides in 23 countries in Latin America and the Caribbean, 2010 or latest available year**[12]

Source: Small Arms Survey Database

Little has been written on armed violence in Cuba and the reasons for the extremely low level of firearm homicides are not well understood. Community policing may play a role in social control and in keeping homicide rates down (Kruger, 2007). One report emphasizes that, since the 1959 revolution, gang violence has been 'exported' to the United States. Before 1959, Cuba was a major hub for organized crime, but now Cuban organized crime groups reportedly operate mainly in the United States (Galeotti, 2006). Another report finds that organized crime in Cuba is tightly controlled by the government, and that related activity levels are relatively low (Stratfor, 2008). Still, many questions remain as to why Cuba's experience is so markedly different from that of the rest of Latin America and the Caribbean.

Trends in homicide rates

In contrast to global trends, a number of countries in Latin America and the Caribbean exhibit increasing homicide rates. Research shows that 'the analysis of global trends in homicide rates is hampered by the lack of time-series data in many countries, especially in Africa,' (UNODC, 2011, p. 25); nevertheless, available data reveals that in most countries in Asia and Europe, homicide rates are decreasing. UNODC finds that between 2005 and 2009, homicide rates increased mainly in countries that already suffered from high levels, including countries in Latin America and the Caribbean; meanwhile, 'in 101 countries with low homicide rates—mainly located in Europe and Asia—and in 17 countries with medium homicide rates, they decreased in the same period' (UNODC, 2011, p. 24).

Figure 1.3 shows the trend of average national homicide rates between 1995 and 2010 in each of the three sub-regions of Central America, the Caribbean, and South America. Each country was given equal weight within its sub-region; in criminal justice research, this method is used to generate a composite (sub-regional) rate (UNODC, 2011). The figure highlights that the three sub-regions had similar average national homicide rates in 1995: somewhere between 15 and 20 per 100,000 population. Since then, the trends have moved in very different directions. While Central America and the Caribbean show an increase in homicide rates, the average in South America decreased slightly.

It is important to note that all three sub-regions are highly diverse and that, among specific countries, there are considerable variations in patterns and trends of homicide rates. In fact, the sub-regional trends seen in Figure 1.3 often reflect a significant change in the national homicide rate of one or two countries in one sub-region, such as Honduras, Trinidad and Tobago, or Colombia.

Figure 1.4 presents the changes in national homicide rates between 1995 and 2010 (or the earliest and latest reported year within this time period), by sub-region. It shows that, on average, more countries in Central America and the Caribbean experienced an increase, rather than a decrease, in homicides. In contrast, in South America, there is a balance between countries with increasing and decreasing homicide rates. Overall, the country with the greatest change in homicide rates between 1995 and 2010 was Honduras; between 1999 (the earliest for which data is available) and 2010, the national homicide rate rose from 42.0 to 81.9 per 100,000.

Figure 1.3 **Trends in national homicide rates in 34 countries in Central America, the Caribbean, and South America, 1995–2010**[14]

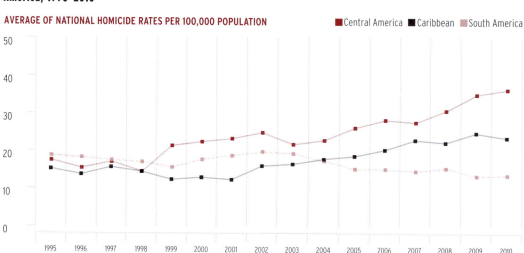

Source: Small Arms Survey Database

LATIN AMERICA AND THE CARIBBEAN 15

Figure 1.4 **Changes in national homicide rates in 24 countries in Central America, the Caribbean, and South America, 1995–2010**[15]

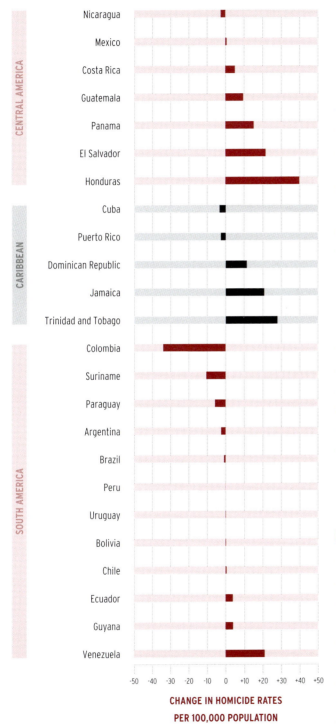

CHANGE IN HOMICIDE RATES PER 100,000 POPULATION

Source: Small Arms Survey Database

Map 1.1 displays the information shown in Figure 1.2 geographically. It presents a cross-country comparison of homicide rates on the basis of data from 2010 or the latest available year and the percentage of homicides that were carried out with firearms.

Together with the trend data shown in Figure 1.4, the map reveals certain patterns. While a number of countries suffer from very high and increasing homicide rates as well as very high proportions of firearm homicides, others have low, stable, or decreasing homicide rates, and a lower proportion of firearm homicides. El Salvador, Guatemala, Honduras, Jamaica, and Venezuela all exhibit homicide rates of more than 30 per 100,000, rates that have been rising since 1995, and proportions of firearm homicides above 70 per cent. The proportion of firearm homicides in Brazil, Colombia, Panama, and Puerto Rico also exceeds 70 per cent. In contrast, Argentina, Chile, Cuba, Peru, Suriname, and Uruguay all have homicide rates below 10 per 100,000, decreasing or stable rates since 1995, and proportions of firearm homicides below 60 per cent.

UNPACKING PATTERNS AND TRENDS

Many different factors result in certain countries—such as El Salvador, Guatemala, Honduras, Jamaica, and Venezuela—getting 'caught in the crossfire' of firearm homicides.[16] This section aims to shed light on some of these factors. A breakdown of national homicide rates between 1995 and 2010 and the percentage of firearm homicides for countries in the region can be found in the annexe to this chapter. The following

Map 1.1 **National homicide rates and proportion of firearm homicides in 23 countries in Latin America and the Caribbean, 2010 or latest available year**[17]

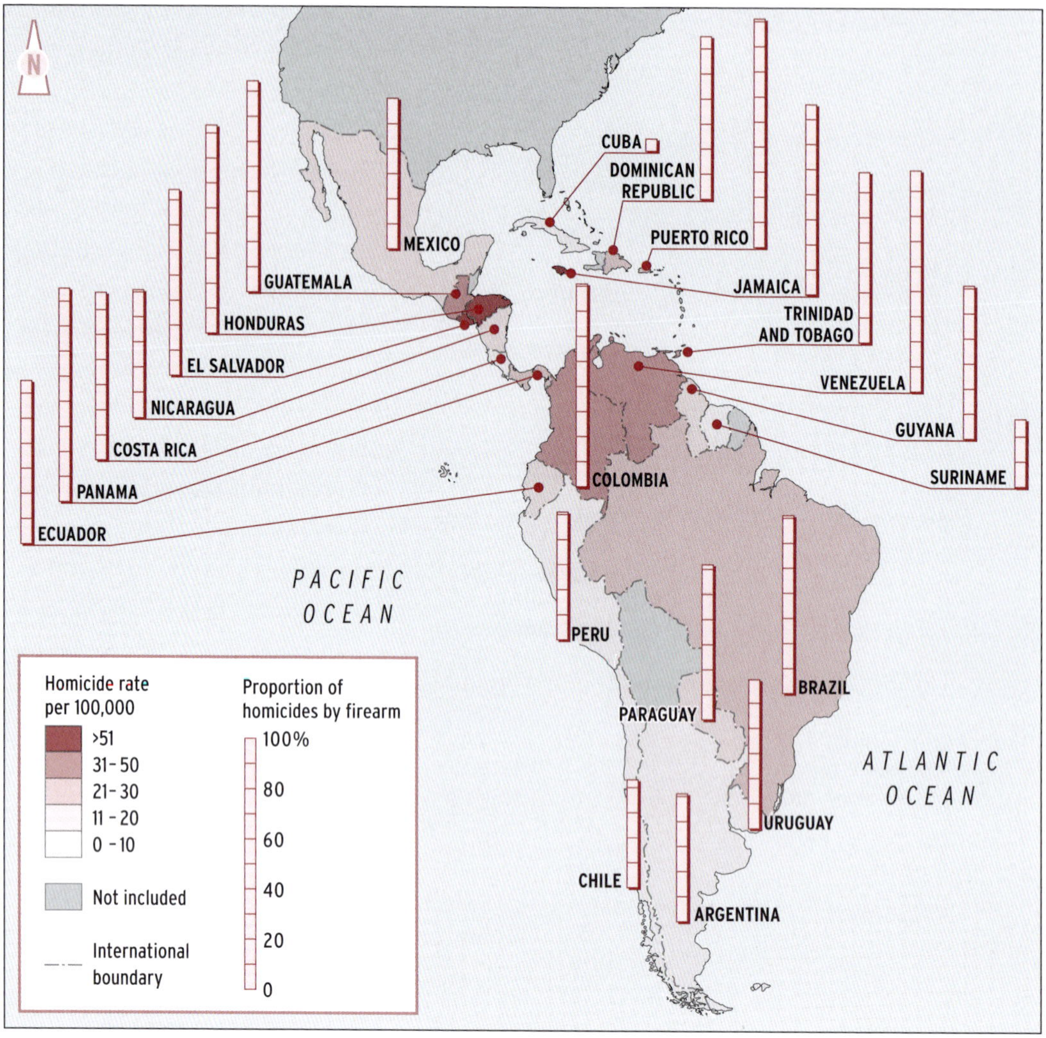

Source: Small Arms Survey Database

analysis considers the 23 countries highlighted on Map 1.1. The aim of this section is not to provide a comprehensive description of characteristics of firearm homicides in each country, but to shed light on selected patterns and trends.

Mexico

At the national level, the homicide rate in Mexico in 2010 was 18.6 per 100,000 (see Annexe 1.1). Yet while some parts of the country are almost unaffected by armed violence, others suffer from very high homicide rates. Moreover, some states have experienced a dramatic increase in homicides in recent years (see Figure 1.5). A case in point is Chihuahua, a state of more than three million inhabitants in the northern part of the country. In only three years, the intentional homicide rate in Chihuahua increased more than fivefold, from 19 in 2007 to 103 in 2010 (ICESI, n.d.).[18]

Figure 1.5 **Homicide rate in Chihuahua, Mexico, compared to other states, 1997-2010**

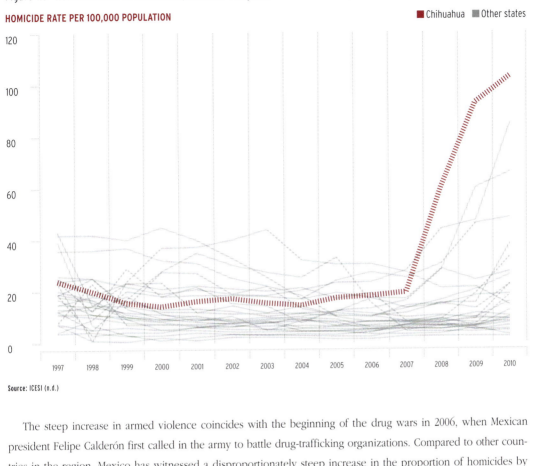

Source: ICESI (n.d.)

The steep increase in armed violence coincides with the beginning of the drug wars in 2006, when Mexican president Felipe Calderón first called in the army to battle drug-trafficking organizations. Compared to other countries in the region, Mexico has witnessed a disproportionately steep increase in the proportion of homicides by firearms (see Annexe 1.1). One study confirms that law enforcement and the military are not only facing more firearms, but also more militarized weaponry (Turbiville, 2010, p. 124). Another highlights that 'the AK-47 assault rifle has gained a bad reputation in recent years as the weapon of choice of the Mexican cartels' (Ortiz, 2011).[19]

The Northern Triangle

In the northern part of Central America, **El Salvador**, **Guatemala**, and **Honduras** form the so-called Northern Triangle. The three countries suffer not only from the highest homicide rates in the entire region, but also from increasing homicide rates. In all three countries, firearms are used in the vast majority of homicides (see Annexe 1.1).

The Northern Triangle is highly affected by large-scale drug trafficking. Situated mid-way between the cocaine-producing countries of South America and consumers in North America, Central American drug-trafficking organizations have been active since the 1970s. Yet, whereas the previous decades saw a comparatively stable situation with a 'pax mafiosa', the past ten years have witnessed a crackdown on drug-trafficking organizations in Colombia and the rise of powerful Mexican cartels that compete for control in Central America (Bosworth, 2010). The drug war in Mexico since 2006 has further destabilized the region, as the Mexican cartels increasingly move south (DRUG VIOLENCE). In May 2001, the International Crisis Group estimated that about 500 members of the Mexican drug cartel Los Zetas were active in Guatemala. The May 2011 massacre of 27 farmers in northern Guatemala is described as a drug deal gone bad between Los Zetas and another cartel (ICG, 2011, p. 2).

Drug-trafficking organizations in the Northern Triangle frequently cooperate with *maras*, youth gangs active in all three countries. The *maras* were formed in Los Angeles by Central Americans, many of whom sought refuge from the civil wars in El Salvador (1980–92) and Guatemala (1960–96). In the mid-1990s, US immigration law was tightened and, between 1998 and 2005, the United States repatriated almost 46,000 convicted gang members, with 90 per cent of the deportees sent to El Salvador, Guatemala, and Honduras, where they gradually displaced the local gangs, known as *pandillas* (Jütersonke, Muggah, and Rogers, 2009). The transnational transposition of US gang culture may be causing more frequent and more brutal violence 'due to the fact that it is less embedded within a local institutional context than traditional Central American *pandilla* culture, and therefore less rule-bound and constrained' (Rodgers, Muggah, and Stevenson, 2009, p. 9).

In a recent study on crime and violence in Central America, the World Bank (2011) finds that the availability of firearms is another important factor in firearm homicides in the Northern Triangle. While it is not known how many guns entered Central America during the civil wars, one study suggests that up to two million AK-47s were delivered during the last days of the cold war (Agozino et al., 2009, p. 295). El Salvador and Honduras were the largest recipients of weaponry from the US government in the 1980s and early 1990s, El Salvador because of the war against the communist guerrillas and Honduras because it was the primary base of operations for the US-backed Nicaraguan resistance known as the Contras (Godnick, Muggah, and Waszink, 2002, p. 5).

> Drug-trafficking organizations in the Northern Triangle frequently cooperate with *maras*.

The incomplete disarmament process after the civil wars resulted in the continuing presence of wartime firearms in the region. For example, although 3,000 combatants were demobilized in Guatemala after the peace accord in 1996, only 1,800 firearms were returned (IEPADES, 2006b, p. 12). Certainly not all firearms in circulation today can be traced backed to the civil wars. Indeed, the World Bank finds that between 2000 and 2006, the value of imported firearms in Guatemala almost tripled, from about USD 3 million to USD 8 million. It further estimates that 2.8 million firearms are currently circulating throughout the three countries, the great majority of them illegally owned (World Bank, 2011, p. 20).

Southern Central America

Compared to the countries of the Northern Triangle, the three countries in the southern part of Central America—**Costa Rica**, **Nicaragua**, and **Panama**—have significantly lower homicide rates (see Annexe 1.1). The relative efficiency of their justice systems may be one reason why. The World Bank indicates that Costa Rica and Panama are the only countries in the Central America sub-region whose governance indicators tend to rank in the upper half of the global sample, partly explaining the lower homicide rates (World Bank, 2011, p. 11).

Like El Salvador and Guatemala, Nicaragua also has a history of civil war (1972–91) and is known for the presence of youth gangs. It appears to suffer from lower levels of gang violence than its northern neighbours, however. While most refugees from El Salvador and Honduras went to Los Angeles, where some of them formed the *maras*, many Nicaraguans left for Miami, where gangs were either African-American or Cuban-American and generally did not accept Nicaraguans.[20] To this day, Nicaragua's predominant youth gangs are the home-grown *pandillas*. Rodgers, Muggah, and Stevenson (2009) suggest that their presence—and the absence of *maras*—may help explain why Nicaragua suffers lower levels of violence than El Salvador and Guatemala. Nevertheless, evidence shows that not only *maras*, but also *pandillas*, have become involved in drug trafficking and dealing over the past decade.[21]

Although on a lower scale than the countries in the Northern Triangle, Costa Rica, Nicaragua, and Panama all suffer from increasing homicide rates. In addition, Panama experienced the second-highest proportion of firearm homicides in 2010 in Latin America and the Caribbean, after Puerto Rico. Research has shown that a negative relationship generally exists between the percentage of homicides committed with firearms and the proportion of homicides solved by the police (Geneva Declaration Secretariat, 2011, p. 102). There is thus concern that, with an increase in gun violence in Panama, law enforcement's ability to clear homicide cases may decline, which can create an atmosphere of impunity regarding gun homicides. As the World Bank notes:

These worries are compounded by fears of contagion from their three more violence-prone northern Central American neighbours and the prospect that they too could become havens for the drug trafficking that drives the high crime rates in the [Northern Triangle] (World Bank, 2011, p. 1).

The volume of drug seizures in Panama has already surpassed that of other Central American countries. Although this may also be due to the efficiency of the law enforcement system, it remains an indicator of active trafficking (World Bank, 2011, p. 12).

The Caribbean

Trends in homicides in many Caribbean countries are difficult to assess, since many of the islands have populations of less than 500,000. Of all the Caribbean countries reviewed in this chapter (see endnote 6), only five countries have a population of more than 500,000. Cuba and Puerto Rico are described in the previous section; the data for the other three countries—**the Dominican Republic**, **Jamaica**, and **Trinidad and Tobago**—shows that they have all witnessed increases in homicide rates. The situation in Jamaica is unique in the sense that homicide rates have consistently been higher than in any other country in the sub-region since 1995.[22] In all three countries, the majority of homicides are committed with firearms (see Annexe 1.1).

Agozino et al. (2009) find that a central reason for the high proportion of firearm homicides in the Caribbean is the 'weaponization' of society, meaning that the use of guns has become increasingly embedded in significant sectors of society. This trend is especially pronounced within youth gangs. Data on homicides in Jamaica from 1998 to 2002 shows that gangs disproportionately use firearms when committing violent crimes. While around 70 per cent of all murders

A police investigator walks towards the body of a woman killed during a shooting rampage carried out by gunmen in Rockhall, Kingston, on election day, October 2002. © Andrew Winning/Reuters

involved guns in 2002, 94 per cent of gang- and drug-related murders involved firearms (Lemard and Hemenway, 2006).[23] A recent study on gangs and guns in Trinidad and Tobago also reveals higher frequencies of gun use among gangs. It finds that, while homicides due to blunt, sharp, or other instruments remained comparatively stable over the years, those committed with firearms increased. This increase in gun-related homicides coincides with an increase in the number of urban gangs (Townsend, 2009, p. 18).[24]

Because relatively few gun homicides are solved in Trinidad and Tobago, impunity for gun violence may also be a factor; Agozino et al. show that in 2006 the clearance rate for non-gun homicides was more than 62 per cent, whereas it was only slightly more than 20 per cent for gun homicides (Agozino et al. 2009, p. 292). Impunity surrounding gun homicides in a country with high levels of violence and frequent gun crime can result in cycles of retaliatory attacks.

Colombia and Venezuela

Colombia and Venezuela both show high homicide rates and high proportions of firearms used in homicides, but the trends are very different. Colombia—having started at a much higher level in 1995—experienced the most significant decrease in homicide rates among all countries in Latin America and the Caribbean. In contrast, homicide rates in Venezuela, starting at a lower level, have increased significantly (see Annexe 1.1).

A masked gunman points his handgun towards student opponents of President Hugo Chávez during a shootout at Venezuela's Central University, Caracas, November 2007. © Gregorio Marrero/AP Photo

Since the 1990s, Colombian cities have introduced numerous measures to address gun violence; these play an important role in the reduction in homicides (see Box 1.2). Yet, after several years of steep drops in violence—mainly between 2002 and 2006—the rate of decline in homicide rates slowed down significantly. Restrepo and Aponte (2009) refer to this phenomenon as the 'glass floor' of homicides (*piso de cristal de los homicidios*). Furthermore, while national homicide rates largely decreased, the proportion of firearm homicides remained almost stable. This development stands in contrast to one of the main observations of this chapter, namely that the proportion of homicides committed with firearms tends to increase and decrease in parallel with homicide rates. Although the trend is not visible at the national level in Colombia, it is on a sub-national level (see Figure 1.6).

While the homicide rate of Colombia decreased significantly, the rate in Venezuela showed a steady increase since 1995 (see Annexe 1.1).[25] In addition, some argue, policies that resulted in reductions in Colombia simultaneously drove illicit actors to neighbouring countries, where there was less pressure from the military and police. In this sense, Colombia's relative success may have created new problems for other countries, such as Venezuela.[26]

In addition to such dynamics, however, a number of factors independent of Colombia may explain the rise in Venezuelan homicide rates. Romero (2010) finds that high and increasing income inequality, resulting from economic reforms of the 1980s and 1990s, is one of the reasons for the surge in violence in Venezuela. He points

Box 1.2 Reductions in firearm homicides in Colombian cities

In the 1990s, a number of Colombian cities witnessed extremely high homicide rates. The three cities most affected were Bogotá, Cali, and Medellín. In 1991, homicide rates in Medellín peaked at 266.3 per 100,000. In Cali, the homicide rate in 1994 stood at 121.5, and in 1993 the peak in Bogotá was 80.9. Since then, the homicide rates have dropped significantly in all three cities.

In Cali, homicides dropped from 121.5 in 1994 to 84.6 in 1997. The drop is believed to be related to a set of violence reduction interventions implemented by Cali's then mayor, Rodrigo Guerrero. The initiatives were reproduced in other cities in Colombia, including Bogotá (starting in 1997) and Medellín (starting in 2005).[27] As shown in Figure 1.6, implementation of the initiative did not correlate directly with major reductions in homicides in the cities; other measures may also have been relevant in bringing about the reductions.

A number of gun control measures were part of the urban violence reduction efforts in the 1990s. With an estimated 5.9 guns per 100 persons, civilian gun ownership (legal and illegal) in Colombia is low by global standards (Karp, 2007). There are two primary reasons for this low level. First, the Colombian state enforces strong regulation of civilian arms possession. Second, the criminal market for guns is tightly controlled by the non-state groups involved in armed conflict (guerrillas and paramilitaries) and by drug-trafficking organizations. Still, the vast majority of weapons circulating in Colombia are probably illegal (Small Arms Survey, 2006, p. 219).

In response to the challenge posed by illicit weapons, the government introduced bans on carrying firearms and implemented firearm collection programmes. Villaveces et al. (2000) observe that the enforced ban on carrying firearms on weekends after paydays, on holidays, and on election days in Cali (1993-94) and Bogotá (1995-97) contributed to a significant drop in homicide rates in both cities.

Homicide rates in Colombian cities may also have dropped in response to the 2003-04 demobilization and disarmament of the paramilitary group Autodefensas Unidas de Colombia (United Self-Defence Forces of Colombia, AUC) and other 'hard' security measures that complemented the municipal violence prevention policies.[28]

Despite the downward trends beginning in 2003, homicide rates in Cali and Medellín started to increase again from 2007 and 2008, as Figure 1.6 shows. Recent reports suggest that the demobilization of the AUC led to a fragmentation of the criminal hegemony over the drug trade. Many of these new groups work alongside and against each other in the interests of the drug trade. The changing power structures of drug-trafficking organizations as a result of the demobilization of the AUC, as well as mafia-related violence, are thought to be important factors influencing the recent increase in urban homicides, especially in Medellín.[29]

Figure 1.6 **Homicide rates and proportion of firearm homicides in Bogotá, Cali, and Medellín, 1990-2009**

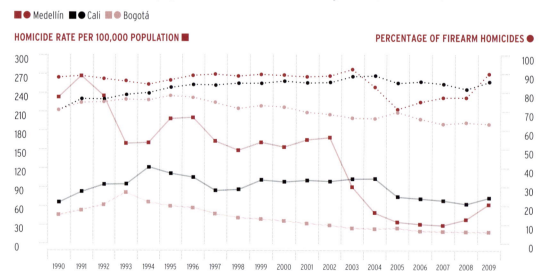

Source: elaboration based on Aguirre and Restrepo (2010, p. 272); DANE (n.d.)

out that, 'while many Latin American economies are growing fast, Venezuela's has continued to shrink, [and] the gap between rich and poor remains wide' (Romero, 2010).[30] Romero also observes an increasing atmosphere of impunity, with a continuing collapse of the old political and social order and a judicial system that has grown increasingly politicized (Romero, 2010).

In line with one of the overall trends identified in this chapter, the proportion of firearm homicides in Venezuela has increased in parallel with rising homicide rates (see Annexe 1.1). Civilian gun ownership is widespread, although there is great uncertainty about the actual number of legal and illegal guns in circulation, estimated at between 1.6 and 4.1 million guns, the vast majority of which are illegal firearms (Karp, 2009, p. 56).

The Venezuelan military and security services serve as important sources of illegal firearms (Karp, 2009, p. 57). Unlike most of the world's militaries, Venezuela's armed forces are growing, and the number of military weapons is increasing. In what many analysts see as another potentially destabilizing factor, Venezuela will soon begin to manufacture its own Kalashnikov-pattern rifles with the assistance of a Russian arms company (InSight, 2010).[31] There is concern that the new rifle will begin entering the black market (Tulyakov, 2010).

Brazil

In 2010, Brazil's homicide rate stood at 25.6 per 100,000. A majority of these homicides (71.3 per cent) were committed with firearms (see Annexe 1.1). Homicide rates have remained relatively stable since 2000 (the earliest year for which data is available). At the sub-national level, this picture is more complex. Violence is unevenly distributed, with pockets of very high levels of violence. High rates of firearm-related youth and gang violence remain characteristics of Brazil's larger cities, where 'young people live in juvenile subcultures that glorify the warrior ethos and

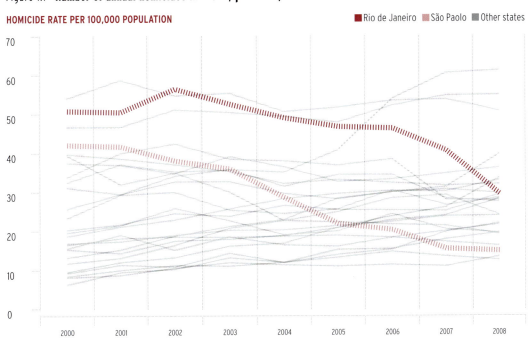

Figure 1.7 **Number of annual homicides in Brazil, per state, 2000–08**

Source: elaboration based on FBSP (2010, p. 34)[32]

the assertion of masculinity through the power expressed in threats based on the use of firearms' (Imbusch, Misse, and Carrión, 2011, p. 115). At the same time, sub-national data for 2000–08 shows that a number of cities, including Rio de Janeiro and São Paulo, experienced major reductions in homicide rates (see Figure 1.7).

De Souza et al. (2007) suggest that a range of gun control measures introduced in Brazil from 2003 onwards have helped reduce the toll of violent deaths and hospitalization. In October 2003, the Brazilian government passed a gun law that:

> *sought to control the flow of firearms into the country, made it illegal to own guns that are not registered or to carry guns outside of one's home or business, instituted background checks for gun purchases, and raised the minimum age for gun purchase to twenty-five* (de Souza et al., 2007, pp. 575–76).

The subsequent reduction in homicides is most pronounced in the city and state of São Paulo, where the number of annual homicides dropped from more than 15,000 in 2001 to slightly more than 6,000 in 2007 (FBSP, 2010, p. 34). Goertzel and Kahn (2009) attribute the 'great São Paulo homicide drop' to more effective policing methods, including strict enforcement of the gun control legislation. There was consistent investment in policing, with more and better training, more resources for prevention, and a special focus on the most affected areas and populations. In addition, between 2000 and 2010, police in São Paulo seized more than 200,000 illegal guns and gun holders surrendered more than 130,000 guns in buy-back campaigns (Mizne, 2011).

The subsequent reduction in homicides is most pronounced in the city and state of Sao Paolo.

The Southern Cone

The sub-region in Latin America and the Caribbean that is least affected by gun violence is the Southern Cone (Cono Sur). This group of four countries in South America—**Argentina**, **Chile**, **Paraguay**, and **Uruguay**—all show comparatively low homicide rates and comparatively lower proportions of firearm homicides (below 60 per cent).[33] With the exception of Paraguay, they have homicide rates between 5.3 and 6.1 per 100,000 population. In addition, all four countries show either stable or decreasing homicide rates. Of the four, Paraguay—having had higher initial rates in 1995—witnessed the most significant decrease (see Annexe 1.1).

A number of researchers have analysed homicide trends in Argentina. Spinelli et al. (2006) find that in Buenos Aires in 2002—the year in which firearm homicides peaked—the firearm-related deaths were often connected to various aspects of organized crime, such as drug trafficking and arms smuggling. At the same time, Spinelli, Macías, and Darraidou (2008) link the reduction of homicides after 2002 to a number of socio-economic trends—such as reduced income inequality and poverty (measured in terms of families living under the poverty line) and increases in gross national product—rather than strict gun control measures.

In early 2007, the Argentine gun buy-back programme—the Programa Nacional de Entrega Voluntaria de Armas de Fuego—was introduced. Individuals who handed in a firearm received ARS 100–140 (USD 32–45), depending on the type of firearm. The programme netted 104,000 firearms and 750,000 units of ammunition throughout the country between 2007 and 2009 (Fleitas, 2010, p. 26).

Whether it had a positive influence on homicide reduction remains under debate. Lenis, Ronconi, and Schargrodsky (2010), for example, find that while it successfully reduced the number of accidental deaths, the programme did not influence the number of homicides.

TYPES AND ORIGINS OF FIREARMS USED IN HOMICIDES

The *type* of firearm involved in a homicide can be identified through ballistic investigation. When a shot is fired, the barrel leaves specific marks on the bullet. This 'fingerprint' can be used to identify the gun that fired the bullet. Increasingly, countries in Latin America use the Integrated Ballistics Identification System (IBIS), which can identify, correlate, and match projectiles and shell casings with the weapons that fired them.[34] As with real fingerprints, successful matches depend on the extensiveness of bullet markings and whether a specific gun barrel has been entered into the database.

If a crime weapon has not been recovered, matching projectiles to barrels is more challenging. This point is of concern in Bogotá, where law enforcement failed to recover firearms from perpetrators in an estimated 95 per cent of all homicide cases in 2002. If the projectile is not too damaged, ballistic examination can usually determine at least the calibre. Yet in the vast majority of cases in Bogotá, recovered bullets examined at the forensics lab did not provide sufficient information to allow for the identification of firearms because the database did not contain details about any projectiles previously fired from the same gun (León-Beltrán and Forero, 2004, p. 5).

In the absence of administrative statistics on *types* of firearms, smaller, localized datasets of guns seized by police can be examined. Yet caution must be used when treating data on firearm confiscations as a proxy for the weapons used in homicides. As Aguirre et al. (2009) observe, for example, confiscated firearms in Colombia can serve as a proxy for the nationwide demand for illicit firearms, but it is impossible to identify a direct relationship between seized firearms and their use in crime. Many of the firearms seized by police are not related to violent crimes but are linked to firearms regulation infractions or other minor offences (Aguirre et al., 2009, p. 15). A study of firearm seizures between 1973 and 2006 in Argentina confirms that only in a very small percentage (1.3 per cent) could the firearm be traced back to a homicide (Dreyfus, 2007a, pp. 42–43).

Despite these limitations, data on firearms seized in combination with specific case studies on the types of firearms involved in homicides help shed light on the types and origins of firearms involved in crime and violence in Latin America and the Caribbean.

Pistols and revolvers are the weapons most frequently used in crimes.

Types of weapons

Gun seizure data reveals that in Latin America and the Caribbean, pistols and revolvers are the weapons most frequently used in crimes. For example, of some 3,000 guns that were confiscated by the National Civilian Police of **Guatemala** in 2006, 45 per cent were pistols, 27 per cent revolvers, 24 per cent shotguns, and 11 per cent craft guns (IEPADES, 2006a, p. 38). In addition, most of the weapons collected by the Government of **Panama** in weapon recovery campaigns are revolvers, rifles, and pistols. The seized sub-machine guns, hand grenades, rifles, and muskets are far fewer in number (FLACSO, 2007, p. 317).

A study on **Jamaica** paints a similar picture. In 2009, 50.6 per cent of the 569 firearms seized after a criminal event were pistols and 21.8 per cent revolvers. Homemade weapons (15.3 per cent), sub-machine guns (3.5 per cent), and shotguns (3.3 per cent) made up a smaller portion. Officials in Jamaica suggest that pistols may play an even greater role than the statistics reveal. The overwhelming majority of casings recovered from Jamaican crime scenes are for 9 mm semi-automatic pistols (Leslie, 2010, p. 34).

The proportion of homemade guns seized both in Jamaica and in Guatemala is significant. Research suggests that craft guns are common among youth gangs across Latin America and the Caribbean. Godnick, Muggah, and Waszink (2002) observe that youth gangs in Central America have been assembling makeshift pistols that are made of bedsprings and metal tubing. Some of the craft guns are sophisticated imitations of .22 mm, .38 mm, and 9 mm pistols. While these arms—which often contain parts and material from other guns—are probably more expensive than the most basic craft guns, they are still cheap, easy to use and dispose of, and difficult for authorities to trace (Godnick, Muggah, and Waszink, 2002, p. 8). A study focusing on the characteristics and practice of gangs in **Ecuador** reveals that revolvers stand out in gang members' preferences. Above all, they favour six-shot .38 mm revolvers. According to the same study, this firearm was also available in a craft version on the black market for around USD 35 (Loor, 2004, p. 25).

Data on gun seizures must be interpreted with caution. It is possible that criminal groups are especially watchful of expensive weapons, such as machine guns or assault rifles, which may partly explain why the police seize relatively few such firearms. Yet, while data on types of firearms used in homicides confirms that pistols and revolvers predominate, it also shows that craft guns are used. The police in **Costa Rica** produce regular detailed reports on criminal violence that include data on the types of firearms used in homicides (see Table 1.1). Information on the type of firearm was provided in 276 cases of the 338 homicides committed with a firearm in 2009. Revolvers were used in 143 cases, pistols in 133. No automatic weapons or assault rifles could be identified. It is possible that a part of the 62 firearms that were not identified were, in fact, automatic weapons or assault rifles. In that case, however, pistols and revolvers would still remain the most common weapons involved in homicides.

Seized weapons are displayed before being destroyed at a police facility in Panama City, October 2010. © Arnulfo Franco/AP Photo

Table 1.1 Firearm-related homicides and firearm types used in Costa Rica, 2003–09

	2003	2004	2005	2006	2007	2008	2009
All homicides	300	280	338	351	369	512	525
Firearm homicides	156	164	196	217	226	349	338
Firearm-related homicides (%)	52.0	58.6	58.0	61.8	61.3	68.2	64.4
Type of firearm identified in firearm homicides							
Pistols	70	79	102	128	118	190	133
Revolvers	82	14	74	72	99	136	143
Automatic and assault	4	12	24	17	12	0	0

Sources: Policía Judicial de Costa Rica (2006; 2007; 2008; 2009)

Table 1.2 Firearm types used in homicide and suicide cases, Honduras, 2005–08

Type of firearm	2005		2006		2007		2008	
	Number	%	Number	%	Number	%	Number	%
Pistol	314	37.6	391	38.6	415	38.4	420	37.2
Revolver	242	28.9	279	27.6	322	29.8	327	29.0
Craft gun	107	12.8	125	12.4	141	13.0	146	12.9
Light flintlock musket	103	12.3	134	13.2	68	6.3	73	6.5
Shotgun	30	3.6	34	3.4	34	3.1	39	3.5
Rifle	n/a	n/a	32	3.2	24	2.2	30	2.7
Sub-machine gun	n/a	n/a	17	1.7	15	1.4	21	1.9
Carbine	n/a	n/a	n/a	n/a	2	0.2	8	0.7
Other	40	4.8	n/a	n/a	60	5.6	65	5.8
Total	**836**	**100**	**1,012**	**100**	**1,081**	**100**	**1,129**	**100**

Sources: IUDPAS (2006; 2007; 2008; 2009)

Pistols and revolvers are also the predominant firearms used in homicides (and suicides) in **Honduras**. The University Institute on Democracy, Peace, and Security runs a violence monitoring project that provides quarterly and annual data on injuries and mortality at the national level in Honduras (see Table 1.2). In 2008, a total of 1,448 gun homicides and 316 gun suicides were recorded. Out of these deaths, 1,129 cases provided sufficient ballistic information to determine the type of firearm. Pistols are the guns most commonly used to commit homicides and suicides (37.2 per cent), followed by revolvers at 29.0 per cent (IUDPAS, 2009, pp. 1, 12). Craft guns are also widely used in Honduras; 12.9 per cent of all firearms recovered in homicides or suicides in 2008 were craft guns.

A study from **Ecuador** in 2005 highlights that, in the cases where bullets could be extracted from victims' bodies at morgues, the vast majority of ballistic examinations identified 9 mm bullets (121 out of 147), which are usually used in pistols, while the rest involved .38 mm bullets (26 out of 147), which are more commonly used in revolvers (Zárate, 2006, p. 108).

These examples suggest that pistols and revolvers—and, to a lesser extent, craft guns—are the guns most commonly used in firearm homicides in Latin America and the Caribbean, rather than machine guns and assault rifles. At the same time, the literature suggests that drug-trafficking organizations are making increasing use of assault rifles and machine guns. The International Crisis Group reports that drug traffickers in **Guatemala** often carry assault weapons, including AK-47s, AR-15s, M-16 rifles, grenades, and even RPGs, Russian-type rocket launchers (ICG, 2011).

There is an emerging body of research on firearms used by drug-trafficking organizations in **Mexico**. The US Bureau of Alcohol, Tobacco, Firearms and Explosives indicates that, while drug-trafficking organizations in Mexico had primarily used .38 mm handguns for some time, cartel members and enforcers have since developed a preference for more powerful firearms. The most common now include the Colt AR-15 (.223-calibre assault rifle), AK-47-pattern weapons (7.62-calibre assault rifle), and pistols produced by the Belgian company FN Herstal (5.57-calibre pistols) (Hoover, 2008). Although precise numbers are not available, another report reveals that military-style firearms, such as 7.62 mm AK-type assault rifles, AR-15-type semi-automatic rifles, and handguns with high calibres (such as .38, .45, and 9 mm pistols), are increasingly seized from Mexican drug cartels (US GAO, 2009, p. 17).

Origins of weapons

Latin America has a long tradition of gun production, with some manufacturers tracing their history back many decades.[35] Brazil has the largest arms industry in the region, followed by Argentina. Firearms are also produced by private or government-owned industries in Bolivia, Chile, Colombia, Ecuador, Mexico, Paraguay, Peru, and Venezuela (Small Arms Survey, 2004, p. 17). While most of the production is intended to equip the military and law enforcement institutions, some of the production is for private use. Research shows that, '[w]ith the important exceptions of major exporters led by Argentina, Chile, Mexico, and above all Brazil, [Latin America's] small arms producers tend to be niche manufacturers, serving captive local markets' (Small Arms Survey, 2004, p. 26).

Table 1.3 highlights the makes of the revolvers and pistols used in homicides in Bogotá, **Colombia**, in 2002. It shows that

Table 1.3 **Firearm types used in homicides in Bogotá, 2002**

Type of gun	Number of recovered guns	Per cent
Revolver Smith & Wesson	42	37.8
Revolver Llama	28	25.2
Revolver Ruger	11	9.9
Revolver Colt	5	4.5
Pistol Browning	5	4.5
Pistol Beretta	4	3.6
Pistol Walther	3	2.7
Pistol Smith & Wesson	3	2.7
Revolver Taurus	2	1.8
Pistol Taurus	2	1.8
Pistol Jericho	2	1.8
Rifle AK-47	2	1.8
Pistol Glock	1	0.9
Pistol Sig Sauer	1	0.9
Total	**111**	**100**

Source: León-Beltrán and Forero (2004, p. 6)

the most common type of firearm was the US-manufactured Smith & Wesson revolver, followed by the local Llama revolver. The Llama revolver is a product of the government-owned manufacturer INDUMIL, which primarily equips the Colombian army and the national police, although it also supplies the domestic market and exports some weapons (Small Arms Survey, 2006, p. 218). Since the data covers only a few months, however, the number of weapons is small and the figures were not officially released. Whether the predominant use of Smith & Wesson and Llama revolvers is a general pattern in homicides in Colombia remains an open question. Whether these weapons are legally or illegally owned also remains unknown.

As described in Box 1.2, Colombia has strict gun laws, and the vast majority of weapons circulating in the country are probably illegal. One study of the illegal gun market in Bogotá reveals that in 2006, pistols and revolvers could be bought for as little as USD 260. Machine guns, such as the M-60, could be purchased for around USD 10,000, or half the market price. The same study shows that criminals in Colombia increasingly rent firearms to avoid prosecution and enforcement of gun restrictions (Aguirre et al., 2009, p. 7).

Box 1.3 Firearms seized in Rio de Janeiro, 1951–2003

Table 1.4 shows that the bulk of the firearms confiscated by the police in Rio de Janeiro between 1993 and 2003 were revolvers (60.6 per cent), while pistols constituted 26.7 per cent (Dreyfus and Marsh, 2006, p. 23). At the same time, the data suggests a rise in the number of weapons with larger firepower, including sub-machine guns, machine guns, rifles, and carbines (Dreyfus, 2007b, p. 13).

Analysing the producers, Bandeira and Bourgois (2006) find that most of the weapons (76.6 per cent) are Brazilian-made, 35.0 per cent by Taurus and 21.0 per cent by Rossi (p. 12). Rivero (2005) shows that only one-fifth of Brazilian-made weapons (19 per cent) were registered at some point, whereas the rest (81 per cent) was never legally registered. In those cases, either the legal gun owners committed crimes or the guns had been diverted from their legal owners (Rivero, 2005, p. 206).

Rivero further demonstrates that a significant number of the high-power firearms were types that are typically for exclusive use by law enforcement or the military, such as the Brazilian-made .38 mm Taurus and Imbel pistols, Taurus 9 mm pistols, and pistols produced by Smith & Wesson (United States), Ruger (United States), and Norinco (China) (Rivero, 2005, p. 220). This suggests that many weapons are diverted from Brazilian government-controlled stockpiles to the illicit market. Dreyfus confirms this finding, showing that 18 per cent of firearms that have been traced back to their origins stem from official institutions, including the military police and the army (Dreyfus, 2007b, p. 22).

In some cases, guns were allegedly exported via **Paraguay** but never really left Brazil, only to be released on the local illicit market. Paraguay has long functioned as a hub for illicit procurement of weapons, both foreign- and Brazilian-made (Dreyfus and Marsh, 2006, p. 44; Dreyfus, 2007b, p. 27). For a long time, the country had lax regulations on firearms purchase. This eventually changed, first with the US ban on firearm exports to Paraguay in 1996 and, in 2002, with the adoption of tougher regulations in Paraguay. Among other measures, the purchase of firearms by foreigners was explicitly prohibited.[36] Fieldwork carried out in 2000 and 2005 showed that, despite these regulations, a substantial number of firearms imported until the mid-1990s could still be purchased in Paraguayan gun shops (Dreyfus and Marsh, 2006, pp. 36–37).

In addition, there is evidence of a flourishing illicit arms trade in the 'Tri-Border Area' between Argentina, Brazil, and Paraguay. In 2003, a report by the US Federal Research Division provided evidence that the area serves as a haven for illicit activities ranging from counterfeiting to intellectual property theft and money laundering; it has been identified as a central point for the laundering of funds from drug trafficking and for the funding of terrorist organizations (Hudson, 2003).

The data on weapons seized in Rio de Janeiro was influential at the policy level. It spurred the Parliamentary Commission of Inquiry to investigate arms trafficking in Brazil. A coalition of non-governmental organizations[37] also used the data to inform the discussion and the campaign advocating the adoption of the national disarmament statute (gun law) in December 2003. The new law introduced the prohibition for civilians to purchase firearms (except in at-risk professions) and the centralization at the federal level of permits and weapons registration (FLACSO, 2007, p. 122).

The presence of illicit gun markets is often heavily linked to the illicit drug trade. Studies show that Colombia 'has a massive illegal market for weapons, with a complex network of buyers and sellers—many of them driven by the armed conflict and narco-trafficking' (Small Arms Survey, 2006, p. 217). Colombia's conflict and drug trade not only attract illegal firearms but also create pools of illicit guns that flow into neighbouring countries, Central America, and the Caribbean. Research on armed violence in **Jamaica** suggests that criminals commonly smuggle weapons such as AK-47s and M-16s from South and Central American countries—including from Colombia, Honduras, and Venezuela—into Jamaica, although many of the illicit guns originate in the United States (Leslie, 2010, p. 41).

In his research on **Honduras**, Bosworth (2010) observes that, while most drug-trafficking transactions are made in cash, payment in arms also occurs. Honduran authorities report that Mexican drug cartels such as Los Zetas move high-calibre firearms, mostly stolen from security forces in Mexico and Guatemala, into Honduras in exchange for cocaine.

Other reports indicate that Honduras has become the sub-regional centre of the booming business in illegal guns. More than 850,000 weapons are estimated to be circulating in the country. Of those, only 258,000 weapons are

Table 1.4 Types of firearms captured by police in Rio de Janeiro, 1951-2003

Type of firearm	1951-80		1981-92		1993-2003	
	Number	%	Number	%	Number	%
Revolver	16,868	69	23,197	77	22,402	60
Pistol	3,495	14	3,802	12	9,865	26
Garrucha (single- or double-barrel single-shot handgun)	378	1	1,080	3	1,617	4
Fuzil (assault rifle)	7	0	38	0	1,579	4
Carabina (carbine)	196	0	627	2	466	1
Metralhadora (machine gun)	12	0	97	0	442	1
Arma tiro a tiro (firearm without deposit or charger)	3,433	14	953	3	264	0
Submetralhadora (sub-machine gun)	8	0	77	0	192	0
Escopeta (shotgun)	4	0	43	0	25	0
Garruchão (long version of single- or double-barrel shotgun)	17	0	56	0	23	0
Artesanal (craft gun)	0	0	5	0	22	0
Rifle	17	0	27	0	11	0
Mosquetão[38]	2	0	14	0	9	0
Bazooka	0	0	4	0	9	0
Caneta (pen gun)	1	0	2	0	4	0
Laça granada (grenade launcher)	0	0	0	0	2	0
Lança rojão (rocket launcher)	0	0	0	0	1	0
Other	6	0	19	0	16	0
No information	1	0	5	0	10	0
Total	**24,445**	**100**	**30,046**	**100**	**36,959**	**100**

Source: Rivero (2005, p. 217)

officially registered. As one study notes: 'Huge stockpiles of military weapons, often poorly guarded and controlled by corrupt officials[,] are ending up in the hands of criminal gangs and drug cartels [that operate] throughout Central America and beyond' (Graham, 2011). The Honduran newspaper *El Heraldo* recently reported on an investigation into the whereabouts of approximately 3,000 guns collected between the years 2002 and 2006 (Cullinan, 2011).

The diversion of firearms from military and law enforcement institutions, as well as private security companies, is reportedly not uncommon.[39] A study on state military surplus weapons and ammunition in South America finds that three-quarters of the region's surplus is located in Argentina and Brazil (Karp, 2009). The same study observes that in **Brazil** poor stockpile security and corruption leads to widespread firearms diversion from law enforcement agencies into criminal hands (Karp, 2009, p. 31). The Oversight Division for Arms and Explosives in Rio de Janeiro maintains a registry of weapons seized between 1951 and 2003 in the context of criminal activities, illegal possession, and irregular use (such as permit lapses or alcohol consumption when carrying a firearm). The registry, which was cleaned to retain only crime weapons, shows that diversion from official institutions is not unusual (see Box 1.3).

In **Argentina**, cutbacks in the army in the 1980s and 1990s generated a surplus of at least 400,000 small arms and light weapons.[40] A report on surplus arms in South America suggests that many of these weapons have been diverted. Indeed, a significant number of semi- and fully automatic weapons originating from the Argentine army have been seized from criminals in Argentina and neighbouring countries (Karp, 2009, p. 27). Beyond these, most of the arms involved in crime in Argentina are Argentine-made. Of all the arms seized between 1992 and 2001 in the city of Mendoza after a crime, nearly 74 per cent were Argentine-made. Overall, 37.5 per cent of the arms seized were .22- and .32-calibre revolvers made by the Argentine firm Pasper (Appiazola, 2001, pp. 3–7).[41]

Despite such information, further research is needed on the origins and the legal status of firearms and ammunition, the dynamics of the illicit markets, and the mechanism of firearms involvement in armed violence.

The relationship between the availability of firearms and homicide rates is not straightforward.

MAPPING AN AGENDA FOR FUTURE RESEARCH

As discussed above, numerous factors may explain why the relationship between high homicide rates and the high proportion of firearm homicides is especially pronounced in a significant number of countries in Latin America and the Caribbean. These include the availability of firearms, the presence of organized crime and certain kinds of youth gangs, and impunity regarding firearm homicides. This section explores these factors and highlights gaps in our knowledge that further research might usefully address.[42]

Availability of firearms

The widespread availability of firearms in Latin America and the Caribbean is often cited as an important cause of high levels of armed violence. Yet the relationship between the availability of firearms and homicide rates is not straightforward. In 2007, the Small Arms Survey estimated the civilian firearms possession rates in 178 countries in the world (both legally and illegally owned firearms). A comparison of global possession rates shows that—despite the inclusion of estimates on illegal firearms—Central America, the Caribbean, and South America have comparatively low possession rates. Central America had an average rate of 6.8 per 100, the Caribbean stood at 4.7, and South America at 12.1. In comparison, Western Europe—a part of the world with far lower proportions of firearm homicides—has an estimated average possession rate of 24.9 per 100.[43]

These figures do not differentiate between legal and illegal firearms. In the context of firearm homicides, the availability of illegal firearms may be the more relevant indicator. As highlighted earlier, actual and potential perpetrators may have access to illicit firearms through black markets or as a result of diversion from state institutions. More research is needed to establish whether homicides are generally committed with legal or illegal firearms in Latin America and the Caribbean.

The illicit use of firearms depends not only on their availability but also on other factors, including measures to prevent civilian and criminal access to guns. In most cases it is extremely difficult to evaluate the effects of such measures on levels of armed violence. The impact of 'softer' measures, such as community awareness-raising efforts and gun buy-back programmes, is equally difficult to assess. A recent study on the effects of bans on carrying firearms in Colombian cities confirms that such restrictions contributed to a drop in the number of gun injuries (both fatal and non-fatal). Yet the study also finds that the positive effects largely depended on the efficiency of the law enforcement institutions tasked with enforcing the ban and confiscating guns (Restrepo and Villa, 2011). Whether a tightening of civilian gun regulations can reduce homicide rates is a question that calls for further research.

Little is known about the relationship between the presence of youth gangs and homicide rates.

The presence of youth gangs

Analysis of police statistics across different countries in the Americas, Asia, and Europe demonstrates a close link between the proportion of homicides committed with firearms and the proportion of homicides related to youth gangs (Geneva Declaration Secretariat, 2011, p. 100). In comparison to countries in Asia and Europe, states in the Americas show not only a significantly higher proportion of homicides committed with firearms but also a significantly higher share of all homicides that are attributed to gang violence.

A recent report by UNODC (2011) draws attention to the fact that, in the Americas, young men are much more frequently killed with firearms than in other parts of the world. In the 46 countries in the Americas for which UNODC has available data, a 15–34-year-old man is about six times more likely to be killed with a firearm than with a knife. In contrast, he would be almost as likely to be killed with a knife as with a firearm in the 17 Asian countries under review (UNODC, 2011, p. 40). According to UNODC, this demographic analysis suggests a higher involvement of youth gangs—which are typically composed of young men—in firearm homicides in the Americas.

Despite these claims, it remains unclear which gangs are actually involved in homicides, to what extent, and in what context. While limited evidence is available, there is no national-level breakdown of 'gang presence' or longitudinal data, such as 'increased gang presence'. A recent World Bank report states that:

> *while gangs are doubtless a major contributor to crime in El Salvador, Guatemala and Honduras, the very limited evidence indicates that they are responsible for only a minority share of violence; multiple sources suggest that perhaps 15 percent of homicides are gang related* (World Bank, 2011, p. ii).

Little is known about the relationship between the presence of youth gangs and homicide rates. Systematic data collection and analysis are required to explore possible links between gang membership and the perpetration of homicide.

Organized crime

Research suggests that the extent to which gang members and other perpetrators possess and use firearms varies according to the general availability of and obstacles to buying an illicit gun, as influenced by the presence of a black market or the ease of firearm smuggling (Small Arms Survey, 2010, p. 111). This chapter highlights that firearms smuggling and black markets, on the one hand, and organized crime and drug traffickers, on the other, frequently form two sides of the same coin.

Agozino et al. (2009) explain that the illicit drug and weapons markets are related in several ways. First, guns are required to ensure contracts in an illicit market where there are often no other guarantors. The availability of guns among those involved in the drug market leads to their increased use, such as when drug deals 'go bad'. But once the guns have arrived in local communities, they may also be used in situations not directly connected with the drug market (Agozino et al., 2009, p. 293).

In its recent report on crime and violence in Central America, the World Bank (2011) concluded that the principal driver of violence is, in fact, the illegal drug trade, outranking other possible factors, such as the prevalence of youth gangs, the availability of firearms, and the legacy of conflict. Yet little is still known about the ties between the illegal drug trade, the availability of illicit firearms, and the use of firearms to commit homicides. Comprehensive analysis of the context of homicides, the types of perpetrators, and the type and origin of the instrument involved is still hampered by a lack of data.

Firearm homicides and impunity

As suggested in previous sections, firearms homicides are less likely to be solved than homicides committed with knives or other instruments. One reason for this is that—in contrast to homicides committed with knives or other bladed weapons—firearms do not require physical contact between the victim and the offender (Addington, 2006; Riedel, 2008). As one study points out:

> *Many countries have limited forensics capacity for detailed ballistics analysis, and the lack of close contact between the victim and the offender means that few types of physical evidence (such as the offender's hair, blood, or fingerprints) are left behind* (Geneva Declaration Secretariat, 2011, p. 102).

In addition, firearm homicides that are related to gang activities and drug-trafficking organizations are even less likely to be solved, as witnesses are often reluctant to come forward due to a fear of reprisals.

These findings imply that there may be a connection between high rates of homicide, high rates of homicides related to organized crime, frequent use of firearms, and low police and law enforcement performance in solving gun homicides. Indeed, '[c]ountries showing this combination of factors risk entering a spiral of increasing violence and impunity' (Geneva Declaration Secretariat, 2011, p. 88). In the case of Guatemala, impunity in relation to firearm homicides and organized crime also has to do with the fact that drug traffickers have more money, more firepower, and more mobility than the police—especially the ability to disperse and regroup quickly (ICG, 2011). As summed up by a Guatemalan police officer: 'Here we are with our *pistolitas,* and they have automatic rifles' (ICG, 2011, p. 10). The question of whether there are direct causal relationships between impunity around organized crime, weak law enforcement capacities, and high firearm homicides calls for further research.

CONCLUSION

Many countries in Latin America and the Caribbean experience significantly higher homicide rates than other parts of the world. This chapter finds that the region also suffers from significantly higher proportions of homicides committed with firearms. In places with high or increasing levels of violence, the firearm—an effective instrument for committing homicides—is used even more frequently.

At the same time, the chapter shows that there are significant differences among countries in Latin America and the Caribbean. A number of them—mainly in the Southern Cone—exhibit homicide rates below 10 per 100,000, with firearms used in fewer than 60 per cent of cases. At the other end of the spectrum are countries with homicide rates of more than 30 and even exceeding 80 per 100,000, with firearm homicides accounting for up to 90 per cent of all cases. These include the three countries in the Northern Triangle—El Salvador, Guatemala, and Honduras—as well as Colombia and Venezuela.

A literature review suggests that pistols and revolvers are the weapons most frequently used in firearm homicides in Latin America and the Caribbean. At the same time, the trafficking and use of heavier firearms, such as assault rifles or sub-machine guns, is a feature of large-scale drug trafficking organizations. Evidence suggests that the majority of firearms used in crimes are illegal, yet little is known about the legal status of firearms used in armed violence in Latin America and the Caribbean.

While the chapter has shed light on the patterns and distribution of homicides and firearm homicides in the region, further analysis is needed to refine our understanding of this complex phenomenon. Of the topics requiring attention, the most pressing include the factors underpinning homicides, including firearm homicides. Illuminating perpetrators' access to guns, especially through illicit markets, is also an important area for assessing—and ultimately addressing—armed violence in the region.

LIST OF ABBREVIATIONS

AUC	Autodefensas Unidas de Colombia (United Self-Defence Forces of Colombia)
IBIS	Integrated Ballistics Identification System
IML	Instituto de Medicina Lega (Institute of Forensic Medicine, El Salvador)
UNODC	United Nations Office on Drugs and Crime

ANNEXE

Online annexe at <http://www.smallarmssurvey.org/de/publications/by-type/yearbook/small-arms-survey-2012.html>

Annexe 1.1. Trends in firearm homicides, 1995-2010

Annexe 1.1 comprises 23 graphs showing trends in homicide rates between 1995 and 2010 and the percentage of firearm homicides for Argentina, Brazil, Chile, Colombia, Costa Rica, Cuba, the Dominican Republic, Ecuador, El Salvador, Guatemala, Guyana, Honduras, Jamaica, Mexico, Nicaragua, Panama, Paraguay, Peru, Puerto Rico, Suriname, Trinidad and Tobago, Uruguay, and Venezuela.

ENDNOTES

1 For an overview of the research on violence in Latin America and the Caribbean, see Imbusch, Misse, and Carrión (2011).
2 The terms 'firearms', 'arms', and 'guns' are used interchangeably in this chapter.
3 The International Classification of Disease code for 'assault' is X85–Y09; it excludes injuries due to legal intervention and operations of war (WHO, n.d.).
4 Interview with an IML representative conducted by Matthias Nowak, San Salvador, 23 August 2011.
5 In Colombia, for example, three different sources provide data on homicides: 1) the National Police, 2) the National Administrative Department of Statistics (DANE), which produces public health data, and 3) the National Institute of Legal Medicine and Forensic Sciences. This study makes use of police data. The police report lower homicide rates than do the other two sources, but they provide the longest time series of homicide rates (see Aguirre, Moscoso, and Restrepo, 2011).
6 This chapter groups countries in line with UN geographical regions (UNSD, n.d.a). For purposes of this chapter, the 34 countries under review are categorized as follows. *Central America* is composed of Belize, Costa Rica, El Salvador, Guatemala, Honduras, Mexico, Nicaragua, and Panama.

The *Caribbean* comprises Anguilla, Antigua and Barbuda, the Bahamas, Barbados, Cuba, Dominica, the Dominican Republic, Grenada, Jamaica, Puerto Rico, Saint Kitts and Nevis, Saint Lucia, Saint Vincent and the Grenadines, and Trinidad and Tobago. *South America* contains Argentina, Bolivia, Brazil, Chile, Colombia, Ecuador, Guyana, Paraguay, Peru, Suriname, Uruguay, and Venezuela.

7   Anguilla, Antigua and Barbuda, Bolivia, and Saint Lucia are not included in this calculation due to a lack of data on the proportion of firearm homicides; the Bahamas, Barbados, Belize, Dominica, Grenada, Saint Kitts and Nevis, and Saint Vincent and the Grenadines are excluded because they have populations below 500,000 (UN, n.d.b). In countries with such small populations, small changes in the number of homicides can lead to large changes in the rate, which makes it difficult to generate reliable rates per 100,000 population for a specific year within an acceptable confidence interval.

8   Elaboration of Geneva Declaration Secretariat (2011, p. 99).

9   A breakdown of national homicide trends and the percentage of firearm homicides in relation to the total homicide rate can be found in the annexe to this chapter.

10 Of the total of 34 countries (see endnote 6) Anguilla, Antigua and Barbuda, Bolivia, and Saint Lucia are excluded from Figure 1.1 due to a lack of data on the proportion of firearm homicides.

11 Trends in countries in Latin America and the Caribbean are analysed in more detail in the next section.

12 See endnote 7.

13 Interview with a representative of the Ministry of Public Health conducted by Katherine Aguirre, Havana, Cuba, August 2011.

14 See endnote 6.

15 Countries with a population of less than 500,000 are excluded (Anguilla, Antigua and Barbuda, the Bahamas, Barbados, Belize, Dominica, Grenada, Saint Kitts and Nevis, Saint Lucia, and Saint Vincent and the Grenadines), see endnote 7.

16 See UNODC (2007).

17 The 23 countries are Argentina, Brazil, Chile, Colombia, Costa Rica, Cuba, Dominican Republic, Ecuador, El Salvador, Guatemala, Guyana, Honduras, Jamaica, Mexico, Nicaragua, Panama, Paraguay, Peru, Puerto Rico, Suriname, Trinidad and Tobago, Uruguay, and Venezuela; see endnote 7.

18 See also Geneva Declaration Secretariat (2011, p. 64).

19 For a more detailed review of the characteristics of drug-related violence and of the types of firearms used in Mexico, see the chapter on drug violence in this volume (DRUG VIOLENCE).

20 Author interview with Dennis Rodgers, 21 December 2011.

21 See Rodgers, Muggah, and Stevenson (2009) and Rodgers (2006).

22 For literature on Jamaica, see, for example, Clarke (2006) and USAID (2007).

23 Cited in Leslie (2010, p. 33).

24 The research, undertaken in 2009, also finds that police efforts to reduce gun violence are sometimes mitigated by the government's direct financial support to urban gangs. As Townsend notes: 'In exchange, come election days, these gangs have been frequently called upon to turn out loyal supporters and physically menace would-be opposition voters' (Townsend, 2009, p. 15).

25 For literature on violence in Venezuela, see Briceño-León (2006) and Leon (2010).

26 Author interview with William Godnick, public security coordinator, United Nations Regional Centre for Peace, Disarmament and Development in Latin America and the Caribbean (UN-LiREC), 9 December 2011.

27 For a review of the effect of multi-institutional efforts to prevent and reduce crime and violence in urban centres in Colombia, see Guerrero (1999), Villaveces et al. (2000), and Aguirre and Restrepo (2010).

28 See Spagat (2008) and Rozema (2008).

29 See Pachico (2011) and Ortiz and McDermott (2011).

30 That said, the World Bank reports that inequality in Colombia is greater than in Venezuela; see World Bank (n.d.).

31 The factory is due to open in 2012 (*Universal*, 2011).

32 The population statistics are based on IBGE (2008).

33 On firearms in Argentina, Paraguay, and Uruguay, see Fleitas (2006).

34 For example, in May 2011, Mexican and Belizean government officials inaugurated a training course in ballistic investigation for members of the Belize Police Department. The training included an introduction to IBIS (Embassy of Mexico in Belize, 2011).

35 On the production of firearms in Latin America, see Klare and Andersen (1996) and Small Arms Survey (2004).

36 See Dreyfus et al. (2010).

37 The coalition included Viva Rio in Rio de Janeiro, Convive in Brasília, and the Instituto Sou da Paz in São Paulo, among others (FLACSO, 2007).

38 *Mosquetão* refers to an old type of rifle with a short barrel.

39 Latin America stands out as the region in which private security companies are the most heavily armed. They are a source not only of legal demand for firearms and ammunition, but also of diversion to illicit actors (Small Arms Survey, 2011, p. 114).

40 See UNGA (1997) for what weapons are included in the small arms and light weapons categories.

41 The United States is the second-most represented country of origin, accounting for 5.5 per cent of the weapons seized (Appiazola, 2001, pp. 3–7).

42 While this chapter has touched on socio-economic and structural factors that may affect levels of crime and violence in Latin America and the Caribbean—including economic inequality and overall levels of poverty—further exploration of these themes is beyond its scope. For more analysis of the main underlying factors of high crime rates in the region—high inequality, low police presence, and low incarceration rates—see Soares and Naritomi (2010).

43 Based on Karp (2007).

BIBLIOGRAPHY

Addington, Lynn A. 2006. 'Using National Incident-Based Reporting System Murder Data to Evaluate Clearance Predictors.' *Homicide Studies*, Vol. 10, No. 2, pp. 140–52.

Agozino, Biko, et al. 2009. 'Guns, Crime and Social Order in the West Indies.' *Criminology and Criminal Justice*, Vol. 9, pp. 287–305.

Aguirre, Katherine, Manuel Moscoso, and Jorge Restrepo. 2011. *¿Qué hay detrás de las diferencia de los datos de homicidios en 2009?* 11 July.
<http://blog.cerac.org.co/%C2%BFque-hay-detras-de-las-diferencia-de-los-datos-de-homicidios-en-2009>

— and Jorge Restrepo. 2010. 'El control de armas como estrategia de reducción de la violencia en Colombia: pertinencia, estado y desafíos.' *Revista Criminalidad*, Vol. 52, No. 1, pp. 265–84.

—, et al. 2009. *Assessing the Effect of Policy Interventions on Small Arms Demand in Bogotá, Colombia*. Documentos de CERAC No. 14. Bogotá: Conflict Analysis Resource Center (CERAC). December.

Appiazola, Martín. 2001. *Contra los mitos: un análisis estadístico de armas secuestradas en Mendoza, Periodo 1992–2001*. September.
<http://www.martinappiolaza.com/2001/09/informe-contra-los-mitos-un-analisis.html>

ATF (United States Bureau of Alcohol, Tobacco, Firearms and Explosives). 2010. 'Puerto Rico: January 1, 2010–December 31, 2010.' Washington, DC: ATF, Department of Justice. <http://www.atf.gov/statistics/download/trace-data/2010/2010-trace-data-puerto-rico.pdf>

Bandeira, Antonio and Josephine Bourgois. 2006. *Armas de fuego: ¿Protección? ¿O Riesgo?* Stockholm: Foro Parlamentario sobre Armas Pequeñas y Ligeras.

Bosworth, James. 2010. *Honduras: Organized Crime Gaining Amid Political Crisis*. Washington, DC: Woodrow Wilson Center for Scholars.

Briceño-León, Roberto. 2006. 'Violence in Venezuela: Oil Rent and Political Crisis.' *Ciencia e Saúde Colectiva*, Vol. 11, No. 2, pp. 315–25.

Clarke, Colin. 2006. 'Politics, Violence and Drugs in Kingston, Jamaica.' *Bulletin of Latin American Research*, Vol. 25, No. 3, pp. 420–40.

Cullinan, Jeanna. 2011. '3,000 Guns "Disappear" in Honduras.' 10 November.
<http://insightcrime.org/insight-latest-news/item/1827-3000-guns-disappear-in-honduras>

DANE (Departamento Administrativo Nacional de Estadística). n.d. 'Defunciones no fetales.' Bogotá: DANE.
<http://www.dane.gov.co/index.php?option=com_content&view=article&id=788&Itemid=119>

Davis, Diane. 2006. 'The Age of Insecurity: Violence and Social Disorder in the New Latin America.' *Latin American Research Review*, Vol. 41, No. 1, pp. 178–97.

Dreyfus, Pablo. 2007a. 'Dime con qué armas andan, y te diré qué campaña quieres.' In Khatchick DerGhougassian, ed. *Las armas y las víctima: Violencia, proliferación y uso de armas de fuego en la provincia de Buenos Aires y la Argentina*. Buenos Aires: Universidad de San Andrés.

—. 2007b. 'Armas pequenas e leves: controle do tráfico ilegal no caso do Brasil.' IV Conferência do Forte de Copacabana Rio de Janeiro. 16 November.

— and Nicholas Marsh. 2006. *Tracking the Guns: International Diversion of Small Arms to Illicit Markets in Rio de Janeiro*. Rio de Janeiro and Oslo: Viva Rio, Institute for Studies on Religion, and International Peace Research Institute, Oslo.
<http://www.prio.no/upload/943/Tracking%20the%20guns_All.pdf>

—, et al. 2010. *Small Arms in Brazil: Production, Trade, and Holdings*. Special Report No. 11. Geneva: Small Arms Survey, Viva Rio, and Institute for Studies on Religion.

Embassy of Mexico in Belize. 2011. 'Mexico Provides Training on Ballistics to Belizean Police.' 10 May.
<http://portal.sre.gob.mx/belice_eng/index.php?option=news&task=viewarticle&sid=174>

FBSP (Fórum Brasileiro de Segurança Pública). 2010. *Anuário do Fórum Brasileiro de Segurança Pública*.
<http://www2.forumseguranca.org.br/node/24104>

FLACSO (Facultad Latinoamericana de Ciencias Sociales). 2007. *Armas pequeñas y livianas: una amenaza a la seguridad hemisférica*. San José, Costa Rica: FLACSO.

Fleitas, Diego M. 2006. *El problema de las armas de fuego en el Cono Sur: Los casos de Argentina, Paraguay, y Uruguay*. Documento de Trabajo No. 1. San José, Costa Rica: Facultad Latinoamericana de Ciencias Sociales (FLACSO).

—. 2010. *La Seguridad Ciudadana en Argentina y su relación con el Contexto Regional*. Buenos Aires: Facultad Latinoamericana de Ciencias Sociales (FLACSO).

Galeotti, Mark. 2006. 'Organized Crime Gangs Pose Threat to Cuban Development.' *Jane's Intelligence Review*, Vol. 18, No. 2.

Geneva Declaration Secretariat. 2011. *Global Burden of Armed Violence 2011: Lethal Encounters*. Cambridge: Cambridge University Press.

Godnick, William, Robert Muggah, and Camilla Waszink. 2002. *Stray Bullets: The Impact of Small Arms Misuse in Central America*. Occasional Paper No. 5. Geneva: Small Arms Survey.

Goertzel, Ted and Tulio Kahn. 2009. 'The Great São Paulo Homicide Drop.' *Homicide Studies*, Vol. 13. November, pp. 398–410.

Graham, Ronan. 2011. 'Honduras Guns Feeding Central America's Arms Trade.' 12 August.
<http://insightcrime.org/insight-latest-news/item/1398-honduras-guns-feeding-central-americas-arms-trade>

Guerrero, Rodrigo. 1999. 'Programa Desarrollo, Seguridad y Paz, DESEPAZ de la Ciudad de Cali.' Rio de Janeiro: Inter-American Development Bank.
<http://idbdocs.iadb.org/wsdocs/getdocument.aspx?docnum=362232>

Hoover, William. 2008. 'Statement of William Hoover, Assistant Director for Field Operations, Bureau of Alcohol, Tobacco, Firearms and Explosives before the United States House of Representatives Committee on Foreign Affairs Subcommittee on the Western Hemisphere.' 7 February.
<http://foreignaffairs.house.gov/110/hoo020708.htm>

Hudson, Rex. 2003. *Terrorist and Organized Crime Groups in the Tri-Border Area (TBA) of South America*. Washington, DC: Federal Research Division of the Library of Congress. July.

IBGE (Instituto Brasileiro de Geografia e Estatística). 2008. 'Projeção da População do Brasil por sexo e idade: 1980–2050—Revisão 2008.'
<http://www.ibge.gov.br/lojavirtual/fichatecnica.php?codigoproduto=9066>

ICESI (Instituto ciudadano de estudios sobre la inseguridad—Citizen Institute for Studies on Insecurity). n.d. 'Estadísticas oficiales.'
<http://www.icesi.org.mx/estadisticas/estadisticas_oficiales.asp>
ICG (International Crisis Group). 2011. *Guatemala: Drug Trafficking and Violence*. Crisis Group Latin America Report No. 39. 11 October.
IEPADES (Instituto de Enseñanza para el Desarrollo Sostenible). 2006a. *Control de Armas de Fuego: Manual para la Construcción de la Paz por la Sociedad Civil*. Guatemala City: IEPADES.
—. 2006b. *El tráfico ilícito de armas en Guatemala*. Guatemala: IEPADES.
Imbusch, Peter, Michel Misse, and Fernando Carrión. 2011. 'Violence Research in Latin America and the Caribbean—A Literature Review.' *International Journal of Conflict and Violence (IJCV)*, Vol. 5, No. 1, pp. 87–154.
InSight. 2010. 'Venezuela to Manufacture Kalashnikov Rifles.' 3 December.
<http://insightcrime.org/insight-latest-news/item/305-venezuela-to-start-producing-kalashnikov-rifles>
IUDPAS (Instituto Universitario de Democracia, Paz y Seguridad). 2006. *Observatorio de la Violencia: mortalidad y otros*. Boletín Enero–Diciembre 2005. Tegucigalpa, Honduras: IUDPAS.
—. 2007. *Observatorio de la Violencia: mortalidad y otros*. Boletín Enero–Diciembre 2006. Tegucigalpa, Honduras: IUDPAS.
—. 2008. *Observatorio de la Violencia: mortalidad y otros*. Boletín Enero–Diciembre 2007. Tegucigalpa, Honduras: IUDPAS.
—. 2009. *Observatorio de la Violencia: mortalidad y otros*. Boletín Enero–Diciembre 2008. Tegucigalpa, Honduras: IUDPAS.
Jütersonke, Oliver, Robert Muggah, and Dennis Rodgers. 2009. 'Gangs, Urban Violence, and Security Interventions in Central America.' *Security Dialogue*, Vol. 40, pp. 373–97.
Karp, Aaron. 2007. 'Completing the Count: Civilian Firearms Online—Annexe 4: Civilian Gun Ownership for 178 countries, in Descending Order of Averaged Firearms.' In Small Arms Survey. *Small Arms Survey 2007: Guns and the City*. Cambridge: Cambridge University Press.
—. 2009. *Surplus Arms in South America: A Survey*. Working Paper No. 7. Geneva: Small Arms Survey and the Conflict Analysis Resource Center (CERAC). August.
Klare, Michael and David Andersen. 1996. *A Scourge of Guns: The Diffusion of Small Arms and Light Weapons in Latin America*. Washington, DC: Federation of American Scientists.
Kruger, Mark. 2007. 'Community-Based Crime Control in Cuba.' *Contemporary Justice Review*, Vol. 10, No. 1, pp. 101–14.
Lemard, Glendene and David Hemenway. 2006. 'Violence in Jamaica: An Analysis of Homicides, 1998–2002.' *Injury Prevention*, Vol. 12, pp. 15–18.
Lenis, David, Lucas Ronconi, and Ernesto Schargrodsky. 2010. *The Effect of the Argentine Gun Buy-back Program on Crime and Violence*. Berkeley: University of California, Berkeley. <http://socrates.berkeley.edu/~raphael/IGERT/Workshop/PEVAF_September_27_2010.pdf>
Leon, Daniel. 2010. *The Political Economy of Violence: The Case of Venezuela*. Boca Raton, Florida: Dissertation.com.
León-Beltrán, Isaac de and Luz Janeth Forero. 2004. *Una descripción de las armas de fuego homicidas en Bogotá para el año 2002 y una propuesta para aumentar el costo del servicio de homicidio*. Documento 27. Bogotá: Fundación Método. June.
Leslie, Glaister. 2010. *Confronting the Don: The Political Economy of Gang Violence in Jamaica*. Occasional Paper No. 26. Geneva: Small Arms Survey.
Loor, Kleber, ed. 2004. *Pandillas y Naciones de Ecuador—Alarmante realidad, tarea desafiante: de víctimas a victimarios*. Rio de Janeiro: Children in Organized Armed Violence (COAV). <http://www.coav.org.br/publique/media/Report%20Equador.pdf>
Mizne, Denis. 2011. *Security Issues in LatAm: Challenges and Opportunities Facing the Region*. Presentation at the 6th Annual LatAm CEO Conference. New York, 18–19 May. <http://www.itauconference.com/downloads/itau-denis-mizne.pdf>
Ortiz, Andres and Jeremy McDermott. 2011. 'Murder in Colombia: Reduction and Shift into the Cities.' 24 January.
<http://www.insightcrime.org/insight-latest-news/item/465-murder-in-colombia-reduction-and-shift>
Ortiz, Ildefonso. 2011. *Weapon of Choice: The AK-47's Price, Reliability Garner Fans on Both Sides of the Law*. 13 September.
<http://m.themonitor.com/news/reliability-54703-law-weapon.html>
Pachico, Elyssa. 2011. 'Mafia War Feared in Cali, as Rastrojos Face New Competition.' 28 March.
<http://www.insightcrime.org/insight-latest-news/item/719-mafia-war-feared-in-cali-if-rastrojos-face-new-competition>
Policía Judicial de Costa Rica. 2006. *Casos y personas fallecidas por homicidio doloso en Costa Rica durante el 2006*.
<http://www.poder-judicial.go.cr/planificacion/Estadisticas/Anuarios/anuariopoliciales2006/principal.htm>
—. 2007. *Casos y personas fallecidas por homicidio doloso en Costa Rica durante el 2007*.
<http://www.poder-judicial.go.cr/planificacion/Estadisticas/Anuarios/anuariopoliciales2007/principal.htm>
—. 2008. *Casos y personas fallecidas por homicidio doloso en Costa Rica durante el 2008*.
<http://www.poder-judicial.go.cr/planificacion/Estadisticas/Anuarios/Anuario_Policial_2008/index.htm>
—. 2009. *Casos y personas fallecidas por homicidio doloso en Costa Rica durante el 2009*.
<http://www.poder-judicial.go.cr/planificacion/Estadisticas/Anuarios/Anuario_Policial_%202009/elementos/principal.html>
Restrepo, Jorge and David Aponte, eds. 2009. *Guerra y violencias en Colombia: herramientas e interpretaciones*. Bogotá: Editorial Pontificia Universidad Javeriana.
— and Edgar Villa. 2011. 'Do Bans on Carrying Firearms Work for Violence Reduction? Evidence from a Department-Level Ban in Colombia.'
<http://www.depeco.econo.unlp.edu.ar/cedlas/ien/pdfs/meeting2011/papers/Villa.pdf>
Riedel, Marc. 2008. 'Homicide Arrest Clearances: A Review of the Literature.' *Sociology Compass*, Vol. 2, Iss. 4, pp. 1145–64.
Rivera-Lyles, Jeannette. 2007. 'Airlines' Lax Rules Fuel Lethal Trade.' *Orlando Sentinel*. 18 March.
<http://www.orlandosentinel.com/news/nationworld/orl-asecprguns18031807mar18,0,1670758.story>
Rivero, Patricia S. 2005. 'O mercado ilegal de armas de fogo na cidade do Rio de Janeiro: Preços e simbologia das armas de fogo no crime.' In Rubem César Fernández, ed. *Brasil: as armas e as victimas*. Rio de Janeiro: Institute for Studies on Religion.
Rodgers, Dennis. 2006. 'Living in the Shadow of Death: Gangs, Violence and Social Order in Urban Nicaragua, 1996–2002.' *Journal of Latin American Studies*, Vol. 38, Iss. 2, pp. 267–92.

—, Robert Muggah, and Chris Stevenson. 2009. *Gangs of Central America: Causes, Costs, and Interventions*. Working Paper 23. Geneva: Small Arms Survey.

Romero, Simon. 2010. 'Venezuela, More Deadly Than Iraq, Wonders Why.' *The New York Times*. 22 August.
<http://www.nytimes.com/2010/08/23/world/americas/23venez.html?pagewanted=all>

Rozema, Ralph. 2008. 'Urban DDR Processes: Paramilitaries and Criminal Networks in Medellín, Colombia.' *Journal of Latin American Studies*, Vol. 40, pp. 423–52.

Small Arms Survey. 2004. *Small Arms Survey 2004: Rights at Risk*. Oxford: Oxford University Press.

—. 2006. *Small Arms Survey 2006: Unfinished Business*. Cambridge: Cambridge University Press.

—. 2010. *Small Arms Survey 2010: Gangs, Groups, and Guns*. Cambridge: Cambridge University Press.

—. 2011. *Small Arms Survey 2011: States of Security*. Cambridge: Cambridge University Press.

Soares, Rodrigo R. and Joana Naritomi. 2010. 'Understanding High Crime Rates in Latin America: The Role of Social and Policy Factors.' In Rafael Di Tella, Sebastian Edwards, and Ernesto Schargrodsky, eds. *The Economics of Crime: Lessons for and from Latin America*. Chicago: University of Chicago Press, pp. 19–55.

de Souza, Maria de Fátima Marinho, et al. 2007. 'Reductions in Firearm-Related Mortality and Hospitalizations in Brazil after Gun Control.' *Health Affairs*, Vol. 26, No. 2, pp. 575–84.

Spagat, Michael. 2008. *Colombia's Paramilitary DDR: Quiet and Tentative Success*. Bogotá: Centro de Recursos para el Análisis de Conflictos (CERAC).

Spinelli, Hugo, Guillermo Macías, and Victoria Darraidou. 2008. 'Procesos macroeconómicos y homicidios: Un estudio ecológico en los partidos del Gran Buenos Aires (Argentina) entre los años 1989 y 2006.' *Salud colectiva*, Vol. 4, No. 3, pp. 283–99.

Spinelli, Hugo, et al. 2006. 'Firearm-related Deaths and Crime in the Autonomous City of Buenos Aires, 2002.' *Ciencia e Saúde Colectivo*, Vol. 11, No. 2, pp. 327–38.

Stratfor. 2008. 'Organized Crime in Cuba.' 16 May.

Townsend, Dorn. 2009. *No Other Life: Gangs, Guns, and Governance in Trinidad and Tobago*. Working Paper No. 8. Geneva: Small Arms Survey.

Tulyakov, Ivan. 2010. 'Venezuela Launches Official Production of Third-Generation Kalashnikov Rifles.' 14 July.
<http://english.pravda.ru/world/americas/14-07-2010/114222-kalashnikov-0/>

Turbiville, Graham H. Jr. 2010. 'Firefights, Raids, and Assassinations: Tactical Forms of Cartel Violence and Their Underpinnings.' *Small Wars and Insurgencies*, Vol. 21, No. 1. March, pp. 123–44.

UNGA (United Nations General Assembly). 1997. *Report of the Panel of Governmental Experts on Small Arms*. A/52/298. 27 August.
<http://www.un.org/Depts/ddar/Firstcom/SGreport52/a52298.html>

UNSD (United Nations Statistics Division). n.d.a. 'Composition of Macro Geographical (Continental) Regions, Geographical Sub-regions, and Selected Economic and Other Groupings: Geographical Region and Composition.' <http://unstats.un.org/unsd/methods/m49/m49regin.htm#ftnb>

—. n.d.b. 'UNData: Country Profiles.' <http://data.un.org/CountryProfile.aspx?crName=Antigua%20and%20Barbuda>

Universal, El (Caracas). 2011. 'Venezuela iniciará la producción de Kalashnikov en 2012.' 17 September.
<http://www.eluniversal.com/2011/09/17/venezuela-iniciara-la-produccion-de-kalashnikov-en-2012.shtml>

UNODC (United Nations Office on Drugs and Crime). 2007. *Crime and Development in Central America: Caught in the Crossfire*. Mexico City: UNODC.

—. 2011. *Global Homicide Report*. Vienna: UNODC.

—. n.d. 'UNODC Homicide Statistics.' <http://www.unodc.org/unodc/en/data-and-analysis/homicide.html>

USAID (United States Agency for International Development). 2007. *Guns, Gangs, and Governance: Towards a Comprehensive Gang Violence Prevention Strategy*.

US GAO (United States Government Accountability Office). 2009. *Firearms Trafficking: U.S. Efforts to Combat Arms Trafficking to Mexico Face Planning and Coordination Challenges*. GAO-09-709. June.

Villaveces, Andrés, et al. 2000. 'Effect of a Ban on Carrying Firearms on Homicide Rates in 2 Colombian Cities.' *Journal of the American Medical Association*, Vol. 283, No. 9, pp. 1205–09.

World Bank, 2011. *Crime and Violence in Central America: A Development Challenge*. Washington, DC: World Bank.

—. n.d. 'GINI Index.' <http://data.worldbank.org/indicator/SI.POV.GINI>

WHO (World Health Organization). n.d. 'International Statistical Classification of Diseases and Related Health Problems: 10th Revision.'
<http://apps.who.int/classifications/apps/icd/icd10online/>

Zárate, Milton G. 2006. *Delincuencia organizada transnacional y el uso de armas de fuego: Nueva amenaza para la seguridad hemisférica y sus desafíos para combatirla (caso Ecuatoriano)*. Thesis. Buenos Aires: Universidad del Salvador.

ACKNOWLEDGEMENTS

Principal author

Elisabeth Gilgen

Contributors

Katherine Aguirre and Matthias Nowak

Mexican soldiers burn drugs seized during military operations, Tijuana, December 2011.
© Guillermo Arias/Xinhua/Eyevine/Redux

When Business Gets Bloody
STATE POLICY AND DRUG VIOLENCE

2

INTRODUCTION

The term 'drug violence' can evoke a variety of images and contexts—from interpersonal aggression by addicts and turf wars among corner dealers in retail settings to full-blown militarized confrontations among powerful, heavily armed cartels. While Latin America is home to the full spectrum of drug violence, a number of countries in the region suffer acute, destabilizing armed violence featuring large, well-armed drug trafficking operations in conflict with one another and state forces.

Mexico is the extreme case. Since President Felipe Calderón called in the army to wage an all-out war on the country's drug cartels in December 2006, more than 47,000 lives have been lost in a maelstrom of violence (Cave, 2012). While the government's crackdown has fragmented the cartels into smaller organizations, many splinter groups have proven just as violent as their predecessors. In fact, cartel violence has only grown in intensity, lethality, and brazenness since the crackdown. In addition to spiralling violence inside the country, the fragmentation of the Mexican cartels now threatens to alter the dynamics of the drug trade landscape in El Salvador, Guatemala, and Honduras.

Rio de Janeiro presents another facet of the drug violence landscape. In this city, where prison-based drug syndicates have held territorial control over the favelas (shantytowns) for more than two decades, state security forces began a new programme in 2008 to retake and then occupy favelas with long-term community-oriented police forces. In contrast to previous approaches in Rio and elsewhere in Latin America, the programme prioritizes the most violent of syndicates and aims not to eradicate the illicit drug trade but to reduce the worst of drug-related violence and re-establish state authority.

This chapter reviews recent empirical trends and theoretical explanations of drug violence in Latin America, with a focus on armed violence between organized actors—particularly cartels and prison-based syndicates—and state forces in Mexico, the Northern Triangle of Central America, and Brazil. Among the chapter's findings:

- The onset and intensity of systemic drug violence on the scale seen in Mexico and Rio de Janeiro are highly variable and sensitive not only to drug trafficking and market structure, but also to state anti-narcotics policy and enforcement.
- In Mexico, a blanket crackdown has led to numerous arrests and fragmented some of the larger cartels; however, violence both among cartels and between cartels and the state has risen dramatically and continuously since President Calderón brought in the Mexican army to combat drug trafficking in late 2006.
- Mexican cartels—responding in part to the crackdown in Mexico—are establishing footholds in Central America, especially Guatemala and Honduras, destabilizing local relations among 'native' organized crime groups and threatening to overrun weak police and armed forces.
- In Rio de Janeiro, the state has regained control over more than 20 favelas, including some of the city's largest, from the prison-based drug trafficking syndicates that previously controlled them. These syndicates appear to be

shifting from strategies of armed dominion and confrontation to non-violent low-level dealing. But it is too early to tell if this systematic 'pacification' programme will result in sustained reductions in armed violence.

The chapter is divided into four main sections. The first presents an overview of the actors, state policies, and types of violence associated with the drug trade. The second section focuses on Mexico and the complex factors that are driving the escalation of violence there. The third section turns to Central America, especially the 'Northern Triangle' countries (El Salvador, Guatemala, and Honduras), where Mexican drug-trafficking organizations seem to be making important inroads, with potentially disastrous implications for security in the region. The fourth section examines the state's systematic 'pacification' of drug syndicate-controlled slums in Rio de Janeiro. The chapter ends with some concluding reflections.

DRUG VIOLENCE IN THE AMERICAS: ACTORS, VIOLENCE, AND STATE RESPONSES

The so-called 'drugs–violence nexus' has been a topic of active research and debate for decades, especially in high-consumption countries such as the United States and the United Kingdom.[1] The focus of much of this research is on violence related to drug use, as well as on the dynamics of retail drug markets. Such violence can be immensely costly to society, but it seems qualitatively different from the sort of organized armed attacks being carried out, for example, by Mexico's drug cartels. While Latin America is certainly not free from drug consumption and related types of violence, it has witnessed episodes of extreme internecine fighting among large, powerful trafficking organizations, and, at times, brazen anti-state violence that evokes comparisons to civil war. Latin America is also unique in the sense that the production and transhipment of drugs (towards retail markets in the United States, Europe, and other wealthy destinations) often outstrip retail and consumption as the most important drug-related economic activities. This section examines the various actors who populate different drug market sectors and how they produce numerous types of violence; it then surveys policy responses by states and how these may curtail or aggravate violence.

Actors

The international drug trade involves a host of different actors engaged in numerous illicit activities. One basic distinction can be made between consumers and traffickers; as detailed below, violence stemming from consumption and state responses to it are qualitatively different from that arising from the dynamics of drug dealing. Yet drug trafficking operations also vary significantly in size, type, and degree of organization. Categories are slippery; the term 'gangs' can mean anything from street-corner outfits with a handful of members to powerful prison-based networks such as Rio de Janeiro's Comando Vermelho, Central America's *maras,* and Southern California's Mexican Mafia (Skarbek, 2011; Lessing, 2010). Larger organizations run the gamut from business-oriented, organized crime operations—such as Colombia's Cali Cartel in the 1990s and the traditional organized crime families of Central America—to more violent but non-ideological groups such as Pablo Escobar's Medellín Cartel or Mexico's Los Zetas, to revolutionary and ideological insurgencies that have parleyed their control of drug-plant cultivation areas into a permanent source of revenue, such as the Revolutionary Armed Forces of Colombia (FARC) or Afghanistan's Taliban.

One useful criterion for categorization is the identification of a market sector within which organizations operate; although groups may be active in more than one sector—or even diversify into and from other criminal or insurgent activities—their market sector offers a rough guide to their size, tactical capacity, and incentives.

Producers

Most illicit drugs begin as plants and so require cultivation over significant land areas. Typically, peasant or indigenous populations carry out plantation and harvest, under the protection or duress of armed groups, which extract a large proportion of the rents. The comparative advantage for this sector is territorial control over rural areas, rather than trafficking expertise or contacts, so production is a common point of entry into the drug trade for armed insurgencies and paramilitary groups. Thus, in Colombia, the once strongly ideological FARC is now deeply involved in coca leaf production, and in Afghanistan the Taliban have shifted away from a *fatwa* against poppy cultivation to offering protection to growers (Labrousse, 2005; Felbab-Brown, 2005). In Mexico, where rural insurgency has had far less success, 'pure' wholesaling trafficking organizations, such as the Sinaloa cartel, directly control some marijuana- and poppy-growing regions. Where production is legal or decriminalized, such as in Bolivia (coca) or California (marijuana), there is less need for armed protection from state forces, so producers can operate as small businesses or collectives.

Transhipment and smugglers

This category comprises groups that process and move large quantities of drugs towards retail markets. Many of the largest and most infamous drug-trafficking organizations belong in this category, including the powerful 'cartels' of Colombia and Mexico. While never quite living up to the textbook definition of a cartel, they do represent networks of dealers who join forces to increase profits, and their history is often one of tentative pacts pockmarked with periods of infighting. International smugglers occupy the most lucrative position in the supply chain, buying coca or opium at farm-gate prices and transporting the drugs towards and across the US border at a total mark-up on the order of 2,000 per cent (Reuter and Greenfield, 2001, p. 167). The trade is specialized, requiring contacts with suppliers as well as distributers or retailers in the destination market; it also frequently involves high-level corruption of border and customs officials in multiple countries. These factors help make this one of the most concentrated sectors of the drug trade, while the sheer profitability gives these few organizations immense resources. Within destination countries, wholesale markets appear less concentrated and more like a shifting panorama of autonomous dealers (Reuter and Haaga, 1989).

> International smugglers occupy the most lucrative position in the supply chain.

Retailers

Retail drug trafficking is a highly risky enterprise. It involves a large number of illegal transactions, and dealers must be accessible to consumers without drawing police attention. At the same time, it is highly profitable; once heroin and cocaine have entered the United States, the price per pure gram grows roughly tenfold between wholesale and the final street purchase (Reuter and Greenfield, 2001, p. 166). The result is a competitive and unstable market structure. There may be incentives to expand or even 'corner' entire markets, yet as retailing organizations grow, they become easier and more appealing targets for state repression and predation by rivals and can easily fragment when a leader is arrested or killed. The overall result is that retail sectors tend to be characterized by small trafficking operations—frequently street gangs with local ties to the neighbourhood or slum in which they operate (Dorn, Murji, and South, 1992; Hagedorn, 1994).

While it is difficult for any one retail organization to grow and remain dominant over a large region, a different model of agglomeration involving prison gangs has arisen in some places, with the potential to alter retail markets fundamentally. By consolidating control over inmate life and propagating through entire prison systems, these gangs can project power outward, onto members of street gangs who anticipate future imprisonment (Lessing, 2010). This coercive power can be used to organize local retailers into broader, regional syndicates. For example, the powerful Mexican Mafia prison gang wields a kind of 'governance' over the Latino street gangs of Southern California, defining turf and setting limits on internecine conflict (Skarbek, 2011). In Rio de Janeiro, the Comando Vermelho prison gang successfully dominated the regional retail market by imposing a system of mutual aid among affiliated local retail outfits (Amorim, 1993). The prison-based governance mechanism has proven resilient to state repression, allowing Rio's retailers to amass man- and firepower on the scale of large, wholesaling 'cartels' (Lessing, 2008).

Violence

Goldstein's classic tripartite typology of drug violence usefully distinguishes 'systemic' violence, associated with the machinations of the drug trade itself, from 'economic–compulsive' violence—which users commit during property crimes to obtain funds to purchase drugs—and 'psychopharmacological' violence, accidental or 'irrational' violence due to the effect of the drugs themselves (Goldstein, 1985). This typology was developed in the context of research on US drug markets. Not surprisingly, it thus has a strong focus on consumption as opposed to production and smuggling; the latter two types of violence are, by definition, exclusive to consumers.[2] Goldstein himself found economic–compulsive violence to be very rare and the majority of psychopharmacological violence to be alcohol-related and non-lethal (Goldstein, 1997, pp. 116–17). Later studies also found systemic violence to be of greater importance in generating armed violence, even in wealthy countries (MacCoun, Kilmer, and Reuter, 2003, pp. 68, 72).

At the global level, it is clearly drug traffickers—not users—who are responsible for the bulk of the violence, in particular the kind of militarized conflict seen in Brazil, Colombia, and Mexico. It is thus useful to update this typology by further distinguishing among subtypes of systemic violence. As Reuter (2009) notes, such subdividing can take many forms; the approach taken here is to attempt to identify underlying mechanisms and, where applicable, to 'cross-tabulate' them with the market sector.

For example, one prime cause of systemic violence is the fact that drug-trafficking organizations, as illegal operations, have no recourse to the legal system to resolve disputes or enforce contracts; violence often fills this gap (Reuter, 2009, p. 275). In retail sectors, this can provoke violence by dealers

Police investigators work at a crime scene where the bodies of seven men were found alongside three banners threatening rival gangs, Ciudad Juárez, Mexico, November 2008.
© Alejandro Bringas/Reuters

against users who fail to pay their debts; a retailer may need to make an example of a delinquent debtor in order to recover a payment or maintain a credible threat over other users. In the production and transhipment sectors, this type of violence can be directed at other trafficking organizations in the supply chain—as well as civilian populations in areas under their 'control'—to prevent the police from being informed (Akerlof and Yellen, 1994). Larger organizations may even use violence against their own members to punish disobedience or skimming (Reuter, 2009, p. 175).

The logic of turf wars is somewhat different. In this case, trafficking organizations use violence to appropriate territory, clientele, routes, or other assets from rivals. This type of violence is more likely to occur between drug-trafficking organizations in the same market sector; examples include battles among street gangs over corners in US urban areas, the internecine conflict among Mexican cartels over smuggling routes, and drug syndicates' invasions of rival favela territory in Rio de Janeiro. That said, actors from one sector may try to 'invade' another sector; in particular, the profitability of retail operations can lead smugglers and wholesalers to attempt takeovers of retail markets.

These two types of violence—extra-legal justice and turf war—are, in a sense, endemic to the drug trade, though they vary depending on the ability of trafficking groups to resolve disputes peacefully and strike stable truces. Anti-state violence by drug-trafficking groups, by contrast, is relatively rare, and not at all a 'fact of life' of the drug trade. In the United States, for instance, it is extremely rare for traffickers to kill police officers. Even smugglers and wholesalers generally avoid attacking security agents for fear of triggering retaliatory repression by the state. As cases such as Brazil, Colombia, and Mexico show, however, when powerful trafficking organizations do adopt a confrontational strategy, the ensuing conflict can be long-lived and the consequences dire.

The mechanisms underlying cartel–state conflict are not well studied or well understood. In ongoing work, Lessing (2011) identifies several plausible channels. One key logic is the use of violence in the negotiation of bribes, as captured by Pablo Escobar's infamous, repeated offer to Colombian officials: *'plata o plomo'* (silver or lead—that is, bribe money or the assassin's bullet). In retail markets this type of violence is exceptional (since killing police officers is likely to draw enormous scrutiny and repression), but when it does occur, it tends to be limited to street police who are already corrupt. The practice can be much more widespread in the producer and transhipment sectors, where drug-trafficking organizations may not only be negotiating bribes with higher-level officials such as judges, investigators, commanding police and army officers, and politicians, but also often have sufficient 'reach' to attempt to corrupt or intimidate state actors who are not 'on the take'.[3]

Plata o plomo-style violence is aimed primarily at state enforcers—security forces and judicial personnel—in an attempt to keep them from enforcing state policies. A slightly different form of anti-state violence occurs when large drug-trafficking organizations attempt to force state leaders themselves to change a piece of formal drug policy; examples are extradition in Colombia, the deployment of federal troops in Mexico, and prison conditions in Rio de Janeiro (Snyder and Durán Martínez, 2009, pp. 81–82; *Reforma*, 2010a; Penglase, 2005). Since traffickers seek to influence decision-makers rather than enforcers in such cases, they are likely to resort to acts of violence that foment a sense of crisis and generate political costs. During the Escobar period in Colombia (1984–1993), this took the form of terror tactics such as car bombs and elite kidnappings; in Rio de Janeiro, common tactics include bus burnings and forced closure of business districts. In Mexico, some controversy exists over what, if any, incidents truly qualify as terrorism, but certain actions seem to fit this logic, including the Familia Michoacana cartel's massive, probably coerced, street protests against federal troops by residents in 2010 and a coordinated attack on federal agents, coupled with a call for direct talks with President Calderón, in 2009 (*Reforma*, 2010a; Gómez, 2009; *El Universal*, 2009b).

> Once cartels opt for anti-state aggression, the motive to limit inter-cartel violence is weakened.

A crucial interaction is that between turf war and anti-state violence. Most researchers agree that Mexico's violence began as inter-cartel fighting; indeed, it was the upsurge in this violence that Calderón sought to address by calling in the army in late 2006. By that point, Mexico's cartels had amassed significant arsenals and in some cases semi-privatized standing armies.[4] It is plausible that such an accumulation of firepower contributed to the cartels' decisions to take an aggressive stance against state forces.

Yet the onset of cartel–state conflict might in turn escalate inter-cartel fighting. Retail outfits rely on anonymity and invisibility; they thus have strong incentives to keep internecine violence 'hidden' and at a low intensity, but even larger smuggling operations have strong motives to avoid drawing unwanted attention from authorities. Once the latter opt for anti-state aggression, however, the motive to limit the intensity of inter-cartel violence is severely weakened. Potential examples of this effect include the brutal bombing campaign carried out by the usually restrained Cali cartel against Pablo Escobar in the wake of his frontal attack on the state in the late 1980s, as well as the accelerating pace of gruesome murders and mutilations in Mexico since 2006.

This points to the danger of systemic violence in producer and transhipment countries with powerful and resource-rich smuggling cartels—and in places like Rio de Janeiro, where retail markets are dominated by powerful prison-based drug syndicates with cartel-sized arsenals. These large, powerful drug-trafficking organizations can afford to engage in prolonged campaigns of aggression against rivals, state forces, and civilian groups; even if they do not literally threaten the viability of the national state, they can induce levels of violence and insecurity bordering on civil war.

A second channel of interaction can occur if the state 'maxes out' its repressive capacity, especially if it is trying to fight all groups at once. As state forces get stretched thin, the chance that any one criminal group will be effectively suppressed falls. This situation can actually induce new groups to form, or to diversify into other criminal activities in which armed violence was previously seen as too risky. Such diversification may be occurring in Mexico, where new forms of armed criminality are on the rise. For example, in April 2010, an unidentified 'commando unit' of at least 30 gunmen and a dozen vehicles blocked off an entire section of downtown Monterrey and systematically kidnapped targets who were staying in two expensive hotels (*El Universal*, 2010); in June of that year, another commando unit took over a state-owned oil well, where it held workers hostage for weeks (*Reforma*, 2010b; 2010c).

State policy and response

While the drug trade—including the violence associated with it—is often cast as a problem for states to 'deal with' or respond to, it is important to recognize that states play a key role in defining the rules of the game. The legal status of drug consumption, sale, and trafficking; official policies on sentencing, surveillance, and extradition; the institutional structure and capacity of police and other state forces; and operational decisions such as where and when to apply repressive force all fundamentally shape the incentives and, ultimately, the actions of drug traffickers. The existence of coffee shops in the Netherlands, where adults may legally purchase marijuana and hashish, as well as cannabis dispensaries in California, are reminders of how drug markets are fundamentally shaped by state policy. In production and transhipment markets, this point is sometimes overlooked. When epidemics of drug use and violence arise, governments may be called on to respond, but that response hardly occurs in a vacuum; rather, it is a shift in policy by what is by any measure the most important actor in the drug trade.

This section looks at policy and responses in the Latin American context, with a focus on violence rather than drug use. Systemic violence is likely to be particularly responsive to state repressive policy—that is, decisions about how and when to apply repressive force against drug traffickers. Whereas psychopharmacological violence is accidental or 'irrational' and is likely to rise and fall with drug use (Blumstein, 1995), systemic violence is organized and strategic, arising when traffickers find it more profitable or otherwise advantageous to use violence than not. Such strategic decisions depend on complex interactions among drug-trafficking organizations and state enforcers; they are not a simple function of the size or profitability of the drug trade (see Box 2.1). This is particularly true of the large-scale violence—including anti-state violence—that shook Colombia in the Escobar era and is currently afflicting Mexico.

States play a key role in defining the rules of the game.

The starting point for many Latin American countries—blessed with low drug addiction rates and cursed with other, more serious security problems—has been a laissez-faire approach to drug trafficking. For much of the late 1970s and early 1980s, for example, the Colombian government more or less turned a blind eye to cocaine trafficking. It was during this period that the large 'cartels' formed and Pablo Escobar rose to public prominence as a wealthy benefactor and, eventually, an elected congressman. Central American governments also allowed international smugglers to operate free from significant interference for years, and although Brazil has waged a militarized war against syndicates in its urban areas, these retail markets represent only a fraction of the total flow of cocaine that wholesalers have for decades moved through the country towards European markets—with relatively little fanfare (Dowdney, 2003, pp. 41–42).

A variation on the laissez-faire approach is a government-negotiated pact. Although no official would call it that, such arrangements were common in Mexico through the 1990s, under the rule of the then hegemonic Partido Revolucionario Institucional (Institutional Revolution Party, PRI). The PRI dominated Mexican society and politics by dividing up territory, markets, and benefits among friendly groups, which then became dependent supporters of the PRI; Mexico's cartels were no exception. Such an arrangement has the benefit of giving the state some leverage over cartels but may also involve deep and high-level corruption.

These approaches can be appealing, at least in the short run, to governments of transhipment countries, since smuggling itself does not cause immediate, observable damage to the countries where it takes place; indeed, it generates employment and immense profits. But over time it tends to corrupt police, bureaucrats, and even elected officials, potentially generating diplomatic and economic pressure from consumer countries, particularly from the United States, to increase state repression of the drug trade. Such factors can lead to a policy of active repression of the drug trade—a crackdown—which often involves the creation or expansion and deployment of specialized police or armed forces, as well as new legal and judicial instruments such as seizure or extradition laws.

Box 2.1 Policy v. profits: what drives large-scale drug conflict?

The relationship between the size and profitability of the drug market on the one hand and violence on the other is unclear, both theoretically and empirically. In the case of large-scale conflict, both among large drug cartels and between cartels and the state, however, it seems unlikely that an increase (or decrease) in the flow of drugs is driving the violence.

While the true size of drug markets is difficult to measure, even the imperfect estimates of price and production have told a relatively straightforward story over the past 30 years. They document a steadily expanding supply and falling prices–probably in response to growing demand–from the mid-1980s throughout the 1990s, followed by a long period of relatively stable prices and production levels, despite massive eradication efforts throughout the Andean region. These trends stand in contrasts to the abrupt onset of intense anti-state violence in Colombia during the 'narco-terror' period (roughly 1984-93), when Pablo Escobar, powerful head of the Medellín cartel, waged an all-out war of intimidatory violence against the Colombian state (see Figure 2.1).

While it is true that the market for cocaine was expanding during Colombia's narco-violence decade, the onset of cartel-state violence is more plausibly linked to Escobar's expulsion from Congress by then attorney general Rodrigo Lara Bonilla, whom Escobar promptly murdered, than to shifts in drug demand or supply. Similarly, the abrupt end of open conflict between drug cartels in 1993 was surely more a product of Escobar's death at the hands of US-aided police forces and the subsequent fragmenting of Colombia's two mega-cartels into myriad successors than a sudden change in supply and demand. In fact, Colombia only became a major producer of cocaine in the period immediately following Escobar's death, when cartels had splintered and essentially abandoned confrontation strategies in favour of anonymity and cooptation (Lessing, 2011).

More recently, around the time of the onset of cartel-state conflict in Mexico, US cocaine prices began to rise. As Figure 2.2 shows, however, the relationship between market conditions and violence is subtle and inconstant, and the causality is unclear. During the onset of violence, from 2007 to 2008, US prices rose in startlingly tight correlation with both the official Mexican drug-related homicide count (0.95 correlation) and the tally kept by *Milenio*, a national newspaper (0.92 correlation). Thereafter, prices remained flat, while homicides continued to increase sharply, yielding extremely weak correlations (0.20 and 0.18, respectively).[5]

One plausible explanation for this stark shift is that in the short run, the Mexican crackdown and attendant violence created uncertainty, shortages, and perhaps even hoarding or speculation in US retail drug markets, but that, as dealers adapted to a more fluid and redundant supply chain, prices levelled off.[6] Similarly, some analysts attribute price spikes in the 1980s to the onset of the narco-violence period in Colombia, noting that these were ephemeral and did not affect the overall downward trend (Caulkins and Reuter, 1996, p. 7).

Whatever the factors involved, it is hard to make a convincing case that profitability is driving drug violence when there is not even a stable correlation between violence and prices. Of course, if there were no demand for drugs, and hence no retail market, there would be no powerful drug cartels in Mexico to fight a drug war. So drug demand and drug flows can be thought of as a necessary precondition for drug violence. Similarly, other factors often cited as drivers of drug violence in Mexico–such as socioeconomic and demographic conditions, institutional weakness, culture, and geographic location–are simply too static or slow-moving to explain the pace of the violence epidemic, though they may have contributed as intensifying factors.

What did change in 2006 was the Mexican government's antinarcotics policy: President Felipe Calderón called in the army to combat the drug cartels, and the cartels responded with violence. Yet even the relationship between government crackdowns and drug violence is not clear-cut. In Mexico, the big increase in repression occurred during the initial rollout of the army, in late 2006 and early 2007, but the sharpest increase in violence, including direct attacks on the army itself, took place in 2010 (see Figure 2.3).

DRUG VIOLENCE 49

Figure 2.1 **The cocaine market v. periods of cartel-state conflict, 1982–2010**

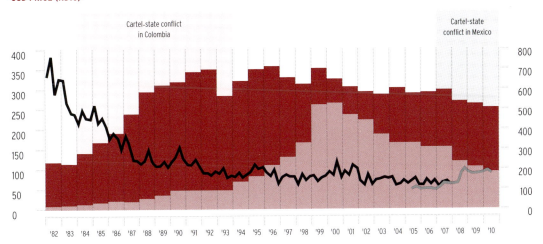

Notes: Production values for 1982-85 and 2010 are estimates. The US Office of National Drug Control Policy (Fries et al., 2008) has produced a long time series of USD price per pure gram data for US purchases for various weight categories; the <2 grams captures typical retail purchases. More recent data is only available from NDIC (2009; 2010; 2011), which is not disaggregated by weight.

Sources: NDIC (2009; 2010; 2011); ODCCP (1999; 2000a; 2000b); UNODC (2003, 2010); Fries et al. (2008)

Figure 2.2 **Drug-related homicides in Mexico v. US street price of cocaine, 2007–10**

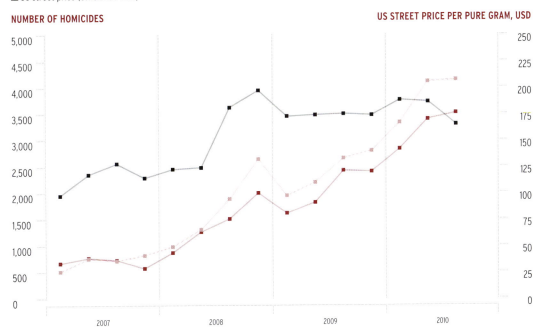

Sources: Milenio (2010a; 2010b; 2011); NDIC (2011); SNSP (2011)

Figure 2.3 **Mexico: crackdowns v. anti-state attacks, December 2006–2011***

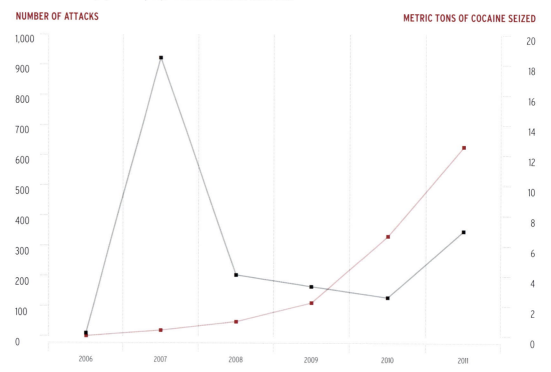

Notes: The figure for attacks in 2011 is a yearly estimate based on official data for the period 1 January–14 June. The figure for metric tons seized reflects official data for all of 2011.
Sources: SEDENA (2011); Aranda (2011)

Frequently, such crackdowns are rhetorically cast as declarations of war, with leaders demonizing traffickers and characterizing any form of negotiation or quarter as a form of surrender. But a critical question is whether the state applies the same amount of repressive force to all drug-trafficking organizations, irrespective of their behaviour—what this chapter refers to as a *blanket crackdown*—or whether the state metes out repression in proportion to the amount and severity of violence employed by traffickers—referred to here as *proportional response*. While policy-makers rarely make this distinction, recent research has emphasized the fact that blanket crackdowns fail to generate disincentives for violence and may in fact spur drug-trafficking organizations on to greater violence (Kleiman, 2011; Guerrero-Gutiérrez, 2011, p. 71; Lessing, 2011).

Proportional response necessarily implies directing less repressive force against less violent traffickers, which many traditional policy-makers deride as unacceptably 'tolerating' criminal activity. Critics identify it with the government-negotiated pact approach or focus on the apparent absence of social sanction. Its advocates portray proportional response as the application of *harm reduction* principles to problems of public security.[7] Originating in the public health realm, harm reduction aims to minimize the negative health and social effects of drug use, rather than reducing consumption itself. The no-questions-asked needle exchange is the paradigmatic harm reduction programme, now widely implemented as a means of reducing HIV/AIDS transmission and other transmissible diseases, even in countries with traditionally hard-line stances on drugs.

Harm reduction has taken hold as an approach to limiting ill effects flowing from the consumption of drugs, and policing strategies focused on the most violent drug retailers appear to have been successfully implemented in US cities such as Boston and New York (Kennedy et al., 2001; Zimring, 2011). Nevertheless, allowing drug producers, smugglers, and retailers to operate as long as they avoid the most harmful activities—armed violence—remains a hard sell. Classic harm-reduction advocacy views addicted drug users as victims rather than criminals, a case that cannot easily be made about drug dealers, particularly large-scale trafficking organizations with militarized arsenals. But some governments, driven to find new approaches to drug violence in the face of massive human costs, are shifting the primary objective of policing away from the total eradication and interdiction of drugs towards the minimization of associated violence. As discussed below, Rio de Janeiro's 'pacification' approach stands out as a clear case of explicitly proportional response.

MEXICO

By some measure, Mexico's drug war is now the most violent sub-national conflict of the century, even compared with civil wars (see Figure 2.4).[8] To many observers, the violence in Mexico is unsurprising given the underlying conditions: enormous profit margins, organized cartels angling for primacy, a ready supply of weapons from the United States, widespread police corruption, weak institutions (especially in border areas), and few legitimate economic opportunities that might keep youths from entering the drug trade. While these ingredients do seem like a natural recipe for violence, they have all been characteristics of Mexico for years, if not decades. The violence, on the other hand, has been

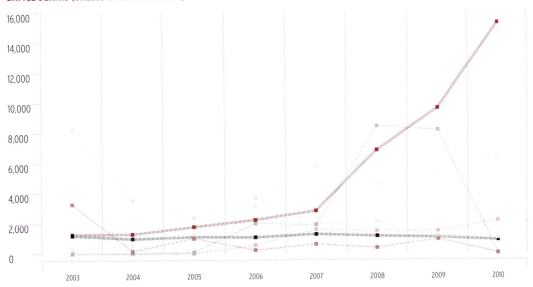

Figure 2.4 **Deaths in civil wars and drug conflicts, 2003-10**

Source: UCDP (2011); ISP (2012); SSP-Rio (2003); Reforma data cited in Ríos and Shirk (2011)

remarkably precipitous, with annual drug-related homicides now eight times higher than they were in 2005. As Figure 2.4 shows, the explosion of violence in Mexico has been just that: an abrupt break with the past.

Background

Drug trafficking in Mexico has deep roots; marijuana and opium production in the state of Sinaloa began in the 19th century, while the army's involvement in anti-narcotics dates back to the 1940s (CNN México, 2011; Astorga, 1999). Through the late 1980s, however, drug trafficking was largely limited to production and export of marijuana and, to a lesser extent, opiates to the United States; the army only intermittently engaged in eradication of plantations in areas of difficult access. Since then, two key changes have taken place. First, US law enforcement cracked down heavily on the Colombian cartels' Caribbean trafficking routes, inducing them to rely increasingly on Mexican trafficking organizations that could transport shipments overland at US border crossings such as Juárez, Laredo, and Tijuana (Meiners, 2009). As an increased share of the cocaine trade shifted to Mexico, the value of these border crossings increased, as did the size and firepower of the Mexican cartels that controlled them.

Second, Mexico's political system underwent a sea change, as the ruling PRI slowly lost its grip on power, first losing local and state elections, then control of Congress, and, finally, in 2000, the presidency, for the first time in 70 years (Magaloni, 2005). The PRI's penetration into and control over virtually all aspects of Mexican social and economic life seem to have extended to the drug trade,

Soldiers carry the coffins of six members of Mexico's Army, who were found decapitated in Chilpancingo, Mexico. © Claudio Cruz/AP Photo

and probably contributed to the relatively stable and non-violent relationship among cartels (Snyder and Durán Martínez, 2009, p. 73).

From the late 1990s through the first non-PRI presidency, that of Vicente Fox (2000–06), an inter-cartel turf war slowly arose. Cartels began to invest in armament and build private armies; a particularly ominous development was the Gulf cartel's corrupting and subcontracting of a US-trained special forces unit, the Grupo Aeromóvil de Fuerzas Especiales, as the cartel's private militia. The group came to be known as Los Zetas and later broke with the Gulf cartel to become an autonomous, and extremely violent, cartel. Although drug-related violence in this period looks extremely mild in retrospect, the number of homicides did double over the course of the Fox administration. Soon after a narrow and violently contested election victory, President Calderón (2006–present) declared 'war' on the drug trade, ordering the army to crack down on cartels in urban areas and along major land routes. When the cartels reacted with increased levels of armed violence, including attacks on state forces, the government essentially doubled down, arguing that by capturing and killing cartel leaders it would eventually break up the organizations into fragments too small to fight the state (Guerrero-Gutiérrez, 2011).

Recent trends

To date, the Calderón government's 'headhunting' strategy has yielded the arrest or death of some 28 major traffickers, which in turn has succeeded in fragmenting the Mexican drug trade. By one count, the number of principal cartels has grown from six to 16 (Guerrero-Gutiérrez, 2011, pp. 11, 64). Rather than abating, however, the violence has not only increased but also *accelerated* (see Figure 2.5).

Moreover, the trend is not limited to inter-cartel violence; a similar rate of increase can be seen in attacks on army soldiers. Between December 2006 and September 2011, 174 public officials were killed, including more than 30 mayors

Figure 2.5 **Cartel fragmentation and drug-related violence, 2006-11**

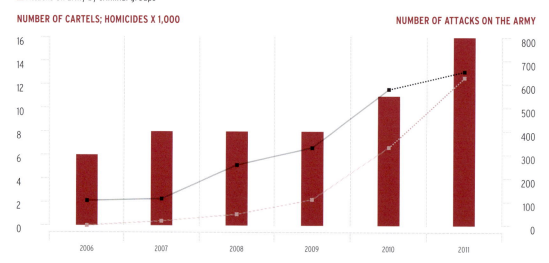

Note: The 2011 figures for homicides and attacks are yearly estimates based on data for the period January–June.

Sources: Reforma data cited in Guerrero-Gutiérrez (2011, p. 45); Aranda (2011); Ríos and Shirk (2011)

(Guerrero-Gutiérrez, 2011, p. 45). In short, splintering has not made cartels less willing or able to attack the state. This is also reflected in the rising number of seizures of small arms; particularly troubling is what appears to be an increasing cartel preference for military-style firearms and hand grenades over handguns (US Embassy in Mexico, 2009; see Figure 2.6). Such a preference for military-type weapons—which are costly to acquire, in both monetary and logistical terms—is an indication of the strong incentives cartels face to engage in brazen forms of armed violence.

Violence has also spread out from the border and hotspots in drug-trafficking regions such as Sinaloa and Michoacán to more central locales such as Acapulco, Cuernavaca, and Monterrey. Moreover, Los Zetas, now an autonomous cartel, along with other armed criminal groups with limited or unclear ties to the principal drug cartels, have increasingly expanded into other criminal activities such as human trafficking, kidnapping, mass extortion, and theft of petroleum from the state-owned firm PEMEX (Guerrero-Gutiérrez, 2011, pp. 32–34). As the number of cartels has grown, so has their variety. Table 2.1 presents a four-way typology of cartel types.

In the face of the sharp spike and spread of violence, the Calderón administration has dug in, consistently defending its blanket approach and seeking to 'lock in' the policy beyond the end of Calderón's term in 2012 (Archibold, Cave, and Malkin, 2011). In practice, however, the policy may be moving towards a more focused, proportional response. A recent study shows that the majority of arrests from 2006 to 2010 were

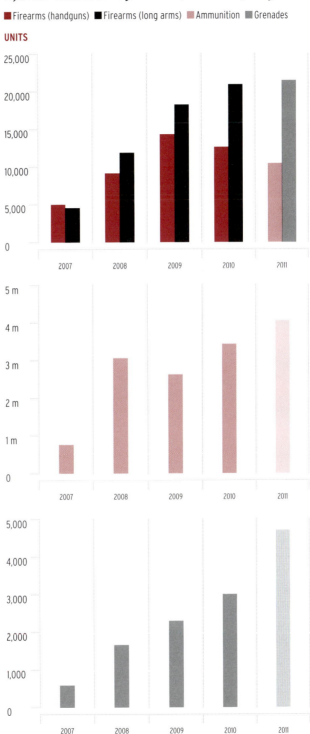

Figure 2.6 **Seizures of illegal firearms, ammunition, and grenades**

■ Firearms (handguns) ■ Firearms (long arms) ■ Ammunition ■ Grenades

Note: The 2011 figures are estimates based on data for January–June 2011.
Source: PGR (2011, p. 278)

Table 2.1 Typology of Mexican cartels

Category	Description	Organizations
National cartels	Operate throughout the country, including at entry and exit points; they seek to expand and diversify into activities such as human trafficking and oil theft.	Sinaloa, Los Zetas, and Gulf cartels
'Toll collector' cartels	Control border crossings and charge other cartels fees to pass through; they are largely confined to border areas and have difficulty diversifying.	Tijuana and Juárez cartels
Regional cartels	Operate within fixed areas; they exercise limited control over routes passing through their territory.	Los Caballeros Templarios and Pacífico Sur cartels
Local organizations	Splinters of larger cartels that frequently diversify into other illicit activity to make up for lost drug turf.	La Resistencia, La Mano con Ojos, Los Incorregibles, La Empresa, La Nueva Administración, La Nueva Federación para Vivir Mejor, and Cártel Independiente de Acapulco, among others

Source: Guerrero-Gutiérrez (2011, pp. 28, 31-37)

of members of the Gulf and Zeta cartels (then only recently split); US officials attributed this to Los Zetas' more brazen behaviour (Burnett, Peñaloza, and Benincasa, 2010). In 2011, the apparent focus on Los Zetas intensified, with a dedicated military operation ('Northern Lynx') and several high-level arrests and busts (Gómora, 2011; Corchado, 2011). The government, however, continues to publicly reject any departure from a blanket approach, equating proportional response with corrupt pacting, and collusion with the Sinaloa cartel in particular. In the words of then national security spokesman Alejandro Poiré, on the president's blog:

> *The federal government does not favour any criminal organization; it weakens them all systematically without distinction. To benefit any criminal group [. . .] is to validate the outdated argument that crime can be 'managed'* (Poiré Romero, 2011).[9]

The other traffic: US arms into Mexico

To what extent do Mexican cartels use US firearms against one another and the state? Separating facts from rhetoric and misinformation has sometimes proven difficult.

A number of preliminary points can be made. First, Mexican gun control laws are among the strictest in the world. There is only one retail gun outlet in the entire country, run by the National Defense Secretariat, where citizens who have passed a rigorous screening and licensing process may purchase a single low-calibre handgun and a box of ammunition each; since 2006, the store has sold about 6,500 guns per year (Booth, 2010). In comparison, the history, culture, and political system of the United States have produced far more permissive firearm laws. In the United States, civilians can readily obtain firearms—including, in some states, assault weapons—on large retail and secondary markets. Non-existent or partial state-level owner licensing and gun registration regulations in many states frequently make purchases difficult to trace.

Second, tens of thousands of 'crime guns' seized in Mexico have been traced back to US manufacture or points of sale (Goodman, 2011); nevertheless, it is difficult to pinpoint the exact share of cartels' weapons that originated in the United States (Stewart, 2011).

Third, as discussed below, there is evidence that the expiration of the US Assault Weapons Ban (AWB) in 2004 led to an increase in (not necessarily drug-related) gun violence in Mexican municipalities near the US border, apparently by increasing the supply of the previously banned weapon types (Dube, Dube, and García-Ponce, 2011; Chicoine, 2011). At the same time, it is not clear what effect a new AWB or stricter US gun laws in general would have on the current violence in Mexico, as cartels have access to other sources of firearms, especially the arsenals of Central America (Stewart, 2011).

US federal gun regulations: the AWB and background checks

In 1994, Congress passed and US president Bill Clinton signed the federal Assault Weapons Ban. The AWB prohibited civilian sales of so-called 'semi-automatic assault weapons', defined in the law to include 19 named firearms and copies of those firearms, plus other weapons with at least two specified physical characteristics from a list of features (Dube, Dube, and García-Ponce, 2011, pp. 5–6). The law also banned the sale and possession of magazines with a capacity of more than ten rounds of ammunition. The AWB featured a sunset clause, providing for its expiration after ten years.

Two new studies find that the expiration of the ban in 2004 led to an increase in violence in Mexico. Chicoine (2011) estimates that the expiration accounts for at least 16.4 per cent of the increase in homicide rates in Mexico between 2004 and 2008. Dube, Dube, and García-Ponce (2011), cautious not to conflate the potential effects of firearms supply on armed violence with the complex dynamics of cartel-related conflict in the Calderón period, focus exclusively on the 2004–06 period. To isolate the effect of the expiration of the AWB, they make use of the fact that California retained a state ban on the weapons while Texas and Arizona did not. They find that Mexican municipalities along the Texas and Arizona borders experienced an additional 40 per cent increase in homicides relative to those along the California border; they estimate that at least 158 additional deaths per year were attributable to the AWB expiration.

> Two new studies find that the 2004 expiration of the AWB led to an increase in violence in Mexico.

On the other hand, it is unclear to what extent the AWB restricted access to automatic weapons. Because of the law's reliance on a list of specific characteristics, largely cosmetic in nature, gun producers found ways around the ban by manufacturing 'post-ban' sports versions. As a 2001 review in *Gun World* magazine notes: 'In spite of assault rifle bans [. . .] the Kalashnikov, in various forms and guises, has flourished [...] with more models, accessories and parts to choose from than ever before' (VPC, 2003). The reality seems to be that incomplete bans such as the AWB make obtaining weapons only marginally more difficult for the highly motivated; unfortunately, this group may include actors with criminal intent.

Controversies: the 90 per cent and Operation Fast and Furious

In 2008, a top official of the US Bureau of Alcohol, Tobacco, Firearms and Explosives (ATF) testified before Congress that more than 90 per cent of firearms seized in or on their way to Mexico originated in the United States (ATF, 2008). A subsequent US Government Accountability Office report based on ATF data puts this figure at 87 per cent (US GAO, 2009, p. 15). The 90 per cent figure was widely circulated and discussed, and Mexican government officials incorporated it into their talking points as a crucial reason for the violence. However, this estimate was based on a small sample of the 30,000 firearms seized from criminals in Mexico in 2008. Of the 7,200 (24 per cent) guns that Mexican authorities submitted to the ATF for tracing, only about 4,000 (13 per cent) were successfully traced. Of these 4,000, 87 per cent had US origins (Stewart, 2011).

Mexican President Calderón glances behind after unveiling a banner reading 'No More Weapons' in the border city of Ciudad Juárez, Mexico, February 2012.
© Alfredo Guerrero/Mexico Presidency/Reuters

The relatively small number of traced guns is due to the difficulty of tracing weapons for which buyer and registration information is missing, or which have changed hands multiple times. But a sampling bias also prevents the drawing of broad conclusions based on the trace data. The guns submitted to the ATF for tracing would normally have carried some indication of US origin (such as markings denoting US manufacture or import into the United States). Ultimately, the fact that US origins were confirmed for 3,480 of the 30,000 firearms seized by Mexican authorities reveals little about the true relative contribution of US guns to Mexican crime.

Another controversy concerns an ATF 'gunwalking' operation that deliberately—and without the Mexican government's knowledge—permitted illegal purchases and transfers to Mexico by 'straw purchasers'[10] of some 2,500 firearms, including hundreds of assault weapons, between 2009 and 2011. The rationale of the operation—known as 'Fast and Furious'—was to trace illicit weapon flows to higher-ups in drug cartels. It was later discovered that 'walked guns' had been used in violent crimes, including the December 2010 murder of a US Border Patrol agent. The resulting scandal damaged US–Mexican relations and led to the reassignment or replacement of most of the top ATF officials who were directly involved, with Congressional investigations continuing into 2012 (Attkisson, 2011a; 2011b; 2011c).

The role of Fast and Furious firearms in Mexican cartel violence is not well understood. Individual cases have come to light, however, such as a May 2011 cartel attack that forced a government helicopter into an emergency landing. The operation has yielded indictments for some 20 alleged arms traffickers, but overall it has been a major setback for the ATF and its wider Project Gunrunner programme, which is designed to stop US-sourced gun flows to Mexican cartels and to improve tracing practices (Attkisson, 2011a; 2011b; 2011c).

The US gun market in perspective

The ease of access among Mexican (not to mention US) cartels to firearms, particularly high-powered rifles, is extremely troubling and has prompted the Mexican government and civil society to call for better controls. At the same time, analysts and policy-makers should not exaggerate the importance of supply-side US gun control measures on Mexican cartel violence. The US gun market existed for decades before the current explosion of violence; it was thus probably not the primary trigger. Nor would limiting US inflows alone necessarily reduce the violence by any meaningful degree. The cases of Brazil and Colombia suggest that when traffickers demand firearms, supply arises from somewhere to fill it.

The explosion of Mexican drug violence is generating intense demand for firearms by groups with enormous cash flows and close contact with corrupt officials in Mexico and suppliers of illicit small arms in Central America and overseas. This suggests that a key component of any successful violence-reduction strategy must aim to reduce cartels' incentives to employ armed violence, thus minimizing demand for firearms (Atwood, Glatz, and Muggah, 2006).

CENTRAL AMERICA'S NORTHERN TRIANGLE

The countries of Central America's Northern Triangle—El Salvador, Guatemala, and Honduras—are not strangers to armed violence and drug trafficking. While the region has largely put the spectre of civil war behind it, crime and murder rates have been among the highest in the world for some time. The US deportation of more than 45,000 convicts to Central America between 1998 and 2005 contributed to the transformation of the region's numerous street gangs into *mara* gang networks involved in multiple criminal enterprises, and especially extortion rackets (Jütersonke, Muggah, and Rodgers, 2009, p. 308; Cruz, 2010; LATIN AMERICA AND THE CARIBBEAN). The region suffers not only from perennial challenges to economic development, but also what has been called a 'Hobbesian trap'—a concentrated economic elite that prefers to rely on private security than to shore up weak state institutions through taxes (*Economist*, 2011). Against this backdrop, the prospect of an 'invasion' by Mexico's powerful cartels—partially in response to the official crackdown—and their potential collusion with the region's *maras* have arisen as major security concerns. These issues have spurred regional summits and led the United States to appropriate USD 248 million in funding for the region between 2008 and 2010, as part of its Mérida Initiative and a new Central America Regional Security Initiative (Seelke, 2010, p.1).

A soldier stands next to a message written in blood following the massacre of 27 people by members of Los Zetas at a ranch in Petén, Guatemala, May 2011. © Moises Castillo/ AP Photo

Background

Central America's history, like much of Latin America's, is characterized by clashes between countervailing forces of reform—of both society and the state, sometimes classically liberal, sometimes radical—and the conservatism of traditional, highly concentrated elites. An exacerbating factor is the political fragmentation of the region, which may have been useful to colonial rulers and elites but contributed to a legacy of small national states with perennially low capacity, high levels of corruption, and severe vulnerability to foreign influence and intervention (Mahoney, 2001, pp. 11–45).

These antecedents contributed to the region's descent into civil war and political violence during the cold war. In Guatemala, a US-backed coup in 1954 replaced a democratically elected leftist government with a military regime, eventually leading to a 36-year civil war (1960–96) that left more than 200,000 dead. El Salvador also fought a bloody civil war (1980–92) that killed some 75,000 and sent hundreds of thousands of refugees to the United States, among other places. Honduras did not experience a civil war but suffered under a repressive regime in the 1980s and served as a staging area for raids by US-backed Contra rebels into neighbouring Nicaragua (USAID, 2006).

Such armed conflicts leave varied, long-lasting scars. One important legacy that scholars are only recently beginning to explore systematically is how civil conflict can foster the growth of corruption, organized crime, and armed violence associated with the drug trade. Though peace came to Central America in the 1990s, the region retained an enormous stock of war-making capital, both physical—weapons and military equipment—and human; in El Salvador

Table 2.2 Mexican cartel activity in Central America

Dates	Event
March 2008	Los Zetas cartel attacks one of Guatemala's largest drug gangs, killing 11.
November 2008	A battle between a local ally of the Sinaloa cartel and Los Zetas in Huehuetenango, Guatemala, leaves as many as 60 dead.
March 2009	After threatening the national head of police, Los Zetas threaten to kill Guatemalan president Álvaro Colom.
December 2009	The Honduran drug czar, Col. Julian Aristides González, is assassinated days after publicly disclosing the discovery of clandestine airstrips in the drug-trafficking region of Olancho.
February 2010	Honduran police find an encrypted note from leaders of the Barrio 18 gang, indicating that Los Zetas have paid for the assassination of Security Minister Oscar Álvarez.
December 2010 & February 2011	The Guatemalan government declares and renews a state of siege in the Alta Verapaz department in response to cartel-related violence.
May 2011	Los Zetas massacre 27 people at a ranch in Petén, Guatemala, leaving an admonition literally written with blood for a leading Guatemalan drug trafficker. In response, the government declares another state of siege and deploys the army to the region.

Sources: Dudley (2011, pp. 31-32); El Universal (2009a); ICG (2011, pp. 2-3)

alone, some 45,000 combatants were demobilized (USAID, 2006, p. 50). Combined with poor economic conditions and weak state institutions (two other legacies of civil war), these factors have contributed to historically high rates of violence in the region. But they also constitute valuable 'resources' for the international drug trade, in particular the Mexican cartels, whose turf war and ongoing conflict with the state have led them to seek out both man- and firepower.

Guatemala's *Kaibiles* are a case in point. Created by the military government in the 1970s, this elite special forces battalion suffered extreme, brutal training and went on to commit atrocities against civilians during the civil war (USBCIS, 2000). While attempts were made at reform in the wake of the reconciliation process, deserters from the group have increasingly been found working as hired guns for Mexican cartels. Indeed, they are thought to be training Mexican soldiers who have deserted the army to join the cartels' ranks (Stratfor, 2006).

Further important factors in the region are ingrained corruption and endemic low state capacity. State budgets cannot compare in size to potential drug profits; indeed, the conservative estimate of USD 20 billion for the share of drug revenues controlled by Mexican cartels dwarfs Honduras' USD 12 billion gross domestic product (Meiners, 2009). At the same time, officials who refuse to cooperate with traffickers have little in the way of protection, as the recent murder of the Honduran drug czar illustrates (see Table 2.2). Such imbalances make the *plata o plomo* approach—a combination of generous bribes and violent threats—a very effective way to intimidate state officials. This not only facilitates the trafficking of drugs, but also allows cartels illicit access to the region's poorly guarded arms stocks; the US Embassy in Mexico estimates that 90 per cent of military-origin seized weapons in that country came from Central American military stocks (US Embassy in Mexico, 2009).

Violence and crime

According to the UN Development Programme, 'Central America is the most violent region of the world, with the exception of those regions where some countries are at war or are experiencing severe *political* violence' (UNDP, 2009, p. 10). Even this qualification may no longer be necessary; a new estimate based on data covering 2004–09 puts Central America ahead of Southern Africa as the region with the highest violent death rate (29.0 v. 27.4 per 100,000), almost four times the global average of 7.9 per 100,000. This violence is heavily concentrated in the Northern Triangle region. El Salvador has the highest violent death rate of any country in the world (61.9), while Honduras and Guatemala rank fourth and seventh, respectively (Geneva Declaration Secretariat, 2011, pp. 51–60). Trends in homicide rates show a similar pattern, with the Northern Triangle countries suffering from rates two to three times that of Panama, Nicaragua, and Costa Rica—and Belize occupying a middle position (LATIN AMERICA AND THE CARIBBEAN).

Interestingly, rates of property crime and overall victimization are not particularly high in the region, but rather at or below the average for Latin America (World Bank, 2011, p. 2). This is consistent with the idea that much of the violence is 'systemic', associated with the workings of the drug trade, as opposed to related to more atomistic and economic factors affecting overall crime. The World Bank's study of violence in the region cites an unpublished econometric study that finds that 'controlling for other factors, drug-trafficking hotspots have murder rates more than double those in areas of low trafficking intensity' (World Bank, 2011, p. 21). Other typical risk factors, such as a high population of young men, also correlate with higher levels of violence. Of course, correlation is not causation; it is possible that drug traffickers are drawn to violent or poorly policed regions. Indeed, the report notes that the region as a whole has suffered from high murder rates since at least the 1960s, prior to the expansion of the drug trade and much of the region's civil war violence (World Bank, 2011, p. 22). What is clear is that the region suffers from a confluence of high overall homicide rates and particularly acute violence in zones where drug-trafficking and youth gangs are common.

Spillover and balloon effects

Central America has, for decades, served as a stopover for illicit drug flows making their way from the Andean countries to the United States. In this respect, the region is attractive both geographically, with access to the Caribbean, the Pacific, and the land route through Mexico, and institutionally, for its generally weak law enforcement and criminal justice institutions and imperfect state control over national territory. From the beginning of the cocaine boom in the late 1970s through the mid-1990s, Central American organized crime families—and, for a time in Panama, military governor Manuel Noriega—provided transportation and logistical services for large Colombian and Mexican cartels (Dudley, 2011, pp. 25–28). Yet the region's role was limited during the period, as most shipments went through the Caribbean directly to the United States. The mid-1990s crackdown that blocked these routes and fomented the rise of Mexico's cartels raised the profile of Central America as well, but direct aerial and maritime routes from Colombia to Mexico still predominated well into the 2000s.

In yet another example of the 'balloon effect'—in which repression in one area displaces cartels and violence to another—the onset of the recent cartel–state conflict in Mexico seems to be leading cartels to develop land routes that pass directly through Central America (Meiners, 2009). Today, an estimated 90 per cent of cocaine that enters the United States passes through Central America (World Bank, 2011, p. 12). At the same time, the intensity of the fighting in Mexico and the diversification of its cartels into arms and human trafficking have created increased demand for weapons, soldiers, sources of revenue, and border territory. This spillover or contagion effect has made the Northern Triangle, and particularly Mexico's southern border with Guatemala and Belize, an important locus for armed criminal activity.

> The onset of cartel-state conflict in Mexico may be leading cartels to develop land routes through Central America.

While the presence of Mexican cartels in Central America dates back at least to the early 1990s, when Sinaloa cartel leader Joaquín 'El Chapo' Guzmán was arrested in Guatemala, it is only in the last few years that they have begun to establish significant operational footholds in the region (Dudley, 2011, p. 28). This process appears to be bringing them into direct competition with traditional, domestic criminal organizations, as well as state forces, both corrupt and honest. Most of the recent serious violent events have taken place in Guatemala and, to a lesser extent, in Honduras.

Thus far, three interrelated trends can be perceived in Central America. First, Mexican cartels are directly attacking traditional organized crime syndicates for turf, upsetting what appears to have been relatively peaceful relations among the latter. Second, Mexican cartels are battling each other for supremacy in key trafficking regions, particularly near the Mexican border. Third, cartels are engaging in limited anti-state intimidation and violence linked to attempts to corrupt officials (Dudley, 2011; Meiners, 2009). A crucial trend for analysts to monitor is whether anti-state violence intensifies into the kind of open clashes and unilateral attacks on state forces seen in Mexico. This requires tracking not only total levels of violence, but distinguishing among modalities, tactics, and groups involved.

Gangs

Gangs have an enormous presence in the Northern Triangle, with roughly 70,000 members from more than 900 gangs (World Bank, 2011, p. ii). Over the past two decades, partially in response to mass deportation of Central American gang members from the United States, these small, traditional, local gangs have become increasingly consolidated as 'cliques' of larger *maras*, fluid criminal networks largely controlled from within the prison system. Relying on the ability to coordinate outside activity, some of these prison-based *maras* now operate large-scale extortion and protection rackets (Lessing, 2010, p. 164); in El Salvador, they apparently control a lucrative market for crack cocaine in low-income neighbourhoods (Dudley, 2011, p. 44). To date, however, gangs account for only a small share of violence, in the range of 15 per cent of homicides (World Bank, 2011). No clear pattern of systematic partnership with or integration into the Mexican cartels now operating in Central America has yet emerged.

One reason may be that whereas El Salvador is in many ways the 'heart' of *mara* culture, Mexican cartel activity has focused on Guatemala and Honduras. It may also be that *maras* make unattractive partners for large cartels (Dudley, 2011, p. 42). Whereas domestic organized crime families may have experience in the drug trade or access to corrupt officials, and the Guatemalan *kaibiles* have extensive military training and discipline, *mara* members—typically poor, local youths who lack training and experience—may seem unreliable and relatively unskilled. It is certainly the case that Los Zetas tend to recruit from the ranks of current and former soldiers, reflecting their military background. However, the Sinaloa cartel, like many of the traditional drug cartels of Mexico, has often relied on street gangs, 'subcontracting' out acts of armed violence and intimidation (Guerrero-Gutiérrez, 2010). Moreover, the linkages between *mara* leaders in Central America and affiliated cliques operating in the United States could be of great interest to Mexican cartels seeking greater control over retail markets. Clearly, this issue merits continued scrutiny.

Implications

The drug trade in Central America and especially the Northern Triangle is in flux—and the twin problems of violence and corruption have the potential to worsen considerably in the coming years. While it is important to shore up weak institutions and attempt to establish basic human security in the region, sending outmanned security forces into combat with powerful cartels could produce Mexico-like results, leading to desertions, corruption, and brazen cartel–state clashes.

Recent developments in Central America appear to be the most recent example of the 'balloon effect', in which repression in one geographic area pushes drug trafficking activity onto an alternate route. As drug flows increase along that new route, official corruption inevitably grows, with illicit rents making their way into state enforcers' pockets. Worse, armed violence can erupt, not only among cartels but also against state forces, especially when governments try to crack down. While Mexico would be relieved to have its cartels shift the bulk of their activities to Central America, such a shift would not solve anything at the regional level.

US aid and expertise in law enforcement, justice, and intelligence operations in the region might bring some improvements in the institutional effectiveness of these countries, but this input is unlikely to eliminate drug trafficking in the region and could even push cartels into the relatively low-violence countries of south Central America. US policy-makers often argue that the war on drugs cannot be abandoned lest cartels run wild, sowing violence and corruption. But state anti-trafficking policies also influence the extent and modalities of violence and corruption. Moreover, repressive crackdowns and violent cartel 'blowback' in upstream markets appear to have only marginal and temporary effects on drug flows into the United States. What Central America needs is a regional approach that attempts to channel drug flows in a way that does a minimum of harm to institutionally weak states and developing economies, while creating incentives for cartels to minimize armed violence.

RIO DE JANEIRO

Background

Whereas most retail drug markets are competitive and fragmented, including in most of Brazil's other large cities, a handful of powerful, prison-based trafficking syndicates have dominated Rio de Janeiro's retail drug trade, and the favelas out of which it operates, for more than 25 years (Lessing, 2008). These groups, born as prison gangs, expanded outward in the 1980s to take control of most of the city's nearly 1,000 favelas, holding off police invasions with militarized force, and inevitably retaking communities once police had left. This stalemate seemed intractable, a 'worst-of-all-worlds' equilibrium, in which police corruption was thoroughgoing, police operations were incredibly lethal but strategically ineffective, and nearly one million favela residents continued to live under the control of non-state actors. The situation was so grim that when police-linked paramilitary groups known as *milicías* began to seize control of outlying favelas to establish their own exploitive rule, Rio's then mayor publicly declared them to be the lesser of two evils (Bottari and Ramalho, 2006).

Rio State Governor Sérgio Cabral began his first term in 2007 with a traditional hard-line stance that led to an increase in lethal confrontations between police and the syndicates; this approach fit into a longer pattern of oscillation in state policy between periods of benign neglect and lethal crackdowns (see Figure 2.7).[11] A June 2007 joint operation by police and federal armed forces in the Complexo do Alemão favela, the stronghold of Rio's largest and oldest trafficking syndicate, the Comando Vermelho (CV), was significant. While more than 1,300 troops and officers laid siege to the favela for a month, 'traffickers circulated, carrying assault rifles and pistols [. . .] and impudently pointed them at federal soldiers, challenging them to a confrontation in the favela' (Costa, 2007).[12] Finally, troops invaded the area, leaving 19 civilians dead (all allegedly drug traffickers) and yielding some arms seizures but no arrests. Though the operation—which officials then touted as the largest ever in Brazil's history—drew severe criticism from human rights organizations and even the United Nations for its use of excessive lethal force (Alston, 2007), it nonetheless left the favela under the dominion of the CV (Costa et al., 2007).

A fireman hoses down a truck set on fire by members of the Comando Vermelho, Rio de Janeiro, Brazil, November 2010. © Buda Mendes/Getty Images

Figure 2.7 **Rio de Janeiro: civilians killed by police in alleged 'self-defence'**[13]

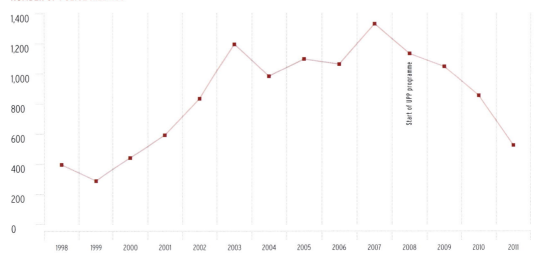

Note: Figures are for the state of Rio de Janeiro.
Source: author's calculations based on SSP-RJ (2003) and ISP (2012)

Following the 2007 invasion, however, the state introduced a new public security approach centred on the use of *unidades de polícia pacificadora* ('pacification' police units, or UPPs). One by one, police special forces teams cleared favelas of drug traffickers and handed them over to the community-oriented UPP battalion that was specially recruited and trained for permanent deployment in that favela. As discussed below, the UPP programme represents a clear departure from Rio's traditionally violent and repressive approach to policing the favelas. One crucial aspect of this approach is that the UPPs' stated mission is not to eliminate drug trafficking itself, but rather to bring an end to the presence of armed drug traffickers through targeted response and subsequent occupation of favelas.

As the state prepared to invade Alemão in 2010, there were reasons for both hope and fear. The UPP programme had thus far been a success; by announcing ahead of time which favelas were to be occupied, the state had given traffickers time to clear out, and there had been no major confrontations. At the same time, traffickers expelled from cleared favelas had been regrouping in Alemão; if the CV was going to make a stand, it would surely be there. In the end, the invasion generated fewer casualties than expected, and, more importantly, it left Alemão firmly in the hands of the state for the first time in more than a generation, ending what Rio's Public Security Secretary and architect of the UPP programme José Beltrame called the 'belief in the invincibility' of Alemão (Carneiro, 2010). The event marks a turning point in the history of Rio's drug war.

This section explores the UPP programme in more detail, discusses recent trends in drug-related violence, and considers some of the side effects and threats to the long-term success of the new approach in Rio.

The UPP approach

The UPP strategy consists of four steps.[14] First, the state announces the next community to be cleared of drug syndicates. While an exact date is not given, announcements come far enough in advance to give traffickers a chance to flee. Second, law enforcement floods the favela with overwhelming force, usually by the elite Batalhão de Operações

Policiais Especiais (Special Police Operations Battalion, BOPE), yet sometimes with additional support from the military, especially armoured vehicles. Third, the favela community is cleared and the UPP installed. For a month or so after the occupation, the BOPE remains in the community, conducting searches for arms and drugs and preventing attempts by trafficking syndicates to retake the community. In the transition, a UPP proximity policing unit is installed, usually in a new building built to purpose. Fourth, the BOPE and army withdraw, leaving the UPP in charge (Stahlberg, 2011).

Rio's conventional ostensive police force, the *polícia militar* (military police), has long been structured and trained to use primarily repressive tactics. The UPPs, on the other hand, are drawn from fresh recruits who have received training in human rights, preventive policing, and community relations. A large proportion of UPP officers are women, and all of them receive bonuses, intended to raise the profile of work that traditional, hard-line officers have tended to view as 'soft'. In some communities, the process of trust-building is reinforced with a 'Social UPP' component—a range of social programmes aimed at residents, some of them administered by or with UPP officers, others simply appropriating the UPP 'brand'. At the time of writing, 19 UPP battalions had been installed in communities cleared of drug syndicates and *milícias*, with plans to pacify a total of 40 favelas by the 2014 World Cup (de Aquino, 2011).

Few of the ideas underlying the UPP programme are new. The concept draws on basic community- and proximity-policing practices and lessons learned from the experiences of Medellín and Bogotá, Colombia (Stahlberg, 2011, p. 7).[15] Yet much of the programme's core approach was directly adapted from earlier alternative policing projects in Rio. Key among these was the pioneering 1999 Mutirão pela Paz (Mobilization for Peace) project, which combined police occupation of a target favela with community outreach and a raft of new social programmes (Soares, 2000, pp. 280–84). Another such project was the innovative successor programme Grupamento de Policiamento em Áreas Especiais (Policing in Special Areas Unit), installed on a pilot basis in a group of favelas in 2001 (Huguet, 2009); the idea was to establish a permanent police base in the favelas that would not attempt to eradicate the drug trade itself, but rather its armed presence.[16]

> The idea was not to eradicate the drug trade itself, but rather its armed presence.

This stance is central to the UPP strategy. As asserted by the UPP programme's public face, Secretary of Public Security José Beltrame, 'We cannot guarantee that we will put an end to drug trafficking nor do we have the pretension of doing so' (Phillips, 2010). When confronted by journalists with footage of drugs being sold in a UPP-occupied favela, Beltrame did not apologize for the UPP's central harm-reduction approach:

The basic mission was to disarm the drug dealers and bring peace to the residents. The footage doesn't appear to show anyone armed. [. . .] I can't guarantee there is no drug dealing going on, in some dark corner, in a place as large as City of God.… That positive outcome is worth infinitely more than the sale of a half dozen packets [of cocaine] (Araújo, 2010).[17]

Similarly, Beltrame openly defends the UPP policy of letting traffickers flee prior to pacifications in order to avoid confrontations:

What difference does the arrest of a drug lord make to the life of people who live in a given community? [. . .] Will it reduce crime rates? Arresting drug lords is important, but it isn't the most important thing. Without territory, they are much less 'lords' than they were before' (Bastos, 2011a).[18]

Three factors set the UPP programme apart from previous interventions: its city-wide scale; its systematic approach of pacifying one favela at a time and announcing future operations in advance; and the unusually strong support it is receiving, not only from a rare political alignment that spans the municipal, state, and federal levels of government, but also from business elites and the media giant *Globo*. These factors have allowed the UPP programme, perhaps the first large-scale attempt to address drug trafficking and violence using a proportional response and harm-reduction approach, to remain politically tenable—an important development in its own right.

Assessing the results

The pacification effort and the attendant social investment projects in many of the pacified favelas have changed the face of Rio de Janeiro. Most recently and saliently, state forces retook Rocinha, the largest of Rio's favelas and in many ways the most prominent, abutting the single wealthiest neighbourhood in the city. After more than a generation of dominion by armed traffickers, the state was able to reclaim the enormous area in only two hours, and without firing a shot (Briso and Cerqueira, 2011). The entire South Zone—home to most of the city's tourist attractions and wealthy residents—is now essentially free of trafficking-dominated favelas, and large-scale public projects are bringing residents into the fabric of formal city life.

There are strong indications that Rio's syndicates, particularly the CV, have shifted from confrontation to strategic retreat and non-violent trafficking. The most prominent

Policemen patrol the Rocinha favela as part of an operation to install a UPP, Rio de Janeiro, November 2011. © Sergio Moraes/Reuters

is the taking of Complexo do Alemão with a minimum of fighting in 2010, after the bloody and failed attempt in 2007. Perhaps even more surprising to observers of Rio's drug war was the wholly bloodless pacification of Rocinha and its neighbour Vidigal, once the stronghold of one the CV's principal rivals. In fact, drug traffickers seem to be disarming in anticipation of pacifications. This occurred in São Carlos favela, site of the first post-Alemão UPP installation (Costa, Moura, and Daflon, 2011); it also seems to be happening in Maré, a very large complex of favelas that is not even scheduled to be pacified.[19]

Despite these qualitative changes, the effect of the UPP programme on crime and violence has not been rigorously quantified for a number of reasons. Because UPPs are not assigned randomly to favelas, and because the installation of a UPP in one favela could have an effect on other non-UPP favelas, comparisons of crime rates between UPP and non-UPP communities are hard to interpret.[20] Access to crime data is also a problem. Recently, Rio's Public Security Secretary began releasing crime data for the first 13 favelas to receive UPPs, touting falling crime numbers and the fact that in the first half of 2011, all but two UPP favelas reported no homicides (Cândida and Ramalho, 2011). While these certainly seem like strong results, there is no comparable data available for non-UPP favelas, nor does the published data cover periods prior to pacification, making a meaningful comparison impossible. Longer time series are available for administrative areas that include multiple favelas and surrounding neighbourhoods, but these inevitably group together both UPP and non-UPP favelas.

While state-wide data shows a 28.7 per cent decline in homicides between 2006 and 2009, these figures have recently been called into question. A rigorous econometric examination of mortality data from the federal Instituto de Pesquisa Econômica Aplicada (Institute for Applied Economic Research) shows that over this same period, 'deaths of unknown motive' have been rapidly increasing in Rio but not elsewhere in Brazil, suggesting that some homicides have been 'hidden' in this reporting category (Cerqueira, 2011). The study's 'corrected count' estimate increases the number of homicides in 2009 alone by some 3,100 (more than 60 per cent of the total) and suggests that the 'true' homicide rate has fallen by only 3.6 per cent over the period (Roman, 2011). The study and its strong implications of deliberate manipulation of homicide data have put the government on the defensive (Bottari and Leite, 2011), further complicating efforts to assess the effectiveness of the UPP project.

Ultimately, reducing rates of crime and violence is one of the primary goals of the pacification project, and achieving this aim will be crucial to winning the acceptance and approval of residents. To date, no such reductions have been clearly demonstrated. In fact, as discussed below, the state may initially be less effective than the long-entrenched drug lords at providing local security. However, by re-establishing a monopoly on the use of force and by apparently inducing a shift in traffickers' strategies from fighting to hiding, the programme has greatly reduced its own future cost. São Carlos was, in a sense, already pacified when the troops arrived, and Maré appears to have 'pacified itself' before an official operation was ever announced. The more smoothly these operations go, the more credibility the leadership has to openly defend and promote proportional response. Yet this progressive and innovative approach has a number of risks associated with it, as discussed below.

Potential threats and looming problems
Milícias: Rio's home-grown paramilitaries

The CV and its rival prison-based syndicates are not the only armed groups plaguing Rio. The so-called *milícias*, paramilitary groups comprised largely of rogue police officers, have established armed control over hundreds of favelas and outlying areas, taken over illicit markets and extortion rackets, terrorized the communities they control, and infiltrated the political system through armed clientelism and electioneering (Freixo et al., 2008; Cano, 2008; Hidalgo and Lessing, 2009). The *milícias*, weakening the state from within its own security apparatus, and with far more political power and coercive reach than favela-bound traffickers, are, according to Beltrame, a graver threat today than drug syndicates (de Aquino, 2011). Recent examples include the brutal murder by *milícia*-linked police of corruption-fighting judge Patricia Acioli (Nogeira, 2011); meanwhile, a flurry of death threats led Rio State Assemblyman Marcelo Freixo—who has spearheaded efforts to investigate the *milícias*—to flee the country at the invitation of Amnesty International (Loureiro, 2011).

The UPP programme seems to be aimed more at retaking territory from syndicates than from *milícias*. Of the more than 20 favelas pacified to date, only one was controlled by *milícias* prior to its pacification, which was itself an unplanned response to the widely publicized torture of journalists by the local *milícia* group (de Freitas, 2009).[21] Moreover, the steady progress of the pacification process and the severe blow dealt to the CV with the invasion of Alemão seems to have benefitted the *milícias*, who now control more favelas than the CV (Paes Manso, 2010). *Milícia* power is also extending inward, into the state's security forces; a study by the Rio Police internal affairs office finds that in every one of the 18 police battalions located in the city of Rio, as well as special forces such as the BOPE, at least one officer was involved with a *milícia* group (Monken, 2009). There have been denunciations of extortion and other *milícia*-like activities by UPP troops, and Beltrame has admitted that UPP penetration by the *milícias* is to be feared (Bastos, 2011a).

The danger of *milícia* infiltration within the highly trained BOPE corps is particularly troubling, especially in view of the case of Mexico's Grupo Aeromóvil de Fuerzas Especiales. That elite, US-trained anti-drug force deserted en masse, becoming Los Zetas, today the most violent cartel in Mexico (Manwaring, 2009, p. 19). Unless carefully overseen and managed, elite corps like the BOPE can quickly be transformed from an asset into a highly destabilizing and destructive force.

Balloon effects

Another important issue is the so-called 'balloon effect', the suggestion that by forcing traffickers out of Rio proper, the UPP programme is merely pushing them to migrate to new areas, in effect moving crime and violence around from one region to another. This is a particularly acute concern given the UPP violence-reducing policy of allowing traffickers to flee rather than cornering them. For example, it was widely suspected that many traffickers from the first-pacified favelas had migrated to the CV's headquarters of Vila Cruzeiro and Alemão, and the flight of dozens of armed traffickers from those two favelas was actually filmed by news helicopters. Consequently, many residents and authorities now fear 'crime migration'—the invasion by Rio's drug traffickers of other municipalities and even states. Federal highway police, for example, stepped up patrols in neighbouring states the day after Alemão was retaken (Gomes and Cymrot, 2011).

> One concern is funding for the UPP programme once the 2014 World Cup and 2016 Olympics have passed.

Just as determining the actual benefits of the UPPs on the communities that received them is difficult, it is virtually impossible to know with certainty whether and where balloon effects have actually occurred. However, some crime migration is consistent with early observations. One analysis—which compares the first trimesters of 2009 (when only two UPPs had been installed) and 2011 (by which time 16 UPPs had been installed)—finds that in four of Rio's largest neighbouring municipalities, arrest rates rose two to four times faster than in Rio itself (Bastos, 2011b). Another study reveals that, in the wake of the retaking of Cruzeiro and Alemão, seasonal robbery rates fell in the area surrounding these favelas, but rose sharply in the region of Mangueira, where the fleeing traffickers are believed to have first regrouped (*R7*, 2011). News stories from remote towns in the interior of Rio de Janeiro State make clear that both residents and authorities attribute increases in certain local crime rates to migrating traffickers from UPP favelas in Rio, but the causal link remains conjectural. Further research is needed to clarify the relationship between crime levels in adjacent areas (Werneck, 2010).[22] The government, meanwhile, acknowledges that the balloon effect may occur but has argued that it is not a major concern because only the top drug lords are able to migrate, leaving the foot soldiers behind. Other experts argue that denying the existence of the balloon effect would be to 'deny the obvious' (Bastos, 2011b).

Intra-community destabilization and long-term financing

Another concern relates to the intra-community effects of the UPP programme and its long-term prospects. Most favelas have been under the armed dominion of drug traffickers for decades; syndicates frequently come to provide public goods for the communities they dominate. In particular, they establish a rough social order in which transgression is likely to be discovered and harshly punished (Leeds, 1996). When that source of parallel power is removed, it can leave a vacuum in the daily life of favelas. This vacuum is particularly acute in communities that have not yet received UPP units, since the BOPE and army soldiers who occupy the community in the medium term are not trained to play the kind of mediation roles that traffickers once did and lack the information structures necessary to administer street-level social order. This can lead to increases not only in reported property crime, but also in domestic violence, both of which tended to be very effectively prohibited, or at least hushed up, by traffickers (*Povo do Rio,* 2012).[23] An additional concern, especially for favela residents, is that the government will not be able to maintain the funding necessary to administer or expand the UPP programme once the 2014 World Cup and the 2016 Olympic Games have passed.

In a sense, both of these concerns reflect the same underlying problem: for decades, the state outsourced the monopoly on the use of force and the provision of social order to non-state actors willing to work for free, or even to pay bribes to corrupt police for the privilege (Werneck, 2011). Taking over those functions, even in just a handful of the city's 1,000 favelas, will be a long, slow, costly task.

CONCLUSION

'Systemic' drug violence—associated with drug traffickers and cartels as opposed to consumers—is a complex phenomenon influenced by the structure of illicit markets as well as state policies and institutions. Violence involving small drug-retailing outfits such as street gangs can be common, lethal, and hard to reduce, but it is also limited by the size and resources of such groups, and by their desire to maintain anonymity. In particular, retail drug organizations rarely attack state forces. Far more troubling, if rarer, is violence involving the large, articulated trafficking syndicates and cartels. Such drug-trafficking organizations have directly engaged state forces, as well as rivals, in brazen armed conflict, leading to extensive 'drug wars'—armed conflicts with destructive consequences similar to civil wars, but of a very different nature.

Although quantitative data is scarce, the case studies reviewed in this chapter illuminate the challenges, effects, and potential unintended consequences of state efforts to contain and reduce drug cartel violence in Latin America. In Mexico, President Calderón's 2006 crackdown, designed to fall on all cartels roughly equally, appears to have instead triggered a rapid explosion of violence. The effects of cartel fragmentation are being felt throughout the country and in Central America. Six years into an exhausting and brutal conflict with tens of thousands of casualties, the blanket response approach may be giving way to a more proportional focus on the most deadly Mexican cartels. Even sharper shifts in policy are likely if, as many predict, the PRI regains the presidency in the 2012 election.

In Rio de Janeiro, a proportional-response approach aimed at curbing the most violent drug-trafficking activity has returned control of many favelas to the state after more than 20 years. But it is not yet clear whether this is translating into a reduction in overall violence, nor whether criminal networks are simply relocating. Maintaining control is also a long-term, complex, and costly proposition, and is as much about state service provision as violence prevention. It remains to be seen whether state and federal authorities are prepared to stay the course. Yet the shift in approach seems to have altered the logic of cartel–state conflict that was in effect for decades; as such, it merits sustained scrutiny to better assess its true effectiveness and possible application to other settings.

LIST OF ABBREVIATIONS

ATF	Bureau of Alcohol, Tobacco, Firearms and Explosives
AWB	Assault Weapons Ban
BOPE	Batalhão de Operações Policiais Especiais (Special Police Operations Battalion)
CV	Comando Vermelho
FARC	Fuerzas Armadas Revolucionarias de Colombia (Revolutionary Armed Forces of Colombia)
PRI	Partido Revolucionario Institucional (Institutional Revolution Party)
UPP	*Unidade de polícia pacificadora* ('pacification' police unit)

ENDNOTES

1. See, for example, Dorn, Murji, and South (1992); Goldstein (1985); and MacCoun, Kilmer, and Reuter (2003).
2. Traffickers could, of course, commit this kind of violence, but they would do so as consumers.
3. Author interview with a retired Colombian justice official, Cartagena, Colombia, 27 December 2010.
4. Los Zetas were originally a kind of private security brigade employed by the Gulf cartel.
5. All correlation coefficients are based on the author's calculations.
6. Author communication with Peter Reuter, 25 September 2011.
7. See, for example, Global Commission on Drug Policy (2011, p. 14).
8. The most violent armed conflict of the 21st century has been the Eritrea-Ethiopa border war, with between 70,000 and 100,000 battle deaths from May 1998 to May 2000.
9. Author's translation.
10. Straw purchasers are individuals who are authorized to buy firearms for personal use but who unlawfully buy them with intent to transfer them to drug-trafficking organizations or other criminals.
11. In Brazil, formal responsibility for policing and public security policy rests almost entirely with governors and state legislatures. The federal government can provide key support, whether financial, political, or logistical (such as assistance from the armed forces) but can also remain largely disengaged. Since municipalities generally lack their own police forces, mayors have little direct power to affect policing policy; still, their support or opposition can have an important impact on the success of state-level policies.
12. Author's translation.
13. When police killings of civilians in the course of patrols or operations are reported, they are automatically placed in this category—*auto de resistência*, or 'act of resistance'. While the majority of these incidents probably occur during armed confrontations, they do not reflect a judicial determination and there is ample evidence that at least some of these killings represent summary executions by police against subdued, fleeing, or unarmed opponents. See, for example, Cano (1997).
14. Stahlberg (2011) provides a detailed overview of the UPP programme.
15. For more on the municipal measures implemented in Colombia, see Chapter 1 on Latin America and the Caribbean in this volume.
16. The programme's designer and commanding officer describe these rules as 'don't walk around openly armed, don't sell drugs near schools, and don't employ children'. Author interview with Maj. Antonio Carballo Blanco, Rio de Janeiro, June 2003.
17. Author's translation.
18. Author's translation.
19. Author interviews with the director of a Maré-based social service NGO, Rio de Janeiro, 14 January 2012, and with a Maré-based field researcher, Rio de Janeiro, 22 December 2011.
20. In statistical terms, there might be selection bias or contamination. If favelas are selected for receiving UPPs even partly on the basis of crime and violence rates, then observed differences between UPP and non-UPP favelas could be due to the selection process rather than the 'treatment' of receiving a UPP. 'Contamination'—pacification in one favela driving changes in crime elsewhere—could happen directly, because traffickers migrate from a pacified favela to a neighbouring, non-UPP favela, or indirectly, because the UPP programme is lowering crime rates everywhere. Either channel would violate the so-called stable-unit treatment value assumption (SUTVA).
21. As of this writing, there were 19 UPPs in place and one scheduled for inauguration in April 2012, serving some 27 communities. In some cases, multiple smaller favelas were pacified in a single operation and are served by a single UPP. In addition, Vila Cruzeiro and the Alemão complex of favelas were pacified in 2011; they are currently patrolled by army and non-UPP police troops, with no date for UPP implantation set.
22. The main obstacle to assessing the balloon effect is accurately documenting where traffickers flee.
23. Author interview with a community activist, Complex do Alemão, 29 March 2011.

BIBLIOGRAPHY

Akerlof, George and Janet Yellen. 1994. 'Gang Behavior, Law Enforcement, and Community Values.' In Henry Aaron, Thomas Mann, and Timothy Taylor, eds. *Values and Public Policy*. Washington, DC: Brookings Institution, pp. 173–209.

Alston, Philip. 2007. 'Press Statement by the Special Rapporteur of the UN Human Rights Council on Extrajudicial, Summary or Arbitrary Executions.' Brasília: United Nations Human Rights Council. 14 November.

Amorim, Carlos. 1993. *Comando Vermelho: a História Secreta do Crime Organizado*. Rio de Janeiro: Editora Record.
<http://www.scribd.com/doc/6784630/Comando-Vermelho-A-Historia-Secreta-Do-Crime-Organizado-Carlos-Amorim>

Aranda, Jesús. 2011. 'Ataques al Ejército cobran la vida de 253 soldados desde 2006.' *La Jornada* (Mexico). 1 August.
<http://www.jornada.unam.mx/2011/08/01/politica/014n1pol>

Araújo, Vera. 2010. 'Feira de drogas resiste à UPP da Cidade de Deus.' *O Globo*. 2 July.
    <http://oglobo.globo.com/rio/mat/2010/07/01/feira-de-drogas-resiste-upp-da-cidade-de-deus-917044669.asp>
Archibold, Randal, Damien Cave, and Elisabeth Malkin. 2011. 'Mexico's President Works to Lock In Drug War Tactics.' *The New York Times*. 15 October.
Astorga, Luis. 1999. *Drug Trafficking in Mexico: A First General Assessment*. UNESCO Discussion Paper No. 36.
    <http://www.unesco.org/most/astorga.htm>
ATF (Bureau of Alcohol, Tobacco, Firearms and Explosives). 2008. 'Statement of William Hoover.' 7 February.
    <http://www.atf.gov/press/releases/2008/02/020708-testimony-atf-ad-hoover-sw-border.html>
Attkisson, Sharyl. 2011a. 'Gunrunning Scandal Uncovered at the ATF.' CBS News. 23 February.
    <http://www.cbsnews.com/stories/2011/02/23/eveningnews/main20035609.shtml?tag=contentMain;contentBody>
—. 2011b. 'Major ATF Phoenix Shake-up after "Gunwalker."' CBS News. 17 May. <http://www.cbsnews.com/8301-31727_162-20063716-10391695.html>
—. 2011c. '"Gunwalker" Guns Linked to Helicopter Shooting.' CBS News. 6 June. <http://www.cbsnews.com/8301-31727_162-20069270-10391695.html>
Atwood, David, Anne-Kathrin Glatz, and Robert Muggah. 2006. *Demanding Attention: Addressing the Dynamics of Small Arms Demand*. Geneva: Small Arms Survey and Quaker United Nations Office.
    <http://www.smallarmssurvey.org/fileadmin/docs/B-Occasional-papers/SAS-OP18-Demand.pdf>
Bastos, Marcelo. 2011a. 'Secretário de Segurança admite temer que policiais das UPPs atuem como milicianos.' *R7*. 12 May.
    <http://noticias.r7.com/rio-de-janeiro/noticias/secretario-de-seguranca-admite-temer-que-policiais-das-upp-atuem-como-milicianos-20110512.html>
—. 2011b. 'UPPs provocam aumento de prisões e denúncias fora do Rio.' *R7*. 21 June.
    <http://noticias.r7.com/rio-de-janeiro/noticias/upps-provocam-aumento-de-prisoes-e-denuncias-fora-do-rio-20110621.html>
Blumstein, Alfred. 1995. 'Youth Violence, Guns, and the Illicit-Drug Industry.' *Journal of Criminal Law and Criminology*, Vol. 86, No. 1, pp. 10–36.
    <http://www.saf.org/lawreviews/blumstein1.htm>
Booth, William. 2010. 'Only One Gun Store, but No Dearth of Weapons.' *Washington Post*. 29 December.
    <http://www.washingtonpost.com/wp-dyn/content/article/2010/12/28/AR2010122803644.html?hpid=topnews>
Bottari, Elenlice and Renata Leite. 2011. 'Governo estadual contesta pesquisa que questionou queda nos homicídios.' *O Globo*. 25 October.
    <http://oglobo.globo.com/rio/mat/2011/10/25/governo-estadual-contesta-pesquisa-que-questionou-queda-nos-homicidios-925661481.asp>
— and Sérgio Ramalho. 2006. 'Milícias avançam pelo corredor do Pan 2007.' *O Globo*. 12 December.
    <http://oglobo.globo.com/rio/governo-estadual-contesta-pesquisa-que-questionou-queda-nos-homicidios-2896711>
Briso, Caio, and Sofia Cerqueira. 2011. 'A retomada de São Conrado.' *Veja Rio*. 23 November.
    <http://vejario.abril.com.br/edicao-da-semana/sao-conrado-rocinha-upp-646869.shtml>
Burnett, John, Marisa Peñaloza, and Robert Benincasa. 2010. 'Mexico Seems to Favor Sinaloa Cartel in Drug War.' National Public Radio (United States). 19 May. <http://www.npr.org/2010/05/19/126906809/mexico-seems-to-favor-sinaloa-cartel-in-drug-war>
Cândida, Simone, and Sérgio Ramalho. 2011. 'ISP: áreas de UPPs têm redução de crimes.' *O Globo*. 14 September.
    <http://oglobo.globo.com/rio/mat/2011/09/14/isp-areas-de-upps-tem-reducao-de-crimes-925359728.asp>
Cano, Ignacio. 1997. *Letalidade da Ação Policial no Rio de Janeiro*. Rio de Janeiro: Instituto de Estudos da Religião.
—. 2008. 'Seis por Meia Dúzia? Um Estudo Exploratório do Fenômeno das Chamadas "Milícias" no Rio De Janeiro.' In *Segurança, Tráfico e Milícias no Rio de Janeiro*. Rio de Janeiro: Justiça Global. <http://www.cancun2003.org/download_pt/Milicias_SeisporMeiaDuzia.pdf>
Carneiro, Júlia Dias. 2010. 'Operação acaba com crença em 'invencibilidade' do Alemão, diz Beltrame.' BBC Brasil. 28 November.
    <http://www.bbc.co.uk/portuguese/noticias/2010/11/101128_rio_beltrame_jc.shtml>
Caulkins, Jonathan and Peter Reuter. 1996. 'The Meaning and Utility of Drug Prices.' *Addiction*, Vol. 91, No. 9, pp. 1261–64. <http://www.publicpolicy.umd.edu/uploads/cms/faculty/reuter/Working%20Papers/The%20Meaning%20and%20Utility%20of%20Drug%20Prices.pdf>
Cave, Damien. 2012. 'Mexico Updates Death Toll in Drug War to 47,515, but Critics Dispute the Data.' *New York Times*, 11 January.
    <http://www.nytimes.com/2012/01/12/world/americas/mexico-updates-drug-war-death-toll-but-critics-dispute-data.html>
Cerqueira, Daniel. 2011. 'Mortes Violentas Não Esclarecidas e Impunidade no Rio de Janeiro.' Rio de Janeiro: Instituto de Pesquisa Econômica Aplicada. <http://www2.forumseguranca.org.br/files/MortesViolentasNaoEsclarecidaseImpunidadenoRiodeJaneiro.pdf>
Chicoine, Luke. 2011. *Exporting the Second Amendment: U.S. Assault Weapons and the Homicide Rate in Mexico*. Working Paper. Notre Dame: University of Notre Dame. <http://nd.edu/~lchicoin/Chicoine_AWB_Mexico.pdf>
CNN México. 2011. 'El "Triángulo Dorado", resguarda la droga entre el frío clima y la pobreza.'
    <http://mexico.cnn.com/nacional/2011/11/17/el-triangulo-dorado-resguarda-la-droga-entre-el-frio-clima-y-la-pobreza>
Corchado, Alfredo. 2011. 'Arrest Signals Targeting of Zetas.' *Dallas Morning News*. 5 July.
Costa, Ana Cláudia. 2007. 'Quatro mortos e quatro feridos no segundo dia de operação no Alemão.' *O Globo*. 14 June.
    <http://oglobo.globo.com/rio/mat/2007/06/14/296166611.asp>
—, Athos Moura, and Rogério Daflon. 2011. 'Polícia ocupa Complexo de São Carlos para instalação de Unidades de Polícia Pacificadora (UPPs).' *O Globo*. 6 February.
<http://extra.globo.com/noticias/rio/policia-ocupa-complexo-de-sao-carlos-para-instalacao-de-unidades-de-policia-pacificadora-upps-1013906.html>
—, et al. 2007. 'Megaoperação no Alemão deixa 19 mortos.' *O Globo*, 27 June. <http://oglobo.globo.com/rio/mat/2007/06/27/296546114.asp>
Cruz, José Miguel. 2010. 'Central American Maras: From Youth Street Gangs to Transnational Protection Rackets.' *Global Crime*, Vol. 11, No. 4, pp. 379–98.
    <http://dx.doi.org/10.1080/17440572.2010.519518>

de Aquino, Ruth. 2011. 'José Mariano Beltrame: "A milícia hoje me preocupa mais que o tráfico".' *Epoca*. 16 October.
    <http://revistaepoca.globo.com/Sociedade/noticia/2011/10/jose-mariano-beltrame-milicia-hoje-me-preocupa-mais-que-o-trafico.html>

de Freitas, Guedes. 2009. 'Estado inaugura UPP no Batan e planeja outras ocupações.' *Diario Oficial do Governo do Estado do Rio de Janeiro*. 18 February.
    <http://governo-rj.jusbrasil.com.br/politica/1734583/estado-inaugura-upp-no-batan-e-planeja-outras-ocupacoes>

Dorn, Nicholas, Karim Murji, and Nigel South. 1992. *Traffickers: Drug Markets and Law Enforcement*. New York: Routledge.

Dowdney, Luke. 2003. *Children of the Drug Trade*. Rio de Janeiro: 7 Letras.

Dube, Arindrajit, Oeindrila Dube, and Omar García-Ponce. 2011. 'Cross-border Spillover: U.S. Gun Laws and Violence in Mexico.'
    <https://files.nyu.edu/od9/public/papers/Cross_border_spillover.pdf>

Dudley, Steven. 2011. 'Drug Trafficking Organizations in Central America: *Transportistas*, Mexican Cartels and Maras.' In Cynthia Arnson and Eric Olson, eds. *Organized Crime in Central America: The Northern Triangle*. Washington, DC: Woodrow Wilson Center.

Economist. 2011. 'The Drug War Hits Central America.' 14 April. <http://www.economist.com/node/18560287>

El Universal (Mexico). 2009a. '"Los Zetas" amenazan a presidente guatemalteco.' 2 March. <www.eluniversal.com.mx/nacion/166126.html>

—. 2009b. 'Presunto líder de La Familia llama al diálogo.' 15 July. <http://www.eluniversal.com.mx/notas/612528.html>

—. 2010. 'Comando irrumpe en hoteles de Monterrey y se lleva a 7 personas.' 22 April. <http://www.eluniversal.com.mx/estados/75605.html>

Felbab-Brown, Vanda. 2005. 'The Coca Connection: Conflict and Drugs in Colombia and Peru.' *Journal of Conflict Studies*, Vol. 25, No. 2.
    <http://journals.hil.unb.ca/index.php/JCS/article/view/489/824>

Freixo, Marcelo, et. al. 2008. *Relatório final da Comissão Parlamentar de Inquérito destinada a investigar a ação de milícias no âmbito do estado do Rio de Janeiro*. Rio de Janeiro: Asembléia Legislativa do Estado do Rio de Janeiro.
    <http://www.marcelofreixo.com.br/site/upload/relatoriofinalportugues.pdf>

Fries, Arthur, et al. 2008. *The Price and Purity of Illicit Drugs: 1981–2007*. Alexandria: Institute for Defense Analysis.

Geneva Declaration Secretariat. 2011. *Global Burden of Armed Violence 2011: Lethal Encounters*. Geneva: Geneva Declaration Secretariat.

Global Commission on Drug Policy. 2011. *War on Drugs: Report of the Global Commission on Drug Policy*.
    <http://www.globalcommissionondrugs.org/Report>

Goldstein, Paul. 1985. 'The Drugs/Violence Nexus: A Tripartite Conceptual Framework.' *Journal of Drug Issues*, Vol. 39, pp. 143–74.
    <http://www.drugpolicy.org/docUploads/nexus.pdf>

—. 1997. 'The Relationship between Drugs and Violence in the United States of America.' In United Nations International Drug Control Programme. *World Drug Report*. Oxford: Oxford University Press, pp.116–21.

Gomes, Luiz Flávio and Danilo Cymrot. 2011. 'UPPs, prevenção situacional e o deslocamento do crime.' Rio de Janeiro: Instituto de Pesquisa e Cultura. 10 February. <http://www.ipclfg.com.br/prevencao-do-crime-e-upps/upps-prevencao-situacional-e-o-deslocamento-do-crime/>

Gómez, Francisco. 2009. 'Inédita narcoembestida.' *El Universal* (Mexico). 12 July. <http://www.eluniversal.com.mx/nacion/169639.html>

Gómora, Doris. 2011. '"Lince Norte" daña estructura "zeta".' *El Universal* (Mexico). 5 August. <http://www.eluniversal.com.mx/nacion/187762.html>

Goodman, Colby. 2011. *Understanding Data and Evidence on U.S. Firearms Seized in Mexico*. Testimony before the Government Reform and Oversight Committee, US House of Representatives. 30 June.

Guerrero-Gutiérrez, Eduardo. 2010. 'Pandillas y cárteles: La gran alianza.' *Nexos*. <http://www.nexos.com.mx/?P=leerarticulov2print&Article=73224>

—. 2011. *Security, Drugs, and Violence in Mexico: A Survey*. Mexico City: Lantia Consultores.
    <http://iis-db.stanford.edu/evnts/6716/NAF_2011_EG_(Final).pdf>

Hagedorn, John. 1994. 'Neighborhoods, Markets, and Gang Drug Organization.' *Journal of Research in Crime and Delinquency*, Vol. 31, No. 3, pp. 264–94.
    <http://jrc.sagepub.com/content/31/3/264.abstract>

Hidalgo, F. Daniel and Benjamin Lessing. 2009. *Politics in the Shadow of the Drug War*. Paper prepared for the XXVIII International Congress of the Latin American Studies Association. Rio de Janeiro, Brazil.

Huguet, Clarissa. 2009. *Drug Trafficking, Militias, Police Violence and State Absence*. Paper prepared for the 10th Berlin Roundtables on Transnationality. Berlin, Germany, March. <http://www.irmgard-coninx-stiftung.de/fileadmin/user_upload/pdf/urbanplanet/collective_identities/Huguet_Essay.pdf>

ICG (International Crisis Group). 2011. 'Guatemala: Drug Trafficking and Violence.' 11 October.
    <http://www.crisisgroup.org/en/regions/latin-america-caribbean/guatemala/139-guatemala-drug-trafficking-and-violence.aspx>

ISP (Instituto de Segurança Pública—Public Security Institute). 2012. 'Dados Oficiais.' <http://www.isp.rj.gov.br/Conteudo.asp?ident=150>

Jütersonke, Oliver, Robert Muggah, and Dennis Rodgers. 2009. 'Gangs, Urban Violence, and Security Interventions in Central America.' *Security Dialogue*, Vol. 40, Nos. 4–5, pp. 373–97. <http://sdi.sagepub.com/content/40/4-5.toc>

Kennedy, David, et al. 2001. 'Reducing Gun Violence: The Boston Gun Project's Operation Ceasefire.' *National Institute of Justice Research Report*
    <https://www.ncjrs.gov/pdffiles1/nij/188741.pdf>

Kleiman, Mark. 2011. 'Surgical Strikes in the Drug Wars.' *Foreign Affairs*, September/October.

Labrousse, Alain. 2005. 'The FARC and the Taliban's Connection to Drugs.' *Journal of Drug Issues*, Vol. 35, No. 1, pp. 169–84.
    <http://www.archido.de/index.php?option=com_docman&task=doc_view&gid=1388>

Leeds, Elizabeth. 1996. 'Cocaine and Parallel Polities in the Brazilian Urban Periphery: Constraints on Local-Level Democratization.' *Latin American Research Review*, Vol. 31, No. 3, pp. 47–83. <http://links.jstor.org/sici?sici=0023-8791%281996%2931%3A3%3C47%3ACAPPIT%3E2.0.CO%3B2-L>

Lessing, Benjamin. 2008. 'As Facções Cariocas em Perspectiva Comparativa.' *Novos Estudos–CEBRAP*, Vol. 80, pp. 43–62.
    <http://www.scielo.br/scielo.php?script=sci_arttext&pid=S0101-33002008000100004&nrm=iso >

—. 2010. 'The Danger of Dungeons: Prison Gangs and Incarcerated Militant Groups.' In Small Arms Survey. *Small Arms Survey 2010: Gangs, Groups, and Guns.* Geneva: Small Arms Survey.
<http://www.smallarmssurvey.org/fileadmin/docs/A-Yearbook/2010/en/Small-Arms-Survey-2010-Chapter-06-EN.pdf>

—. 2011. *The Logic of Violence in Criminal War: Cartel–State Conflict in Mexico, Colombia, and Brazil.* Paper presented at Violence, Drugs and Governance: Mexican Security in Comparative Perspective Conference. Stanford University, Palo Alto, California, 3 October.
<http://iis-db.stanford.edu/evnts/6716/LessingThe_Logic_of_Violence_in_Criminal_War.pdf>

Loureiro, Cláudia. 2011. 'Deputado Marcelo Freixo, do RJ, deixará o país após ameaças de morte.' *G1.* 31 October.
<http://g1.globo.com/rio-de-janeiro/noticia/2011/10/deputado-marcelo-freixo-do-rj-deixara-o-pais-apos-ameacas-de-morte.html>

MacCoun, Rob, Beau Kilmer, and Peter Reuter. 2003. 'Research on Drugs–Crime Linkages: The Next Generation.' In Henry Brownstein and Christine Crossland, eds. *Toward a Drugs and Crime Research Agenda for the 21st Century.* Washington, DC: National Institue of Justice, pp. 65–95.
<https://www.ncjrs.gov/pdffiles1/nij/194616c.pdf>

Magaloni, Beatriz. 2005. 'The Demise of Mexico's One-Party Dominant Regime.' In Frances Hagopian and Scott Mainwaring, eds. *The Third Wave of Democratization in Latin America: Advances and Setbacks.* New York: Cambridge University Press, pp. 121–48.
<http://www.stanford.edu/group/polisci/faculty/magaloni/thedemiseofmexicosoneparty.pdf>

Mahoney, James. 2001. *The Legacies of Liberalism: Path Dependence and Political Regimes in Central America.* Baltimore: Johns Hopkins University Press.

Manwaring, Max G. 2009. *A New Dynamic in the Western Hemisphere Security Environment: The Mexican Zetas and Other Private Armies.* Washington, DC: Strategic Studies Institute, United States Army War College.

Meiners, Stephen. 2009. 'Central America: An Emerging Role in the Drug Trade.' Stratfor.
<http://jotl.wordpress.com/2009/03/28/central-america-an-emerging-role-in-the-drug-trade-by-stephen-meiners/>

Milenio. 2010a. 'Noviembre, el mes con menos muertes diarias.' 1 December.

—. 2010b. 'Un ejecutado cada hora durante 2009.' 2 January.

—. 2011. 'Declina en agosto número de ejecuciones ligadas al "narco".' 1 September.

Monken, Mario Hugo. 2009. 'Todas unidades policiais do RJ têm pelo menos um miliciano.' *Jornal do Brasil.* 7 August.
<http://www.jb.com.br/rio/noticias/2009/08/07/todas-unidades-policiais-do-rj-tem-pelo-menos-um-miliciano/>

NDIC (National Drug Intelligence Center). 2009. 'National Drug Threat Assessment 2009.' Washington, DC: United States Department of Justice.
<http://www.justice.gov/ndic/pubs31/31379/index.htm>

—. 2010. 'National Drug Threat Assessment 2010.' Washington, DC: United States Department of Justice. <http://www.justice.gov/ndic/pubs38/38661/>

—. 2011. 'National Drug Threat Assessment 2011.' 2011-Q0317-001. Washington, DC: United States Department of Justice.
<http://www.justice.gov/dea/concern/18862/index.htm>

Nogeira, Italo. 2011. 'Police officers accused of killing judge in Rio.' Folha.com. 13 September.
<http://www1.folha.uol.com.br/internacional/en/dailylife/974384-police-officers-accused-of-killing-judge-in-rio.shtml>

ODCCP (United Nations Office for Drug Control and Crime Prevention). 1999. *Global Illicit Drug Trends 1999.* New York: ODCCP.
<http://www.unodc.org/unodc/en/data-and-analysis/WDR.html>

—. 2000a. *Global Illicit Drug Trends 2000.* New York: ODCCP. <http://www.unodc.org/unodc/en/data-and-analysis/WDR.html>

—. 2000b. *World Drug Report 2000.* New York: ODCCP. <http://www.unodc.org/unodc/en/data-and-analysis/WDR.html>

Paes Manso, Bruno. 2010. 'Milícias ocupam mais favelas no Rio que facção.' *O Estado de S. Paulo.* 11 November.
<http://www.estadao.com.br/noticias/impresso,milicias-ocupam-mais-favelas-no-rio-que-faccao,638198,0.htm>

Penglase, R. Ben. 2005. 'The Shutdown of Rio de Janeiro. The Poetics of Drug Trafficker Violence.' *Anthropology Today,* Vol. 21, No. 5, pp. 3–6.
<http://www.blackwell-synergy.com/doi/abs/10.1111/j.0268-540X.2005.00379.x>

PGR (Procuraduría General de la República—Office of the Mexican Attorney-General). 2011. *Quinto Informe de Labores.* Mexico City: PGR.
<http://www.pgr.gob.mx/Temas%20Relevantes/Documentos/Informes%20Institucionales/5o%20Informe%20de%20Labores%20PGR.pdf>

Phillips, Tom. 2010. 'Rio de Janeiro Police Occupy Slums as City Fights Back against Drug Gangs.' *Guardian.* 4 April.
<http://www.guardian.co.uk/world/2010/apr/12/rio-de-janeiro-police-occupy-slums>

Poiré Romero, Alejandro. 2011. 'El quinto mito: El Gobierno Federal favorece a Joaquín "El Chapo" Guzmán y al grupo criminal del Pacífico.' Blog de la Presidencia. 4 July.
<http://www.presidencia.gob.mx/el-blog/el-quinto-mito-el-gobierno-federal-favorece-a-joaquin-el-chapo-guzman-y-al-grupo-criminal-del-pacifico/>

Povo do Rio. 2012. 'Rocinha vira terra de ninguém.' 9 January.

R7. 2011. 'Ocupação do Alemão e da Penha provoca migração de crimes para outras áreas da cidade, aponta ISP.' 24 May. <http://noticias.r7.com/rio-de-janeiro/noticias/ocupacao-do-alemao-e-da-penha-provoca-migracao-de-crimes-para-outras-areas-da-cidade-aponta-isp-20110524.html>

Reforma. 2010a. 'Arma "La Familia" protesta callejera.' 13 December. <http://www.lacronicademexico.com/index.php?option=com_content&view=article&id=1870:arma-la-familia-protesta-callejera&catid=39:narcotrafico&Itemid=2>

—. 2010b. 'Toma Narco Gigante-1.' 10 June.

—. 2010c. 'Chocan por pozo Ejército y sicarios.' 26 June.

Reuter, Peter. 2009. 'Systemic Violence in Drug Markets.' *Crime, Law and Social Change,* Vol. 52, No. 3, pp. 275–84.
<http://www.springerlink.com/content/p0p3453k2x8m3482/fulltext.pdf>

— and Victoria Greenfield. 2001. 'Measuring Global Drug Markets.' *World Economics,* Vol. 2, No. 4, pp. 159–73.

— and John Haaga. 1989. *The Organization of High-level Drug Markets: An Exploratory Study*. Santa Monica: RAND.
<http://www.rand.org/pubs/notes/N2830.html>

Ríos, Viridiana and David Shirk. 2011. *Drug Violence in Mexico: Data and Analysis Through 2010*. San Diego: Trans-Border Institute.
<http://justiceinmexico.files.wordpress.com/2011/03/2011-tbi-drugviolence.pdf>

Roman, Clara. 2011. 'A violência está maquiada.' *Carta Capital*. 25 October.
<http://www.cartacapital.com.br/politica/violencia-no-rio-de-janeiro-esta-maqueada>

SEDENA (Secretaría de la Defensa Nacional–Mexican National Defense Secretariat). 2011. 'Combate al narcotráfico.'
<http://www.sedena.gob.mx/index.php/actividades/combate-al-narcotrafico/3426-combate-al-narcotrafico>

Seelke, Clare Ribando. 2010. *Mérida Initiative for Mexico and Central America: Funding and Policy Issues*. Washington, DC: Congressional Research Service. <http://books.google.com/books?id=ZufBDBi3u6MC&pg=PA2&lpg=PA1&ots=s0IKC3XrGk#v=onepage&q&f=false>

Skarbek, David. 2011. 'Governance and Prison Gangs.' *American Political Science Review*, Vol. 105, No. 4, pp. 702–16.
<http://polisci2.ucsd.edu/pelg/skarbek_governance_prison_gangs.pdf>

SNSP (Sistema Nacional de Seguridad Pública–Mexican National Public Security System). 2011. 'Base de datos de fallecimentos.'
<http://200.23.123.5/Documentos/baseFallecimientosRD.xlsx>

Snyder, Richard and Angelica Durán Martínez. 2009. 'Drugs, Violence, and State-Sponsored Protection Rackets in Mexico and Colombia.' *Colombia Internacional*, Vol. 70, pp. 61–91.

Soares, Luiz Eduardo. 2000. *Meu casaco de general: 500 dias no front da segurança pública do Rio de Janerio*. Rio de Janeiro: Companhias das Letras.

SSP-RJ (Secretaria de Estado de Segurança Pública do Governo do Rio de Janeiro–Public Security Secretariat of the State of Rio de Janeiro). 2003. 'Anuário Estatístico do Núcleo de Pesquisa e Análise Criminal.'

Stahlberg, Stephanie Gimenez. 2011. 'The Pacification of Favelas in Rio de Janeiro.' Palo Alto, CA: Stanford University.
<http://cddrl.stanford.edu/publications/23471>

Stewart, Scott. 2011. 'Mexico's Gun Supply and the 90 Percent Myth.' Stratfor. 12 February.

Stratfor. 2006. 'Kaibiles: The New Lethal Force in the Mexican Drug Wars.' 25 May.

UCDP (Uppsala Conflict Data Program). 2011. 'Battle-related Deaths Dataset Version 5.0.' <http://www.pcr.uu.se/research/ucdp/datasets/>

UNDP (United Nations Development Programme). 2009. *Opening Spaces to Citizen Security and Human Development: Human Development Report for Central America, 2009–2010*. <http://hdr.undp.org/en/reports/regional/latinamericathecaribbean/irdhc-2009-2010-summary.pdf>

UNODC (United Nations Office on Drugs and Crime). 2003. *Global Illicit Drug Trends 2003*. Vienna: UNODC.
<http://www.unodc.org/unodc/en/data-and-analysis/WDR.html>

—. 2010. *World Drug Report 2010*. Vienna: UNODC. <http://www.unodc.org/unodc/en/data-and-analysis/WDR.html>

USAID (United States Agency for International Development). 2006. 'Central America and Mexico Gang Assessment.' Washington, DC: Bureau for Latin American and Caribbean Affairs, Office of Regional Sustainable Development, USAID.

USBCIS (United States Bureau of Citizenship and Immigration Services). 2000. 'Guatemala: Kaibiles and the Massacre at Las Dos Erres.' Washington, DC: USBCIS. <http://www.unhcr.org/refworld/docid/3ae6a6a54.html>

US Embassy in Mexico. 2009. 'Scenesetter for the Firearms-Trafficking Conference, April 1–2.' Cable to US Secretary of State. Reference ID 09MEXICO880. Mexico City. 25 March. <http://wikileaks.org/cable/2009/03/09MEXICO880.html>

US GAO (United States Government Accountability Office). 2009. 'Firearms Trafficking: U.S. Efforts to Combat Arms Trafficking to Mexico Face Planning and Coordination Challenges.' Washington, DC: GAO. 19 June. <http://www.gao.gov/products/GAO-09-709>

VPC (Violence Policy Center). 2003. '"Officer Down": Assault Weapons and the War on Law Enforcement.' Washington, DC: VPC.
<http://www.vpc.org/studies/officeone.htm>

Werneck, Antônio. 2010. 'Migração de bandidos do Rio faz violência explodir em Friburgo.' *O Globo*. 18 September.

—. 2011. 'Traficante Nem diz que metade do seu faturamento ia para policiais.' *O Globo*. 11 November.
<http://oglobo.globo.com/rio/mat/2011/11/10/traficante-nem-diz-que-metade-do-seu-faturamento-ia-para-policiais-925781139.asp>

World Bank. 2011. *Crime and Violence in Central America: A Development Challenge*. Washington, DC: World Bank.

Zimring, Frank. 2011. 'How New York Beat Crime.' *Scientific American*, Vol. 305, No. 2, pp. 74–79.

ACKNOWLEDGEMENTS

Principal author

Benjamin Lessing

A boy recovers in hospital from a bullet wound that has left him unable to recognize anyone and with no sensation on his left side, eastern Chad, 2007. © Corbis

A Matter of Survival
NON-LETHAL FIREARM VIOLENCE

INTRODUCTION

Not all gunshots kill. Many victims survive. This may sound like good news, but the consequences of firearm injuries can be severe. Treatment and recovery place a heavy burden on survivors, their families, communities, and society. Non-lethal firearm violence—often representing narrowly avoided homicide—is far more widespread than firearm death worldwide. Improved knowledge of the incidence and patterns of non-lethal firearm violence would clarify the overall burden of armed violence on society and underpin the development of effective responses. Yet our current understanding of non-lethal firearm injuries is limited, hampered by a lack of data.

This chapter reviews available data on the incidence of non-lethal firearm violence, focusing on interpersonal assaults committed in non-conflict settings.[1] It includes an overview of estimates for countries in which data collection is relatively robust. It also highlights the need for improved incidence and trend monitoring. The main findings indicate that:

- Worldwide, at least two million people—and probably many more—are living with firearm injuries sustained in non-conflict settings over the past decade. Their injuries generate considerable direct and indirect costs, such as those incurred through treatment, recovery, and lost productivity.
- Available data suggests that shooting victims in countries with lower overall levels of firearm violence have a better chance of surviving their injuries.
- Whether a firearm injury leads to severe disability or death is influenced by firearm type, ammunition velocity, and calibre, as well as the availability and quality of medical care, among other factors.
- Robust data on non-lethal firearm violence is still relatively uncommon, and collected data rarely conforms to standardized coding protocol, limiting its comparability. The use of simple forms and relatively inexpensive injury surveillance techniques would greatly improve available information.

This chapter has three main sections. The first introduces the concept of non-lethal firearm violence and associated terminology, explaining how the type of firearm and ammunition and the availability of medical care influence the survivability of gunshot injuries. The second section reviews data sources for non-lethal firearm violence, presents available information on the incidence of and trends in firearm-related injuries, and provides some estimates of direct and indirect costs. The third section reviews sample injury surveillance systems and highlights some of the challenges to improving data collection on non-lethal firearm violence. The conclusions offer reflections on how surveillance efforts might be improved, taking into account the importance of assisting developing countries in establishing well-designed data collection systems.

NON-LETHAL GUN VIOLENCE IN PERSPECTIVE

In the shadow of homicide

Tracking progress towards armed violence reduction requires accurate data to establish baselines and measure trends. Homicide represents a relatively robust indicator, since data is more readily available and presents fewer comparability problems than any other crime or violence indicator. A number of authoritative reports that assess international trends in violence make almost exclusive use of lethal violence data, including the *Global Burden of Armed Violence* and the *World Development Report* (Geneva Declaration Secretariat, 2011; World Bank, 2011). The extensive use of homicide statistics as a violence indicator has created additional demand for such data. Over the past few years both public health and criminal justice sources have become more accessible and complete. For example, the United Nations Office on Drugs and Crime (UNODC) recently published a *Global Study on Homicide* and the World Health Organization (WHO) is providing statistics on cause of death in its *Global Burden of Disease* (UNODC, 2011a; WHO, 2008).[2]

The use of mortality statistics as an indicator of violence levels is partly a function of the relative ease in obtaining, reporting, and comparing deaths. Criminal justice and public health authorities across countries and cultures tend to treat homicides using broadly similar definitions and concepts. Public health reporting requirements may raise the pressure to collect and report comprehensive data on causes of death, helping to ensure that all or most deaths that occur are captured. Over and above legal obligations, fatal outcomes tend to be viewed as extremely serious across all cultures, thus entailing an ethical obligation to report.

In contrast, various definitions and counting rules are used to capture non-lethal firearm incidents, depending on the nature and characteristics of the injuries as well as the needs and purposes of the data-collecting entity. Depending on the severity of the injury, there may be fewer legal obligations to report non-fatal injuries, even intentional ones. Data is typically obtained through local, city, or state injury surveillance or monitoring systems that are capable of recording the context and mechanism of injuries, such as emergency department admissions. Ideally, the data is coded according to WHO's uniform International Classification of Diseases (ICD) system, now in its tenth revision

Figure 3.1 **Number of armed violence monitoring systems collecting data on different indicators (multiple responses), n=20, multiple responses**

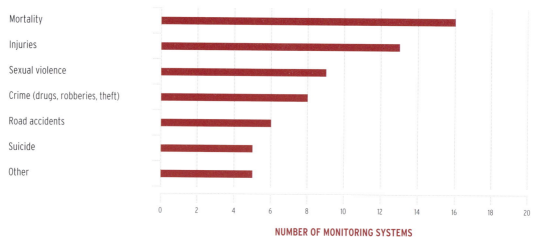

Source: adapted from Gilgen and Tracey (2011, p. 30)

(WHO, n.d.b). In practice, health practitioners in many countries view the time and management requirements necessary to apply the ICD-10 codes as excessively burdensome. Thus, while affiliated hospitals may use a unified data collection system, collection and coding methods are highly variable within and across countries (see Box 3.1).

Police incident reports and victimization surveys provide additional sources of information. They can include details about the nature and circumstances of injuries, including on weapons used, as well as important qualitative information from victims about their experiences. But they are equally marked by a lack of uniformity in methods and reporting.

These factors result in wide gaps in non-lethal violence data, with a notable North–South divide. Given that developing countries are often deeply affected by armed violence, bridging that data gap must become a priority if violence reduction policies are to respond effectively to local conditions. A Geneva Declaration study on 20 armed violence monitoring systems ('observatories') from 13 countries reveals that mortality statistics are the most frequently collected (see Figure 3.1). Data collection on non-fatal injuries is less common.

Quantifying severity

The ICD-10 system allows for public health data to identify whether injuries were caused by interpersonal violence (assault), whether a firearm was used, and whether injuries are *serious* or *slight*. Under this classification, a serious injury leads a patient to be admitted to hospital; a slight injury is one for which a patient can be treated in the emergency

Box 3.1 Questions of terminology

Armed violence is defined here as 'the intentional use of illegitimate force (actual or threatened) with arms or explosives, against a person, group, community, or state, that undermines people-centred security and/or sustainable development' (Geneva Declaration Secretariat, 2008, p. 2).

Interpersonal violence, which is committed by one or more persons against another or others, is distinguished from *self-directed violence,* or *self-harm* (suicides and suicide attempts). It includes subcategories that clarify the relationship between perpetrators and victims and the setting where the violence occurs, as follows:

- *family* or *domestic violence* includes abuse or maltreatment of relatives–including children and elderly family members–usually (though not exclusively) in the home;
- *intimate partner violence* involves perpetrators who are current or former intimate partners of the victim; and
- *community violence* includes gang or youth violence, rape or sexual assault by strangers, and violence in institutional settings, between individuals who are not related (Krug et al., 2002, pp. 6-7).

While common usage of the term 'injury' suggests non-fatal outcomes alone, injuries are in fact a leading cause of death in many countries–whether due to motor vehicle collisions, intentional violence, natural disasters, or other causes. *Injury* thus indicates acute bodily trauma; serious injuries may cause death days, months, or even years after they were inflicted–creating complications for data collection and reporting.

A *fatal injury* is one from which the victim dies either immediately or after treatment. ICD-10[3] stipulates that if the patient dies within 30 days of the incident, the case should be classified as fatal (Butchart et al., 2008, pp. 12-13). For the criminal justice system, depending on the country, a case of assault may be reclassified as a homicide if the victim dies up to one year after sustaining injury.[4] Thus, a victim who dies from his wounds more than 30 days but less than one year after being injured will not necessarily be recorded as a fatality in public health statistics while being counted as a homicide in criminal justice statistics.

The terms *victim* and *survivor,* often used to refer to those who have suffered a violent injury, reflect the attitudes and priorities of those who use them. 'Victim', mostly used in the criminal justice system, is a label that may carry a stigma of helplessness; the use of the term 'survivors' highlights injury as a form of oppression or lived experience as opposed to a medical condition. 'Disability', the term and concept, describes a person's interaction with society rather than his or her attributes (WHO and World Bank, 2011, p. 4; WHO, n.d.a, p. 1).

Sources: Krug (2002); Buchanan (2011)

department and then released. It should be noted, however, that many slight injuries—and some serious ones—are never reported to or treated in health facilities.

Weapons and ammunition

The severity of a gunshot injury—and the likelihood of death or permanent impairment—is affected by the technical specifications of the ammunition used, including bullet size, type of tip (such as hollow-tipped, pointed, or round-nose), velocity, and 'flight pattern'. These factors influence bullet trajectory through the body and the subsequent damage to tissue, organs, and bones.

In general, the higher the bullet velocity, the more likely the injury is to be lethal. For example, handguns have slower velocity projectiles than rifles and tend to cause less severe injuries. A study in Nigeria's Gombe State finds that the majority of patients presented wounds caused by low-velocity gunshots fired from locally assembled firearms (Ojo, 2008). Another study in Kano, Nigeria, also indicates that the firearms used in assaults in the country are mainly low-velocity handguns (Mohamed et al., 2005, p. 298). The latter study notes that the higher fatality rate observed in the United States in comparison to Nigeria may reflect the use of high-calibre, high-velocity pistols in the United States.

In addition to bullet speed, the rate of fire is an important factor affecting the severity of injuries. Semi-automatic and automatic pistols, whose rate of fire exceeds those of single-shot rifles and repeating revolvers, are likely to cause greater injury. Indeed:

The increased use of semi-automatic weapons [in the United States] has resulted in changed wounding patterns with an increased number of bullet wounds per incident per body and a subsequent higher mortality (FICAP, 2009, p. 8).

Bullets lacerate and crush tissue and bones in the direct path of the projectile, also causing what is known as 'cavitation'. When a bullet enters the body, a temporary vacuum is opened for a few thousandths of a second behind it. The pressure applied by the temporary cavity on surrounding tissues and organs provokes injuries far from the bullet path; these can be hard to detect, particularly in soft organs. This pressure is also capable of fracturing bones several centimetres from the bullet track (Prokosch, 1995, pp. 18–19; Waters and Sie, 2003, p. 121). The greater the speed of the bullet, the larger the initial cavity; a large cavity may be 30 to 40 times the diameter of the bullet. After the bullet has gone through tissue, the temporary cavity disappears, leaving a lasting cavity or wound track.

Depending on the type of ammunition and other factors, the projectile may 'tumble' into the body (known as 'yaw'), further increasing the wound cavity. If the bullet fragments, each piece will follow a distinct path, thereby multiplying the damage (Prokosch, 1995, pp. 191–92).

Access to medical care

The physical location of the injury largely determines the types of consequences. Injuries to the extremities often result in fractures that may lead to haemorrhages, infections, amputation, or permanent trauma due to joint or bone deformation. Brain and spinal cord injuries can cause irreversible damage such as paralysis, sexual dysfunction, limited movement, seizure disorders, incontinence, and severe facial disfigurations. Abdominal gunshot wounds may require specialized surgical skills to preserve vital organs. Gunshot wounds to the head have the highest risk of being lethal; similarly, suicide attempts with firearms—in which the gun is most frequently aimed at the head—are most likely to result in death (Vyrostek, Annest, and Ryan, 2004, fig. 21).

A major factor influencing whether an injury is fatal is the accessibility of emergency and trauma care services. Prompt access to qualified medical services may determine not only the victim's chance of survival, but also the

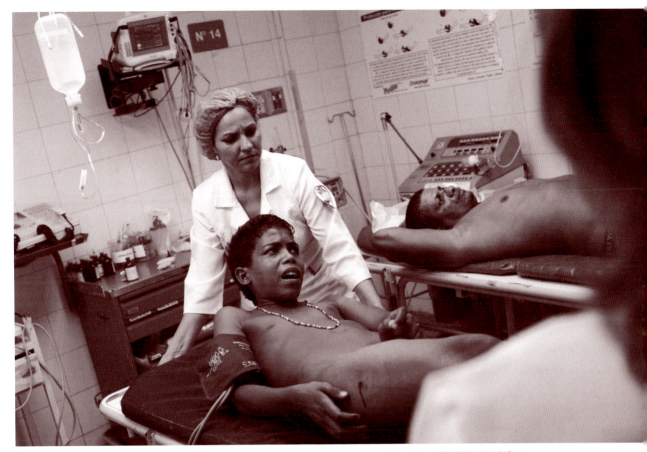
A 14-year-old boy with a gunshot wound to his leg is treated by medical personnel in the emergency room of a hospital near the Petare slum in Caracas, Venezuela, November 2009. © Carlos Garcia Rawlins/Reuters

long-term prognosis and the likelihood of disability. The speed of initial service may also make a difference, as the treatment received in the first hours after the injury may determine whether the patient will die or survive, and, in the latter case, the subsequent quality of life. This factor is particularly significant in rural and low-income areas where only a minority of patients reach the hospital by ambulance. A study conducted in Nairobi, for instance, finds that only 7.7 per cent of firearm injury survivors were brought to hospital by ambulance and only a quarter arrived within one hour of the incident (Hugenberg et al., 2007, p. 416).

Sometimes the knowledge and training of emergency service providers is more decisive than the availability of sophisticated equipment or technology (Hofman et al., 2005, p. 14).

Acute care for gunshot victims may also depend on the existence and enforcement of legislation requiring that medical personnel *immediately* report all victims of firearm incidents to the police. This type of regulation is sometimes aimed at ensuring that injured criminals do not escape through the medical system. In some circumstances, a lack of coordination and fear may create tension between physicians and law enforcement officials. This problem has been noted in Nigeria.[5] The strict application of these kinds of laws is likely to generate delays in the provision of medical care to victims of gun violence and increased mortality rates, especially where police and treatment centres are far from each other (CLEEN, 2010, p. 6).

ASSESSING THE SCALE AND SCOPE OF NON-LETHAL FIREARM VIOLENCE

Intentional firearm violence

In the context of efforts to assess and reduce the incidence of armed violence, this chapter focuses on intentional injuries committed with a firearm (firearm assaults), as compared to self-inflicted and unintentional injuries or wounds caused by law enforcement officers in the course of their work (that is, 'legal intervention').[6] This section examines the characteristics and extent of non-lethal firearm violence as it can be estimated from a variety of sources, such as public health, law enforcement, and victimization surveys.

What proportion of firearm violence is due to intentional assaults? It varies significantly by context. In low-income countries a large proportion of firearm-related deaths occur as a consequence of interpersonal violence, mostly in urban settings. In upper middle- and high-income countries, suicides represent the largest proportion of firearm-related deaths (WHO, 2001, p. 3). The case of the United States is illustrative. Figure 3.2 shows the distribution of firearm injuries by outcome (fatal and non-fatal) and intent, also highlighting self-directed harm (suicide and attempted suicide). Suicide represents the majority (60 per cent) of firearm deaths, whereas intentional assaults account for 37 per cent. The majority (67 per cent) of non-fatal firearm-related injuries are caused by intentional interpersonal violence. Unintentional injuries make up more than one-quarter of non-fatal events, while attempted suicides account for only five per cent. The distinction between intentional and unintentional may sometimes be blurred, as in the case of injuries caused by stray bullets (see Box 3.2).

Figure 3.2 **Distribution of fatal and non-fatal firearm injuries in the United States, by intent, 2009**[7]

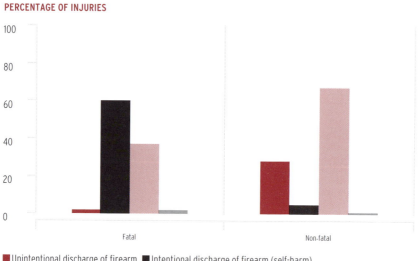

Note: Totals are not equal to 100 due to rounding.
Sources: CDC (n.d.a); Kochanek et al. (2011)

US Representative Gabrielle Giffords and her husband light a candle during a one-year memorial vigil for the victims and survivors of the shooting incident in which she was critically injured, Tucson, Arizona, January 2012. © Matt York/AP Photo

Box 3.2 Stray bullets and celebratory shootings

A clear-cut distinction between 'intentional' or 'unintentional' may be difficult to apply to injuries caused by stray bullets, though they are almost always categorized as unintentional injuries in both health and crime statistics. An analysis of the circumstances of 284 cases of injuries caused by stray bullets in the United States finds that the majority of events (59 per cent) were related to violence (see Figure 3.3); accidents related to firearm maintenance, shooting sports, and celebratory gunfire accounted for 18 per cent of cases combined (Wintemute et al., 2011). The study could not identify the circumstances of 23 per cent of stray bullet-related injuries.[8]

Stray bullets are relatively common in violence-affected contexts and represent a serious concern in many Latin American countries. A survey of media reports and National Police data in Colombia reveals that stray bullets injured at least 1,200 men and almost 700 women between 2001 and 2011, mostly in urban areas (CERAC, 2011). In the municipality of Rio de Janeiro, Brazil, stray bullets were responsible for 9.4 per cent of non-fatal firearm injuries in the first six months of 2011 (Teixeira, Provenza, and Oliveira, 2011, p. 9).

In many countries, people express excitement by firing a gun in the air at weddings, parties, and around New Year celebrations, but also at funerals to show respect for the dead. Such celebratory shootings, while not meant to be violent, do result in casualties. In 2008–09, they represented 5 per cent of all stray bullet injuries in the United States (see Figure 3.3). During New Year celebrations at the end of 2010 in the Philippines, stray bullets injured 30 people, three of them fatally (Suerte Felipe, 2011; BBC, 2011). That same year in Italy, eight people were injured and one killed by stray bullets (*Corriere della Sera*, 2011).

An extensive body of research is devoted to injuries caused by traditional or celebratory shootings in different contexts and cultures, including the specific types of injury they incur (commonly low-velocity head injuries).[9] In particular, these studies focus on injuries among women and children, who represent a sizeable portion of victims of this type of incident. A study conducted in a South African hospital finds that the majority (42 per cent) of children who were treated for firearm injuries over the past decade had been hit by stray bullets (see Box 3.3).

Figure 3.3 Circumstances of injuries caused by stray bullets in the United States, March 2008–February 2009

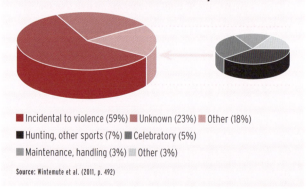

- Incidental to violence (59%)
- Unknown (23%)
- Other (18%)
- Hunting, other sports (7%)
- Celebratory (5%)
- Maintenance, handling (3%)
- Other (3%)

Source: Wintemute et al. (2011, p. 492)

Members of the Philippine national police force display their taped-up gun barrels in Manila, an exercise intended to prevent stray gunfire during New Year celebrations, December 2007. © Jay Directo/AFP Photo

In the United States, the use of firearms to commit suicide is thus a key factor affecting the proportions and consequences of firearm injuries. When firearms are used to attempt suicide, death occurs in approximately 85 per cent of the cases, a higher rate than suicides attempted using other methods; the death rate is also higher than for intentional assaults with firearms (Vyrostek, Annest, and Ryan, 2004, figs. 20, 21). In developing countries, in contrast, intentional assaults are likely to represent the majority of both fatal and non-fatal firearm-related injuries.

Data sources and trends
Public health data

Injury data typically originates in medical services, which are well placed for capturing the number of patients treated for firearm-related injuries. Although any hospital could collect detailed information on patients being treated for injuries, the identification of different causes, including firearms used, and regular mechanisms for sharing, compiling, and analysing cases at the aggregate level are rare.

The relative burden of firearm-related injuries compared to other types of injuries—such as those sustained in road accidents, falls, and fires—may depend on a range of factors, including the overall levels of violence in the area, firearm availability, and law enforcement presence. Figure 3.4 shows the distribution of injury diagnoses in five hospitals across Uganda in 2004–05 (GBI, 2010). Almost a quarter of all cases were classified as violence-related, with 3 per cent attributed to gunshots and 21 per cent to blunt force or bladed instruments.

Public health sources provide the bulk of statistical data that can be used to assess the extent and trends of non-lethal firearm injuries. For this reason, the analysis presented below relies primarily on public health data.

Figure 3.4 **Distribution of injury diagnoses in five Ugandan hospitals, 2004–05[10]**

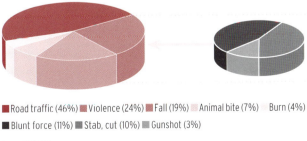

■ Road traffic (46%) ■ Violence (24%) ■ Fall (19%) ■ Animal bite (7%) ■ Burn (4%)
■ Blunt force (11%) ■ Stab, cut (10%) ■ Gunshot (3%)

Source: GBI (2010)

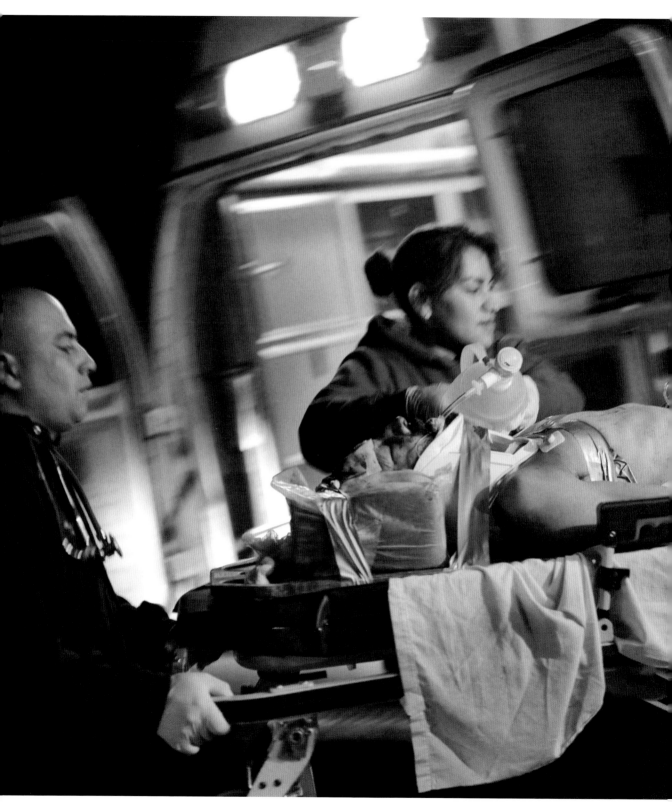
A man is rushed to hospital with two bullet wounds following an incident in Nido de las Aguilas, Tijuana, Mexico, March 2008.
© Washington Post/Getty Images

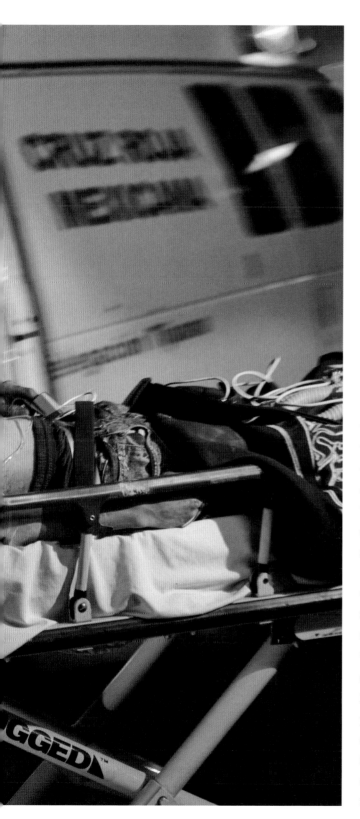

Victimization survey data

Victimization surveys represent another source of information on non-lethal firearm violence. They typically ask respondents about their personal experiences as victims, or sometimes as witnesses, of violence. Most capture information on crimes committed with firearms. Data may cover incidents in which guns were used to threaten or coerce a victim, types of firearms involved, and physical and psychological consequences for victims and witnesses of violence.

Because incidents of firearm violence are statistically rare, the margin of error in victimization surveys is very large, and they rarely connect different types of weapons with types of outcomes. For example, a recent survey carried out in Liberia shows that in 38 per cent of all reported cases of violence, the victim was injured by an instrument (the type is not identified), while in 4.4 per cent of all cases the incident resulted in the death of the victim (Gilgen and Murray, 2011, p. 8). Victimization studies suggest that, on average, approximately a third of victims of all crimes suffer (non-fatal) physical injuries.[11]

The International Crime Victims Survey (ICVS) includes data on incidents involving firearms in 38 countries and cities. Figure 3.5 shows the percentage of survey respondents who were held at gunpoint during robberies and assaults in the eight cities that rank highest in these categories, over the five years preceding the survey. In three of the cities—Johannesburg, Rio de Janeiro, and São Paulo—nearly one in every ten respondents was the victim of a robbery in which the offender had a firearm.

Figure 3.5 **Prevalence of respondents who were victims of armed robberies and assaults with firearms, for the five years preceding the survey**

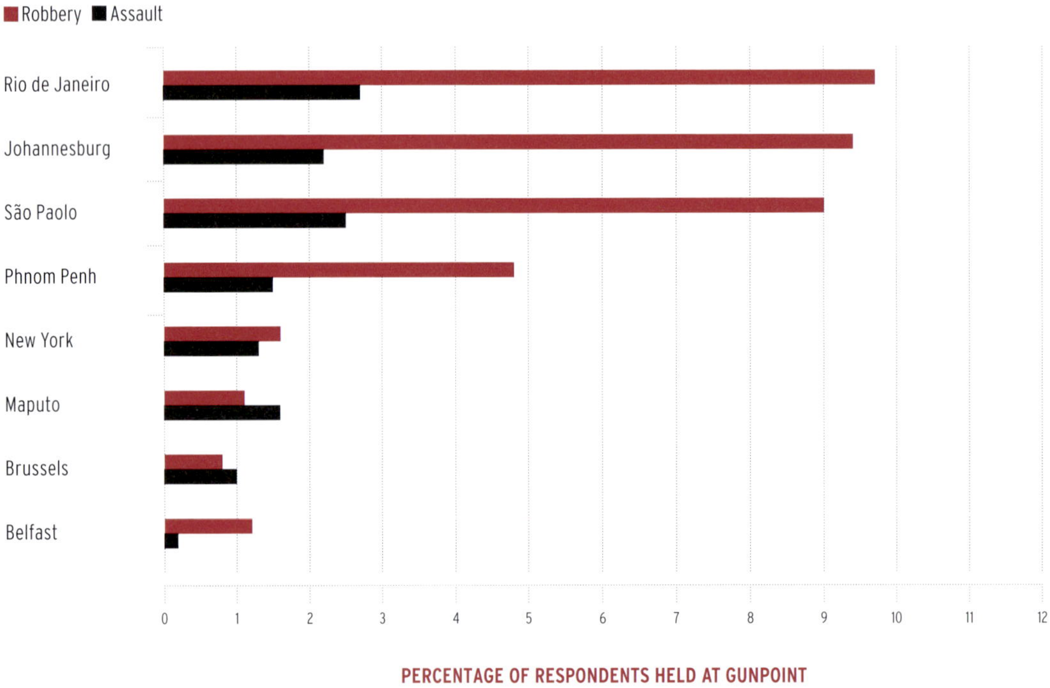

■ Robbery ■ Assault

PERCENTAGE OF RESPONDENTS HELD AT GUNPOINT

Note: The eight cities in this graph ranked highest for robbery and assault among the 38 countries and cities included in the ICVS 2004-05.
Source: van Dijk, van Kesteren, and Smit (2007)

Surveys provide insight into the significant variation in the use of firearms in non-fatal crimes across countries and cities. The 2008 national victimization survey in Mexico finds that more than half of the victims of all crimes were confronted with firearms (Juárez, 2010, p. 16, fig. 7). In Kenya, a survey conducted in 2011 among 2,400 households reveals that slightly more than a third of those who were victims of crime or violent encounters faced a firearm, with 15 per cent of victims noting that the assailants had a handgun, and 17 per cent reporting an automatic weapon, such as an AK-47 (see Figure 3.6). The high percentage of incidents involving automatic weapons is alarming as it indicates their widespread distribution over the territory and their involvement in crime (Small Arms Survey and KNFP, 2012).

Figure 3.6 **Weapons identified by victims of crime or violence in Kenya, 2011**[12]

■ Bladed weapon (25%)
■ Automatic weapon (such as AK-47) (17%)
■ Handgun (pistol or revolver) (15%)
■ Military equipment (1%)
■ No weapon was used (19%)
■ Crude or traditional weapon (15%)
■ Unknown (7%)
■ Rifle or shotgun (1%)

Source: Small Arms Survey and KNFP (2012, fig. 2.14)

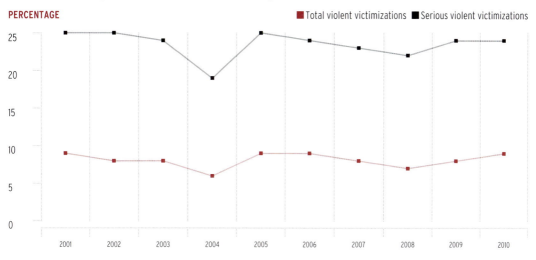

Figure 3.7 **Percentage of violent victimizations involving firearms in the United States, 2001-10**

Source: Truman (2010, p. 8, figure 7)

Irrespective of the different levels of violence in the 38 countries and cities surveyed, the ICVS finds that the proportion of incidents involving firearms was higher in urban areas and was, on average, 12 per cent for robberies and 5 per cent for assaults (van Dijk, van Kesteren, and Smit, 2007, pp. 76, 80). In the United States the proportion of non-fatal firearm-related crime remained relatively stable in the period 2001–10, accounting for between 6 and 9 per cent of total violent crime and 20–25 per cent of serious crime[13] (see Figure 3.7). In 2010 firearms were involved in approximately a quarter of cases of rape, robbery, and aggravated assault; firearms were most likely to be used in the robberies (29 per cent of cases) (Truman, 2010, p. 8 and table 4).

Despite its limitations, victimization survey data is useful for assessing the overall extent of firearm violence and for supplementing public health data. This is particularly true for understanding non-physical consequences, such as serious psychological stress to victims, family members, and friends. These effects can be felt even when a firearm is only used to threaten.

Putting the data together: lethal v. non-lethal violence

What are the consequences of firearm violence? The proportion of firearm incidents that result in death varies across different contexts. The Small Arms Survey has examined data on non-lethal firearm injuries, selecting data representing intentional violence or assault from approximately 28 countries and territories, a relatively small sample in comparison to homicide databases.[14] A range of factors influences the poor availability of comparable non-fatal injury statistics. Much of the data is collected locally (for example by a few hospitals or a group of cities or provinces) using widely diverging methodologies and information collection systems. For this reason, among others, data is seldom nationally representative or comparable across countries.

One way of analysing data on lethal and non-lethal firearm violence is to estimate national 'case fatality rates'[15]—the number of cases with a lethal outcome divided by the total number of lethal and non-lethal cases. This concept is used in epidemiology to provide a rough indicator of the proportion of persons who do not survive a specific type of disease or injury over a specific period of time, with the objective of reducing the proportion through improved medical services, prevention programmes, and other interventions.

Figure 3.8 Non-fatal firearm injuries and firearm homicides in 26 countries, latest available year

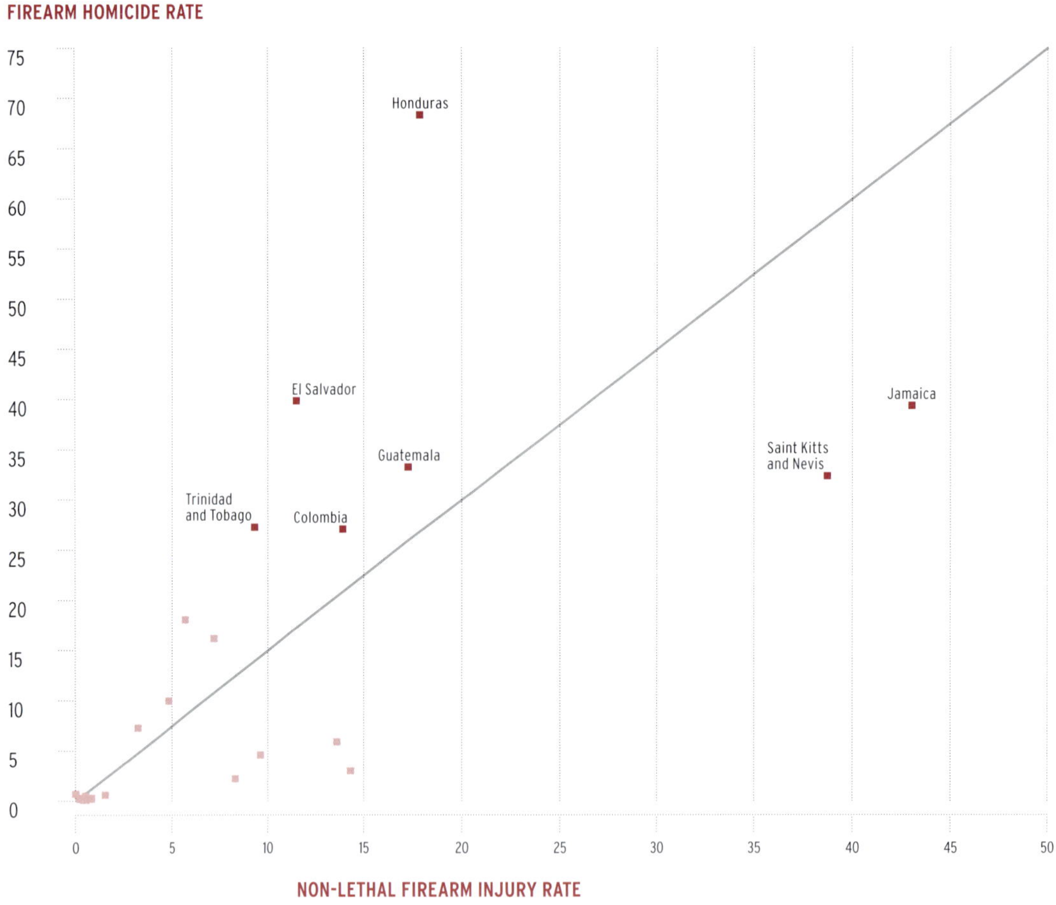

FIREARM HOMICIDE RATE

[Scatter plot with x-axis "NON-LETHAL FIREARM INJURY RATE" (0 to 50) and y-axis "FIREARM HOMICIDE RATE" (0 to 75). Labeled points include: Honduras (~17, 70), El Salvador (~11, 41), Jamaica (~43, 40), Guatemala (~17, 34), Saint Kitts and Nevis (~39, 33), Trinidad and Tobago (~9, 28), Colombia (~14, 28).]

Source: Small Arms Survey (2011); UNODC (2011b)

Figure 3.8 compiles injury data from the 2011 Small Arms Survey non-lethal firearm injury database and firearm homicide data from UNODC (2011b). The estimates must be regarded as tentative, as data on fatal and non-fatal injuries originate from different sources and data collection systems, may not be representative of the same populations, and reflect different time periods. Yet the data suggests that the higher a country's firearm homicide rate, the higher its case fatality rate for all firearm violence. In the 26 countries for which relevant data is available,[16] there is a correlation between the rate of firearm homicides and non-lethal firearm injuries (0.689, N=26).

A more rigorous comparison of case fatality rates would require compatible data series and counting methods, which exist for only a handful of countries.[17] Data from the United States and the UK (England and Wales), for example, yields a case fatality rate of close to 20 per cent, or approximately four non-fatal cases for every death.[18] But there is significant variation in the case fatality rate across countries and in the same country at different points in time. A study carried out in Kano, Nigeria, documents eight non-fatal firearm injuries treated for every firearm death, for a case fatality rate of 11 per cent, much lower than that observed in the United States and the UK (Mohammed et al., 2005, p. 298).

Figure 3.9 **Trends in non-fatal firearm injuries and firearm homicides in seven selected countries, 2004-09 (index year 2004=100)**

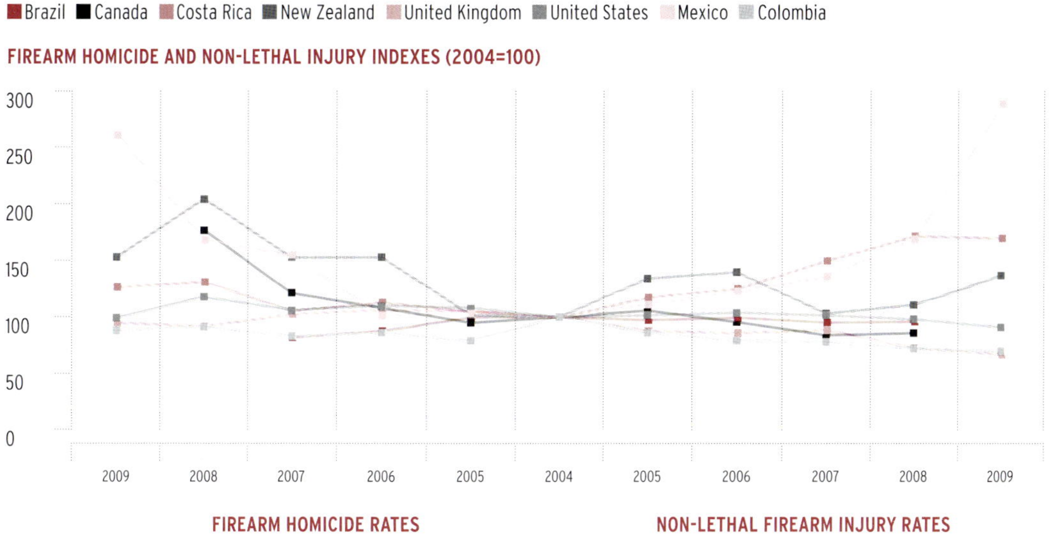

■ Brazil ■ Canada ■ Costa Rica ■ New Zealand ■ United Kingdom ■ United States ■ Mexico ■ Colombia

FIREARM HOMICIDE AND NON-LETHAL INJURY INDEXES (2004=100)

FIREARM HOMICIDE RATES   NON-LETHAL FIREARM INJURY RATES

Source: Small Arms Survey (2011); UNODC (2011b)

Among the countries and territories shown in Figure 3.8, few have comprehensive longitudinal series that can be used for trend analysis. Based on available data from the United States, a study demonstrates that the lethality of serious assaults in the country dropped dramatically between 1960 and 1999 (Harris et al., 2002). Figure 3.9 shows an example of changes in national rates of lethal and non-lethal injuries for 2004–09 in seven countries for which relevant data is available. It highlights trends in firearm homicide rates on the left side and non-lethal firearm injuries on the right. In three of seven countries—namely Costa Rica, Mexico, and New Zealand—both lethal and non-lethal injuries increased over the observed period (LATIN AMERICA AND THE CARIBBEAN; DRUG VIOLENCE).

Based on the data presented in this section, it is possible to generate an average global case fatality rate of 48 per cent for intentional, non-conflict firearm injuries, or approximately one non-fatal injury for every fatal injury. In the worst-case scenario, if every gunshot were fatal (a 100 per cent case fatality rate), there would be no survivors. Countries such as Brazil, Colombia, and Mexico, which all show higher rates of firearm homicide, show a case fatality rate of around 70 per cent (LATIN AMERICA AND THE CARIBBEAN; DRUG VIOLENCE). If the correlation were to hold globally, gunshot victims in countries with higher overall levels of firearm violence could be considered less likely to survive their injuries; by contrast, the lower the overall levels of firearm violence, the better the chances that a shooting victim will survive his or her injury.

The 2011 edition of the *Global Burden of Armed Violence* estimates that there are 396,000 intentional non-conflict homicides per year (Geneva Declaration Secretariat, 2011, p. 43). The 2008 edition concludes that approximately 60 per cent of all homicides worldwide are committed with firearms (Geneva Declaration Secretariat, 2008, p. 67); UNODC sets the proportion lower, at approximately 42 per cent globally (UNODC, 2011a, p. 10). Applying these two estimates as low and high limits (42–60 per cent) to the total annual number of homicides generates between 166,000 and 238,000 firearm homicides per year. A calculation based on the average case fatality rate would put the number of non-fatal firearm assaults at the same level or higher. Assuming that trends are stable, and not taking into consideration the potentially reduced life expectancy for firearm injury survivors, these calculations suggest that an estimated 2 million people worldwide are living with firearm injuries sustained in non-conflict settings over the past decade. This is a conservative estimate; in many countries the number of survivors is increasing, suggesting that the number of persons living with the consequences of firearm injuries is much higher. Assuming a ratio of 3:1 (three non-lethal injuries for every death), as is often cited for the United States, an estimated 500,000 to 750,000 people are injured by firearms every year.

Assessing the cost of firearm injuries

Direct medical costs for firearm injuries, including hospital stays, diagnostic procedures, surgery, and blood products, are substantial and often exceed the costs of treating other injuries and medical emergencies (Norberg et al., 2009, p. 443). But they represent only a portion of the total costs to the victim and society.

Research was carried out in the United States in the 1990s, when the firearm violence epidemic was at its peak, to assess the overall cost of firearm injuries. One study estimates that direct and indirect costs exceeded USD 20 billion in 1990, of which USD 1.4 billion represented direct medical costs (Max and Rice, 1993, p. 171). Another study, focusing exclusively on medical costs, estimates the mean cost per injury at about USD 17,000, which includes hospitalization (as victims who survive firearm injuries frequently require multiple rehospitalizations) and subsequent medical treatment spread over a victim's lifetime. Based on the number of firearm injuries in the US in 1994, the study estimated a total cost of USD 2.3 billion. The study finds that approximately three-quarters of these costs were borne for gunshot injuries due to violence (Cook et al., 1999, p. 453).

The local impact of the costs related to armed violence depends on the rate of firearm injuries and average income levels. A study carried out in Jamaica reveals that firearm-related injuries accounted for approximately 16 per cent of all injuries in 2006 but roughly 75 per cent of total direct medical costs for fatal injuries, 53 per cent of direct medical costs for serious injuries, and 6 per cent of direct medical costs for slight injuries (Ward et al., 2009, p. 448). A 2005 study observes that the average cost of treating one serious firearm injury was 13 times greater than the South African government's annual per capita expenditure on health (Allard and Burch, 2005, p. 591). In Kenya, a study based on a six-month surveillance of medical treatment of gunshot injuries in Nairobi finds that the average hospital bill was approximately USD 225—more than six times the monthly income of someone living below the poverty line (Hugenberg et al., 2007, p. 415).

According to a WHO typology, a comprehensive assessment of direct costs of firearm violence would include expenses linked to policing and imprisonment, legal services, foster care, and private security (Butchart et al., 2008, p. 7, table 1). Tangible indirect costs include loss of productivity, lost investments in social capital, and higher insurance costs; a broad range of intangible indirect costs may also be taken into account, such as reductions in or limitations on health-related quality of life (pain and suffering, both physical and psychological), job opportunities, access to schools and public services, and participation in community life.

CHALLENGES TO NON-FATAL INJURY SURVEILLANCE

As noted above, systematic monitoring of non-lethal firearm violence presents a series of challenges. Obstacles may be particularly difficult to overcome in areas where violence is pervasive, and where surveillance is thus most needed. But while a North–South divide with respect to non-lethal violence surveillance exists, the systems in place in many industrialized countries are often incomplete as well. This section describes some efforts in developed and developing countries to monitor non-fatal firearm injuries—and the main roadblocks they face.

Sample surveillance systems

The 'gold standard' of non-lethal firearm violence surveillance involves the systematic generation of detailed health records in emergency departments and hospital admissions. This type of system is extremely rare, however.

In practice, health-based surveillance systems are almost never comprehensive or complete. The United States is one of the few countries with a relatively sophisticated, nationwide non-fatal injury surveillance system that captures firearm violence. The National Electronic Injury Surveillance System (NEISS) is operated by the US Consumer Product Safety Commission, which collects data on injuries treated in a nationally representative sample of 66 hospital emergency departments.[19] The Commission itself does not release data on firearm-related injuries, but the National Center for Injury Prevention and Control of the US Centers for Disease Control and Prevention (CDC) accesses, analyses, and publishes firearm data. The data is used to monitor progress towards the government's goal of achieving a 10 per cent reduction in the national rates of non-fatal firearm injuries from their 2007 levels by 2020—a reduction from 20.7 to 18.6 per 100,000.[20]

The US system captures a range of information for non-fatal firearm injuries treated in emergency rooms, including age, sex, race or ethnicity of the victim; intent of injury (unintentional, self-harm, assault, legal intervention, undetermined intent); primary body region affected; and place of occurrence (such as home, public place, street, school).[21] Data is coded according to WHO's ICD-10 categories.

But even this relatively advanced system has limitations. For example, estimates for non-fatal firearm injuries treated in hospital emergency departments can be only provided at the national level, not at the state and local levels. It does not capture outpatient or clinic (non-hospital) visits. Full coding remains a challenge; even though the system is designed to capture information on the race or ethnicity of the injured person, the type of firearm used (such as a handgun, rifle, or shotgun), and the victim–suspect relationship, this information is rarely coded. Figure 3.10 compares information from the Inter-University Consortium for Political and Social Research and the Federal Bureau of Investigation regarding the type of firearms used in cases of fatal and non-fatal injuries. The type of firearm used was not coded in 68 per cent of all cases of non-fatal injuries,[22] as opposed to only 17 per cent of fatal cases (FBI, 2010). Approximately three-quarters of lethal firearm injuries are caused by handguns, whereas the distribution by type of firearm causing non-fatal injuries is largely unknown (though handguns probably also represent the majority in non-fatal cases).

Figure 3.10 **Non-fatal firearm injuries treated in hospital emergency departments and homicide victims in the United States, by type of firearm, January 2006–December 2008**

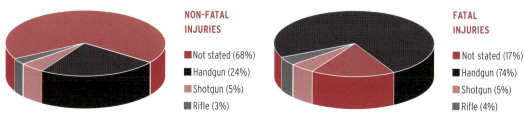

NON-FATAL INJURIES
- Not stated (68%)
- Handgun (24%)
- Shotgun (5%)
- Rifle (3%)

FATAL INJURIES
- Not stated (17%)
- Handgun (74%)
- Shotgun (5%)
- Rifle (4%)

Source: ICPSR (2010, p. 28); FBI (2010)

A small number of other, mainly Northern countries have promising systems in place that capture non-fatal firearm injuries. For example, the Netherlands monitors the numbers and rates of hospital admissions for non-fatal firearm assaults[23] for in-patients in hospitals and clinics across the country. The Centraal Bureau voor de Statistiek produces publicly available annual reports on both rates of non-fatal firearm assaults per 10,000 population for men and women and the average number of days of treatment per injury. The type of weapon and the victim–offender relationship is not captured, however (CBS, n.d.).

Key challenges

> Very few countries have comprehensive data collection systems for firearm injuries.

Most other countries have far weaker systems, if they have them at all. According to an initial survey conducted by the Small Arms Survey, some form of injury data is collected in approximately 60 countries, but the vast majority provide little or no data on non-fatal violence, are unable to disaggregate data according to weapon type, and do not specify intentionality (Pavesi, 2011, pp. 6–8). Among the most significant obstacles to better surveillance are a lack of comprehensiveness and standardization, non-representative sampling, and data entry and computerization problems, as discussed below.

Comprehensive and standardized injury surveillance. Ideally, hospital-based injury data collection systems would document firearm-related injuries within the framework of all-injury data collection systems. The primary advantage of a comprehensive system is the potential for widespread standardization in coding injuries, the instruments that cause them, and intentionality; such standardization would provide common definitions and details on injury context. Yet to date, only a fraction of countries have made progress towards comprehensive injury surveillance.

WHO's ICD-10 system provides a universally applicable scheme for coding non-fatal firearm violence, but its application is far from universal. Simple forms and questionnaires for the purpose of injury surveillance have been designed on the basis of WHO guidelines. For example, a model form developed in 2007 by the Central America injury project of the Pan American Health Organization and the CDC has been slightly modified to fit local needs in Colombia, El Salvador, and Nicaragua (Zavala et al., 2007, p. 435).

Mortality vs. injury surveillance. The non-fatal component is missing in many surveillance systems, which typically capture information on instruments used for fatal injuries only. For example, the National Injury Mortality Surveillance System (NIMSS) in South Africa represents the only means to estimate the extent of firearm violence in the country (see Box 3.3).

Representative sampling. Since no system can capture every injury at every hospital, non-fatal injury surveillance relies on a statistical sampling of cases to generate representative data (whether at the city, state, provincial, or national level). For example, the design used by the NEISS system is a stratified probability sample of all US hospitals that have at least six beds and provide 24-hour emergency services. These hospitals are divided into four strata based on their size, plus one children's hospital stratum (Hootman et al., 2000, pp. 268–69; CDC, n.d.b). With statistical analysis, the sample data can thus be extrapolated to the national level. Yet in countries with fewer hospitals, where injuries cluster significantly in one geographical area, or where injuries are not seen in hospital emergency departments—as is the case in many rural, underdeveloped areas—generating a representative sample can be challenging. Furthermore, violence levels are not among the elements taken into account for sampling, thus areas with either very high or very low levels of violence may be included in the sample and generate a bias in the analysis.

Box 3.3 Measuring fatal and non-fatal firearm violence in South Africa

In the years leading up to and immediately following the fall of the Apartheid regime in 1994, levels of gun violence in South Africa increased dramatically, leading to a national debate about guns and armed violence in the country. While many civil society and some political stakeholders called for the passage of new, comprehensive national civilian gun regulations, there was a clear need for better data on gun homicides and non-fatal gun injuries.

During this period, the public health community initiated fatal injury surveillance projects to help ascertain the distribution of violence and injury deaths and to identify injury control priorities. This approach was first demonstrated in Cape Town in the mid-1990s (Lerer, Matzopoulos, and Phillips, 2007); since 1998, the mortuary-based National Injury Mortality Surveillance System has collated data from mortuaries across South Africa. The NIMSS serves three interest groups: the forensic medical services, the crime prevention and justice community, and violence and injury prevention agencies (NIMSS, 2004, pp. 2-3).

This data was particularly instructive for tracking year-to-year homicide rates from 2000 onwards, as well as the decline in gun deaths following the implementation of the national Firearms Control Act No. 60, which was passed in 2000 and came into full effect in 2004 (South Africa, 2000).[24] Longitudinal analysis has demonstrated the importance of sustained and consistent injury mortality reporting to understand trends in violent injuries.

In 2003 the NIMSS annual report included city-specific chapters, and the following year mortality rates were calculated for the period 2001 to 2004 for four cities—Cape Town, Durban, Johannesburg, and Pretoria—alerting researchers to a substantial decrease in firearm homicides relative to non-firearm homicides (NIMSS, 2004; 2005). Subsequent reports suggest that this reduction was sustained through 2007 (NIMSS, 2007; 2008).

Data from two mortuaries serving Cape Town shows that changes in the number of homicides processed at these two mortuaries were attributable to fluctuations in the number of firearm homicides, while levels of non-firearm homicides remained stable. The Cape Town data revealed that there had been a consistent year-to-year increase in firearm homicide beginning in 1994, when firearms were involved in just 28 per cent of all homicides processed at the two mortuaries, to 49 per cent in 2002, after which there was a substantial decline. According to police statistics, rates of common assaults and assaults with the intent to inflict grievous bodily harm also began steep declines in 2003 (SAPS, 2011, p. 3).

There is no comprehensive surveillance system for non-fatal firearm injuries in South Africa. Monitoring does occur in specific settings, however. One such example is the Red Cross War Memorial Children's Hospital (RXH) in Cape Town, the country's only dedicated paediatric trauma unit for children under the age of 13. The hospital provides secondary and tertiary services for all hospitals in the Western Cape region, which has a population of 4.5 million.

The Childsafe South Africa programme, established in 1987, has maintained a database of all patients treated at the RXH trauma unit since 1991 (Childsafe South Africa, n.d.). The data includes demographic markers, details of the cause of injury, age at injury, an injury severity score, and details of the outcome. The database has produced two reports on paediatric firearm injuries treated in the hospital (for the years 1991-2000 and 2001-10), covering 441 cases in total (including lethal and non-lethal cases). Figure 3.11 shows the number of non-lethal firearm injuries in children admitted to the hospital between 1991 and 2010.

The studies find that from 2000 to 2010 (169 patients), most children (80 per cent) were hit by a single bullet. The majority (42 per cent) were hit by stray bullets; 14 per cent were hit intentionally by an adult; 3 per cent were hit intentionally by another child; 2 per cent of children were shot while they were playing with a gun. In 14 per cent of the cases the gunshot was accidental. Four children presented with fatal injuries. During the same period, 33 children (under the age of 12) died immediately after being shot and were sent directly to a mortuary.

In analysing the longitudinal data provided by Childsafe, Hutt et al. (2004) find that there was a significant decline in injuries over the period during which the Firearms Control Act was introduced, passed, and implemented (2000-04). While levels have increased slightly since then, they remain significantly lower than they were before the act was introduced.

These findings suggest that the introduction of gun control legislation has led to a reduction in firearm injuries, including those caused by accident and stray bullets.

Source: Kirsten and Matzopoulos (2011); Campbell et al. (2011)

Figure 3.11 **Annual number of non-lethal gunshot wounds in children attending the RXH, 1991–2010**

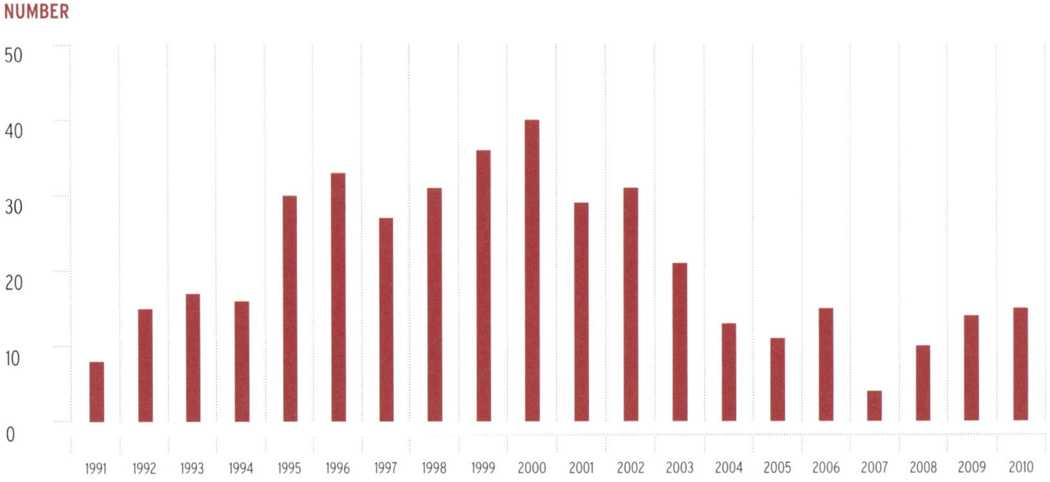

NUMBER

Source: Childsafe South Africa (n.d.); correspondence from Sebastian van As, medical doctor and board member, Childsafe South Africa, 6 October 2011

Data entry and computerization. In most developing countries, information is recorded using pen and paper, subject to the training and capacity of dedicated staff, who often operate in very challenging conditions. Figure 3.12 shows a record of admission in Liberia based on a very simple data collection scheme (Winnington, 2011b). In such cases, data is far less likely to conform to international standardized codes and is more difficult to collate and analyse. Still, the regular application of such methods ensures that crucial information can be captured so that it may help provide early assistance to victims.

Making do: one-off surveys

These multiple challenges mean that in most parts of the world, data on non-fatal firearms violence is primarily generated at the local level, if at all, through non-representative surveys, which are often supported by external funding from donor governments or philanthropic organizations that seek to promote an evidence-based approach to armed violence prevention. Once external backing ends, these initiatives are rarely continued.

A case in point is a survey of violence-related injuries in three hospitals in Timor-Leste, which was funded by the Australian Agency for International Development for the period 2006–08 and made possible

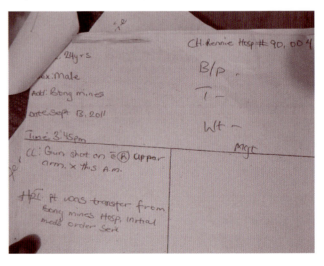

Figure 3.12 **Example of a hospital admission record, C.H Rennies Hospital, Kakata, Upper Margibi County, Liberia, 2011**

Courtesy of Dr A. Winnington, IPPNW (New Zealand)

NON-LETHAL VIOLENCE 99

Box 3.4 Small arms victimization and disability: Jonglei, South Sudan

Jonglei is the largest state in South Sudan, with a population of approximately 1.36 million. Despite the 2005 peace agreement between the Sudan People's Liberation Army and the Government of Sudan, and South Sudan's secession in 2011, Jonglei remains highly insecure. In addition to experiencing low-level resource-based conflict, the state is at the epicentre of ongoing militia violence (HSBA, 2011).

As modern health facilities are rare, access to health services in Jonglei State is extremely poor. The injured who do reach a hospital or clinic are treated in facilities that are short on equipment, medicines, and skilled medical staff. Those who do not are frequently treated by traditional healers and bonesetters who are often unable to help, and sometimes introduce infections and further complications.

Bor State Hospital, the only referral hospital in Jonglei State, was run by Médecins sans Frontières-Belgium until 2009, when the organization withdrew after a security-related incident. Since then, the level of services in Bor Hospital has deteriorated. At this writing, most hospital staff did not have formal qualifications and some were illiterate and thus incapable of producing patient documentation; hospital equipment was substandard or broken. However, the hospital has an outpatient department as well as surgical, orthopaedic, medical, paediatric, and maternity wards—and a physical rehabilitation unit was recently added with the support of Handicap International.

Data in Tables 3.1 and 3.2[25] shows that firearm-related injuries resulting from the use of small arms and light weapons represented 7.2 per cent of all cases over the period under review (2008-11), while landmine and unexploded ordnance incidents represented 0.4 per cent. Small arms and light weapons incidents increased as a proportion of all injuries in 2009. The typology of disability caused by gunshot incidents is diverse, with 26 per cent of victims left with permanent deformities or chronic loss of joint motion, and 18 per cent with fractures; 17 per cent required amputation.

Source: Dejito and Turton (2011)

A 20-year-old woman recovers from a gunshot wound to her arm, and a spear wound to her back, at a hospital in Akobo, Jonglei state, August 2009.
© Peter Martell/AFP Photo

Table 3.1 Typology of injury, Bor and Twic East Counties, Jonglei, 2008-11

Cause of injury	Year				Total	Percentage of total
	2008	2009	2010	2011 (8 months)		
Congenital	30	10	29	1	70	1.72
Landmine or unexploded ordnance	5	3	9	1	18	0.44
Small arms and light weapons	79	100	61	52	292	7.20
Burns	15	50	31	15	111	2.74
Other injuries*	640	681	972	253	2,546	62.74
Illness**	184	78	107	79	448	11.04
Unknown	101	37	0	435	573	14.12
Total	1,054	959	1,209	836	4,058	100.00

Notes:
* 'Other injuries' include road accidents, domestic violence, falls, and animal bites.
** 'Illness' includes stroke, diabetes, malaria, typhoid, polio, and other illnesses characterized by sudden weakness or paralysis, high fever, and/or convulsions. There is often no proper diagnosis because of the technical and professional limitations of the health facilities.

Source: Dejito and Turton (2011)

Table 3.2 Consequences of small arms disability, Bor and Twic East Counties, Jonglei, 2008-11

Disability	Year				Total	Percentage of total
	2008	2009	2010	2011 (8 months)		
Deformity and contracture	29	23	16	8	76	25.76
Paralysis and weakness	8	5	1	0	14	4.75
Amputation	19	15	13	2	49	16.61
Fracture	5	11	23	15	54	18.31
Other physical*	14	0	0	4	18	6.10
Wound**	4	46	10	24	84	28.47
Total	79	100	63	53	295	100.00

Notes:
* 'Other physical' refers to blindness and visual impairment, deafness and auditory impairment, and speech impairment.
** 'Wound' refers to both fresh and infected wounds.

Source: Dejito and Turton (2011)

through a cooperative agreement with the East Timorese Ministry of Health. It generated substantive findings about non-fatal firearm injuries but also highlighted some of the core problems of monitoring non-fatal injuries in developing countries. The report concludes that emergency department data 'is currently not sufficiently robust or systematically recorded to provide a reliable picture of interpersonal violence in Timor-Leste society' (TLAVA, 2009, p. 1). The same could be said for many other countries across the developing world. Box 3.4 discusses a recent targeted survey conducted in Jonglei, South Sudan, an area highly affected by gun violence but without adequate health facilities to treat injuries, and with little or no injury monitoring capacity.

When hospital and survey data is inaccessible or nonexistent, news reports can provide another option for documenting non-fatal firearm injuries. In 2010, the Small Arms Survey conducted a retroactive review of media-reported incidents of armed violence, both fatal and non-fatal, in Yemen, covering the period from 1 September 2008 to 31 August 2009. The study identifies 199 separate incidents of armed violence involving 728 intentional deaths and 734 non-fatal injuries. Media reports were often detailed enough to capture whether incidents resulted from political or social conflict, whether they were criminally motivated or related to domestic violence, and whether they were intentional (YAVA, 2010, p. 2). Yet numerous caveats apply to the use of media reports to monitor violence.[26] In particular, journalist access to high-risk areas can be severely limited, and local and national interests may exert control over reporting, creating sampling bias. In many cases, news reporting is biased towards urban events, while under-representing rural areas. News stories also often fail to capture crucial details about the circumstances of violent events.

CONCLUSION

What happens after a bullet hits a body? The impact and consequences of armed violence cannot be measured exclusively by counting the number of people killed. Most victims survive, but there are still serious gaps in our knowledge of trends and patterns of firearm injuries that do not result in death, as well as of the long-lasting consequences experienced by gun violence survivors.

The good news is that a tentative research agenda is emerging. The integration of statistics from various sources has already enhanced our picture of firearm violence at the local, national, and regional levels. The Global Burden of Injuries project has begun developing analytical tools to produce better estimates of the extent of violence-related injuries using a variety of sources (GBI, n.d.); its continuation would advance relevant knowledge. Where possible, data should distinguish injuries caused by intentional violence from other types of injury. In particular, the use of WHO injury surveillance protocols should be further expanded.

Yet estimation techniques are a weak substitute for emergency room surveillance systems, which remain rare. Developing, supporting, and sustaining hospital-based surveillance systems may create extra work for beleaguered medical staff, but the value in doing so is undeniable—not only for administrative and planning purposes, but also for improved pre-hospital and emergency care, and for the design, targeting, and monitoring of prevention and control strategies. Injury surveillance systems that capture non-fatal injuries also represent important entry points for donors focused on violence prevention.

Until non-fatal injuries are systematically monitored and the data is made available to researchers and policy-makers, an accurate picture of the full impact of gun violence on societies will remain elusive. Far from being an abstract need, expanding the evidence base is critical to identifying, developing, and evaluating promising prevention measures. As of 2012, however, most incidents of gun violence in non-conflict settings—some hundreds of thousands of cases per year—still go unrecorded.

LIST OF ABBREVIATIONS

CDC	Centers for Disease Control and Prevention
ICD	International Classification of Diseases
ICVS	International Crime Victim Survey
NEISS	National Electronic Injury Surveillance System
NIMSS	National Injury Mortality Surveillance System
RXH	Red Cross War Memorial Children Hospital, Cape Town
UNODC	United Nations Office on Drugs and Crime
WHO	World Health Organization

ENDNOTES

1   According to the Geneva Declaration, nine out of ten violent deaths occur in non-conflict settings (Geneva Declaration Secretariat, 2011, p. 1).
2   It should be noted that many African and Asian countries lack reliable homicide data from either the criminal justice or public health systems.
3   ICD-10 codes X93–X95 capture the incidence of intentional firearm injuries (WHO, n.d.b).
4   Criminal justice systems in different countries may apply different rules. For an example of counting rules for homicides in the UK, see UK Home Office (2011, p. ii).
5   Section 4(2) of the Robbery and Firearms (Special Provisions) Act reads as follows: 'It shall be the duty of any person, hospital or clinic that admits, treats or administers any drug to any person suspected of having bullet wounds to immediately report the matter to the police' (Nigeria, 1984). The law imposes severe penalties for violations, including imprisonment for physicians and the closing down of clinics (Nigeria, 1984, s. 4(4)).
6   ICD-10 code Y35.0 captures any injury sustained as a result of an encounter with a law enforcement official, whether on-duty or off-duty. It includes injuries sustained by law enforcement officials, suspects, and bystanders involving firearm discharge (WHO, n.d.b).
7   This graph reflects data obtained from CDC (n.d.a) for non-fatal injuries caused by firearms in 2009. Based on the category 'all intents' (which yields the total), the chart shows the unintentional, self-harm, and assault categories, while the 'undetermined, other' category represents the difference between the sum of the latter three and the total.
8   The 284 cases were selected from 510 cases that had been published in media reports in the 12 months from March 2008 to February 2009. Of the 284 stray bullet injuries, 20 per cent were fatal (Wintemute et al., 2011).
9   See, for example, Incorvaia (2007) and Özdemir and Ünlü (2009).
10 Categories in Figure 3.4 reflect the International Classification of External Causes of Injuries; for more details, see ICECI Coordination and Maintenance Group (2004).
11 See Truman (2010) and Alvazzi del Frate (1998, pp. 65–66).
12 The number of victims was 398, or 21.1 per cent of a sample of 1,884 respondents.
13 Serious crime includes rape and sexual assault, robbery, and aggravated assault (Truman, 2010, p. 2).
14 The Small Arms Survey database on non-lethal firearm injuries includes primarily health statistics. In the absence of public health information for any country, it contains crime statistics that conform to a variety of classifications and definitions (such as non-fatal firearm injury, non-lethal violence, non-fatal shooting, non-fatal physical assault, assaultive injury, serious injury, and gunshot wound). The GBAV 2011 database on homicide contains data from 199 countries and territories (Geneva Declaration Secretariat, 2011); the UNODC database on homicides committed with firearms covers 116 countries (UNODC, 2011b).
15 The concept of 'case fatality rate' is based on analysis of CDC data (CDC, n.d.a), which uses the following formula: fatal injuries / [fatal + non-fatal injuries] * 100.
16 The countries are: Australia, Belgium, Brazil, Canada, Chile, Colombia, Costa Rica, Denmark, Egypt, El Salvador, Estonia, Guatemala, Honduras, Israel, Jamaica, Latvia, Mexico, Netherlands, New Zealand, Nicaragua, Panama, Saint Kitts and Nevis, Trinidad and Tobago, United Kingdom (England and Wales), United Kingdom (Northern Ireland), and United States of America. Firearm homicide data is not available for two of the 28 countries in the Small Arms Survey non-lethal firearm injury database.
17 For example, some hospital admissions do not make a distinction between *serious* and *slight* injuries. The countries in which these hospitals operate thus record a higher number of injuries than those whose hospitals only capture serious injuries. Consequently, these countries will register an apparent lower proportion of fatalities, resulting in a lower case fatality rate. Furthermore, in some settings, especially in developing countries, it may be difficult to obtain and match data from different sources, such as numbers of victims who died at the scene, patients treated by private hospitals, and police records, which often lack contextual details (Ojo et al., 2008, p. 6).
18 Data covers 2009 in the United States and 2011 in the UK, as calculated based on Small Arms Survey (2011) and UNODC (2011b). This ratio is more favorable than the 3:1 ratio (three non-lethal injuries for every one death) frequently cited for the United States; see, for example, Annest et al. (1995, p. 1751).
19 Author correspondence with Lee Annest, director, Office of Statistics and Programming, National Center for Injury Prevention and Control, CDC, 30 November 2011. The number of participating hospitals and emergency departments has changed since the initiative began. See ICPSR (2009, p. iii).
20 See IVP-31 of the Healthy People 2020 Objectives (DHHS, n.d., p. 14).
21 Author correspondence with Lee Annest, director, Office of Statistics and Programming, National Center for Injury Prevention and Control, CDC, 30 November 2011.

22 Author correspondence with Lee Annest, director, Office of Statistics and Programming, National Center for Injury Prevention and Control, CDC, 30 November 2011.
23 Non-fatal firearm assaults are ICD-9 code E965.
24 For a fuller discussion of the passage of the Firearms Control Act, its implementation phasing, and documentation of its impacts, see Small Arms Survey (2008, pp. 186–99).
25 Data in Tables 3.1 and 3.2 covers only the patients who accessed medical services. The data was extracted from Handicap International's database in Jonglei in August 2011 and represents a sample of approximately 4,000 cases admitted to hospital in the period 2008–11.
26 For a discussion of the challenges of using media reports to document armed violence, see Small Arms Survey (2005, pp. 235–38).

BIBLIOGRAPHY

Allard, Denis, and Vanessa Burch. 2005. 'The Cost of Treating Serious Abdominal Firearm-related Injuries in South Africa.' *South African Medical Journal*, Vol. 95, No. 8. August, pp. 591–94.

Alvazzi del Frate, Anna. 1998. *Victims of Crime in the Developing World*. UNICRI Publication No. 57. Rome: United Nations Interregional Crime and Justice Research Institute. <http://www.unicri.it/documentation_centre/publications/icvs/_pdf_files/No57/n57.htm>

Annest, Joseph, et al. 1995. 'National Estimates of Nonfatal Firearm-related Injuries.' *Journal of the American Medical Association*, Vol. 273, No. 2. 14 June, pp. 1749–54.

BBC News. 2011. 'Who, What, Why: How Dangerous Is Firing a Gun into the Air?' 22 August. <http://www.bbc.co.uk/news/magazine-14616491>

Buchanan, Cate. 2011. *Focusing on Survivors of Armed Violence: Trends and Opportunities*. Unpublished background paper. Geneva: Small Arms Survey.

Butchart, Alexander. 2008. *Manual for Estimating the Economic Costs of Injuries Due to Interpersonal and Self-directed Violence*. Geneva: World Health Organization. <http://whqlibdoc.who.int/publications/2008/9789241596367_eng.pdf>

Campbell, Nathan, et al. 2011. *Gunshots in Children: What Role Can Society Play to Prevent Them?* Unpublished background paper. Cape Town: Childsafe South Africa.

CBS (Centraal Bureau voor de Statistiek). n.d. 'Ziekenhuisopnamen: gelsacht, leeftijd en diagnose-indeling ICD9: E965 Aanslag m. vuurwapens en explos.' The Hague and Heerlen: CBS. <http://statline.cbs.nl/StatWeb/publication/?DM=SLNL&PA=71860ned&D1=1,3,5,7,8&D2=1,2&D3=22&D4=1168&D5=0,4,9,14,19-28&VW=T>

CDC (Centers for Disease Control and Prevention, National Center for Injury Prevention and Control). n.d.a. 'WISQARS Nonfatal Injury Reports.' Accessed 24 October 2011. <http://webappa.cdc.gov/sasweb/ncipc/nfirates2001.html>

—. n.d.b. 'Data Sources for WISQARS Nonfatal.: 5.2 The National Electronic Injury Surveillance System (NEISS).' Accessed 23 February 2012. <http://www.cdc.gov/ncipc/wisqars/nonfatal/datasources.htm#5.2>

Childsafe South Africa. n.d. 'Statistics/Research.' Cape Town: Childsafe South Africa. Accessed 28 February 2012. <http://www.childsafe.org.za/research.htm>

CLEEN. 2010. *Emergency Response to Victims of Gun Violence and Road Accidents*. Lagos: CLEEN Foundation. <http://www.cleen.org/Emergency%20Response%20to%20Victims%20of%20Gun%20Violence%20and%20Road%20Accidents.pdf>

Cook, Philip, et al. 1999.'The Medical Costs of Gunshot Injuries in the United States.' *Journal of the American Medical Association*, Vol. 281, No. 5. 4 August, pp. 447–54. <http://jama.ama-assn.org/content/282/5/447.full.pdf+html>

Corriere della Sera. 2011. 'Botti di Capodanno, un morto a Napoli e circa 500 feriti in tutta Italia.' 1 January. <http://www.corriere.it/cronache/11_gennaio_01/capodanno-napoli-morto-immondizia-incendi-botti-feriti_9516d4ec-157d-11e0-8bc9-00144f02aabc_print.html>

Dejito, Rezia and James Turton. 2011. *Small Arms Misuse and Victimisation: Jonglei State, South Sudan*. Unpublished paper. September. Lyon: Handicap International.

DHHS (United States Department of Health and Human Services). n.d. 'Injury and Violence Prevention.' Washington, DC: DHHS. <http://www.healthypeople.gov/2020/topicsobjectives2020/pdfs/Injury.pdf>

van Dijk, Jan, John van Kesteren, and Paul Smit. 2007. *Criminal Victimisation in International Perspective: Key Findings from the 2004–2005 ICVS and EU ICS*. The Hague: Boom Juridische Uitgevers. <http://english.wodc.nl/onderzoeksdatabase/icvs-2005-survey.aspx?cp=45&cs=6796>

FBI (Federal Bureau of Investigation). 2010. 'Crime in the United States: Expanded Homicide Data, Table 8' <http://www.fbi.gov/about-us/cjis/ucr/crime-in-the-u.s/2010/crime-in-the-u.s.-2010/index-page>

FICAP (Firearm & Injury Center at Penn). 2009. *Firearm Injury in the U.S.* Philadelphia: FICAP. <http://www.uphs.upenn.edu/ficap/resourcebook/pdf/monograph.pdf>

GBI (Global Burden of Injuries). 2010. 'Injury in Uganda: Architecture of Data Sources.' Poster presented at the Meeting for Estimating the Burden of Injuries in Sub-Saharan Africa, Burden of Injuries in Africa. Swansea, 17–20 September. <https://sites.google.com/a/globalburdenofinjuries.org/africa/meeting_swansea_2010/posters/Poster_Uganda.pdf?attredirects=0>

—. n.d. 'About This Project.' Accessed 16 February 2012. <http://www.globalburdenofinjuries.org/>

Geneva Declaration Secretariat. 2008. *Global Burden of Armed Violence 2008*. Geneva: Geneva Declaration Secretariat.

—. 2011. *Global Burden of Armed Violence 2011: Lethal Encounters*. Cambridge: Cambridge University Press.

Gilgen, Elisabeth and Ryan Murray. 2011. *Reading between the Lines: Crime and Victimization in Liberia*. Liberia Issue Brief No. 2. September. Geneva: Small Arms Survey. <http://www.smallarmssurvey.org/fileadmin/docs/G-Issue-briefs/Liberia-AVA-IB2.pdf>

— and Lauren Traccy. 2011. *Contributing Evidence to Programming. Armed Violence Monitoring Systems*. Geneva. Geneva Declaration Secretariat. <http://www.genevadeclaration.org/fileadmin/docs/general/GD-WP-2011-Contributing-Evidence-to-Programming.pdf>

Harris, Anthony, et al. 2002. 'Murder and Medicine: The Lethality of Criminal Assault 1960 1999.' *Homicide Studies*, Vol. 6, No. 2. May, pp. 128–66. <http://www.wku.edu/~james.kanan/Murder%20and%20Medicine.pdf>

Hofman, Karen, et al. 2005. 'Addressing the Growing Burden of Trauma and Injury in Low- and Middle-Income Countries.' *American Journal of Public Health*, Vol. 95, No. 1. January, pp. 13–17. <http://ajph.aphapublications.org/cgi/reprint/95/1/13>

Hootman, Jennifer, et al. 2000. 'National Estimates of Non-fatal Firearm-related Injuries Other than Gunshot Wounds.' *Injury Prevention*, Vol. 6, pp. 268–74. <http://injuryprevention.bmj.com/content/6/4/268.full.pdf+html>

HSBA (Sudan Human Security Baseline Assessment). 2011. 'George Athor's Rebellion, Jonglei State.' *HSBA Facts & Figures*. Geneva: Small Arms Survey. April. <http://www.smallarmssurveysudan.org/pdfs/facts-figures/armed-groups/southern-sudan/emerging/HSBA-Armed-Groups-Athor.pdf>

Hugenberg, Florian, et al. 2007. 'Firearm Injuries in Nairobi, Kenya: Who Pays the Price?' *Journal of Public Health Policy*, Vol. 28, No. 4, pp. 410–19.

Hutt, John, et al. 2004. 'Gunshot Wounds in Children: Epidemiology and Outcome.' *African Safety Promotion*, Vol. 2, No. 2, pp. 4–14. <http://www.sabinet.co.za/abstracts/safety/safety_v2_n2_a2.html>

ICECI (International Classification of External Causes of Injuries) Coordination and Maintenance Group. 2004. ICECI Version 1.2. Amsterdam and Adelaide: Consumer Safety Institute and Australian Institute of Health ad Welfare National Injury Surveillance Unit. <http://www.rivm.nl/who-fic/ICECIeng.htm>

ICPSR (Inter-University Consortium for Political and Social Research). 2009. *Firearm Injury Surveillance Study, 1993–2008: Study Description*. ICPSR 30543. Ann Arbor: ICPSR.

—. 2010. *Firearm Injury Surveillance Study, 1993–2008: Codebook*. ICPSR 30543. Ann Arbor: ICPSR.

Incorvaia, Angelo, et al. 2007. 'Can a Falling Bullet Be Lethal at Terminal Velocity? Cardiac Injury Caused by a Celebratory Bullet.' *Annals of Thoracic Surgery*, Vol. 83, pp. 283–84. <http://ats.ctsnetjournals.org/cgi/reprint/83/1/283>

Juárez, Mario Arroyo. 2010. *Cuadernos del ICESI 6: Mortalidad por homicidios en México*. Mexico City: Instituto Ciudadano de Estudios sobre la Inseguridad. <http://www.icesi.org.mx/documentos/publicaciones/cuadernos/cuaderno_6.pdf>

Kirsten, Adèle and Richard Matzopoulos. 2011. *Measuring Fatal and Non-fatal Firearm Violence in South Africa*. Unpublished background paper. November. Geneva: Small Arms Survey.

Kochanek, Kenneth, et al. 2011. 'Deaths: Preliminary Data for 2009.' *National Vital Statistics Reports*, Vol. 59, No. 4. 16 March, pp. 1–51. <http://www.cdc.gov/nchs/data/nvsr/nvsr59/nvsr59_04.pdf>

Krug, Etienne, et al., eds. 2002. *World Report on Violence and Health*. Geneva: World Health Organization. <http://whqlibdoc.who.int/publications/2002/9241545615_eng.pdf>

Lerer, Leonard, Richard Matzopoulos, and Rozette Phillips. 1997. 'Violence and Injury Mortality in the Cape Town Metropole.' *South African Medical Journal*, Vol. 87, No. 3. March, pp. 298–301. <http://archive.samj.org.za/1997%20VOL%2087%20Jan-Dec/1-4/Articles/03%20March/3.8%20VIOLENCE%20AND%20INJURY%20MORTALITY%20IN%20THE%20CAPE%20TOWN%20METROPOLE.%20Leonard%20B.%20Lerer,%20Richard%20G.%20Matzop.pdf>

Max, Wendy and Dorothy Rice. 1993. 'Shooting in the Dark: Estimating the Cost of Firearm Injuries.' *Health Affairs*, Vol. 12, No. 4. Winter, pp. 171–85. <http://content.healthaffairs.org/content/12/4/171>

Mohammed, Aminu, et al. 2005. 'Epidemiology of Gunshot Injuries in Kano, Nigeria.' *Nigerian Journal of Surgical Research*, Vol. 7, Nos. 3–4, pp. 296–99. <http://www.ajol.info/index.php/njsr/article/viewFile/12301/15393>

Nigeria. 1984. Robbery and Firearms (Special Provisions) Act. Assented to 29 March. <http://www.aksjlegalresource.com/resource/Laws_of_the_Federation%5CROBBERY%20AND%20FIREARMS%20_SPECIAL%20PROVISIONS_%20ACT.pdf>

NIMSS (National Injury Mortality Surveillance System). 2004. *A Profile of Fatal Injuries in South Africa: Fifth Annual Report of the National Injury Mortality Surveillance System, 2003*. Cape Town: Crime, Violence and Injury Lead Programme, Medical Research Council and University of South Africa. <http://www.sahealthinfo.org/violence/injury2004.htm>

—. 2005. *A Profile of Fatal Injuries in South Africa: Sixth Annual Report of the National Injury Mortality Surveillance System, 2004*. Cape Town: Crime, Violence and Injury Lead Programme, Medical Research Council and University of South Africa. <http://www.sahealthinfo.org/violence/national2004.pdf>

—. 2007. *A Profile of Fatal Injuries in South Africa: Seventh Annual Report of the National Injury Mortality Surveillance System, 2005*. Cape Town: Crime, Violence and Injury Lead Programme, Medical Research Council and University of South Africa. <http://www.sahealthinfo.org/violence/2005injury.htm>

—. 2008. *A Profile of Fatal Injuries in South Africa: Eighth Annual Report of the National Injury Mortality Surveillance System, 2007*. Cape Town: Crime, Violence and Injury Lead Programme, Medical Research Council and University of South Africa. <http://www.mrc.ac.za/crime/nimss07.PDF>

Norberg, Johannes, et al. 2009. 'The Costs of a Bullet: Inpatient Costs of Firearm Injuries in South Africa.' *South African Medical Journal*, Vol. 99, No. 6. June, pp. 442–44.

Ojo, Emmanuel, et al. 2008. 'Gunshot Injuries in a North Eastern Nigerian Tertiary Hospital.' *Internet Journal of Surgery*, Vol. 16, No. 2. <http://www.ispub.com/journal/the-internet-journal-of-surgery/volume-16-number-2/gunshot-injuries-in-a-north-eastern-nigerian-tertiary-hospital.html>

Özdemir, Mevci and Ağahan Ünlü. 2009. 'Gunshot Injuries Due to Celebratory Gun Shooting.' *Turkish Neurosurgery*, Vol. 19, No. 1, pp. 73–76. <http://neurosurgery.dergisi.org/pdf/pdf_JTN_639.pdf>

Pavesi, Irene. 2011. 'Non-lethal Firearm Injuries: Multisource Data Collection.' Unpublished methodological note. 20 November. Geneva: Small Arms Survey.

Prokosch, Eric. 1995. *The Technology of Killing: A Military and Political History of Antipersonnel Weapons*. London: Zed Books.

Roberts, J. E. H. and R. S. S. Statham. 1916. 'On the Salt Pack Treatment of Infected Gunshot Wounds.' *British Medical Journal*, Vol. 2, No. 2904. 26 August, pp. 282–86. <http://www.jstor.org/stable/20305045>

SAPS (South African Police Service). 2011. *Crime Report 2010/2011*. <http://www.saps.gov.za/statistics/reports/crimestats/2011/crime_situation_sa.pdf>

Small Arms Survey. 2005. *Small Arms Survey 2005: Weapons at War*. Oxford: Oxford University Press.

—. 2008. *Small Arms Survey 2008: Risk and Resilience*. Cambridge: Cambridge University Press.

—. 2011. Database on Non-lethal Firearm Injuries. Geneva: Small Arms Survey.

— and KNFP (Kenya National Focal Point). 2012. *Availability of Small Arms and Perceptions of Security in Kenya: An Assessment*. Geneva: Small Arms Survey.

South Africa. 2000. Firearms Control Act No. 60. Assented to 10 April 2001. *Government Gazette*, Vol. 430. Pretoria.
<http://www.saps.gov.za/docs_publs/legislation/juta/Act60of2000.pdf>

Suerte Felipe, Cecile. 2011. 'Robredo Admits Limited Time vs Drive on Indiscriminate Gun Firing.' *Philippine Star*. 4 January.
<http://www.philstar.com/Article.aspx?articleId=645062&publicationSubCategoryId=63>

Teixeira, Paulo Augusto Souza, João Batista Porto de Provenza, and Marcello Montillo Oliveira. 2011. *Bala Perdida*. Rio de Janeiro: ISP.
<http://urutau.proderj.rj.gov.br/isp_imagens/Uploads/BalaPerdida_1sem2011.pdf>

TLAVA (Timor-Leste Armed Violence Assessment). 2009. *Tracking Violence in Timor Leste: A Sample of Emergency Room Data, 2006–08*. Issue Brief No. 4. October. Geneva: Small Arms Survey and ActionAid Australia. <http://www.timor-leste-violence.org/pdfs/Timor-Leste-Violence-IB4-ENGLISH.pdf>

Truman, Jennifer. 2010. *Criminal Victimization 2010*. Washington, DC: Bureau of Justice Statistics. <http://bjs.ojp.usdoj.gov/content/pub/pdf/cv10.pdf>

UK (United Kingdom) Home Office. 2011. *Home Office Counting Rules for Recorded Crime: Violence Against the Person*. April. <http://www.homeoffice.gov.uk/publications/science-research-statistics/research-statistics/crime-research/counting-rules/count-violence?view=Binary>

UNODC (United Nations Office on Drugs and Crime). 2011a. *2011 Global Study on Homicide*. Vienna: UNODC.
<http://www.unodc.org/documents/data-and-analysis/statistics/Homicide/Globa_study_on_homicide_2011_web.pdf>

—. 2011b. 'Homicide Statistics: Homicides by Firearms.' Vienna: UNODC.
<http://www.unodc.org/documents/data-and-analysis/statistics/Homicide/Homicides_by_firearms.xls>

Vyrostek, Sara, Joseph Annest, and George Ryan. 2004. 'Surveillance for Fatal and Nonfatal Injuries: United States, 2001.' *Morbidity and Mortality Weekly Report: Surveillance Summary*, Vol. 53, No. 7. 3 September, pp. 1–57. <http://www.cdc.gov/mmwr/preview/mmwrhtml/ss5307a1.htm#fig21>

Ward, Elizabeth, et al. 2009. 'Results of an Exercise to Estimate the Costs of Interpersonal Violence in Jamaica.' *West Indian Medical Journal*, Vol. 58, No. 5, pp. 446–51.

Waters, Robert and Ien Sie. 2003. 'Spinal Cord Injuries from Gunshot Wounds to the Spine.' *Clinical Orthopaedics & Related Research*, Vol. 408. March, pp. 120–25.

WHO (World Health Organization). 2001. *Small Arms and Global Health: WHO Contribution to the UN Conference on Illicit Trade in Small Arms and Light Weapons*. Geneva: WHO. <http://whqlibdoc.who.int/hq/2001/WHO_NMH_VIP_01.1.pdf>

—. 2008. *Global Burden of Disease: 2004 Update*. Geneva: WHO. <http://www.who.int/healthinfo/global_burden_disease/estimates_country/en/index.html>

—. n.d.a. *Disability and Rehabilitation WHO Action Plan 2006–2011*. Geneva: WHO.
<http://www.who.int/disabilities/publications/dar_action_plan_2006to2011.pdf>

—. n.d.b. 'International Statistical Classification of Diseases and Related Health Problems: 10th Revision.'
<http://apps.who.int/classifications/apps/icd/icd10online/>

— and World Bank. 2011. *World Report on Disability*. Geneva: WHO. <http://www.who.int/disabilities/world_report/2011/en/index.html>

Winnington, Andrew. 2011a. *IPPNW Scoping Study of Liberian Hospital Data: Recommendations to LAVO (Liberian Armed Violence Observatory)*. Unpublished presentation. Somerville, MA: International Physicians for the Prevention of Nuclear War.

—. 2011b. *Tool to Enhance Violence Injury Data Collection in the Emergency Department*. Unpublished training manual. Geneva: Liberian Armed Violence Observatory.

Wintemute, Garen, et al. 2011. 'Stray Bullet Shootings in the United States.' *Journal of the American Medical Association*, Vol. 306, No. 5. 3 August, pp. 491–92. <http://www.ucdmc.ucdavis.edu/vprp/publications/jamastraybullet.pdf>

World Bank. 2011. *World Development Report 2011: Conflict, Security, and Development*. Washington, DC: World Bank.
<http://wdr2011.worldbank.org/fulltext>

YAVA (Yemen Armed Violence Assessment). 2010. *Fault Lines: Tracking Armed Violence in Yemen*. YAVA Issue Brief No. 1. May. Geneva: Small Arms Survey.

Zavala, Diego, et al. 2007. 'Special Section: A Multinational Injury Surveillance System Pilot Project in Africa.' *Journal of Public Health Policy*, Vol. 28, pp. 432–41. <http://www.ippnw.org/pdf/research-multi-jphp.pdf>

Zawitz, Marianne and Kevin Strom. 2000. *Firearm Injury and Death from Crime, 1993–97*. Washington, DC: Bureau of Justice Statistics.
<http://bjs.ojp.usdoj.gov/content/pub/pdf/fidc9397.pdf>

ACKNOWLEDGEMENTS

Principal author
Anna Alvazzi del Frate

Contributors
Emile LeBrun, Cate Buchanan, Adèle Kirsten, Richard Matzopoulos, Irene Pavesi, Reyia Dejito, James Turton, Sebastian van As, Kavi Bhalla, Matthias Nowak

Kazakh Army soldiers march during the opening ceremony of the Steppe Eagle 2011 joint tactical military exercise, at Ili military range outside Almaty, August 2011.
© Shamil Zhumatov/Reuters

Blue Skies and Dark Clouds
KAZAKHSTAN AND SMALL ARMS

INTRODUCTION

In May 2011, two deadly explosions targeted facilities of Kazakhstan's National Security Committee (KNB) in Aktobe Oblast and Astana.[1] One of the attacks was reportedly the first suicide bombing in the country (Lillis, 2011a). Two months later, in the same oblast, authorities conducted a two-week operation in the villages of Shubarshi and Kenkiyak to neutralize an armed group responsible for the killing of two police officers and suspected of religious radicalism. The special units killed nine members of the group and lost two more officers during the intervention (MIA, 2011; Mednikova and Bogatik, 2011). While Kazakhstan is generally perceived as a beacon of stability in an otherwise troubled region, such incidents demonstrate that economic growth and political stability do not render a country immune to home-grown armed violence.

Kazakhstan has been an active participant in the UN small arms process, submitting four national reports on its implementation of the UN Progamme of Action since 2005 (PoA-ISS, 2010). The country has carried out a well-developed set of small arms control initiatives, including strengthened controls over private firearm ownership, large-scale civilian weapons collection, and the destruction of excess stockpiles of arms and ammunition inherited from the Soviet Union (RoK, 2010a). Yet beyond Kazakhstan's own national reporting, the country's small arms control efforts, as well as the nature and significance of the threats they are meant to address, have only rarely been studied.[2] Indeed, the bulk of international attention on small arms issues in Central Asia has been directed at the country's conflict-ridden neighbours.[3]

This chapter presents the findings of research initiated in April 2010 by the Small Arms Survey. The project, supported by the Government of Norway, was designed to assess levels of small arms availability in the country, evaluate the impact of firearms on crime and security, and review government initiatives to address small arms issues. It relies on a variety of research methods, including a nationwide survey of 1,500 households as well as six focus group discussions with members of communities affected by unintended explosions at munitions sites.[4]

The chapter's main findings include the following:

- Household survey results suggest that civilians in Kazakhstan owned between 190,000 and 225,000 firearms in 2010, which translates into a low per capita rate by international standards. Civilian firearm ownership appears more prominent among young men and in urban areas; it also seems to be motivated by a perceived need for protection against criminals.
- Kazakh authorities report having collected and seized more than 60,000 firearms from civilians between 2003 and 2009. They also destroyed at least 20,000 civilian small arms during the same period.
- Kazakhstan's overall positive security outlook is clouded by an increase in crime rates since 2010, as well as recent incidents of armed violence with terrorist, ethnic, and political undertones.

Map 4.1 Kazakhstan

- Although the country's homicide rate has decreased significantly since the 1990s, it remained above the world average, at more than 8 per 100,000, in 2010. The percentage of homicides and robberies committed with small arms has also increased in recent years, but it remains low when compared with rates elsewhere.
- The Ministry of Defence reported the destruction of more than 1.1 million rounds of surplus conventional ammunition between 2003 and 2009 (out of a declared total of 2.5 million). The Ministry also reported destroying about 38,000 state-held small arms and light weapons between 2002 and 2006.
- Kazakhstan has been disproportionately affected by unplanned explosions at munitions sites, with six major incidents known to have occurred since 2001. Focus group discussions with affected communities revealed that the authorities do not organize emergency response training for the civilian population living near ammunition depots.

This chapter is divided into three main sections. It first analyses Kazakhstan's security outlook and discusses some of the main threats to the country's stability. The second section examines civilian firearms—their availability, government controls over them, and their impact on security in Kazakhstan. Lastly, the chapter examines state-held stockpiles, including government efforts to dispose of surplus and the impact of unplanned explosions at munitions sites.

A SAFE HAVEN UNDER THREAT?

Kazakhstan, the world's largest landlocked country, is home to just 16.6 million inhabitants (AS, 2011). A Soviet Republic until December 1991, it now shares borders with China, Kyrgyzstan, the Russian Federation, Turkmenistan, and Uzbekistan. Unlike some of its Central Asian neighbours, Kazakhstan has been spared civil war and ethnic strife, earning it the reputation of a pillar of stability in an otherwise volatile region.[5] This section tests this assessment by reviewing key crime and security indicators and by discussing emerging threats to the country's internal stability. It finds that the overall security situation is satisfactory, although it is becoming more vulnerable to emerging negative trends. While crime rates have decreased markedly since Kazakhstan's independence, the past two years have seen a significant deterioration. Recent acts of religiously motivated terrorism, as well as instances of ethnic and political violence, further cloud the picture.

Context

Kazakhstan has prospered economically over the past decade thanks in large measure to its booming oil and gas industries. Gross national income per capita increased from USD 1,260 in 2000 to USD 7,440 in 2010 (World Bank, 2011). Kazakhstan is also Central Asia's primary recipient of foreign direct investment, with USD 50 billion received since 1991, representing around 80 per cent of total investment in the Central Asia region (Hug, 2011, p. 21). Economic growth has generally contributed to improving socio-economic conditions, particularly in the last ten years. From 2001 to 2009, the proportion of the population with income below the minimum food basket declined from 16.1 per cent to 0.6 per cent, while income inequality decreased by 30 per cent (UN, 2010, pp. 18–19). Unemployment officially stood at a low 5.3 per cent for the third quarter of 2011 (AS, 2011).

> Kazakhstan is becoming more vulnerable to emerging negative trends.

Growing economic significance and political stability have helped turn Kazakhstan into a regional powerhouse with increasing clout in international forums. In 2010, Kazakhstan became the first Central Asian state and former Soviet Republic to chair the Organization for Security and Co-operation in Europe (OSCE). During its presidency, Kazakhstan hosted the OSCE's first heads of state summit in 11 years (Lillis, 2011b; OSCE, 2010). The country was also named chair of the Shanghai Cooperation Organization in 2011 as well as of the Organization of Islamic Cooperation in 2011–12 (RoK, 2011c; MFA, 2011).

Tarnishing these positive features are persistent concerns with governance in the country. President Nursultan Nazarbayev has led Kazakhstan since 1989 and was last re-elected in April 2011 for a five-year term with 95.5 per cent of the vote. Although the turnout rate of 90 per cent was lauded, foreign observers, including the OSCE, expressed concern at the lack of opposition candidates and noted 'serious irregularities' in the electoral process (Lillis, 2011c; OSCE, 2011). Furthermore, Kazakhstan is accorded a score just above mid-level in the United Nations Development Programme's Human Development Index, ranking 66 out of 169 countries (UNDP, 2010). Shortcomings in the country's healthcare and education systems contribute to the modest ranking and stand in sharp contrast to the flattering macro-economic trends mentioned above (ICG, 2011, pp. 28–33). Social tensions also surfaced in 2011 when massive protests by oil and gas workers demanding higher wages affected the national industry for several months (RFE/RL, 2011). In December 2011, a peaceful strike turned violent, leading to pogroms and deadly clashes in the towns of Zhanaozen and Shetpe, in which at least 17 people died after police and security forces fired on a crowd of rioters (Antoncheva, 2011).

Crime

Similar to the macro-economic trends identified above, the crime situation in Kazakhstan has generally improved over the past 15 to 20 years. Despite some major variations from year to year, the total number of crimes recorded in 2010 was much lower than in 1991 (see Figure 4.1). Violent crime has also fallen markedly since independence. Homicides peaked between 1992 and 1998, a period during which they averaged about 2,500 per year, compared with fewer than 1,400 homicides in 2010 (see Figure 4.2).

Recent crime trends are more worrying. After several years of decline, the general crime rate began increasing in 2010 (see Figure 4.1) and kept growing in the first half of 2011. In January–June 2011, law enforcement agencies recorded 80,685 crimes, an 18.6 per cent increase compared with the same period in 2010 (68,053). Regions where the crime rate in the first half of 2011 increased the most compared to 2010 are the capital Astana (a 105.1 per cent increase), Karagandy Oblast (35.5 per cent), Southern Kazakhstan Oblast (33.1 per cent), Almaty city (20.1 per cent),

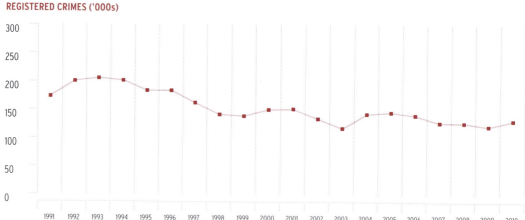

Figure 4.1 **Total recorded crimes in Kazakhstan, 1991–2010**

REGISTERED CRIMES ('000s)

Source: AS (2011)

Figure 4.2 **Intentional homicides in Kazakhstan, 1989–2010**

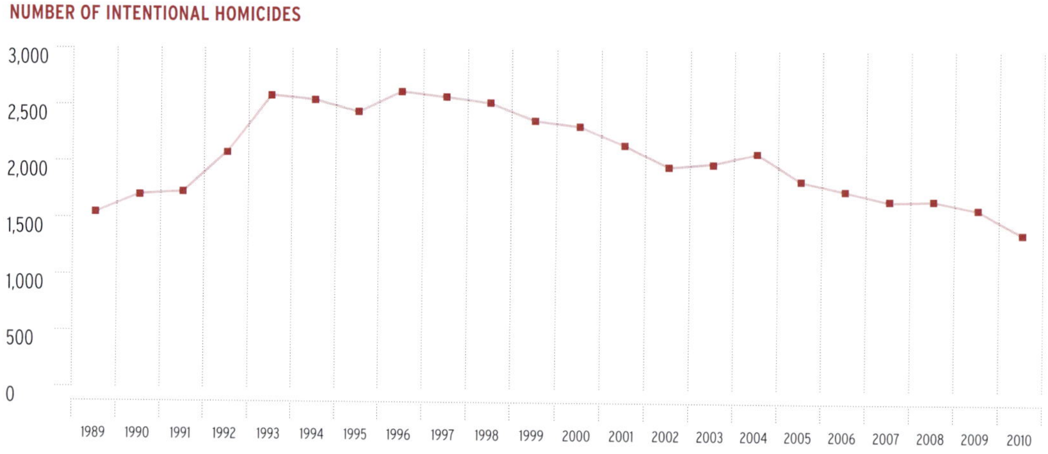

NUMBER OF INTENTIONAL HOMICIDES

Sources: UNICEF (2011) for 1989–2008; PGO (2011a) for 2009 and 2010

Kostanay Oblast (19.2 per cent), Akmola Oblast (17.6 per cent), and Eastern Kazakhstan Oblast (16.9 per cent) (PGO, 2011a; 2011b). Press reports suggest that the increasing crime rate since 2010 is due to a rise in unemployment (Radio Azattyk, 2011). More than three-quarters of the perpetrators arrested over the period January–June 2011 were in fact unemployed, while almost 14 per cent were under the influence of alcohol, and nine per cent were repeat offenders (PGO, 2011a).

Furthermore, despite the important decrease since 1991, Kazakhstan's homicide rate still stood at 8.35 per 100,000 people in 2010—higher than the estimated world rate of 6 per 100,000 and Central Asia's rate of 6.5 per 100,000.[6] Reported homicide rates for other countries in the region are clearly lower: 7.8 per 100,000 in Kyrgyzstan (2009), 1.9 in Tajikistan (2009), 3.8 in Turkmenistan (2006), and 3.0 in Uzbekistan (2007) (UNICEF, 2011). It is unclear whether Kazakhstan's higher rate actually reflects higher homicide levels, or instead more accurate recording and systematic reporting of homicides. Indeed, reporting rates are usually lower in developing countries than in richer states (van Dijk, van Kesteren, and Smit, 2007, p. 17). It may be more appropriate to compare these rates to those of the Russian Federation, a country that ranks just above Kazakhstan in the Human Development Index and whose homicide rate is higher than Kazakhstan's, at 12.5 per 100,000 in 2009 (UNDP, 2010; UNICEF, 2011).

Map 4.2 **Homicide rates per oblast, 2011***

Note: * The homicide rates for the cities Astana and Almaty are not included in the oblast rates. Shaded areas exhibit annual homicide rates above the national rate (8.4 per 100,000 population). Annual homicide rates are calculated based on data available at the time of writing; that is, homicides figures were multiplied by two to provide an annual rate.

Sources: AS (2011); PGO (2011a)

Geographically speaking, as Map 4.2 illustrates, homicide rates in the first half of 2011 were consistently above the national average in eastern and northern Kazakhstan, as well as in oil-producing Western Kazakhstan Oblast. Eastern Kazakhstan Oblast tops the list with an annual homicide rate of 13.4 per 100,000 and also experiences the country's second-highest general crime rate after Almaty city (AS, 2011; PGO, 2011a). Authorities, including then regional police chief and subsequently minister of internal affairs Kalmukhanbet Kasymov, have pointed to alcohol and drug abuse as the main causes of Eastern Kazakhstan's high crime rates. The fact that numerous prisons are located in the east of the country may also explain the phenomenon, as released criminals tend to stay and repeat offences in the oblast in which they were detained (Chernyavskaya, 2010).

Victimization and security perceptions

Results of the household survey conducted for this study reveal a relatively high rate of crime and violent incidents (see Box 4.1). Overall, 5.1 per cent of respondents reported that at least one member of their household had been the victim of a crime or violent encounter over the previous 12 months. Victims suffered injuries in one-third of these cases. The most commonly cited incidents were robberies (40 per cent of reported cases), assaults (34 per cent),

Box 4.1 Household survey on perceptions of security and firearms

In order to measure public perceptions of firearms and security in Kazakhstan, the Small Arms Survey subcontracted the Almaty-based Center for the Study of Public Opinion (CIOM) to carry out a nationwide survey of 1,500 people aged 18–60 (see Annexe 4.1). The survey was carried out in July 2010 using face-to-face interviews at the respondents' homes.

Kazakhstan comprises a total of 14 oblasts that are grouped together in six regions, all of which are characterized by similar geographical, climatic, and economic features. The survey sample covered one oblast per region—Aktobe in the western region, Almaty in the Almaty region, Eastern Kazakhstan in the eastern region, Karagandy in the central region, Pavlodar in the northern region, and Southern Kazakhstan in the southern region—as well as the country's two largest cities, Almaty and Astana. In each surveyed oblast, CIOM randomly selected villages, towns, and cities in order to reflect the rural–urban distribution of the population of the entire region. Interviewers chose households using random selection methods and identified respondents based on age and sex quotas.[7] CIOM verified 23 per cent of interviews through phone calls or onsite supervision. The confidence level of the study is 95 per cent, and the confidence interval is 2.5.

The Small Arms Survey applied statistical weighting to ensure that the final sample's demographic characteristics (such as the distribution of respondents by region, urban vs. rural settings, sex, age, and ethnicity) were commensurate with those of the national population. The weighting also factored in non-response rates. This allows for the extrapolation of results to the national population and for the comparison of findings by region, rural or urban setting, sex, and age group.

Source: CIOM (2010)

Security personnel guard the site of a shootout in the village of Boraldai, near Almaty, December 2011.
© Vladimir Tretyakov/Reuters

threats and intimidation (13 per cent), rape and sexual assaults (5 per cent), and burglaries (4 per cent). The survey also found that almost three-quarters of victims of crime and violence were between 15 and 29 years old. Across all age groups, the sex distribution of victims was 56 per cent women and 44 per cent men. In roughly 60 per cent of cases, crimes were perpetrated in the street or at a public gathering, as opposed to in someone's home. Crimes usually occurred during the day or early evening. In 44 per cent of all cases, respondents said they had not reported the crime to the authorities (CIOM, 2010).

Despite relatively significant crime and victimization rates, survey respondents expressed mixed perceptions about their personal security. On the one hand, when asked about the most serious problems affecting them, they described the security of the members of their household as a relatively minor concern, ranking it fourth behind lack of or inadequate employment, healthcare, and clean water. On the other hand, almost two-thirds of respondents said they were concerned that a member of their household could become the victim of a crime or a violent encounter. Compared to

the national average, a greater proportion of respondents in Astana, Aktobe, Eastern Kazakhstan, and Pavlodar Oblasts expressed concern that members of their household might become victims of crime or violence. In line with the distribution of crime statistics discussed above, Eastern Kazakhstan topped this list with more than 80 per cent of interviewees expressing concern, compared with only 18.5 per cent in Southern Kazakhstan Oblast. While respondents said they felt safe at home and during the day, they described the situation outside and at night in different terms. Indeed, more than 40 per cent of respondents stated that they would feel unsafe or somewhat unsafe after sunset (CIOM, 2010).

Respondents also pointed to greater feelings of insecurity in urban areas. Nineteen per cent of respondents living in rural areas answered that there were no safety or security concerns, compared with just 3 per cent among residents of urban areas. Similarly, 26 per cent of urban respondents identified high crime rates as their main security concern, versus 15 per cent among rural respondents (CIOM, 2010). Figure 4.3 illustrates these differences, showing that, compared to their rural counterparts, about twice as many urban respondents consider their communities unsafe.

Figure 4.3 **Responses to the household survey question, 'Do you consider your city/town/village safe, unsafe, or neither?' (n=1,500), (percentage)**

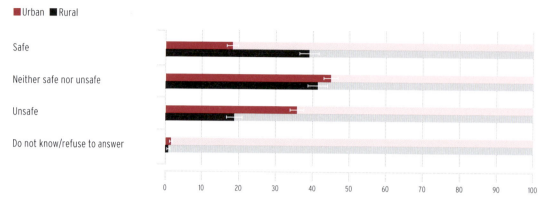

Note: Lines on each bar reveal the corresponding confidence interval. There were significant differences between responses from rural and urban interviewees regarding the extent to which they described their area to be 'safe' or 'unsafe'; such differences were not significant with respect to the response 'neither safe nor unsafe'.

Source: CIOM (2010)

Organized violence

While the threat of large-scale political, ethnic, and terrorist violence in Kazakhstan appears less pronounced than elsewhere in Central Asia,[8] recent incidents have led Kazakh government officials to voice their concern (Lillis, 2011d). The following sections provide examples of organized violence within Kazakhstan itself. They have been grouped according to the political, ethnic, or terrorist nature of the incident, as reported by independent and government sources.

Political violence

Examples of political violence in Kazakhstan include the February 2006 shooting of prominent opposition leader Altynbek Sarsenbayev as well as his bodyguard and driver. They were found on the outskirts of Almaty, all three of them with their hands tied behind their backs and wounds to the back and head. The official investigation concluded that a senate official ordered the assassination of Sarsenbayev, allegedly motivated by long-lasting 'personal enmity', while a former

police officer was found guilty of organizing and carrying out the contract killing with assistance from the KNB's elite Arystan special unit officers (RFE/RL, 2006a; 2006b; 2008). In November 2005, another leading opposition figure—Zamanbek Nurkadilov—was found shot dead in his home. Despite two gunshot wounds to the chest and one to the head, the police officially ruled his death to be a suicide, claiming that no signs of a forced entry had been detected (RFE/RL, 2005).

Ethnic violence

Despite the absence of large-scale ethnic strife in Kazakhstan, security agencies organized robust interventions to suppress a number of localized clashes between ethnic Kazakhs and minority groups.[9] In March 2007, for example, a minor brawl during a game of billiards erupted into a violent shoot-out between ethnic Kazakhs and Chechens in the villages of Malovodnoye and Kazatkom, in Almaty Oblast. The incident resulted in the death of five people, with several others wounded. The authorities managed to stop the violence only after bringing in special police forces, and the area remained cordoned off for several weeks to ensure that order was fully restored. Reports circulated in the press that armed people from other regions of the country intended to come to the scene of these events but were stopped by the police (RFE/RL, 2007; Saydullin, 2007a).

There were also allegations that firearms were used in Mayatas and other villages of the Southern Kazakhstan Oblast when the Kazakh population burned down houses and the property of local Kurds after the arrest of a 16-year-old Kurd suspected of raping a four-year-old Kazakh boy in late October 2007. According to press reports, up to 500 law enforcement personnel were involved in restoring order, and three policemen were injured during the operation (Dzhani, 2007; Saydullin, 2007b; IWPR, 2007).

Terrorist violence

While the revival of Islam in Kazakhstan in the 1990s did not lead to a large-scale emergence of radical religious organizations, some extremist groups have intensified their activities in the country.[10] Compared to the late 1990s and early 2000s, when the majority of criminal cases related to religious extremism

People gather for the funeral of opposition figure Zamanbek Nurkadilov, found shot dead in his home, Almaty, November 2005. © Pavel Mikheev/Reuters

concerned the seizure of prohibited literature, Kazakhstan's security and law enforcement services now report more arrests of members of extremist groups who are in possession of firearms and ammunition and allegedly planning terrorist acts or even the overthrow of the government (USDoS, 2010; Vybornova, 2011).

The years 2010 and 2011 saw a sharp increase in the number of incidents of armed violence that were reportedly related to religious extremism. The 2011 attacks against KNB facilities and the above-mentioned special forces' intervention in Shubarshi and Kenkiyak made most headlines, but several other events of a terrorist nature have also occurred. In April 2011, two suspected religious extremists were killed, and a third was arrested, in a night-time assault on an apartment in Almaty. The suspects, armed with sub-machine guns and grenades, offered strong resistance, wounding 11 special police unit members (Benditskiy, 2011). There have also been several cases of Kazakh nationals being recruited by religious extremists and being killed or arrested by Russian security services for participating in guerrilla activities in the northern Caucasus (Nurseitova, 2011).

Although initially reluctant to label such incidents 'terrorist' in nature, the Kazakh government has recently changed its stance. In August 2011, law enforcement officials for the first time publicly announced that they had foiled a terrorist plot in oil-rich Atyrau Oblast. In a September 2011 speech to Parliament, President Nazarbayev himself acknowledged that the country faced a problem with extremism and pledged to tackle it (Lillis, 2011d).

CIVILIAN SMALL ARMS: UNDER CONTROL, IN DEMAND

This section examines the prevalence and sources of civilian-held firearms, reviews government efforts to control them, and assesses the role they play in insecurity in Kazakhstan. Generally speaking, civilian ownership of firearms is relatively low and appears to be tightly controlled by the authorities. Firearms availability and related insecurity are more pronounced in urban areas, however. Although firearms are rarely used in criminal acts, the proportion of homicides and robberies perpetrated with firearms increased between 2006 and 2010.

A pheasant hunter in Zhetisay, November 2009.
© Carolyn Drake/Panos Pictures

Civilian holdings

General population

About 4.4 per cent of household survey respondents said that someone in their household owned at least one firearm, with an average of five firearms for every 100 households (CIOM, 2010). The types of firearms most frequently owned were hunting rifles (61 per cent of gun-owning households), pistols or revolvers (22 per cent), air pistols (6 per cent), gas pistols (4 per cent), and air guns (4 per cent); the remaining holdings were identified as other firearms (1 per cent) and unspecified firearms (2 per cent) (CIOM, 2010). Applied nationally to Kazakhstan's 4.15 million households, this finding suggests that there are 207,500 privately owned firearms in the country or, taking into account the survey's confidence interval, a range of roughly 190,000–225,000 civilian firearms.[11] This translates into fewer than 1.3 firearms for every 100 Kazakhs, a low rate that places Kazakhstan at 142nd position in international rankings (Small Arms Survey, 2007).

Although household survey respondents can be expected to under-report sensitive issues such as firearm ownership, other sources suggest that 190,000–225,000 is a plausible estimate of the number of firearms in civilian hands in Kazakhstan. Previous studies also found that Kazakhstan's population of 16.6 million is relatively poorly armed by international standards. An earlier estimate offered a range of 100,000–300,000 firearms (Small Arms Survey, 2007). In 2010, the Ministry of Internal Affairs (MIA) reportedly inspected more than 139,000 firearm owners (MIA, 2010a).[12]

Figure 4.4 **Responses to the household survey question, 'How easy or difficult do you think it is to acquire a firearm in your city/town/village?' (n=1,500), (percentage)**

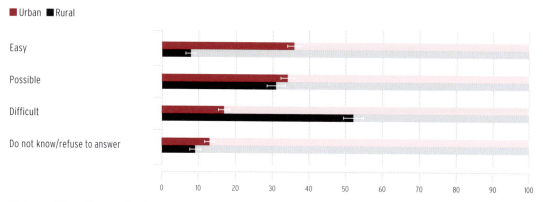

Note: Lines on each bar reveal the corresponding confidence interval.
Source: CIOM (2010)

The household survey also provides insight into motivations for firearm ownership, highlighting differences across urban and rural settings. While only 16 per cent of respondents in rural areas selected criminal intent as one of the top three reasons why people in their communities owned firearms, 74 per cent cited hunting.[13] In contrast, 45 per cent of respondents in urban areas identified criminal intent as one of the top three motivations, while only 37 per cent selected hunting (CIOM, 2010). Responses also suggest that firearms are more accessible in urban settings than in rural areas (see Figure 4.4).

Private security companies

The first private security companies appeared in Kazakhstan in the early to mid-1990s, but it was not until 2000 that the government adopted legislation to regulate them, namely the Law on Security Activity (Bayekenov, 2004; RoK, 2000b). It requires private security companies to obtain a licence for the provision of security services from the Directorate of State Guard Service (under the MIA Committee of Administrative Police and its regional divisions) (MIA, 2004).

According to the Association of Security Organizations of the Republic of Kazakhstan, in 2010 there were about 3,000 private security companies in Kazakhstan that employed nearly 60,000 people, a threefold increase compared to 2001 (Tashimov, 2010; Vasilyeva, 2002). At the same time, the MIA reported that the sector employed some 77,500 guards, 21,500 of whom worked for just two companies, the Kazakhstan Temir Zholy railway operator (14,000) and KazMunaiGas (7,500) (Vesti.kz, 2010; Foster, 2010). These numbers suggest that there are roughly as many private security guards as police officers in Kazakhstan, as the police numbered 69,529 in 2008 (UNODC, n.d.).

The law authorizes private security companies to arm their personnel, but the government introduced some limitations to this right in December 2010, notwithstanding strong opposition from Kazakhstan Temir Zholy and KazMunaiGas (Foster, 2011). Private security companies are now banned from using rifled long- and short-barrelled firearms, and they can only use smooth-bore firearms or barrel-less firearms with non-lethal ('traumatic') cartridges. They are also authorized to use 'electric' weapons (RoK, 2000b; 2010b). Prior to the legislative amendment, private security companies possessed about 7,000 rifled firearms (Foster, 2011).

Foreign security companies are banned from working in Kazakhstan, while foreign legal entities and nationals cannot provide security services or establish private security companies in the country. Recent legislative changes entitle domestic providers of security services to cooperate with foreign security companies in the sharing of experience, advanced training of their personnel, and use of modern security equipment and technology, provided they comply with the legal requirements related to the protection of state secrets (RoK, 2000b; 2010b). The December 2010 legislative amendments also prohibit non-security companies from relying on in-house security operations, in effect requiring them to outsource (Foster, 2011). The only exception is made for security operations established by Kazakhstan's national companies (RoK, 2000b; 2010b).

Sources
Authorized trade

At least 36 companies sell firearms and ammunition in Kazakhstan's civilian market.[14] They sell a variety of pistols (such as the Steyr MA1); bolt-action rifles (such as the Steyr Classic, Elite, and Scout as well as the CZ 452, 527, and 550); and semi-automatic rifles (such as the CZ 858 and Saiga) (Chebotarev, 2010). Few of these companies actually produce civilian weapons or ammunition in Kazakhstan, however. Anna LLP, based in Almaty, is one of the few manufacturers of small arms ammunition for hunting and sports shooting destined for the Kazakh civilian market (Chebotarev, 2010). Kazakhstan Engineering reportedly had a project to set up an enterprise in Petropavlovsk to produce 5.45 mm, 7.62 mm, and 9 mm ammunition, and possibly hunting ammunition. The project had been valued at USD 19 million and production had been expected to start in mid-2007. Kazakhstan Engineering was also reported to have been in talks with Kyrgyzstan about purchasing an ammunition factory in Bishkek, but neither project appears to have materialized (Barabanov, 2008, pp. 32–33). With respect to weapons, the Western Kazakhstan Machine-Building Company (ZKMK, previously known as the Metallist factory) continues to produce a small quantity of hunting rifles (Barabanov, 2008, p. 31; Kenzhegalieva, 2007).[15]

Table 4.1 Value of reported Kazakh small arms imports, by UN Comtrade category, 1992–2008 (in USD)

Types of weapon (UN Comtrade category)	Reported by Kazakhstan	Reported by exporters
Sporting and hunting shotguns (930320)	6,792,782	23,105,081
Small arms ammunition (930630)	5,041,847	5,068,017
Sporting and hunting rifles (930330)	3,030,318	9,128,726
Pistols and revolvers (9302)	2,828,276	1,112,119
Shotgun cartridges (930621)	2,483,837	2,165,181
Military rifles, machine guns, and other (930190)	2,186,057	236,166
Air gun pellets, lead shot, and parts of shotgun cartridges (930629)	995,054	3,534,458
Military weapons (9301)	980,700	22,674
Parts and accessories of shotguns or rifles (930529)	523,721	52,004,510
Parts and accessories of revolvers or pistols (930510)	298,295	364,574
Shotgun barrels (930521)	3,952	48,764
Total	**25,164,839**	**96,790,270**

Source: UN Comtrade (n.d.)

Due to Kazakhstan's relatively low capacity to produce small arms domestically, most weapons and ammunition sold in the country are imported. Available customs data, as reported in the UN Comtrade database, suggests that Kazakhstan imports significantly more small arms than it exports (UN Comtrade, n.d.).[16] Trading partners reported that Kazakhstan imported close to USD 100 million worth of small arms and ammunition (civilian and military) for the period 1992–2008 (see Table 4.1). Yet reported Kazakh exports for the same period amounted to just USD 2 million (see Table 4.1; UN Comtrade, n.d.).[17]

Reported Kazakh imports consist for the most part of sporting and hunting shotguns, rifles, and small arms ammunition—categories usually destined for the civilian market (see Table 4.1). Reports to the UN Register of Conventional Arms confirm this trend, as the largest imports of equipment by quantity for the years 2008–10 concerned hunting weapons (UNODA, n.d.; see Table 4.2). In 2009, Kazakhstan imported a variety of hunting and sporting rifles from countries including Italy (243, 308, 300 Winchester and 30.06 SPR 9.2 x 62), the Russian Federation (MP 161K calibre 22LR, 'Vepr' carbines, 'Saiga', 'Tiger', 'Sobol', 'Korshun', 'CM-2KO', 'Bars-4-1', 'Los', and 'Biathlon-7-3, 7-4'), and the Czech Republic and Germany (various other models) (UNODA, n.d.).

Overall, the available data suggests that Germany and the Russian Federation are the largest and most regular exporters of small arms to Kazakhstan (see Table 4.3). In 2008, Israel declared a large export of parts and accessories of shotguns and rifles (a category usually destined for the civilian market) worth USD 51 million (see Table 4.3; UN Comtrade, n.d.). While this value would make Israel Kazakhstan's top source of arms, UN Comtrade data suggests that Israel is not as regular an arms exporter as Germany or the Russian Federation. Furthermore, the Israeli transfer is not confirmed by data reported by Kazakhstan (UN Comtrade, n.d.).

Table 4.2 Small arms imports as reported by Kazakhstan to the UN Register of Conventional Arms, by category, 2008-10

Exporting country	Year	Type of weapon	Quantity
Germany	2010	Hand-held under-barrel and mounted grenade launchers	18
Czech Republic	2009	Revolvers and self-loading pistols	500
Russian Federation	2009	Revolvers and self-loading pistols	400
United States	2009	Sniper rifles	2
Germany	2009	Sniper rifles of 338 calibre	14
Italy	2009	Hunting rifles	54
Russian Federation	2009	Hunting carbines	1,380
Russian Federation	2009	Hunting and sport weapons	330
Czech Republic	2009	Hunting and sport weapons	1,155
Germany	2009	Hunting and sport weapons	446
Austria	2009	Hunting and sport weapons	50
Austria	2008	Revolvers and self-loading pistols	100
Total			**4,449**

Source: UNODA (n.d.)

Table 4.3 Value of reported Kazakh small arms imports, by exporting country, 1992–2008 (in USD)

Exporting countries	Value reported by Kazakhstan	Value reported by exporters
Armenia	79,399	0
Austria	785,943	929,181
Belarus	219,300	0
Belgium	180,730	0
Brazil	0	8,987
Bulgaria	126,394	0
Canada	127,604	526,349
Cyprus	83,816	799,176
Czech Republic	795,764	1,900,134
Denmark	0	1,376
Finland	151,877	86,895
France	2,200	394,739
Germany	3,752,977	15,590,047
Israel	12,399	51,517,000
Italy	1,999,114	3,678,802
Kyrgyzstan	5,800	495,999
Latvia	0	22,393
Pakistan	0	676
Poland	0	212,399
Russian Federation	15,708,843	15,556,753
Serbia and Montenegro	1,100	0
South Korea	800	0
Spain	188,707	619,291
Sweden	55,910	157,119
Switzerland	54,406	445,282
Turkey	474,071	1,610,490
Turkmenistan	7,800	0
Ukraine	92,511	0
United Arab Emirates	0	98,300
United Kingdom	92,018	1,052,048
United States	164,660	1,086,834
Unspecified	696	0
Total	**25,164,839**	**96,790,270**

Source: UN Comtrade (n.d.)

While the available trade data provides interesting insight, it remains incomplete. As reflected in Tables 4.1 and 4.3, the values of small arms trade reported to UN Comtrade by Kazakhstan differ markedly from those submitted by its trading partners. Overall, the partners have reported exports worth about four times as much as the imports that Kazakhstan has declared. The values of imports Kazakhstan has reported from European states are consistently much lower than the figures given by those countries, with the exception of Belgium and Finland; yet figures for the Russian Federation and Kazakhstan are almost identical (see Table 4.3). Table 4.1 also shows lower reporting by Kazakhstan for sporting and hunting shotguns and, to a lesser extent, rifles.[18]

Illicit sources

Some sources of illicit weapons in Kazakhstan are internal. Diversion of state stockpiles was a major concern in the late 1990s. After the collapse of the Soviet Union, former Soviet military arsenals in Kazakhstan became a source of small arms and ammunition for both the international and the domestic black markets (Vasilyeva, 2010). Among the most high-profile cases of illegal arms deliveries is the 1995 sale of 57 Igla (SA-18) man-portable air defence systems (MANPADS) and 226 9M313 missiles to the former Yugoslavia, then under a UN Security Council arms embargo (Kenzhetaev, 2002). Kazakh authorities also investigated attempts to supply MIG-21 combat aircraft to North Korea in 1999, as well as Mi-8T helicopters to Sierra Leone in 2000, and found that they involved violations of Kazakh legislation (Holtom, 2010, pp. 5–6). There are no openly available statistics on weapons theft from the country's military stockpiles, but in 2008 the Ministry of Defence acknowledged that the armed forces remained one of the sources of illegal firearms and ammunition for the criminal underworld (Severnyy, 2008).

Another source of illicit supply to the domestic small arms market is craft production. The majority of hand-made firearms seized by police in Kazakhstan are actually gas pistols converted to shoot live rounds.[19] In one case in 2008, Russian authorities in Saratov seized an IZh-79 gas pistol that had been modified to shoot live 9 mm rounds. The investigation established that the perpetrator had brought weapons from Oral, Kazakhstan, where he had an accomplice, a worker at the Metallist factory, who had been converting gas and smooth-bore weapons into rifled firearms and equipping firearms with optic sights at home (Kulikov, 2008). According to official statistics, cases of illicit weapons manufacture increased more than threefold between 2006 and 2010 (MIA, 2010c; PGO, 2011a). Little analysis is available on this increase, although it may be partially explained by tightened firearms regulations, which are discussed below.

The extent of cross-border trafficking of firearms into Kazakhstan is also a concern. Central Asia is located on the 'northern' drug route linking Afghanistan to the Russian Federation and Europe (UNODC, 2008, p. 6). The extent to which drugs and firearms trafficking are intertwined in the region remains unclear, although Kazakh authorities have seized firearms in the framework of anti-drug ('Kanal') as well as migration control ('Nelegal') operations. Kanal operations have been taking place regularly since 2003 and are part of a regional effort initiated by the Collective Security Treaty Organization (CSTO), which involves the law enforcement agencies of Armenia, Belarus, Kazakhstan, Kyrgyzstan, the Russian Federation, Tajikistan, and Uzbekistan.

Data on firearms seized during such operations is not available by country but instead for all participating CSTO members. In September 2010, for instance, Kazakhstan, the Russian Federation, and Tajikistan carried out an anti-drug operation named 'Kanal-Yug', resulting in the seizure of more than 1,300 kg of drugs and 220 firearms.[20] In 2009, states participating in Kanal and Nelegal collected a total of 1,881 small arms and 52,759 rounds of ammunition; they also initiated 249 criminal law investigations related to the illegal arms trade.[21] From 16 to 22 November 2010, a larger operation resulted in the seizure of 18 tons of drugs, 1,129 firearms, and 17,000 units of ammunition.[22]

According to press reports, there are several routes for the illicit trafficking of firearms into Kazakhstan, with weapons intended mainly for domestic criminal groups. Kazakh police officials claim that the primary source countries are, in order of importance, the Russian Federation, Tajikistan, and Afghanistan (Isabekov, 2008). The majority of seized Kalashnikov assault rifles and Stechkin pistols originate in the northern Caucasus; Makarov pistols come from Rostov, Krasnodar Kray, Moscow, and St. Petersburg; Kalashnikovs, Makarovs, and Tokarevs with inscribed Latin letters and hieroglyphics are brought from Afghanistan through Kyrgyzstan, Tajikistan, and Uzbekistan—via the routes used for drug trafficking. Organized crime groups also increasingly use European-made pistols such as the German Sig Sauer, the Italian Beretta, and the Czech CZ smuggled from Western Europe and the Russian Federation (Isabekov, 2008).

Events in neighbouring countries can also trigger the funnelling of illicit firearms to Kazakhstan. In April 2010, Kyrgyzstan experienced a violent uprising that resulted in the overthrow of President Kurmanbek Bakiyev as well as the looting of police and military weapons by protesters, prompting Kazakhstan to close its border temporarily (*Kazakhstan Today,* 2010a). More police and military weapons, including pistols, assault rifles, light machine guns, sniper rifles, grenades, and a grenade launcher, were lost during deadly clashes, in June 2010, between ethnic Kyrgyz and Uzbeks in southern Kyrgyzstan. Despite efforts by Kyrgyzstan to recover lost firearms and ammunition, Kyrgyz authorities reported that, as of September 2010, 146 out of 278 firearms and 26,344 out of 43,045 pieces of ammunition that had been lost in June 2010 were still missing (Kylym Shamy, 2011, p. 8).[23] Kazakhstan has expressed concern that some of the lost Kyrgyz arms may find their way across the border (MFA, 2010).[24] Anecdotal reports of arms seizures at the border suggest this concern is justified, although assessing the exact volume of trafficking from Kyrgyzstan is difficult.[25]

Control measures

National legislation

Overall, Kazakhstan imposes tight controls on the acquisition and possession of small arms, including the registration of all privately owned firearms, requirements for their effective storage, and the screening of prospective owners.[26] Kazakh legislation is similar to that of other countries in the region; many elements are holdovers from the Soviet Union. Kazakh legislation has, however, changed over time as part of efforts to address inconsistencies and improve public security.

In December 2010, new legislative amendments were introduced to strengthen the firearms control regime (RoK, 2010b). They include the following:

- a reduction in the number of arms civilians are permitted to own, from five to two;
- obligations for firearm owners to report theft, loss, and unjustified or illegal use of firearms; and
- an obligation for firearm owners to re-register their firearms when moving to a different address (RoK, 2010b).

Kazakh citizens may acquire firearms only for self-defence, sporting, or hunting purposes (RoK, 1998, art. 5).[27] The legal age for owning a firearm in Kazakhstan is 18 years. Applicants for a firearms licence cannot have an outstanding conviction for the commission of 'intent crimes' and cannot have committed two or more specified administrative offences within the same year (RoK, 1998, arts. 19.2, 19.2-1, 19.5). They are also required to provide a medical certificate of good mental health (RoK, 1998, art. 15.3).

If they meet the above requirements, individuals wishing to possess a firearm are required to obtain a number of licences and permits, each subject to a fee. They must pay a fee of about KZT 4,500 (USD 30 or, more precisely, three MRP[28]) for a licence to possess firearms, KZT 760 (USD 5, half an MRP) for a permit to purchase a firearm or to store firearms, and KZT 150 (USD 1, one-tenth of an MRP) for registration and re-registration of every firearm (MIA, 2010b).

> Kazakhstan imposes tight controls on the acquisition and possession of small arms.

124 SMALL ARMS SURVEY 2012

Table 4.4 Weapons collection efforts in Kazakhstan, as reported in national reports on implementation of the UN Programme of Action, 2003–07 and 2009

Action by the state	Type of weapon	2003 KNB	2003 MIA	2004	2005	2006 KNB	2006 MIA	2007	2009	Total all years
Weapons seized and removed	Automatic weapons	9	27		13		15			
	Combination firearms		42		33		50			
	Gas weapons		1,549		1,118	1	1,249			
	Home-made weapons		354		344	4	244			
	Hunting and sawn-off weapons	137								
	Pistols and revolvers	22	467		372	2	354			
	Rifled weapons (unspecified type)	18				10				
	Rifles and carbines		706		532		435			
	Smooth-bore weapons		10,400		7,195	21	6,231			
Total seized and removed		13,731			9,607	8,616		n/a	2,305	34,259
Weapons voluntarily surrendered	Gas weapons		1,270		989		761			
	Home-made weapons		95		82		82			
	Rifled weapons		717		618		636			
	Smooth-bore weapons		6,987		4,348		5,043			
	Other						438			
Total surrendered		9,069			6,037	6,960		n/a	4,453	26,519
Weapons destroyed	Automatic weapons			9	7			22		
	Carbines			41	44			96		
	Combination firearms			20	30			31		
	Gas weapons			467	980			941		

Failure to obtain the above permits and register a firearm is punishable by fines ranging from about KZT 1,500 to 15,000 (USD 10–100) (RoK, 2001, arts. 368, 371). Individuals may only purchase ammunition upon presentation of a valid licence to possess and carry firearms (RoK, 2000a, r. 86).

Once registered gun owners obtain licences, they must:

- undergo training in the handling of small arms based on the curriculum developed by the MIA, if acquiring firearms for the first time (RoK, 1998, art. 15.4);
- register their firearms in an official database maintained by the MIA;
- place their firearms in a safety deposit box, a metallic cabinet, or any other storage device that makes it impossible for other persons to access licensed firearms (RoK, 2000a, r. 91);
- report any theft or loss of firearms in their possession (RoK, 2000a, r. 101); and
- renew their licences and permits in a timely manner.[29]

Kazakh legislation explicitly prohibits civilian possession of short-barrelled shotguns;[30] firearms designed for automatic fire; firearms disguised as other objects; cartridges with armour-piercing, incendiary, or percussion bullets, as well as hollow-point ammunition;[31] and weapons and cartridges that do not meet safety requirements (RoK, 1998, art. 7).[32] While the above suggests that civilian possession of automatic rifles is prohibited, it is not clear whether semi-automatic firearms are allowed for civilian ownership, although such weapons can be seen in the country's gun shops.[33] Firearms held by collectors must be deactivated (RoK, 2000a, r. 47). A weapon holder's licence may be revoked if he or she modifies or converts small arms (RoK, 1998, art. 19).

Enforcement and weapons collection

The Kazakh government has collected significant quantities of small arms and light weapons in the past ten years. Information reported in Kazakhstan's national reports on the implementation of the UN Programme of Action reveals that authorities collected some 60,000, and destroyed more than 20,000, firearms between 2003 and 2009 (see Table 4.4). These are considerable numbers given the country's low level of civilian firearm ownership; they represent about one-third of the estimated total weapons held by civilians in 2010.

Weapon type							Total
Home-made weapons		259				200	
Pistols		170	51		139		
Revolvers		152	48		155		
Rifled weapons (unspecified type)		228			657		
Rifles		312	347			214	
Smooth-bore weapons		4,501	3,605		3,902		
Total destroyed	n/a	**5,708**	**5,563**	n/a	**6,357**	**4,569**	**22,197**

Note: Total figures represent the sum of the different weapon categories as listed in this table. They sometimes differ slightly from the totals provided in the sources. There was no available reporting for the year 2008.

Sources: RoK (2005; 2006; 2008; 2010a)

Table 4.5 Weapon-related crime in Kazakhstan, 2006-10

Type of crime	MIA		Prosecutor General's Office		
	2006	2007	2008	2009	2010
Illegal acquisition, transfer, sale, storage, transportation, or possession of weapons, ammunition, explosive substances, or explosive devices	1,272	1,264	926	1,131	984
Negligent storage of a firearm	9	5	2	3	3
Improper performance of duties related to guarding weapons, ammunition, explosive substances, or explosive devices	0	1	5	1	3
Theft or extortion of weapons, ammunition, explosive substances, or explosive devices	57	51	47	83	59
Violation of weapon handling rules	1	2	1	6	2

Sources: MIA (2010c); PGO (2011a)

Weapons are collected through a combination of voluntary and more forceful schemes. The MIA, as the agency that implements Kazakhstan's firearms legislation, regularly inspects firearms at their places of storage and use, and can seize and destroy unauthorized weapons without compensation (RoK, 1998, art. 30). Table 4.5 provides a breakdown of the different types of weapon-related crimes as recorded in the databases of the MIA and, since 2008, the Prosecutor General's Office.

A 2007 decree established a compensation system for people who surrender arms and ammunition voluntarily to the MIA, with different cash amounts given depending on the type of weapon recovered (RoK, 2007b). All surrendered weapons must be destroyed or dismantled. In 2010, Kazakhstan reported that the government had allocated more than KZT 500 million (USD 3.4 million) in 2008 for the implementation of this legislation, resulting in more than 13,000 firearms being collected (RoK, 2010a, p. 37). As Table 4.5 illustrates, more than 20,000 firearms were also surrendered voluntarily before the adoption of the decree, presumably without any compensation. Official statistics do not reveal

Figure 4.5 Responses to the household survey question, 'In the last two years, have the authorities collected weapons from the population in your city/town/village?' (n=1,500), (percentage)

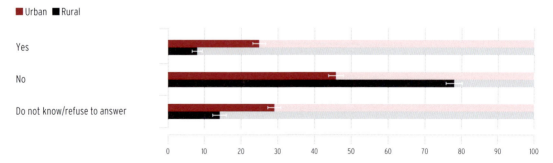

Note: Lines on each bar reveal the corresponding confidence interval.
Source: CIOM (2010)

the number of weapons collected or seized in each of Kazakhstan's 14 oblasts. Results of the CIOM household survey suggest, however, that weapons collection efforts have mainly targeted urban areas (see Figure 4.5).

Kazakh internal affairs agencies also undertake targeted firearm inspections through an operation named 'Karu'. Launched in 1995, the operation aims to prevent and detect crimes committed with firearms, explosives, and explosive devices and also to remove these items from illicit circulation (RoK, 2008, p. 8). Karu operations occur every three months (RoK, 2008, p. 8); the MIA regularly issues press releases to showcase results. In 2010, for instance, authorities inspected more than 139,000 firearm owners and identified more than 5,700 violations of the rules governing the circulation of firearms and 229 violations of hunting rules. A total of 5,196 registered civilian arms were confiscated in the framework of these efforts (MIA, 2010a).

Impact

Overall, firearms do not appear to be a common tool for perpetrating crime in Kazakhstan. In 2006–10 firearms were used in a negligible proportion—between 0.3 and 0.4 per cent—of the total number of crimes recorded in the country (PGO, 2011a). Bladed weapons appear to be more commonly used, although their use decreased from 1 per cent of all crimes in 2006 to 0.5 per cent in 2010 (MIA, 2010c; PGO, 2011a).

The types of crime in which firearms are most frequently used are homicides, robberies, and acts of hooliganism, but their use remains rare overall at below 10 per cent of all cases (see Figure 4.6). In contrast, the proportion of homicides perpetrated with firearms reaches 60 per cent in Latin America and the Caribbean, 24 per cent in Europe, and 22 per cent in Asia.[34] That said, the percentage of homicides and robberies committed with firearms in Kazakhstan increased from 4.1 per cent to 6.9 per cent and from 3.4 per cent to 5.7 per cent, respectively, between 2006 and 2010 (see Figure 4.6). Also noteworthy are recent incidents during which prisoners were able to acquire firearms to help them stage escapes, as well as reports of teenagers getting involved in shootings.[35]

Household survey results suggest that the population does not consider firearms a major problem. While one half of the respondents described firearms as desirable for protection, the other half called them a threat to their safety (CIOM, 2010). More than 60 per cent of respondents nationwide cited personal protection from gangs and criminals

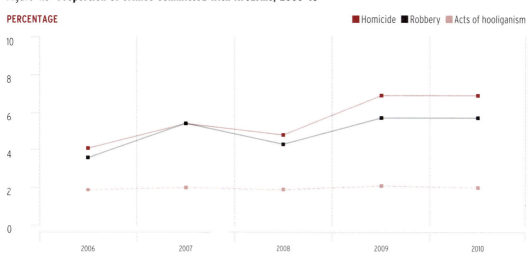

Figure 4.6 **Proportion of crimes committed with firearms, 2006-10**

Source: MIA (2010c); PGO (2011a)

Figure 4.7 **Responses to the household survey question, 'How do you mainly perceive firearms?' (by sex, n=1,497), (percentage)**

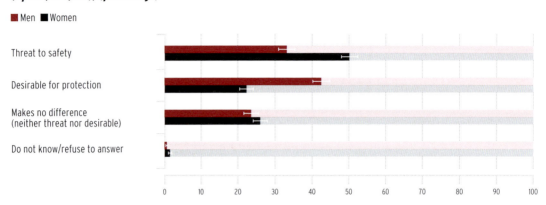

Note: Lines on each bar reveal the corresponding confidence interval.
Source: CIOM (2010)

Figure 4.8 **Responses to the household survey question, 'How do you mainly perceive firearms?' (n=1,498), (percentage)**

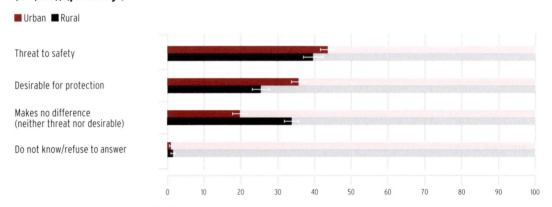

Note: Lines on each bar reveal the corresponding confidence interval.
Source: CIOM (2010)

as one of the top three reasons for gun ownership for people living in their areas. There were significant differences between men and women, however. As Figure 4.7 illustrates, 42 per cent of men said that firearms were desirable for protection, as opposed to 22 per cent of women. Similarly, more than half of women identified firearms as threats to safety, as opposed to just one-third of the men.[36] Younger respondents, as well as people living in urban areas, were also more likely to associate firearms with protection (CIOM, 2010; see Figure 4.8).

STILL SECRET: SMALL ARMS AND THE STATE

This section reviews publicly available information as well as data provided by the Ministry of Defence (MoD) of Kazakhstan on small arms and ammunition currently in state stockpiles. It also examines the number and types of surplus weapons and ammunition destroyed by the country. Lastly, it analyses the threats associated with state stockpiles, focusing on the consequences for local communities of unplanned explosions at munitions sites.

Stockpiles

Kazakh defence, security, and law enforcement agencies include a number of institutions that answer to different ministries, each responsible for its own small arms stockpiles (see Figure 4.9). Kazakh law defines the size of state security forces and their arms stocks as 'state secrets'; there is no public information on their strength or equipment.[37] *Military Balance*, a secondary source, reports the following strength figures for 2010:

- Army: 30,000;
- Navy: 3,000;
- Air Force: 12,000;
- Ministry of Defence: 4,000;
- Government Guard: 500;
- Internal Security: 20,000 (estimate);
- Presidential Guard: 2,000; and
- State Border Protection Forces: 9,000 (estimate) (IISS, 2010, pp. 364–65).

Figure 4.9 **Kazakhstan's defence, security, and law enforcement infrastructure**

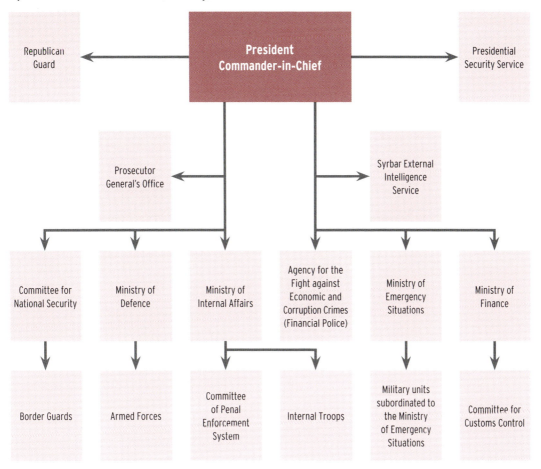

Source: description of Kazakhstan's government agencies in RoK (2011b)

An informed observer judges the size of the army to be closer to 42,000, with the Navy approximately 2,000 strong, as of mid-2010.[38]

No inventory of state security forces' small arms holdings has been made public. Secondary reports indicate there are five main ammunition depots in the country (Ashkenazi, 2010, p. 138). Existing estimates go as high as 550,000–950,000 military small arms and light weapons in the country's armouries, and 200,000–400,000 tons of surplus conventional ammunition (Ashkenazi, 2010, p. 138). The weapons estimates are based on the reported strength of the Kazakh armed forces in the late 1990s and early 2000s (a period during which manpower was at its height), combined with estimated weapons-per-soldier multipliers derived from ratios in other countries in the region.[39] More recently, Kazakhstan's MoD reported a surplus of 2.5 million units of conventional ammunition for 2003, 1.1 million units of which were destroyed by 2009—implying a remaining 1.4 million units of surplus ammunition in 2010 (RoK, 2011a).

For some agencies, including the Financial Police, Customs, as well as the State Courier Service (subordinated to the Prime Minister's Office), national legislation specifies what types of small arms different agencies are to use (see Table 4.6). Secondary sources report that the armed forces' arsenal includes the following small arms and light weapons (Jane's World Armies, 2010):

A group of international OSCE experts examine the condition of conventional ammunition designated for disposal at the Aris Central Arsenal, near Chemkent, June 2005. © Yurii Kryvonos/OSCE

- mortars: 82 mm M37M (150 units), 107 mm M107 (50), 120 mm M-43 (100), 120 mm 2S11 Sani, 120 mm 2S9, Aibat;
- anti-tank weapons: 9K111/AT-4 'Spigot' (200 units), 9P148/AT-5 'Spandrel' (50 units), 9K114 Metis/AT-6 'Spiral', 9K115 Metis/AT-7 'Saxhorn' (24 units), 100 mm T-12 (150 units), RPG-7 (250 units);
- MANPADS: 9K32/9K32M/Strela-2/2M/SA-7a/b 'Grail' (250 units), Strela 3/SA-14 'Gremlin' (50 units), Igla-1/SA-16 'Gimlet';
- pistols: 5.45 mm PSM, 7.62 mm Tokarev, 9 mm Makarov;
- assault rifles: 5.45 mm AK-74, 7.62 mm AKM;
- sniper rifles: 7.62 mm Dragunov;
- sub-machine guns: 5.45 mm AKS-74U;
- light machine guns: 5.45 mm RPK-74;
- general-purpose machine guns: 7.62 mm PKS; and
- heavy machine guns: 12.7 mm DShK.

Sources

Some 50 factories were involved in the production of conventional arms and defence equipment in Kazakhstan before independence from the Soviet Union. By 1995, only 24 military–industrial factories remained operational (Burnashev and Chernykh, 2010). Among them, the state-owned, Oral-based Metallist company was the only one still producing small arms (Chebotarev, 2010; IA, 2004, p. 21). In April 2003, Metallist was incorporated into ZKMK and converted into a producer of oil and gas equipment. The unpredictability of orders for military weapons from both domestic agencies

Table 4.6 Firearms and ammunition assigned to Kazakh state agencies

Government agency	Weapons	Ammunition
Agency for the Fight against Economic and Corruption Crimes (Financial Police)	• 5.45 mm PSM pistol; • 9 mm Makarov pistol; • 5.45 mm Kalashnikov assault rifle, model AKS-74 or AKS-74U; • 7.62 mm Kalashnikov assault rifle, model AKM or AKMS; • 9 mm PP-90 sub-machine gun	• 5.45 mm small-size centre-fire cartridge; • 5.45 × 18 mm pistol cartridge; • 7.62 mm standard bullet and tracer bullet cartridge; • 9 mm pistol cartridge (9 × 18 mm, 9 × 19 mm)
Committee for Customs Control under the Ministry of Finance	• 9 mm Baikal 442 pistol, Makarov pistol, model PM-9; • 12-gauge shotguns, models IZH-81 and Mossberg; • 9 mm Kobalt revolver, also known as RSA/TKB-0216 (Revolver Stechkina-Avraamova, after the names of designers, Stechkin and Avraamov)	• 9 mm ammunition for handguns; • 12-gauge ammunition for shotguns
State Courier Service	• 9 mm Makarov pistol; • 7.62 mm AK gun, model AKMS; • 9 mm Stechkin automatic pistol	n/a

Sources: RoK (2000a; 2003, 2010c)

President Nursultan Nazarbayev looks at a machine gun during a visit to a military site, April 1998. © Shamil Zhumatov/Reuters

and the international market, as well as growing wage arrears and debt, contributed to this conversion (Chebotarev, 2010). Reports also indicate that ZKMK attempts to develop and manufacture new small arms, such as the 9 mm Kobalt revolver and the folded PP-90 sub-machine gun, were not particularly successful (Barabanov, 2008, p. 31).

Kazakhstan's defence manufacturing sector was given new momentum when President Nazarbayev approved a revised military doctrine on 21 March 2007. The doctrine called for the provision of new models of weapons and equipment for the army as well as the modernization and upgrade of existing armaments (RoK, 2007c, p. 16). On 15 May 2009, the president instructed his government to develop a programme to implement the doctrine, including the technological upgrading of the armed forces and arms-producing facilities (RoK, 2009). Official statements indicate that Kazakhstan intends to manufacture arms not only for its armed forces, but also for export to foreign countries (Akhmetova, 2008, p. 4).

On 27 January 2010, Minister of Defence Adilbek Dzhaksybekov revealed that ZKMK had begun producing ammunition for 30 mm BMP-2-mounted machine guns and continued to produce the NSV 12.7 mm 'Utes' heavy machine gun. While meeting with the minister, the plant management expressed ZKMK's intention to expand production to include more small arms ammunition (Parpura, 2010). During the KADEX-2010 arms exhibition, held in Astana in

May 2010, ZKMK announced it had reached an agreement with Israel Military Industries (IMI) to jointly produce a new-generation 'WAVE 300–Tolkyn' small-arms system. This remote-controlled weapon station will reportedly be a combination of the ZKMK-produced NSV 12.7 mm heavy machine gun and the IMI-produced electronic guidance, aiming, and control system. Central Asian states, as well as the Russian Federation, are seen as potential markets for the new weapon (KazTAG, 2010; Kedrov, 2010).

The country's limited military small arms production capacity appears to be reflected in its low levels of military small arms exports. Reported Kazakh exports of these weapons amounted to just over USD 100,000 for the 1992–2008 period (UN Comtrade, n.d.). Kazspetseksport, a state-owned company, is the only entity authorized to export military small arms. The company sells the following types of equipment: 12.7mm NSV machine gun without optical sights; 9 mm PM pistols; 7.62 mm AKM assault rifles; 7.62 mm SKS rifles; 5.45 mm AK-74, AKS-74, and AKS-74U assault rifles; 5.6 mm TOZ-8 and TOZ-17 sports rifles; 26 types of ammunition ranging from 4.5 mm to 23 mm (Chebotarev, 2010).

As domestic production capacity is limited, Kazakhstan is likely to import most of the military small arms it needs for the foreseeable future. According to Defence Vice Minister Ratmir Komratov, the country imports about 70 per cent of its military supplies (of all types) (Moldabayev, 2010, p. 5). Detailed information on military small arms imports is scarce, however. UN Comtrade data as reported by Kazakhstan suggests total imports of military rifles, machine guns, and other light-calibre military weapons amounted to just over USD 3 million for the period 1992–2008, a

Table 4.7 MoD procurement of small arms, parts, and ammunition, 2009

Product	Quantity (units)	Value	
		KZT	USD (rounded)
Small arms and devices for reconnaissance battalions	0	201,989,400	1,383,500
Firing mount for handheld firearms	1	1,700,000	11,600
Sniper rifles	36	126,000,000	863,000
Ammunition, 7.62, 338	4,125	3,300,000	22,600
Large-calibre sniper rifle	2	7,000,000	47,900
Ammunition, 12.7 × 99 mm	1,000	3,500,000	24,000
Closed collimating sight	222	26,640,000	182,500
AK-107 machine gun (Russian Federation)	100	38,000,000	260,300
5.66 mm APS (underwater assault rifle)	44	16,368,000	112,100
5.66 mm APS ammunition for underwater firing	50,000	46,500,000	318,500
5.66 mm ammunition for underwater firing	9,000	8,680,000	59,500
4.5 mm SPP-1 (underwater pistol)	44	12,584,000	86,200
4.5 mm SPS ammunition for underwater firing	50,000	39,500,000	270,500
Installation and acquisition of shotgun rib for sub-machine guns	1,000	15,000,000	102,700

Source: MoD (2009)

relatively small proportion of the country's reported total small arms imports (see Table 4.1). The low value of exports of military small arms to Kazakhstan, as reported by its trading partners, probably reflects the fact that many of the states that regularly export to Kazakhstan, such as the Russian Federation, do not report on military transfers to Comtrade. Kazakh reports to the UN Register on Conventional Arms include some information on military imports. In 2009, for instance, Kazakhstan reported importing sniper rifles from the United States and Germany, as well as grenade launchers from Switzerland (UNODA, n.d.). Imports as reported by Kazakhstan fall well below the actual scale of procurement, however. MoD data, for instance, reveals the procurement of KZT 547 million (USD 3.7 million) worth of small arms and ammunition for 2009 alone (see Table 4.7).

Control measures

The Security Council of the Republic of Kazakhstan's Inter-Departmental Commission on the sale of weapons, military technology, and dual-use goods decides how to deal with surplus weapons and ammunition, based on recommendations made by the MoD's Military–Technical Commission (RoK, 2008, p. 13). Until recently, KazArsenal was the only

Table 4.8 Surplus conventional ammunition destroyed by the Ministry of Defence, 2003-09[40]

Type	2003	2004	2005	2006	2007	2008	2009	Total
37 mm for AZP-39	0	0	0	33,943	9,138	68,823	64,262	176,166
57mm for ZiS-2	0	0	0	2,447	2	21,903	0	24,352
76 mm for ZiS-3	0	0	2,430	46,552	3,292	43,937	0	96,211
85 mm for D-44	8,061	84,126	25,935	52,411	67,679	67,292	0	305,504
85 mm for D-48	0	0	0	9,926	0	0	0	9,926
100 mm for KS-19	0	18,400	0	912	43,828	77,574	0	140,714
115 mm for U-5TS	0	0	0	0	0	4,957	0	4,957
122 mm for D-25TS	0	970	0	0	0	0	0	970
122 mm for M-30	4,127	34,000	50,250	69,692	1,158	6,438	0	165,665
130 mm for M-46	0	0	1,628	0	0	9,553	0	11,181
152 mm for D-20	304	0	0	3,991	17,133	412	0	21,840
152 mm for D-1	3,200	0	0	0	0	0	0	3,200
203 mm for 2S7	0	0	0	0	326	4,761	0	5,087
82 mm mines	0	0	0	0	87,728	4,727	0	92,455
120 mm mines	0	0	0	0	50,777	0	0	50,777
160 mm mines	0	0	0	0	4,952	0	0	4,952
240 mm mines	0	0	0	0	0	1,395	0	1,395
Total	15,692	137,496	80,243	219,874	286,013	311,772	64,262	1,115,352

Source: RoK (2011a, p. 1)

contractor licensed to destroy ammunition in Kazakhstan, but it had no experience dismantling complex ammunition such as MANPADS missiles (NAMSA, 2007, pp. 2–3). KazArsenal's ammunition disposal facilities were located in Kapshagay and Arys, and were reportedly able to destroy up to 250,000 pieces of ammunition per year, depending on the calibre (OSCE, 2005, pp. 4–5). Following the March 2009 explosion at KazArsenal's Arys branch, the government announced the suspension of ammunition disposal activities in Kazakhstan until the causes of the incident were established and improved safety procedures introduced (Kazinform, 2009). On 3 September 2010, the government issued a decree that provided for the creation of a new entity that would be entrusted to process and destroy surplus ammunition (RoK, 2011a, p. 1).

Available information points to substantial progress in the destruction of surplus conventional ammunition by the MoD. In 2003, following an August 2001 ammunition depot explosion in Tokyrau, the Ministry of Defence adopted a 'Comprehensive Programme of Disposal of Conventional Ammunition that Are No Longer Used by the Kazakhstan Armed Forces', which envisioned the destruction of 2.5 million units of conventional ammunition in 2003–07, with 161,000 pieces of calibre 85–152 mm reportedly destroyed in 2003–04. Funding shortages led to an extension of the programme by 7.5 years (OSCE, 2005, p. 3). In December 2009, the MoD announced that 1.5 million pieces of ammunition were to be disposed of (BBC Monitoring/Interfax, 2009). In early 2011, the Ministry informed the Small Arms Survey that it had destroyed 1,115,352 units of conventional ammunition between 2003 and 2009, 939,244 of which were destroyed by detonation (RoK, 2011a, p. 1; see Table 4.8). This suggests that some 1.4 million rounds of ammunition out of the original 2.5 million remained in surplus as of 2010. These figures concern only the Ministry of Defence, however, and not the other institutions listed in Figure 4.9, for which no comparable data is available.

The Ministry of Defence also reported destroying 37,792 units of small arms and light weapons in 2002–06, but there is no information available on remaining levels of surplus arms, if any (RoK, 2011a, p. 2; see Table 4.9). Another report states that the MoD destroyed some 27,723 out of 45,000 weapons earmarked for destruction in 2007 (NAMSA, 2007, p. 2). Kazakhstan has undertaken the destruction of excess stockpiles unilaterally, despite some opportunities for international collaboration. A project designed in partnership with the NATO Ammunition Supply Agency envisioned the destruction of 16,653 small arms and 350 MANPADS missiles and launching tubes in 2007. Officially due to administrative reasons, the project was never implemented, however (NAMSA, 2007, pp. 2–3).

Table 4.9 **Surplus small arms and light weapons destroyed by the Ministry of Defence, 2003-06**

Type	2002	2003	2004	2005	2006	Total
Grenade launchers	0	43	436	566	0	1,045
Rifles and carbines	147	10,937	143	9,586	42	20,855
Machine guns	0	2,960	2,742	2,910	2	8,614
Sub-machine guns/ assault rifles	50	10	0	5,415	0	5,475
Pistols	1,090	573	0	140	0	1,803
Total	1,287	14,523	3,321	18,617	44	37,792

Source: RoK (2011a, p. 2)

Unplanned explosions at munitions sites

Since 2001, six large-scale explosions are known to have occurred at ammunition storage facilities in Kazakhstan. As of late 2011, these explosions made Kazakhstan the 13th most-affected country in the world by such incidents (Small Arms Survey, 2011). The Small Arms Survey, with logistical support from CIOM, conducted focus group discussions in August and September 2010 with members of three affected communities—Ortaderesin (near the August 2001 explosion site of Tokyrau), Arys (the March 2009 incident), and Karaoy (the June 2009 incident) (see Box 4.2). Two focus group discussions with eight participants each were held in each location separately with men and women. Participants were required to have lived in the given location for at least two years prior and since the incident. Discussions focused on respondents' awareness of risks and knowledge of the explosion; impacts such as casualties, property damage, and longer-term consequences on local development and livelihoods; the level of government response; and issues of compensation and preventive measures (see Annexe 4.2).

Risk awareness

In all three locations, the population was aware of the existence of the ammunition depots before the incidents, mainly due to their proximity and because some local residents were employed there as contract workers. In Ortaderesin and Arys, rumours circulated that the depots housed enormous stocks of ammunition and that, if an explosion occurred, it would affect large areas. In Karaoy, residents got used to military exercises that are organized regularly at the two nearby training ranges; some initially mistook the explosion for yet another exercise. In Arys, the explosion occurred on the eve of the Nauryz holiday, and some people first assumed that the military unit was preparing fireworks for celebrations.

Local residents' awareness of the causes of the incidents varies. In Ortaderesin, respondents were not aware of the real cause but put forward several assumptions that surfaced in the press. In fact, the government commission was at the time unable to come to a definitive conclusion as to what had caused the incident (*Kazakhstan Today*, 2002). Respondents in Arys were well aware of the cause of the explosion, partly from press reports, partly from the stories told by fellow residents who had worked at KazArsenal. The investigation, like the press reports that preceded it, concluded that negligence,

---

Box 4.2 Case study background: the Tokyrau, Arys, and Karaoy ammunition depot explosions

The Tokyrau-10 ammunition depot at the Ministry of Defence's military unit 89533, about 50 km east of Balkhash, Karagandy Oblast, exploded on 8 August 2001 (*Kazakhstan Today*, 2001a). According to Sat Tokpakbayev, then minister of defence, the arsenal included 'several tens of thousands of tons' of artillery ammunition, small arms ammunition, and aircraft bombs. The entire stock, mainly left from the Soviet Union's Afghan campaign and partly transferred from Armenia after the 1988 earthquake, was reportedly slated for destruction (*Kazakhstan Today*, 2001b).

The second explosion took place on 20 March 2009 at the ammunition disposal facility of KazArsenal Research and Production Association Ltd., located at military unit 44859 of the Ministry of Defence, 2 km from the town of Arys, Southern Kazakhstan Oblast (about 100 km from Shymkent) (MoE, 2009a). Large ammunition depots were built near Arys before World War II, and ammunition explosions had taken place there before. The press reported that some ammunition and weapons had been transferred to the depots near Arys after the Soviet Army's withdrawal from Afghanistan (*Kazakhstan Today*, 2009b). The Ministry of Defence had subcontracted KazArsenal, a private entity, to dispose of and reprocess out-of-date ammunition for civilian use (Novitskaya, 2009).

The third explosion took place on 8 June 2009 at the ammunition depot of military unit 2466 of the KNB's Border Guard Service, 1 km from the village of Karaoy, Ile District, Almaty Oblast (40 km from Almaty). The unit serves as a field training centre for the KNB Military Institute that trains border guard officers (MoE, 2009b; *Kazakhstan Today*, 2009e; KazTAG, 2009).

violations of safety rules, and a heavy workload at KazArsenal were the main causes of the incident (Novitskaya, 2009). In Karaoy, most respondents said they had little knowledge about the cause, perhaps because of the sensitivity of the topic due to the affiliation of the ammunition depot with the KNB. Investigators concluded that the depot exploded after a contract sergeant accidentally dropped a thunderflash into a box with explosives (KTK, 2009b; Severnyy, 2009); however, some local residents claim that he was used as a scapegoat. In none of the three cases did the authorities explain the causes of the explosions to the local population.

Respondents in one location implied that the explosions might have been related to illegal sales of weapons and ammunition to third parties. Men in Karaoy mentioned a rumour that an inspection of the ammunition depot inventory had been forthcoming and that the explosion had been set intentionally to conceal traces of theft. Participants in other locations were unwilling to discuss this sensitive topic. After the Tokyrau explosion, allegations surfaced in the press that it had been organized to provide a cover-up for theft (Dzhalilov, 2001).

Casualties and impact

Community reaction was the most dramatic in Arys. The incident took place late in the evening and, according to the focus group participants, led to mass fear, shock, and panic rooted in previously existing fears; the situation reportedly resembled a disorganized war-time evacuation. Taking advantage of the situation, taxi drivers and petrol station owners inflated prices, as confirmed by press reports (Novitskaya, 2009). In Ortaderesin, locals were also frightened but did not panic, instead assisting the military personnel and their families and facilitating an organized evacuation. Yet compared to Arys, Ortaderesin is much farther from the explosion site (about 15 km). In Karaoy, some people panicked and left the village, but the majority remained in their houses. Some participants in all three communities initially had the impression that a war had started; residents said the explosions resembled a war-time artillery fire and bombardment.

> Residents said the explosions resembled war-time artillery fire and bombardment.

While not leading to any immediate fatalities among the civilian population of the settlements, the three events differ in terms of the overall damage and casualties they ultimately caused. The Arys explosion resulted in the death of three KazArsenal workers at the site and inflicted injuries on 17 others (two of whom later died in hospital) (*Kazakhstan Today*, 2009c; KTK, 2009a; Novitskaya, 2009). The closeness and intensity of explosions in Arys also resulted in broken windows, knocked-out doors, and cracked roofs and walls. While there was no damage to the property of local residents in Ortaderesin, a teenage scrap metal hunter, a conscript soldier, and five workers were killed and three were wounded by unexploded ordnance in separate incidents between 2003 and 2009.[41] The explosions also destroyed infrastructure, including the military's housing facility, damaged the local railway and power transmission line, and prompted the authorities to limit the water supply for several days.[42] Local residents claimed that, although the clean-up activities were officially terminated, many pieces of ammunition remained scattered in the surrounding area and on the bottom of Lake Balkhash and continued to generate interest from scrap metal hunters. Residents of Karaoy reported no damage to property but also knew that a conscript soldier was killed when a concrete wall of the depot crushed him (MoE, 2009b; Konovalov, 2009; ERA-TV, 2009).

Focus group participants in Ortaderesin and Arys claimed that the explosions have had negative effects on the environment and human health, especially on pregnant women and children, and that they resulted in higher household animal mortality. Some Ortaderesin residents reported that the Tokyrau depot had housed uranium-tipped ammunition and that the incident had led to the radioactive contamination of the surrounding area, but that the authorities had told local residents that there was no radiation in the environment. It appears that the Tokyrau arsenal had indeed contained such ammunition, as reported by Kazakhstan's nuclear scientists, who helped remove highly radioactive ammunition and waste from the incident site and placed them for long-term storage at the former Semipalatinsk nuclear test site (Dmitropavlenko et al., 2010).

Government response

According to respondents, firefighters in all three locations acted quickly to arrive at the site after receiving an emergency alert. During the Tokyrau incident, however, they did not intervene until the explosions subsided because of the risk of injury or death (*Kazakhstan Today,* 2001a; 2001c). Similarly, in Karaoy, emergency responders did not approach the site to contain the blaze because of continuous explosions of ammunition (MoE, 2009b; *Kazakhstan Today,* 2009e). In Arys, firefighters suppressed the fire, risking their lives despite a remaining explosion hazard (*Kazakhstan Today,* 2009d).

The evacuations in the three affected communities were organized in different ways. In Ortaderesin, the local authorities oversaw a relatively smooth evacuation, moving more than 400 civilians and military personnel to Balkhash. Some men remained in the village to take care of their property and cattle. Residents in Arys claim that the authorities did not arrange for any transportation to evacuate the civilian population, while the military unit provided two buses and a car to evacuate its personnel and their families. In Karaoy, residents living in the peripheral streets located close to the incident site were evacuated and accommodated in the local community centre, while others stayed at home or left the village independently.

In all three cases, the authorities closed off the affected areas, putting in place police and military cordons and blocking the roads leading to the towns and the villages. Engineer units were called in to search for unexploded ordnance.

Compensation and prevention

Contrary to their expectations, Ortaderesin residents did not receive any government assistance—unlike the military personnel and their families, who were paid, according to respondents, compensation for their lost property. Residents of Arys claimed that they had been promised they would be compensated for all the damage inflicted by the incident but instead only received insignificant compensation for shattered windows. Focus group participants said that funds that had been allocated for assistance were embezzled by local officials. Respondents in Karaoy did not expect government compensation or assistance since the explosions, despite the damage they had inflicted on military property and housing, did not cause any damage to private property.[43]

The focus groups conducted by the Small Arms Survey also revealed that the authorities in Kazakhstan do not organize emergency response training for the civilian population living near ammunition depots, and that local residents do not know how to behave in emergency situations. Respondents pointed out that there are no designated collection points or evacuation centres such as the Soviet-era bomb shelters, where local residents could gather in the event of an emergency.

Focus group participants in all three locations agreed that, ideally, military facilities, such as ammunition depots and military training ranges, should be located far from human settlements. On the other hand, the majority of participants said they did not strongly object to the location of military facilities near their settlements because they provide much-needed jobs.

CONCLUSION

As of late 2011, the information that the Small Arms Survey was able to gather indicated that security in Kazakhstan had improved since post-Soviet independence. Crime, including homicide, has decreased significantly since the mid-1990s, and criminals appear to use firearms relatively rarely when perpetrating homicides and armed robberies. Civilian respondents put employment, healthcare, and access to water ahead of security when asked about the most serious problems affecting them. Further, the government has a relatively comprehensive set of measures in place to regulate civilian acquisition and possession of small arms.

There are, however, some important caveats to the assumption that Kazakhstan is a secure country in an otherwise unstable region. Threats to Kazakhstan's stability are not limited to events in neighbouring countries; rather, they include a domestic homicide rate that exceeds global and Central Asia averages and a recent increase in the use of firearms in violent crime. Perceptions of insecurity appear to be higher in urban areas, fuelling civilian demand for firearms as a means of self-defence in cities and among young men. Additional negative trends include apparent increases in the illicit manufacture of small arms, and reports of the use of firearms by prisoners and teenagers. The recent surge in terrorist violence on Kazakh territory, combined with prominent cases of ethnic and political violence over the past five years, is especially worrying. While it would be alarmist to speak of an approaching storm, Kazakh skies are not entirely clear.

The six large-scale, unplanned explosions at munitions sites that have occurred in the country since 2001 highlight problems in the management of state stockpiles. The explosions have caused death, injury, and the destruction of private property and public infrastructure. Over the longer term, they have also harmed local environments, livelihoods, and employment. The lack of emergency response training for communities living near depots points to a shortfall of government capacity and will to respond effectively to such accidents. Ensuring the safety and security of state stockpiles, including stores of surplus ammunition, would not only help prevent further accidents, but would also decrease the risk of arms being diverted to unauthorized entities and individuals. While Kazakhstan has taken some unilateral steps in this direction, increased transparency and international cooperation, as is occurring elsewhere, would help the country to benefit from the expanding international knowledge base in this area.

LIST OF ABBREVIATIONS

CIOM	Center for the Study of Public Opinion
CSTO	Collective Security Treaty Organization
IMI	Israel Military Industries
KNB	National Security Committee
KZT	Kazakhstan tenge
MANPADS	Man-portable air defence systems
MIA	Ministry of Internal Affairs
MoD	Ministry of Defence
MRP	*Mesyachniy raschetniy pokazatel* (monthly calculation index)
NISAT	Norwegian Initiative on Small Arms Transfers
OSCE	Organization for Security and Co-operation in Europe
ZKMK	Zapadno-Kazakhstanskaya Mashinostroitelnaya Kompaniya (Western Kazakhstan Machine-Building Company)

ANNEXES

Online annexes at <http://www.smallarmssurvey.org/publications/by-type/yearbook/small-arms-survey-2012.html>

Annexe 4.1. Survey questionnaire

Annexe 4.2. Focus group guide

ENDNOTES

1   Kazakhstan is divided into 14 *oblasts,* or administrative units or regions.
2   The main studies include Ashkenazi (2010) and IA (2004).
3   The Small Arms Survey is no exception, having published studies on Kyrgyzstan and Tajikistan (MacFarlane and Torjesen, 2004; Torjesen, Wille, and MacFarlane, 2005).
4   The Survey also submitted a number of official requests for information to the relevant ministries and discussed its draft report with government representatives during a workshop co-hosted by the Military Strategic Studies Center in Astana in July 2011.
5   See, for instance, Wołowska (2004).
6   AS (2011); Geneva Declaration Secretariat (2011, pp. 51, 119); PGO (2011a); author correspondence with Elisabeth Gilgen, researcher, Small Arms Survey, 6 October 2011.
7   Quotas were based on the sex and age distribution of the relevant region.
8   The 2010 turmoil in Kyrgyzstan is the latest example of such violence. See ICG (2010).
9   See, for example, Brill Olcott (2010); IWPR (2007); von Gumppenberg (2007, pp. 23–25).
10 For additional background on extremist groups in Kazakhstan, see von Gumppenberg (2007, pp. 28–32) and Omelicheva (2011, pp. 82–132).
11 The confidence interval is increased from 2.5 to 3 to take into account the low response rate to this question.
12 In spite of several official requests to the Ministry of Internal Affairs, it was not possible to obtain updated official statistics on registered privately held firearms, which the Ministry considers a confidential matter. Author correspondence with Kazakh Ministry of Internal Affairs, 2 August 2011.
13 According to the Committee for Forestry and Hunting of the Ministry of Agriculture, there were about 85,000 officially registered hunters in Kazakhstan in 2010 (Koemets, Kolokolova, and Kenzhegaliyeva, 2010).
14 See the list of companies at Koramsak (n.d.).
15 The authors extend thanks to Paul Holtom for drawing their attention to these reports.
16 This comparison and the discussion and tables below rest on UN Comtrade data as downloaded by the Norwegian Initiative on Small Arms Transfers (NISAT). The Small Arms Survey sent its analysis of UN Comtrade data to Kazakhstan's Committee for Customs Control on 28 July 2011 with a request for any comments or clarification. The Committee responded that information related to arms transfers was considered a 'state secret' and was therefore confidential (author correspondence with the Committee for Customs Control, 28 August 2011). Note also that Kazakhstan appears to have withdrawn its submissions from the UN Comtrade database in late 2010; this data is still accessible via the NISAT project, which regularly downloads and saves information (see NISAT, n.d.).
17 While this section focuses on civilian-held weapons, data on military weapons is reflected in the tables for informative purposes and to provide a comprehensive overview of available information.
18 On the other hand, Kazakhstan has reported a larger volume of imports of military rifles than have exporters (see Table 1). The fact that some countries do not report exports of military equipment to Comtrade may explain these discrepancies.
19 See, for example, *Kazakhstan Today* (2003; 2010b); Kazinform (2010); and MIA (2010d).
20 Author correspondence with CSTO, 22 February 2011.
21 Author correspondence with CSTO, 11 February 2010.
22 Author correspondence with CSTO, 22 February 2011.
23 Another source reports that in early February 2011 the President of Kyrgyzstan, Roza Otunbayeva, announced that 356 firearms and 63,780 units of ammunition were lost during the June 2010 ethnic violence (P-KR, 2011).
24 Statement by a Kazakh government representative at a workshop held by the Small Arms Survey, Astana, 20 July 2011.
25 On 24 June 2010, for instance, the Zhambyl Oblast police—in cooperation with border guards—detained a Kyrgyz national for attempting to smuggle a bag with a Kalashnikov automatic rifle, a Makarov pistol, a PSN pistol with silencer, and cartridges of 55 different calibres (MIA, 2010e; *Kazakhstan Today,* 2010c).
26 Most of the provisions regulating civilian firearm ownership are laid out in the Criminal Code, in Law No. 339 and Law No. 214-III (RoK, 1997; 1998; 2007a). The government also adopted a number of regulations to implement and enforce this legislation (RoK, 2000a; 2004). The following review is based on these publicly available legal sources.
27 Secondary legislation stipulates that applicants must indicate the relevant reason when applying for a purchasing licence, but that they are not required to provide any evidence of a corresponding need for the weapon (RoK, 2000a, rules 78, 79). Individuals wishing to obtain a licence for the purchase, possession, or carrying of hunting rifles must hold a hunter's certificate (RoK, 1998, art. 15.1; 2000a, r. 78).
28 MRP (monthly calculation index; in Russian, *mesyachniy raschetniy pokazatel*) is a unit used in Kazakhstan to calculate payments such as wages and compensation made under labour legislation as well as taxes, levies, duties, fines, and other fees. As of 1 January 2011, one MRP equalled KZT 1,512 (USD 10).

29 Permits for the possession, storage, and carrying of firearms are valid for five years, after which they can be renewed every five years (RoK, 2000a, rules 87, 90). Licences for purchasing firearms are valid for three months (RoK, 2000a, r. 85).
30 The law defines short-barrelled shotguns as long-bore firearms of barrel length inferior to 500 mm and with a total length inferior to 800 mm, as well as long-bore firearms that can be modified to a length inferior to 800 mm.
31 The law defines hollow-point ammunition as bullets with a displaced centre of gravity.
32 The law also prohibits 'weapons and ammunition that are incompatible with the requirements of forensics', although it does not specify what this means in practice (RoK, 1998, art. 7).
33 Authors' observation, Almaty, September 2010.
34 These rates are based on a selection of countries in each region for which reasonable data was available (Geneva Declaration Secretariat, 2011, p. 100).
35 BNews (2011); Interfax-Kazakhstan (2011); *Kazakhstan Today* (2009a); Kutsay (2010); Shemratov (2010).
36 Interviewed representatives of companies that sell firearms to the civilian market stated that men are their main customers, suggesting that Kazakh men are more attached to firearms than women (Chebotarev, 2010, p. 15).
37 Law prohibits the sharing of such information (RoK, 1999, arts. 11, 14). Anonymous sources indicate that Kazakhstan reports such data as well as arms transfers information within the framework of OSCE information exchange mechanisms. It is not possible to verify this data, however, since national reports to the OSCE are confidential.
38 Author correspondence with an informed source, 3 June 2010.
39 Author correspondence with Aaron Karp, senior consultant, Small Arms Survey, 22 April 2010.
40 Although many types of conventional ammunition are beyond the scope of the Survey's research, this table is included for informative purposes given that it represents unpublished official information of potential interest to other stakeholders.
41 Fomina (2006); ITAR-TASS (2003); *Kazakhstan Today* (2009f); Kazinform (2006).
42 Channel One (2001); Gabchenko (2001); *Kazakhstan Today* (2001a; 2001b; 2001d); ITAR-TASS (2001).
43 Karaoy lies only 1 km from the incident site, suggesting the explosion must have been less severe than in other cases, since the village did not sustain damage.

BIBLIOGRAPHY

Akhmetova, Albina. 2008. 'Squadron, Commence Fire!' *Liter* (Almaty). 16 October.
Antoncheva, Svetlana. 2011. 'Kazakhs Start Criminal Probe into Use of Force during Riots.' Bloomberg. 29 December.
    <http://www.bloomberg.com/news/2011-12-29/kazakhs-start-criminal-probe-into-use-of-force-during-riots-1-.html>
AS (Agency of Statistics of the Republic of Kazakhstan). 2011. Accessed 1 October 2011. <http://www.stat.kz>
Ashkenazi, Michael. 2010. 'Kazakhstan: Where Surplus Arms Are Not a Problem.' In Karp, pp. 133–54.
Barabanov, Mikhail. 2008. 'The Defence Industry of Kazakhstan.' *Eksport vooruzhenyi*. May–June, pp. 28–35.
Bayekenov, Bulat. 2004. 'Not a Peripheral Issue about a Private Detective.' *Kazakhstanskaya Pravda* (Astana). 23 June.
    <http://www.kazpravda.kz/rus/obshtestvo/230604_ne_chastnij_vopros_o_chastnom_detektive.html>
BBC Monitoring/Interfax. 2009. 'Kazakhstan Needs to Scrap about 1.5 m pieces of Ammunition—Defence Ministry.' 9 December.
Benditskiy, Gennadiy. 2011. 'Deadly Force.' *Vremya* (Almaty). 7 April. <http://www.time.kz/index.php?module=news&newsid=20955>
BNews (Kazakhstan). 2011. 'Weapons Were Transferred to the Prisoners in Balkash by Overthrow.' 12 July. <www.bnews.kz/ru/news/post/54410>
Brill Olcott, Martha. 2010. *Kazakhstan: Unfulfilled Promise?* Revised edn. Washington, DC: Carnegie Endowment for International Peace.
Burnashev, Rustam and Irina Chernykh. 2010. *Military Building and Control of Light and Small Arms*. Unpublished background paper. Geneva: Small Arms Survey.
Channel One (Russian Federation). 2001. 'The Fire at the Largest Ammunition Depot in the Republic Continues in Kazakhstan.' 9 August.
    <http://www.1tv.ru/news/techno/113501>
Chebotarev, Andrey. 2010. *Analysis of Small Arms Circulation in Kazakhstan*. Unpublished background paper. Geneva: Small Arms Survey.
Chernyavskaya, Yuliya. 2010. 'Out… and in Again.' *Megapolis* (Almaty). No. 6 (467). 22 February. <http://megapolis.kz/art/Vishel_i_snova_zashyol>
CIOM (Center for the Study of Public Opinion). 2010. *Survey of 1,500 Representative Households in Kazakhstan*. Almaty: CIOM.
Dmitropavlenko, V. N., et al. 2010. 'Utilization of Sources of Radioactive Contamination at Former Military Arsenal in Tokrau Settlement.' *Topical Issues in Radioecology of Kazakhstan*, Iss. 2, pp. 475–90.
Dzhalilov, Adil. 2001. 'Causes of the Fire at the Army Depots Will Be Established in a Year at the Soonest.' *Panorama* (Almaty). No. 34. 7 September.
Dzhani, Feruza. 2007. 'Attacks against Houses of Ethnic Kurds Continue in Kazakhstan.' Ferghana News Agency. 19 November.
    <http://www.fergananews.com/article.php?id=5479>
ERA-TV (Kazakhstan). 2009. 'In the Wake of the Explosion in Karaoy.' 9 June. <http://www.newsfactory.kz/42748.html>
Fomina, Natalya. 2006. 'A Shell Exploded in the Soldier's Hands.' *Novyy Vestnik* (Karagandy). 20 September. <http://www.nv.kz/2006/09/20/4234/>
Foster, Hal. 2010. 'Kazakhstan Strengthens Penalties for Pipeline Oil Rustlers.' Central Asia Newswire. 15 October.
    <http://www.universalnewswires.com/centralasia/viewstory.aspx?id=2015>
—. 2011. 'New Kazakh Law Regulates Private Security Industry and Weaponry.' Central Asia Newswire. 19 January.
    <http://www.universalnewswires.com/centralasia/viewstory.aspx?id=2992>

Gabchenko, Aleksandr. 2001. 'Breath of Death.' *Novosti Nedeli* (Almaty), No. 32. 15 August.
Geneva Declaration Secretariat. 2011. *Global Burden of Armed Violence 2011: Lethal Encounters*. Cambridge: Cambridge University Press.
<http://www.genevadeclaration.org/measurability/global-burden-of-armed-violence/global-burden-of-armed-violence-2011.html>
Holtom, Paul. 2010. *Arms Transfers to Europe and Central Asia*. Background paper. Stockholm: Stockholm International Peace Research Institute. February. <http://www.unidir.org/pdf/activites/pdf7-act508.pdf>
Hug, Adam. 2011. *Kazakhstan at a Crossroads*. London: Foreign Policy Centre. April. <http://fpc.org.uk/fsblob/1334.pdf>
IA (International Alert). 2004. *Small Arms Control in Central Asia*. Eurasia Series No. 4.
<http://undpbangladesh.academia.edu/MPage/Papers/522295/Small_Arms_in_Central_Asia>
ICG (International Crisis Group). 2010. *The Pogroms in Kyrgyzstan*. Asia Report No. 193. 23 August. Bishkek and Brussels: ICG.
<http://www.crisisgroup.org/en/regions/asia/central-asia/kyrgyzstan/193-the-pogroms-in-kyrgyzstan.aspx>
—. 2011. *Central Asia: Decay and Decline*. Asia Report No. 201. Bishkek and Brussels: ICG. 3 February.
<http://www.crisisgroup.org/~/media/Files/asia/central-asia/201%20Central%20Asia%20-%20Decay%20and%20Decline.pdf>
IISS (International Institute of Strategic Studies). 2010. *Military Balance*. London: Routledge.
Interfax-Kazakhstan. 2011. 'Eight People Freed during Special Operation in Balkhash Prison.' 11 July.
<http://www.interfax.kz/?lang=eng&int_id=in_focus&news_id=669>
Isabekov, Rizabek. 2008. 'Where Does Ammunition Come From? From the Forest, Where Else!' *Caravan* (Almaty). 7 March.
<http://www.caravan.kz/article/8926/>
ITAR-TASS. 2001. 'Ammunition Explosions at Military Depots in Karagandy Oblast Ended, while the Fire Continues for the Third Day.' 11 August.
—. 2003. 'One Person Was Killed and Four Others Wounded by the Explosion at One of the Military Bases of Central Kazakhstan.' 5 November.
IWPR (Institute for War & Peace Reporting). 2007. 'Kazakstan: Ethnic Clash a Worrying Sign.' *Reporting Central Asia*, Iss. 517. 24 November.
<http://iwpr.net/report-news/kazakstan-ethnic-clash-worrying-sign>
Jane's World Armies. 2010. 'Kazakhstan.' Berkshire: IHS Global Limited. 5 November.
Karp, Aaron, ed. 2010. *The Politics of Destroying Surplus Small Arms: Inconspicuous Disarmament*. New York and Geneva: Routledge and Small Arms Survey.
Kazakhstan Today. 2001a. 'The Artillery Arsenal of Kazakhstan's Army Is on Fire near Balkhash in Karagandy Oblast.' 9 August.
<http://news.gazeta.kz/art.asp?aid=148944>
—. 2001b. 'The Minister of Defence of the Republic of Kazakhstan: The Fire Will Not End until It Destroys the Entire Stock of Shells Stored at the Military Depots near Balkhash.' 10 August. <http://news.gazeta.kz/art.asp?aid=148968>
—. 2001c. 'The Meeting of Power Agencies Specified Activities Directed at Suppressing the Fire at the Military Depots.' 10 August.
<http://news.gazeta.kz/art.asp?aid=148962>
—. 2001d. 'The Balkhash Directorate for Emergency Situations: The Fire at the Military Depots near Balkhash Have Ended.' 14 August.
<http://news.gazeta.kz/art.asp?aid=149018>
—. 2002. 'The Military Investigative Directorate of the Ministry of Internal Affairs of the Republic of Kazakhstan Continues Its Investigation into the Fire at the Military Depot in Tokyrau.' 18 January. <http://news.gazeta.kz/art.asp?aid=153526>
—. 2003. 'Almaty Police Seized a Gas Pistol Modified to Shoot Live Rounds from a Resident of Eastern Kazakhstan Oblast.' 8 October.
<http://news.gazeta.kz/art.asp?aid=175341>
—. 2009a. 'A Teenager Who Wounded Three Persons in a Street Shooting Has Been Detained in Southern Kazakhstan Oblast.' 24 February.
<http://www.kt.kz/?lang=rus&uin=1133168020&chapter=1153478270>
—. 2009b. 'Two Men Died, Two Went Missing and 16 Others Were Wounded during the Fire at the Military Depots in Arys.' 21 March.
<http://www.kt.kz/?lang=rus&uin=1133168944&chapter=1153480922>
—. 2009c. 'The Territory of the Scientific Production Association KazArsenal Ltd Where the Ammunition Explosions Took Place Is out of the Jurisdiction of the Ministry of Defence of the Republic of Kazakhstan.' 21 March.
<http://www.kt.kz/?lang=rus&uin=1133168944&chapter=1153480933>
—. 2009d. 'The Minister for Emergency Situations of the Republic of Kazakhstan Rewarded Those Who Showed Courage in Suppressing the Fire at the Scientific Production Association KazArsenal Ltd.' 2 April. <http://www.kt.kz/?lang=rus&uin=1133168020&chapter=1153482076>
—. 2009e. 'An Ammunition Depot Located Near the Training Range of Border Guard Troops in Almaty Oblast Is on Fire.' 8 June.
<http://www.kz-today.kz/?lang=rus&uin=1133168944&chapter=1153488540>
—. 2009f. 'Two Military Men Were Wounded in an Explosion in Karagandy Oblast.' 17 June.
<http://www.kt.kz/?lang=rus&uin=1133168944&chapter=1153489475>
—. 2010a. 'MIA of RK: There Are No Cases of Firearm Supplies from Kyrgyzstan to Kazakhstan.' 26 April.
<http://www.kt.kz/?lang=rus&uin=1133168020&chapter=1153515634>
—. 2010b. 'A Top Manager of the Atyrau Munay-Gaz Company Was Murdered in Almaty.' 18 May.
<http://www.kt.kz/?lang=rus&uin=1133168518&chapter=1153517301>
—. 2010c. 'A Resident of Kyrgyzstan in Possession of the Arsenal of Weapons Was Detained in Zhambyl Oblast.' 30 June.
<http://www.kt.kz/?lang=rus&uin=1133168020&chapter=1153520559>
Kazinform. 2006. 'The Commission of the Ministry of Defence of the Republic of Kazakhstan Investigates the Explosion in Balkhash.' 20 February.
<http://www.inform.kz/rus/article/146822>
—. 2009. 'The Head of the Defence Agency of the Republic of Kazakhstan: The Ammunition Disposal Programme Will Be Corrected Taking into Account Safety Improvements.' 26 March <http://www.inform.kz/rus/article/232149>
—. 2010. 'A Woman Attempting to Sneak a Loaded Pistol Was Detained in the Karagandy Airport.' 16 July. <http://www.inform.kz/rus/article/2287006>

KazTAG. 2009. 'No Resident of the Village of Karaoy Was Hurt by the Explosions.' 9 June. <http://www.today.kz/ru/news/kazakhstan/2009-06-09/vzriv5>

—. 2010. 'Kazakhstan Jointly with Israel Intends to Produce a New-generation "WAVE 300–Tolkyn" Small-arms System.' 27 May. <http://www.zonakz.net/articles/4print.php?artid=29559>

Kedrov, Ilya. 2010. 'Israeli–Kazakhstani "Grads" and "Uragans."' *Voyenno-Promyshlennyy Kuryer* (Moscow). No. 23 (339). 16 June. <http://vpk-news.ru/articles/6697>

Kenzhegalieva, Gulmira. 2007. 'From a Ship to a Pistol.' *Caravan* (Almaty). 28 September. <http://www.caravan.kz/article/7767>

Kenzhetaev, Marat. 2002. 'Kazakhstan's Military–Technical Cooperation with Foreign States: Current Status, Structure and Prospects.' *Moscow Defense Brief*, No. 1. <http://mdb.cast.ru/mdb/1-2002/at/kmtcfs/>

Koemets, Yelena, Olga Kolokolova, and Gulmira Kenzhegaliyeva. 2010. 'Barbarian Shooting.' *Caravan* (Almaty). 30 July. <http://www.caravan.kz/article/6079>

Konovalov, Aleksey. 2009. 'It Sounded like a War...' *Vremya* (Almaty). 10 June. <http://www.time.kz/index.php?newsid=10868>

Koramsak. n.d. 'Association Koramsak: Association Members.' Accessed December 2011. <http://www.koramsak.kz/index.php?option=content&task=view&id=17&Itemid=71&ytw=ytw_splitmenu>

KTK (Kazakhstan). 2009a. 'The Number of Casualties of the Tragedy in Arys Has Reached Four.' 20 May. <http://www.ktk.kz/ru/news/video/2009/5/20/4182>

—. 2009b. 'The Military Prosecutor's Office Found a Person Guilty in the Explosion at the Ammunition Depot in Almaty Oblast.' 19 June. <http://www.ktk.kz/ru/news/video/2009/6/19/4544>

Kulikov, Andrey. 2008. 'Arsenal by Transit: A Craftsman from Kazakhstan Was Supplying Criminals with Weapons.' *Rossiyskaya gazeta* (Moscow). 3 April. <http://www.rg.ru/2008/04/03/reg-saratov/oruzhie.html>

Kutsay, Igor. 2010. 'The Mystery of a High-profile Escape.' *Caravan* (Almaty). 12 November. <www.caravan.kz/article/7464>

Kylym Shamy (Centre for Protection of Human Rights). 2011. *Results of Documenting of Data on the Seized/Distributed Firearms, Military Hardware, and Ammunition during Mass Riots in the South of the Kyrgyz Republic.* Bishkek: Kylym Shamy.

Lillis, Joanna. 2011a. 'Kazakhstan: Puzzling Blasts Stir Fears of Islamic Radicalism.' *EurasiaNet's Weekly Digest*. 8 June. <http://www.eurasianet.org/node/63648>

—. 2011b. 'Kazakhstan: Experts Give Astana Mixed Review on OSCE Chairmanship.' EurasiaNet. 13 January. <http://www.eurasianet.org/node/62707>

—. 2011c. 'Kazakhstan: Nazarbayev Landslide Fails to Win over Foreign Observers.' EurasiaNet. 4 April. <http://www.eurasianet.org/node/63221>

—. 2011d. 'Kazakhstan: Astana Confronts Extremist Threat.' *EurasiaNet's Weekly Digest*. 6 September. <http://www.eurasianet.org/node/64133>

MacFarlane, S. Neil and Stina Torjesen. 2004. *Small Arms in Kyrgyzstan: A Small Arms Anomaly in Central Asia?* Occasional Paper No. 12. Geneva: Small Arms Survey. February.

Mednikova, Irina and Alexander Bogatik. 2011. 'Aktobe Situation Raises Concerns of Radicalism, Terror.' Central Asia Online. 11 July. <http://centralasiaonline.com/cocoon/caii/xhtml/en_GB/features/caii/features/main/2011/07/11/feature-02>

MFA (Ministry of Foreign Affairs of the Republic of Kazakhstan). 2010. 'The Statement of the Ministry of Foreign Affairs of the Republic of Kazakhstan Regarding the Situation in the Kyrgyz Republic.' 19 May. <http://portal.mfa.kz/portal/page/portal/mfa/ru/content/press/statement/2010>

—. 2011. 'Kazakhstan and International Organizations: Cooperation between Kazakhstan and the OIC.' 2 June. <http://portal.mfa.kz/portal/page/portal/mfa/en/content/policy/organizations/OIC>

MIA (Ministry of Internal Affairs of the Republic of Kazakhstan). 2004. Order No. 187. 25 March. <http://www.mvd.kz/index.php?p=razdel_more&id5=187&id1=3>

—. 2010a. 'From 12 to 16 April 2010, the Ministry of Internal Affairs Conducted Operative and Preventive Operation "Karu–2010."' <http://www.mvd.kz/eng/index.php?p=razdel_more&id5=6121&id1=5>

—. 2010b. 'Questions and Answers.' <http://www.mvd.kz/index.php?p=conf_group&id_group=73&lang=1&n=2)>

—. 2010c. Official website. <http://www.mvd.kz/>

—. 2010d. 'Speech of K. Zhumanov, Head of the Press Service of the Ministry of Internal Affairs of the Republic of Kazakhstan, at the Weekly Briefing.' 27 January. <http://www.mvd.kz/index.php?p=razdel_more&id5=5150&id1=5>

—. 2010e. 'More than 650 Members of Organized Criminal Groups Were Brought to Justice and Convicted for Committing Various Crimes'. <http://www.mvd.kz/index.php?p=razdel_more&id5=7559&id1=5>

—. 2011. 'Suspects in the Murder of Police Officers Have Been Neutralized in the Aktobe Oblast.' 11 July. <http://www.mvd.kz/index.php?p=razdel&id1=5&year=2011&month=7&day=11>

MoD (Ministry of Defense of the Republic of Kazakhstan). 2009. *Annual Plans for State Procurement of Goods, Works, and Services by the Ministry of Defence.* Astana: Department for Logistical Support of Republic of Kazakhstan's Armed Forces. <http://www.mto.kz/index.php?option=com_content&view=category&layout=blog&id=48&Itemid=74>

MoE (Ministry of Emergency of the Republic of Kazakhstan). 2009a. 'Operational Situation for 21.03.2009.' 21 March. <http://www.emer.kz/conditions/archiv/detail.php?ID=4158>

—. 2009b. 'Explosions of Explosive Devices.' 9 June. <http://www.emer.kz/conditions/archiv/detail.php?ID=4996>

Moldabayev, Dulat. 2010. 'In May KADEX Is Going to Bloom, or the Prospects of Domestic Defence Industry.' *Kazakhstanskaya Pravda* (Astana). 20 January. <http://www.kazpravda.kz/c/1263933918>

NAMSA (North Atlantic Treaty Organization Maintenance and Supply Agency). 2007. *TDY Report: Kazakhstan.* Internal trip report. 8–12 October.

NISAT (Norwegian Initiative on Small Arms Transfers). n.d. Website. <http://www.prio.no/nisat>

Novitskaya, Larisa. 2009. 'Squared Mess.' *OKO* (Astana). 4 September. <http://oko.kz/archive2006/index.php?cont=long&id=1517&year=2009&today=04&month=09>

Nurseitova, Torgyn. 2011. 'Is Training of Suicide Terrorists Possible in Kazakhstan?' KazTAG News Agency. 28 February. <http://www.kaztag.kz/ru/interviews/45449>

Omelicheva, Mariya. 2011. *Counterterrorism Policies in Central Asia*. New York: Routledge.

OSCE (Organization for Security and Co-operation in Europe). 2005. *Report on the OSCE Initial Assessment Visit on SALW and Conventional Ammunition to Kazakhstan, 6–10 June 2005*. Document No. FSC.GAL/68/05. 30 June.

—. 2010. 'Astana Declaration Adopted at OSCE Summit Charts Way Forward.' Press release. 2 December. <http://summit2010.osce.org/en/press_release/node/491>

—. 2011. 'Reforms Necessary for Holding Democratic Elections in Kazakhstan Have Yet to Materialize, Observers Say.' Press release. 4 April. <http://www.osce.org/odihr/76349>

Parpura, Oksana. 2010. 'We Support the Domestic Producer.' Astana: Ministry of Defence of the Republic of Kazakhstan. 26 January. <http://www.mod.gov.kz/mod-ru/index.php?option=com_content&view=article&id=346:2010-02-12-12-11-33&catid=52:2009-06-26-00-57-43&Itemid=41>

PGO (Prosecutor General's Office of the Republic of Kazakhstan). 2011a. Database of the Committee for Legal Statistics and Special Records. <http://www.pravstat.kz>

—. 2011b. 'The Review of the Situation in the Fight against Crime in the Country during 6 Months of 2011.' 11 July. <http://www.pravstat.prokuror.kz/rus/bm/analitik_inf/analit_inf>

P-KR (President of the Kyrgyz Republic). 2011. 'President Roza Otunbayeva Gave Specific Orders Aimed to Ensure Public Security and Strengthen State Borders.' 1 February. <http://www.president.kg/ru/posts/4d6cf5527d5d2e720d0007f0>

PoA-ISS (Programme of Action Implementation Support System). 2010. 'Country Profiles: Kazakhstan.' <http://www.poa-iss.org/CountryProfiles/CountryProfileInfo.aspx?CoI=100&pos=1000>

Radio Azattyk (Kazakh Service of RFE/RL). 2011. 'The Majority of Criminals in Kazakhstan Are Unemployed.' 29 September. <http://rus.azattyq.org/content/criminal_unemployment_kazakhstan/24343521.html>

RFE/RL (Radio Free Europe/Radio Liberty). 2005. 'Kazakh Opposition Figure's Death Ruled Suicide.' 29 November. <http://www.rferl.org/content/article/1063345.html>

—. 2006a. 'Kazakhstan: Authorities Say Suspects Confess to Killing Sarsenbaev.' 20 February. <http://www.rferl.org/content/article/1065973.html>

—. 2006b. 'Kazakhstan: Authorities Insist Personal Enmity Behind Sarsenbaev's Murder.' 27 February. <http://www.rferl.org/content/article/1066199.html>

—. 2007. 'Kazakhstan: Deadly Melee Leaves Unanswered Questions.' 2 April. <http://www.rferl.org/content/article/1075642.html>

—. 2008. 'Kazakhstan: Two Years Later, Opposition Leader's Murder Still Casts Long Shadow.' 13 February. <http://www.rferl.org/content/article/1079473.html>

—. 2011. 'Voice of Kazakh Protest Gaining Strength.' 15 June. <http://www.rferl.org/content/voice_of_kazakh_protest_gaining_strength/24236057.html>

RoK (Republic of Kazakhstan). 1997. Criminal Code. <http://www.zakon.kz/static/ugolovnyy_kodeks.html>

—. 1998. Law No. 339 on State Control over Circulation of Certain Type of Weapons. Assented to 30 December.

—. 1999. Law No. 349-1 on State Secrets. Assented to 15 March.

—. 2000a. Rules on Circulation of Weapons and Their Ammunition. Adopted by Decree of the Government of the Republic of Kazakhstan No. 1176. Assented to 3 August.

—. 2000b. Law No. 85 on Security Activity (as Amended). Assented to 19 October.

—. 2001. Administrative Offences Code (as Amended). Assented to 30 January. <http://www.egov.kz/wps/portal/Content?contentPath=/library2/1_kazakhstan/kr/zakony%20rk/article/kodexy%20rk>

—. 2003. Decree of the Government of the Republic of Kazakhstan No.163 on the Approval of the Standards and List of Weapons and Special Tools that the Personnel of the Financial Police Is Entitled to Use. Assented to 17 February. <http://kazakhstan.news-city.info/docs/sistemsx/dok_ieqkjb.htm>

—. 2004. Decree of the Government of the Republic of Kazakhstan No. 635 on Certain Issues Relating to the Licensing for Development, Production, Repair, Sales, Acquisition, and Exhibition of Combat Small Arms and Their Ammunition. Assented to 8 June.

—. 2005. *Report on the Implementation of the UN Programme of Action*. <http://www.poa-iss.org/CASACountryProfile/PoANationalReports/2005@100@KazakhstanEnglish.pdf>

—. 2006. *Report on the Implementation of the UN Programme of Action*. <http://www.poa-iss.org/CASACountryProfile/PoANationalReports/2006@100@kazakhstan%20%28E%29.pdf>

—. 2007a. Law No. 214-III on Licensing. Assented to 11 January. <http://www.zakon.kz/141150-zakon-respubliki-kazakhstan-ot-11.html>

—. 2007b. Decree of the Government of the Republic of Kazakhstan No. 1299 on Regulations Governing Voluntary Compensated Surrender of Illegally Held Firearms, Ammunition, and Explosives by Citizens. Assented to 26 December.

—. 2007c. Military Doctrine of the Republic of Kazakhstan Adopted by Edict of the President of the Republic of Kazakhstan No. 299. 21 March.

—. 2008. *National Report on the Implementation of the UN Small Arms Programme of Action*. <http://www.poa-iss.org/CountryProfiles/CountryProfileInfo.aspx?CoI=100&pos=1000>

—. 2009. 'Industrial and Technological Development of Kazakhstan for Our Future.' Speech by President Nursultan Nazarbayev. 15 May. <http://www.akorda.kz/ru/speeches/summit_conference_sittings_meetings/v_stwplenie_prezidenta_rk_predsedatelya_ndp_hyp_otan>

—. 2010a. *National Report on the Implementation of the UN Small Arms Programme of Action*. <http://www.poa-iss.org/CASACountryProfile/PoANationalReports/2010@100/PoA-Kazakhstan-2010-E.pdf>

—. 2010b. Law No. 372-IV on Amendments and Additions to Some Legislative Acts of the Republic of Kazakhstan on the Issue of Improving the Tasks of Internal Affairs Agencies in the Area of Ensuring Public Security. Assented to 29 December.

—. 2010c. Decree of the Government of the Republic of Kazakhstan No. 1014 on Approving Categories of Officials of Customs Bodies Authorized to Carry, Store, and Use a Firearm, Lists of Special Tools, and Types of Firearms and Ammunition Used by Officials of Customs Bodies. Assented to 2 October.

—. 2011a. *Letter from the Permanent Mission of Kazakhstan in Geneva Addressed to the Small Arms Survey and Transmitting Information Requested from the Ministry of Defence of Kazakhstan*. 10 January.

—. 2011b. 'Central Executive and Other State Agencies.' <http://ru.government.kz/structure/org>
—. 2011c. 'Kazakhstan's Presidency of the Shanghai Cooperation Organization 2010–2011.' <http://www.sco2011.kz/en/2011/>
Saydullin, Rinat. 2007a. 'Kazakhstan: Who Won the 'Third Chechen War'?' Ferghana News Agency. 19 November.
    <http://www.fergananews.com/article.php?id=5476>
—. 2007b. 'Southern Kazakhstan: A Criminal Conflict Turned into Pogroms and Arsons in the Villages Where Ethnic Kurds Live.' Ferghana News Agency.
    5 November. <http://www.fergananews.com/article.php?id=5452>
Severnyy, Vladimir. 2008. 'Transformatory armii.' *Megapolis* (Almaty), No. 28 (393). 21 July. <http://megapolis.kz/ru/art/Transformatori_armii>
—. 2009. 'Combat Alert-2.' *Megapolis* (Almaty). 22 June. <http://megapolis.kz/art/Boevaya_trevoga2>
Shemratov, Danil. 2010. 'Shooting in Shymkent.' *Caravan* (Almaty). 9 April. <http://www.caravan.kz/article/4560>
Small Arms Survey. 2007. 'Annexe 4: The Largest Civilian Firearms Arsenals for 178 Countries.'
    <http://www.smallarmssurvey.org/fileadmin/docs/A-Yearbook/2007/en/Small-Arms-Survey-2007-Chapter-02-annexe-4-EN.pdf>
—. 2011. 'Unplanned Explosions at Munitions Sites.' 2 November. Geneva: Small Arms Survey.
    <http://www.smallarmssurvey.org/weapons-and-markets/stockpiles/unplanned-explosions-at-munitions-sites.html>
Tashimov, Tulkin. 2010. 'Silnyye stanut silneye.' *Ekspert Kazakhstan* (Almaty). No. 48 (279). 6 December.
    <http://expert.ru/kazakhstan/2010/48/silnyie-stanut-silnee/>
Torjesen, Stina, Christina Wille, and S. Neil MacFarlane. 2005. *Tajikistan's Road to Stability: Reduction in Small Arms Proliferation and Remaining
    Challenges*. Occasional Paper No. 17. Geneva: Small Arms Survey. November.
UN (United Nations in Kazakhstan). 2010. *Millennium Development Goals in Kazakhstan*. <http://www.un.kz/userfiles/1006_oon_mdgr2010_eng.zip>
UN Comtrade (United Nations Commodity Trade Statistics Database). n.d. 'UN Comtrade.' <http://comtrade.un.org/db/default.aspx>
UNDP (United Nations Development Programme). 2010. *Human Development Index and Its Components*.
    <http://hdr.undp.org/en/media/HDR_2010_EN_Tables_rev.xls>
UNICEF. 2011. 'TransMONEE 2011 Database.' Geneva: UNICEF Regional Office for CEECIS. May.
    <http://www.transmonee.org/downloads/EN/2011/tables_TransMonee_2011.xls>
UNODA (United Nations Office for Disarmament Affairs). n.d. Register of Conventional Arms. Tabulated data provided by NISAT.
UNODC (United Nations Office on Drugs and Crime). 2008. *Illicit Drug Trends in Central Asia*. Tashkent: UNODC Regional Office for Central Asia. April.
    <http://www.unodc.org/documents/regional/central-asia/Illicit%20Drug%20Trends_Central%20Asia-final.pdf>
—. n.d. 'Total Police Personnel at the National Level.'
    <http://www.unodc.org/documents/data-and-analysis/Crime-statistics/Criminal_Justice_Resources.xls>
USDoS (United States Department of State). 2010. *International Religious Freedom Report 2010: Kazakhstan*. 17 November.
    <http://www.state.gov/g/drl/rls/irf/2010/148793.htm>
van Dijk, Jan, John van Kesteren, and Paul Smit. 2007. *Criminal Victimisation in International Perspective*. The Hague: Research and Documen-
    tation Centre.
Vasilyeva, Kira. 2010. 'Great Rifle Road.' *New Times*. No. 6, 22 February. <http://newtimes.ru/articles/detail/15951/>
Vasilyeva, Natalya. 2002. 'Sluzhba, kotoraya nas okhranyayet.' *Kazakhstanskaya Pravda* (Astana). 19 October.
    <http://www.kazpravda.kz/rus/obshtestvo/sluzhba_kotoraja_nas_ohranjaet.html>
Vesti.kz. 2010. 'Kazakhstan's Ministry of Internal Affairs Proposed to Change the Law on Security Activity.' 27 May. <http://vesti.kz/society/51999/>
von Gumppenberg, Marie-Carin. 2007. *Kazakhstan—Challenges to the Booming Petro-Economy: FAST Country Risk Profile Kazakhstan*. Working
    Paper 2. Bern: SwissPeace.
    <http://www.isn.ethz.ch/isn/Digital-Library/Publications/Detail/?ots591=0c54e3b3-1e9c-be1e-2c24-a6a8c7060233&lng=en&id=39299>
Vybornova, Galina. 2011. 'Ordinary Jihad.' *Vremya* (Almaty). 24 March. <http://www.time.kz/index.php?module=news&newsid=20724>
Wołowska, Anna. 2004. *Kazakhstan: The Regional Success Story*. Warsaw: Centre for Eastern Studies.
    <http://www.osw.waw.pl/sites/default/files/PRACE_15.pdf>
World Bank. 2011. 'Data and Statistics for Kazakhstan.' Accessed September 2011. <http://www.worldbank.org.kz/WBSITE/EXTERNAL/COUNTRIES/
    ECAEXT/KAZAKHSTANEXTN/0,,contentMDK:20212143~menuPK:361895~pagePK:1497618~piPK:217854~theSitePK:361869,00.html>

ACKNOWLEDGEMENTS

Principal authors

Nicolas Florquin, Dauren Aben, and Takhmina Karimova

Contributors

Center for the Study of Public Opinion (CIOM), Rustam Burnashev, Andrey Chebotarev, Irina Chernykh, Paul Holtom, Ryan Murray, and Nic Marsh

A Somaliland Police Force Special Protection Unit officer provides personal security near Dacarbudhuq, April 2010. © Dominik Balthasar

Between State and Non-state
SOMALILAND'S EMERGING SECURITY ORDER

INTRODUCTION

Just a few years after declaring independence in 1991, Somaliland experienced large-scale armed conflict. Yet, in contrast to south-central Somalia, security in Somaliland was relatively stable in 2011. How has the de facto state managed to achieve a comparatively high level of security despite its recent history of violent conflict and the widespread civilian possession of military firearms?

Some analysts claim that the home-grown disarmament, demobilization, and reintegration (DDR) processes of the early to mid-1990s facilitated improvements in Somaliland's security (Bulhan, 2004, p. 5; Bryden and Brickhill, 2010). Others suggest that the application of traditional systems of governance is the key factor in the post-war decrease in violence (Bradbury, 1997, 2008; Jhazbhay, 2008; Walls, 2009).

This chapter points to other factors and processes that appear to have contributed to improvements in Somaliland's security, distinguishing among different types of violence in the territory and focusing on the changing roles of armed actors in contesting or supporting state authority. In exploring such conflict-related factors, from the national to the local levels, the authors draw on a review of secondary sources, recent survey results, and their own field research in Somaliland. Key findings include the following:

- The overall security situation in Somaliland has improved despite the widespread presence of firearms, including military firearms, in private hands.
- Since the mid-1990s, the resolution of major armed conflicts and the corresponding enhancement of state authority have helped to contain large-scale armed violence in central and western Somaliland and facilitate the establishment of a police force within the territory.
- At the local level, neighbourhood watch groups, working with and under the authority of Somaliland police, are improving security in locations such as Hargeisa and Burao.
- Communal tensions in the form of clan-based violence remain a serious threat to safety and security in Somaliland. Their resolution continues to depend on the integration of all relevant clan groups into the state.

The chapter's first section sketches out Somaliland's recent history, including armed conflicts, and provides a snapshot of security and firearm availability in the territory two decades after its declaration of independence (see Maps 5.1 and 5.2). The following sections examine political, criminal, and communal violence in Somaliland, including factors and processes that appear associated with reductions in violence over the past two decades. The analysis focuses in particular on the historically important city and district of Burao, a location with a turbulent past at the geographical heart of Somaliland. A conclusion recaps the chapter's key arguments.

SETTING THE STAGE: SMALL ARMS AND (IN)SECURITY IN SOMALILAND

In contrast to conflict-ridden south-central Somalia, the northern breakaway Republic of Somaliland is often considered 'an island of relative peace' (Mengisteab, 2009, p. 189) with 'one of the most stable polities in the Horn' (Bradbury, 2008, p. 1).[1] Nevertheless, Somaliland's history has been far from peaceful. The Somaliland state has its roots in a decade-long civil war between the Somali National Movement (SNM) and the regime of Gen. Mohamed Siad Barre. And even in the years following its declaration of independence, Somaliland suffered several bouts of violent conflict.

Following the defeat of Somalia's armed forces in the 1977–78 Ogaden war with Ethiopia, Somalia's military regime became increasingly repressive, leading to the formation of several resistance groups. The SNM, which was among the first armed opposition movements of the early 1980s (see Box 5.1), grew to become a principal challenger to Barre's rule. Towards the end of the decade the guerrilla organization became a mass movement engaged in a fierce civil war with the Somali National Army (SNA). The civil war not only contributed to the removal of the military dictatorship, but also informed the establishment of the Somaliland state, its institutions and identity.[2]

An alliance of clan-based militias, which included the SNM, Somali Salvation Democratic Front, and United Somali Congress, overthrew Barre in January 1991. Four months later, north-western Somalia unilaterally declared independence as the sovereign Republic of Somaliland. In subsequent years, the territory saw numerous rounds of violent conflict between competing political factions largely mobilized along (sub-)clan lines. Yet since the late 1990s Somaliland has followed a trajectory that has diverged sharply from that observed in other parts of the former Somali Republic. Not only has it progressively managed to put major violent conflict behind it, but Somaliland has also built a central administration, including a nascent security apparatus.

Map 5.1 **Somalia**

Awash in arms

With the declaration of independence, the SNM transferred its powers to a two-year transitional government under the leadership of Abdirahman Ahmed Ali 'Tuur' ('hunchback'). Initially, the government's authority hardly reached beyond the capital of Hargeisa (Gilkes, 1993; Spears, 2010, p. 155); as a result, kinship-based militias and *deydey* groups[3] mushroomed. Although they provided security to their own communities, they frequently demanded safe passage fees at provisional roadblocks, contributing to

Map 5.2 **Somaliland**

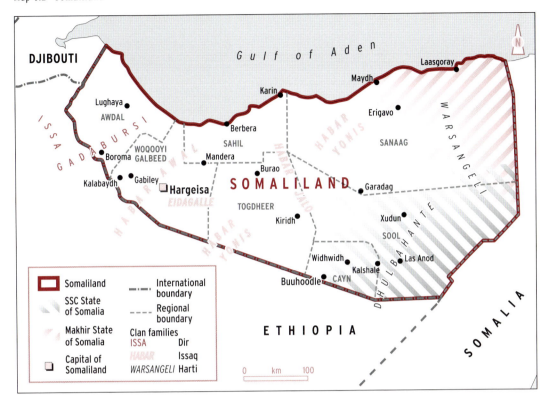

insecurity and political fragmentation in the territory (Bradbury, 2008, p. 88).[4] In order to weaken armed rivals and establish government control, Tuur and his successor conducted several DDR programmes.[5] Yet whereas President Tuur failed to demobilize the militias in the early 1990s, his successor Mohamed Haji Ibrahim Egal was more successful. Shortly after his inauguration in 1993, Egal persuaded most clan militias to surrender their heavy weapons and join the newly formed army (Bradbury, 2008, p. 113).[6] While demobilization was extensive in many areas, the first component of the DDR initiative—disarmament—was only partially implemented. Thus, although DDR successfully took many gunmen off the streets, firearms remained in widespread circulation (Brickhill, 1994; Forberg and Terlinden, 1999).

A recent survey on small arms availability in Somaliland indicates that some 74 per cent of households own small arms, about 73 per cent of which are Kalashnikov-pattern assault rifles (Hughes and Lynge, 2010, p. VIII).[7] Thus, although the possession of heavier weapons is now largely the prerogative of state security forces,[8] assault rifles remain at the disposal of at least every second household in Somaliland. While the public display of guns is not tolerated and is rarely seen in urban areas (Brickhill, 1994, p. 5), and although a gun registration system started in 2006 (Hughes and Lynge, 2010, pp. 43–44), it appears that Somaliland society continues to be heavily armed.

The general picture of weapons holdings in Somaliland is exemplified locally by the case of Burao district, home to Somaliland's second-largest city of the same name. Here, too, it appears that about 80 per cent of all households reportedly own a gun[9] and more than two-thirds of the small arms in private ownership are assault rifles (OCVP, forthcoming a, p. 34). In the city's arms market in June 2011, prices for AK-47s and AKMs ranged between USD 500 and USD 900, while a single 7.62 mm round cost between USD 0.50 and USD 1.00.[10] Yet market prices reflect frequent

Box 5.1 The Somali National Movement

One year after its founding by Somalilanders living in London in 1981, the SNM transferred its headquarters to the Ethiopian border region next to north-western Somalia, from where it launched guerrilla attacks on government installations within Somalia. While the rebel movement initially had very limited military capacity, the SNM turned into a popular mass movement in the late 1980s. Crucial to this transformation was the 1988 decision of dictator Barre to bomb the northern cities of Hargeisa and Burao to rubble in retaliation against members of the Isaaq clan, who comprised the majority of the populations in those cities, because of their alleged support for the SNM. As a consequence, SNM manpower increased massively, transforming its approximately 1,200-strong force into a movement of as many as 50,000 fighters (Flint, 1994, p. 36).[11]

When, in 1988, Barre and Ethiopian leader Mengistu Haile Mariam signed a joint communiqué in which they agreed to cease supporting insurgents operating in each other's territories, the SNM lost its Ethiopian sanctuary. Yet within a month, the SNM managed to establish a presence in rural north-western Somalia. By the time of Barre's overthrow in January 1991, it controlled large swaths of the territory of the former British Somaliland protectorate; it emerged from the civil war as the dominant military force in that part of the crumbling state (Bradbury, 2008, p. 79; Walls, 2009, p. 378).[12]

Despite certain similarities with other Somali rebel movements, the SNM differed significantly from them in numerous respects. First, it was largely self-financed. Rather than receiving massive financial backing from other states or other third parties, the SNM financed most of its activities through donations from the Isaaq community in and beyond Somaliland.[13] Second, the movement's leadership was institutionalized, rather than personalized. Whereas other rebel movements, such as the United Somali Congress or Somali Salvation Democratic Front, were strongly associated with specific individuals– Gen. Mohamed Ali Farah Aideed and Gen. Abdullahi Yusuf Ahmed, respectively–the SNM changed its leadership six times during the course of its decade-long existence (Adam, 1995, p. 76; Bryden, 1999, p. 9). A third difference was that by the time the other Somali rebel movements established themselves in southern Somalia as political organizations in the early 1990s, the SNM was already in the process of handing control over to a transitional civil administration.

fluctuations in the supply of weapons and the demand for them (see Box 5.2). Recent investigations conducted by the Hargeisa-based Observatory of Conflict and Violence Prevention (OCVP) suggest that the availability of firearms within the district of Burao increased between 2009 and 2010 (OCVP, forthcoming a, p. 33).

Evidence shows that the majority of Somaliland's population perceives small arms as a threat to security (Bulhan, 2004, p. 17; Hughes and Lynge, 2010, p. IX). For this reason, the Danish Demining Group (DDG) argues that the private ownership of small arms and light weapons 'will remain one of the big threats to community safety in Somaliland' and that it must be addressed by awareness raising and enhanced gun registration (Hughes and Lynge, 2010, p. 67). Civilian attitudes to guns are complex, however, as suggested by the DDG finding that more than 50

A fighter jet that was used to bomb Hargeisa during the civil war stands as a memorial in the centre of the city. © Andrew McConnell/Panos Pictures

per cent of gun owners report that they hold guns to protect themselves (Hughes and Lynge, 2010, p. 56). In Burao, 38 per cent of respondents claimed to hold a gun for protection (OCVP, forthcoming a, p. 34).

Armed violence and (in)security in Somaliland

Drawing a comprehensive picture of the level and trends of armed violence in Somaliland—classically measured by violent deaths (Geneva Declaration Secretariat, 2011, p. 4)—is virtually impossible due to a lack of data. The first problem is that data for Somaliland is combined with the rest of Somalia in almost all major systematic accounts of armed violence. This is true for direct conflict deaths[14] as well as for homicides.[15] Secondly, even the data available on Somalia

Box 5.2 Small arms in Somaliland[16]

The vast majority of weapons circulating in Somaliland are decades old Kalashnikov-pattern assault rifles, as the import of weapons into the territory is sporadic and modest in comparison to arms deliveries to Puntland and southern Somalia. Small fishing vessels, or *dhows*, intermittently transport arms from Yemen to small ports between Berbera and Bosasso on Somaliland's 800-km northern coastline on the Gulf of Aden. Once weapons are unloaded, arms brokers can easily exploit the mostly ungoverned border between Somaliland and Puntland as a corridor to transport weapons to south-central Somalia, where demand is routinely high due to ongoing conflict.

The regional arms trade does not flow exclusively in one direction, however. At Somaliland's porous and unregulated border with Ethiopia, small numbers of weapons flow in and out. Moreover, arms dealers in Somaliland occasionally purchase arms and ammunition from brokers in Mogadishu, where prices are comparatively low. In February 2011, for instance, at the time that fighting erupted between the Sool, Sanaag, and Cayn (SSC) militia and Somaliland forces, ammunition prices in Mogadishu were exceptionally low due to high rates of diversion from the Transitional Federal Government and affiliated militias to local markets. At USD 0.29 per round by February 2011, 7.62 x 39 mm ammunition was selling for far less in Mogadishu's Bakaara market than in Hargeisa (UNSC, 2011, pp. 42–43).[17] Individuals close to the arms trade in Mogadishu reported that dealers in Somaliland purchased large quantities of ammunition from Mogadishu during that period to meet the spike in demand caused by the fighting.[18]

Somaliland has seen a particular increase in the import of pistols, which have become the weapon of choice for petty criminals despite the fact that, along with their ammunition, they are generally more expensive than military firearms. In late 2010, weathered models of a Czech VZOR 70 and a Chinese Type 54 pistol (copy of the Soviet Tokarev TT-33) were selling in Hargeisa for USD 800 and USD 900, respectively, with their requisite ammunition going for USD 5 per round, more than five times the price of the readily available 7.62 x 39 mm ammunition for Kalashnikov-pattern weapons. A Soviet-produced AKM assault rifle manufactured in 1972 was selling for only USD 550. An arms dealer in Hargeisa explained that the higher price for handguns and their ammunition reflected their scarcity.[19] The Somaliland authorities have intercepted several deliveries of pistols in the past few years. The most recent seizure took place in March 2011, when two Somalilanders were arrested on suspicion of smuggling 37 new Czech pistols from Taiz, Yemen, to Lughaya, a small port west of Berbera.[20]

Demand for weapons in Somaliland changes frequently, primarily in response to fluctuations in criminality—notably armed robbery—and inter-clan disputes in the Sool region. Although state-building and reconciliation efforts have stemmed inter-clan disputes for the most part, clans cling strongly to their weapons out of fear that they may become targets should current inter-clan harmony break down. The unavailability of certain types of ammunition has greatly limited the kinds of weapons that are sold in Somaliland. For instance, since 5.56 x 45 mm ammunition does not circulate widely in East Africa, M-16 rifles, which were once common in Somalia, are rare and, if for sale at all, sell for as little as USD 200.[21]

Somaliland's firearm legislation was amended in 2010 with the adoption of Law No. 39/2010, which requires civilians to obtain a permit to purchase weapons and to register their firearms.[22] With the support of the UN Development Programme (UNDP), the Somaliland Ministry of Interior conducted a large-scale firearm registration campaign across the region between 2006 and 2008, during which some 10,000 weapons were registered. The ministry suspended the exercise prematurely, because of budgetary restraints. Despite these initiatives, the population—particularly in rural areas—does not abide by formal firearm laws, but rather follows traditional customs, which are extremely tolerant of firearm purchase and possession. Moreover, the government does not have the capacity to publicize the new law or provide accessible registration centres to large portions of the population.[23]

A 1992 UN arms embargo prevents the government of Somaliland from legally procuring weapons from beyond its borders. As a result, the government often requires soldiers to bring their own firearms when entering military service. Consequently, the majority of weapons held in state forces closely resemble those in civilian possession. The Somaliland Armed Forces also hold aging high-calibre weapons and tanks left over from the civil war. Unsubstantiated allegations hold that the government of Ethiopia provides weapons to Somaliland on an ad hoc basis as part of a security arrangement that allows Ethiopia unfettered access to Berbera port.[24] Moreover, as Somaliland's relationship with Puntland has grown increasingly antagonistic, both sides have found ways to augment their military arsenals in contravention of UN sanctions.

The emergence of the SSC represents another factor driving demand for weapons in the eastern Sool region of Somaliland. An SSC propaganda video posted on YouTube in July 2011 showed a handful of anti-aircraft guns mounted on 'technicals' (Toyota pick-up trucks with weapon-mounting capabilities), but this does not appear representative of their actual holdings. During the height of fighting in Kalshale in late 2010, for example, the SSC primarily provided fuel and ammunition from a rear base to the feuding clans, as reported by the UN Somalia and Eritrea Monitoring Group (UNSC, 2011, p. 133).

Author: Jonah Leff

Pistols and assault rifles for sale at the central weapons market in Burao, June 2011.
© Dominik Balthasar

as a whole, especially for homicides, is sketchy, incomplete, often based on estimates, and of questionable reliability.[25] That leaves accounts from field researchers, press articles, and occasional survey results, which, even if imperfect, provide at least a glimpse at patterns of armed violence in Somaliland.

The long-term decrease in conflict-related violence

While disaggregated data regarding conflict deaths is not available for Somaliland, some rough annual estimates can be obtained for Somalia as a whole. In combination with qualitative accounts of individual outbreaks of large-scale violence in Somaliland, this data helps to draw a rough picture of the armed confrontations that occurred in Somaliland and consequent casualties over the course of the past two decades.

The conflict data provided by the Uppsala Conflict Data Program and the Stockholm International Peace Research Institute suggests that the high level of conflict deaths reported for Somalia in both the 1990s and the first decade of the 21st century is mainly associated with confrontations in south-central Somalia; it does not indicate any major incidents of armed violence in Somaliland since 1991.[26] The data thus misses the large-scale violence in the territory in 1992 and 1994–95, for which casualties of 1,000 and 2,000–4,000 have been reported, respectively (Bradbury, 2008, pp. 87, 119; IRBC, 1995). Assuming a rather low total figure of 3,000 conflict deaths between 1992 and 1995, the annual average conflict death rate for Somaliland in this period would be 25 per 100,000.[27]

Press articles and research papers also report particular episodes of armed conflict for the years 2003–11. While the terrorist attacks conducted by Al-Shabaab, a radical Islamist militia based in south-central Somalia, reportedly killed 28 people in Hargeisa in 2008 (Somalia Report, 2011), a series of armed confrontations in Kalshale during 2010 and 2011 claimed the lives of between 50 and 100 persons (OCVP, forthcoming e, 2011; UNSC, 2011, p. 130). In Sool region, further clashes between Somaliland and Puntland forces, as well as allied (sub-)clan factions, are not likely to have caused more than 100 deaths since 2007. Even if one assumes a rather high figure of 100 deaths *per year* between 2003 and 2011,[28] the conflict death rate would be no more than 3.3 per 100,000.[29] This points to a significant drop in conflict-related violence from the mid-1990s to the late 2000s.

Recent security trends

Despite the absence of reliable national data, the World Health Organization estimates that the homicide rate in Somalia was 3.3 per 100,000 in 2004 and 1.5 per 100,000 in 2008. While the 2011 edition of the *Global Burden of Armed Violence* is not able to identify any other international sources for homicide rates for Somalia, it does cast doubt on the accuracy of those estimates (Geneva Declaration Secretariat, 2011, p. 57). Meanwhile, the Somaliland administration's efforts to gather crime data have yielded only tentative figures.[30]

Given that 'hard' data on homicide is lacking, household survey data is an important supplement. Several OCVP surveys conducted in numerous Somali districts between 2009 and 2010 as well as the DDG report cited above provide recent data on security perceptions, firearm availability, and certain patterns of armed violence (OCVP, forthcoming a–d; Hughes and Lynge, 2010). Additional data is provided by a recent public opinion survey conducted by the International Republican Institute in Hargeisa (IRI, 2011). While the data does not allow for an in-depth assess-

A signpost declares a 'gun-free space' on the road approaching Burao, July 2007. © Andrew McConnell/Panos Pictures

Table 5.1 Firearm availability and perceptions of safety in Somaliland and Somalia, 2009-10					
	Burao (Somaliland)	Las Anod (Somaliland)	Galkayo (Puntland)		Mogadishu (South-central Somalia)
Perceived improvements in the security situation	Become safer: men: 100%, women: 99% (n=676)[31]	Become safer: men: 75%, women: 68% (n=788)[32]	Become safer: men: 67%, women: 57% (n=662)[33]		Become safer: men: 17%, women: 26% (n=1,382)[34]
Perceived firearm availability trends	More: 73% Same: 16% Fewer: 11% (n=620)[35]	More: 22% Same: 39% Fewer: 39% (n=760)[36]	North: More: 2% Same: 8% Fewer: 91%	South: More: 73% Same: 14% Fewer: 13% (n=536)[37]	More: 65% Same: 26% Fewer: 8% (n=1,336)[38]

Source: OCVP (forthcoming a-d)

ment of the security situation in Somaliland,[39] it does provide some insight into the region's recent security dynamics. Although limited to a 'snapshot' of the security situation between 2009 and 2010,[40] OCVP data from different districts in Somalia allows for comparisons across the central Somaliland district of Burao and districts in other parts of Somalia, such as Galkayo and Mogadishu, and the town of Las Anod in eastern Somaliland (see Table 5.1).

In Burao, nearly all survey respondents reported an improvement in perceived safety between 2009 and 2010 (OCVP, forthcoming a, p. 16).[41] Similarly, the survey conducted by the International Republican Institute in Hargeisa finds that 99 per cent of respondents judged their neighbourhood either 'somewhat safe' (4 per cent) or 'very safe' (95 per cent) (IRI, 2011, p. 11). Findings for Somaliland as a whole are similarly positive. According to the DDG report, more than 50 per cent of all Somalilanders surveyed between August 2008 and August 2009 perceived an increase in security over the previous 12 months (Hughes and Lynge, 2010, p. 13).[42] Another survey, undertaken in 2004, finds that 77 per cent of respondents across Somaliland reported a recent improvement in security at that time (Bulhan, 2004, p. 15).

It has been suggested, however, that perceptions of enhanced safety in the 2000s might be exaggerated due to violent episodes in Somaliland during the 1990s—a circumstance that particularly applies to Burao (Hughes and Lynge, 2010, p. 66; OCVP, forthcoming a, p. 16). Yet, if the civil wars between 1992 and 1995 left a deeper imprint on the public consciousness than more recent events, this further indicates that armed violence has been much less pronounced in the 2000s than it was in the 1990s.

Additional survey results suggest that Burao residents experience significantly higher levels of security than people in all other Somali districts surveyed (see Table 5.2). The OCVP asked respondents to report any knowledge of 'homicides'—which here include both conflict deaths and murder[43]—committed in the past 12 months and about their perception of how often guns were used in attacks. In terms of both violent killings and the use of firearms in attacks, Burao fares significantly better than Las Anod,[44] Galkayo, or Mogadishu. While the homicide rate is unlikely to be zero, as suggested by the results of the OCVP survey presented in Table 5.2, it does seem likely that Burao experienced the least number of murders among the districts surveyed.[45] Further, respondents reported that firearms were only rarely used—in an estimated three per cent of all assaults in Burao—while they are reportedly used in the majority of all attacks in Galkayo and Mogadishu (see Table 5.2). These findings suggest a lower proportion of lethal outcomes in violent assaults in Burao.[46]

Table 5.2 Public perceptions of homicides and firearm use in attacks in Somalia and Somaliland, 2009–10

	Burao (Somaliland)	Las Anod (Somaliland)	Galkayo (Puntland)	Mogadishu (South-central Somalia)
Perceived homicide rate	0% (n=800)[47]	1% (n=800)[48]	6% (n=701)[49]	4% (n=1,588)[50]
Perceived rate of firearms and explosives used in assaults[51]	3% (n=62)[52]	33% (n=98)[53]	57% (n=72)[54]	83% (n=127)[55]

Source: OCVP (forthcoming a–d)

While data on security perceptions and armed violence, including homicides and conflict deaths, is limited, a preliminary analysis supports two tentative conclusions. First, the available accounts suggest that the number of conflict deaths in Somaliland dropped significantly from the mid-1990s to the 2000s. Second, while the data is not sufficient to conclude that all types of lethal violence have decreased in Somaliland over the past 15 years, security in the district of Burao appears relatively stable compared to other Somali districts. Possible explanations for the drop in conflict deaths and the relative security of Burao are discussed below.

The following sections also attempt to explain Somaliland's distinct security environment by distinguishing among *political, criminal,* and *communal* violence in the territory. Although these three forms of violence are often hard to separate in practice, they are analytically distinct. For example, political conflict, understood as an armed challenge to the state, including civil war, generally has a communal dimension in Somaliland's clan-based society, but most local conflicts between sub-clans or lineages do not lead to political mobilization. Likewise, acts of criminal violence such as robbery or murder *can* trigger communal violence between kin groups, but do not necessarily do so. In addition to looking at these distinctions in greater depth, the following sections highlight the security linkages between the national and local levels, including the impact of political conflict resolution on responses to criminal and communal violence. A central focus of this enquiry is the local security situation in Burao.

POLITICAL VIOLENCE

The gradual decline in large-scale political violence is probably the single most important factor in perceived improvements in Somaliland's security. Shortly after its declaration of independence, Somaliland experienced several episodes of large-scale, politically motivated violence. Not unlike in south-central Somalia, these outbursts of civil strife claimed the lives of thousands of people[56] and generated huge numbers of refugees and internally displaced persons.[57] Some 15 years later, political violence is not only far less intense than in the mid-1990s, but it also poses less of a threat to the Somaliland state. The following sections explore some of the factors that contributed to this decline in political violence.

The post-war years

Numerous factors appear to account for the decline in large-scale political violence in post-war Somaliland. One of the key reasons for the divergent trajectories in political violence of Somaliland and south-central Somalia concerns the way the civil war was concluded in each region. In contrast to south-central Somalia, Somaliland experienced one clear military victor. Following the SNM's triumph over the SNA in north-western Somalia, it held a position of military hegemony within the territory it claimed—though not a monopoly, as other clan militias remained active (Bradbury, 2008, p. 79; Walls, 2009, p. 378). While this hegemony was central in suppressing inter-clan violence, it could not prevent major *intra*-clan conflicts, some of which escalated into civil wars.

Shortly after the two-year transitional government of Abdirahman Ahmed Ali Tuur had been inaugurated, it was challenged by opposing factions from within the SNM. Between 1991 and 1993 a number of armed conflicts broke out and nearly led to the military overthrow of the young administration, taking Somaliland to the brink of all-out civil war (Bradbury, 2008, pp. 87–90, 93). In early 1993, in the town of Boroma, Somaliland politicians and traditional authorities held a reconciliation conference that resulted in an agreement on a National Peace Charter. The charter established principles for the preservation of peace and security in Somaliland, such as the decision to establish integrated national security forces and curtail the military powers of the clan militias.[58] While these decisions were contentious, the charter partly helped Tuur's successor, Mohamed Haji Ibrahim Egal, to enhance the state's monopoly over the use of force.[59] The adoption of the charter is thus a second key reason for the abatement of political violence over the medium term.

A third important reason for the decline in political violence was Egal's ability to consolidate political power and strengthen the institutions of the Somaliland state as a result of the armed conflicts of the mid-1990s.[60] The conflicts left the Garhajis sub-clan,[61] which had challenged Egal's rule, politically isolated, partly because Egal successfully portrayed them as opponents of a sovereign Somaliland state.[62] The conflicts also led to the discrediting of certain SNM militia leaders who had tried to restrict and control Egal's policies from within the new government. As Egal stripped important SNM figures, such as Col. Ibrahim Abdillahi 'Dhegaweyne' or Muse Behi Abdi, of military and political power, he integrated several political opponents, such as former SNM chairman Mohamed Mohamoud 'Silanyo', into the government, thus gaining some measure of control over them and, moreover, capturing their popular support base. Egal's consolidation of political power went hand in hand with the fortification of the Somaliland state, including the strengthening of its monopoly over the legitimate use of force. Once the state administration had consolidated its military and political hegemony within the territory it claimed, large-scale political violence diminished—and when it resurfaced, with the conflict involving the SSC, it did so on a much smaller scale.

> Egal's consolidation of political power went hand in hand with the fortification of the Somaliland state.

The simmering pot

Although violent challenges to the state have declined in number and intensity since President Egal consolidated his rule and strengthened state institutions, they have not ended. One recent challenge was the proclamation of the Makhir State of Somalia in Somaliland's north-eastern Sanaag region by members of the Warsangeli clan, who belong to the Harti clan confederation. This aspiring political entity declared in 2007 that it was not part of the Republic of Somaliland but sought an autonomous status within a federal Somalia. The movement did not establish a significant militia, let alone a standing army, and it collapsed roughly one year after its formation.

A more serious challenge to the Somaliland state occurred with the proclamation of the Sool, Sanaag, and Cayn State of Somalia by members of the Dhulbahante clan, who also belong to the Harti clan confederation, in October 2009.[63] In contrast to the Makhir State of Somalia, the SSC managed to establish a military force that has succeeded in challenging state authority in the southern Sool and Cayn regions. Based on the Dhulbahante's political grievances, and fuelled by a long-standing land dispute in the village of Kalshale, the SSC leadership managed to rally some of their kin to their cause.⁶⁴ Largely financed by the Dhulbahante diaspora, the SSC militia clashed with the Somaliland Armed Forces in several ambushes and skirmishes. But as of mid-2011 the rebellion seemed to be drawing to an end after Puntland withdrew political support and the group suffered an internal split, including the defection of its deputy leader to the Somaliland administration (Somaliland Press, 2011d). In June 2011, peace talks between SSC cadres and Somaliland officials in Widhwidh provided further evidence that the conflict might be reaching a peaceful end, although a renewal of violence could not be ruled out (Somaliland Press, 2011c).

It is beyond the scope of this chapter to examine the causes of the recent conflicts in detail, but it appears they arose largely because some of the conditions that led to the end of major armed conflict in the rest of Somaliland, in the mid-1990s, were not met equally throughout the territory. In particular, national state administration has been largely absent from the easternmost areas; the increasing encroachment of the state challenges the Dhulbahante. A second factor in the recent conflicts is the deep-seated feeling of marginalization that the communities of the Harti clan confederation—the Warsangali and Dhulbahante—have felt from the Somaliland state. Their grievances are nurtured by the almost complete absence of basic services in the regions they inhabit and a lack of Harti representation in the government.

Soldiers of the Somaliland Armed Forces patrol the territory of Las Anod, October 2007. © Sergey Vasilyev/WPN

While the recent developments with the SSC show that bargaining among the political elite and the integration of opposition figures into the Somaliland government are still useful in resolving conflict, the state has also increased its coercive capacities. The Somaliland Armed Forces—made up of an estimated 11,000 soldiers, 6,000 of whom are considered active (Menkhaus, 2010, p. 362)—have prevailed, together with local allies, against forces from Puntland and local clans in the conflict over Las Anod.[65] Armed opposition from marginalized groups has thus become an increasingly unpromising endeavour since the mid-1990s, except for group leaders who profit financially from it.[66]

CRIMINAL VIOLENCE AND LOCAL (IN)SECURITY

The decline in large-scale political violence over the past two decades allows criminal violence to come into sharper analytical focus as a distinct phenomenon. As noted above, towns in central Somaliland, notably Burao, have experienced much lower levels of lethal violence than other districts in Somalia. This discrepancy does not appear related to significant differences in firearm availability, which remains relatively high in all parts of Somalia and Somaliland, even though light weapons are also fielded by non-state groups and militias in south-central Somalia. Rather, there are significant variations in the organizational capacities of the perpetrators and in their use of guns in attacks. The absence of organized armed groups and comparatively low levels of criminal violence in western and central Somaliland is mainly due to the relative strength of the Somaliland state and the establishment of a Somaliland Police Force (SLPF). Together with self-organized neighbourhood watch groups, the SLPF has come to play a role in local security provision (see Box 5.3). The following section examines the main perpetrators of criminal violence in Somaliland today, along with the state and non-state actors seeking to combat it.

> Organized armed groups carried out only about three per cent of all attacks.

Perpetrators of criminal violence

In Somaliland's history, criminal violence has frequently been perpetrated by gangs such as the *deydey* groups that robbed civilians in the years following the declaration of independence (Bradbury, 2008, p. 88). Today criminal youth groups are still regarded as threats to local security, committing offences that range from petty crimes, such as mobile phone snatching, to serious assaults, including rape. Yet, as explained below, criminal violence in contemporary Somaliland appears to owe little to gang activity.

Young men appear to be the main perpetrators of criminal violence in urban centres of contemporary Somaliland.[67] The local District Safety Committee (DSC) in Burao considers young criminals a product of rural–urban migration, unemployment, and *qaat*[68] addiction (OCVP, forthcoming a, p. 24).[69] While criminal youths come from a wide variety of backgrounds, they are often school drop-outs with no stable income and poor employment prospects. With *qaat* addiction having become a serious and widespread socio-economic problem in Somaliland (Hansen, 2009), these youths frequently revert to muggings and robberies as a means to finance their *qaat* consumption, reportedly under the umbrella of small gangs in many cases.[70] It has also been suggested that Somaliland youths are prone to mobilization into clan militias in case of communal conflict, and that they represent a pool of easy recruits for other armed groups (OCVP, forthcoming a, p. 24).

However, there is little evidence that gangs or organized groups are the primary perpetrators of youth violence. According to the OCVP district report for Burao, assaults carried out by organized armed groups represent only about three per cent of all attacks and no more than 15 per cent when clan groups are included. In contrast, about 39 per cent of violent attacks are conducted by individuals (OCVP, forthcoming a, p. 22).

Figure 5.1 **Perpetrators of violent attacks in districts in Somalia and Somaliland, 2009–10**

BURAO
■ Attacks by armed groups (15) ■ Attacks by individual and intimate actors (82) ■ Unclear (3)

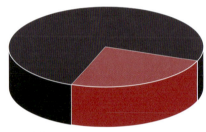

LAS ANOD
■ Attacks by armed groups (25) ■ Attacks by individual and intimate actors (75) ■ Unclear (0)

GALKAYO
■ Attacks by armed groups (47) ■ Attacks by individual and intimate actors (52) ■ Unclear (1)

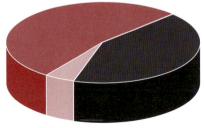

MOGADISHU
■ Attacks by armed groups (53) ■ Attacks by individual and intimate actors (42) ■ Unclear (5)

Sources: OCVP (forthcoming a–d)[71]

One way to evaluate the relative significance of violence by armed actors in Burao, including by gangs and clan groups, is to draw comparisons with other districts in Somalia using OCVP data. If the various organized armed actors are aggregated into one category ('armed groups') and distinguished from individual criminals, friends, relatives, and neighbours ('individual and intimate actors'), the difference between Burao and districts in Puntland and south-central Somalia becomes apparent (see Figure 5.1).

While the vast majority of all attacks in the Somaliland towns of Burao and, for that matter, Las Anod are perpetrated by criminal individuals or persons close to the victim, in Galkayo almost half and in Mogadishu just over half of all assaults are conducted by armed groups. Given Somaliland's recent history of state consolidation and militia demobilization, the relatively low rate of attacks by armed groups is not surprising.[72]

The relative contribution of armed groups to violence correlates with the evidence presented earlier about firearm use. The greater the incidence of attacks by armed groups, the more often firearms are reportedly used in attacks. In Burao, where armed groups appear to commit a relatively limited number of attacks, residents rarely report the use of firearms in attacks. Thus, even though weapons are abundantly present, virtually no non-state groups or individuals use them in attacks, in stark contrast to other districts in Somalia (see Figure 5.2).[73]

Against this background, it appears as though gang violence in Somaliland were exaggerated. DDG's research suggests as much, emphasizing that:

traditional leaders in 15 of the grouped interviews in the 130 communities surveyed [in Somaliland] did say they had a problem with gangs but could not list a single incident occurring in the previous twelve months that had involved them (Hughes and Lynge, 2010, p. 22).

These comparisons should not lead us to underestimate the gravity of criminal violence in Burao. Assaults against women, especially, constitute a serious form of criminal violence in the city and across Somaliland. In fact, women typically report that domestic violence, sexual abuse, and rape are their chief security concerns.[74] According to focus group interviews conducted by the OCVP, sexual assaults take place most often 'in poorly lit areas such as river

Female police officers stand in front of the MaanSoor Hotel, Hargeisa, April 2010. © Dominik Balthasar

162 SMALL ARMS SURVEY 2012

Figure 5.2 **Percentage of attacks committed by armed groups and of attacks in which firearms were used, Somalia, 2009-10**

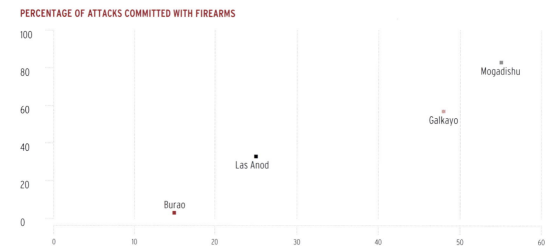

Source: OCVP (forthcoming a–d)[75]

banks and dark alleys' (OCVP, forthcoming a, p. 25). Cases of rape with known perpetrators are often resolved between the elders of the respective families or sub-clans, but women's organizations complain that these agreements are generally to the victims' disadvantage, do not invoke the penal code, and sometimes force a victim to marry her rapist.[76] This practice has, in fact, contributed to a culture of impunity around rape, which some men appear to commit when they cannot afford the costs of marriage.[77]

For these reasons, rape victims tend to turn to the police rather than traditional authorities to report an assault,[78] even though overall satisfaction with police responses to reports of sexual violence is mixed.[79] A Somaliland-based women's organization says the police is not always effective in resolving rape cases but does ask the police, not family or clan elders, to take responsibility for protecting women.[80] Recently, female police officers have been trained to encourage rape victims to report their cases and help to prosecute perpetrators more successfully.

Property crimes, especially theft and burglaries, are another concern shared by many social groups in Burao. In more than a third of property crime cases recorded by the OCVP, victims reported being injured. Like sexual assaults, property crimes are commonly committed at night in poorly lit areas of town.[81] The next section discusses local responses to insecurity on the streets of Burao.

Local security providers

In the wake of its consolidation process and in order to counter criminal activities and provide local safety, the Somaliland state has established a national police force (see Box 5.3). Due to limited state resources, however, other security actors have come to the forefront in the urban centres of central Somaliland, a phenomenon likewise observed in many African states, such as Nigeria, South Africa, and Uganda (Marks and Bonnin, 2010; Baker, 2002; 2008, pp. 101–30). In the near absence of a state, alternative security providers have also become prominent in south-central Somalia (Menkhaus, 2007). However, as Baker shows, the extent to which non-state groups are authorized by or cooperate with the state varies from case to case. In Uganda, for example, close collaboration developed largely under state supervision (Baker, 2008, pp. 129–30). Yet in south-eastern Nigeria, the vigilantes known as the Bakassi Boys became a genuine alternative to the police and regularly evaded attempts by the state to gain control over their activities (Baker, 2002).

In Somaliland, attempts by businesspeople and local communities to provide private protection in Hargeisa and Burao have not led to a competition with state security services. Instead, various patterns of cooperation have emerged, with the state generally assuming a dominant role despite its institutional weakness.

Burglaries and thefts have been identified as the most common forms of property crimes committed in Burao.[82] Large companies have begun to rely on non-state security actors, while also seeking a close relationship with the nascent state authorities. Two cases in point are Dahabshiil, Somaliland's largest money transfer company, and Omar, one of the biggest importers in Somaliland, which delivers imported goods to an increasing number of towns across Somaliland. Dahabshiil's branch in Burao is protected by police forces during the day and private guards at night. Of the 24 armed guards who secure the bank building, nine are local police who earn an extra USD 100 to 150 per month for their service to the company, a significant surplus to their regular monthly salary of USD 60 (see Box 5.3).[83] According to the chief salesman of Omar in Burao, the security situation in Somaliland has dramatically improved since 2000 and now allows the company to reach markets in more remote areas of the territory. Security on the roads and at the company's branches is critical for the business and Omar regularly donates money to Somaliland's police and military to sustain their efforts.[84]

NWGs are mostly providers of, rather than threats to, security.

In recent years, Somaliland has also seen the evolution of neighbourhood watch groups (NWGs) in major cities such as Hargeisa and Burao; these have largely formed as a means of protecting small businesses and private households against burglaries and theft. Based on sub-clan lineages, they are added testimony to the fact that the state's police force is still not wholly effective.[85] Although NWGs in Somaliland are non-state actors, they hardly qualify as 'armed groups'; their members do not typically carry guns but are more frequently equipped with sticks and knives.[86] At night they patrol their communities as well as public marketplaces, where small-scale businesspeople store their goods.[87] Usually standing guard in groups of three to five persons, some NWGs have an overall strength of up to 50 people, divided up and deployed to different streets and areas. Some of the larger groups have clear internal command structures with a chairman, a deputy chairman, and sub-commanders for distinct units.[88] Through a process of confrontation and negotiation between rival NWGs, different groups often divide up the areas they patrol.[89]

Interestingly, some of these groups refer to other NWGs as 'criminal gangs' while seeing themselves as watch groups,[90] an indication of the ambiguous roles some of these groups have taken on. In fact, several local researchers and civil society organizations report that NWGs often include former thieves who have found a more stable income by protecting households and businesses against other thieves.[91] As the NWGs usually charge individual households SOS 500 to 1,000 per night (about USD 5 per month),[92] the financial interest in such work appears clear. They have also attacked non-paying households to coerce them into paying the protection fee.[93] Despite such questionable aspects, many local researchers and NGOs emphasize that NWGs are mostly providers of, rather than threats to, security.

One reflection of NWGs' protective role is that many of them now appear to collaborate closely with the SLPF. Some of the groups include serving police officers, who participate in the patrols or can be reached during the night. Further, the known groups in Burao and Hargeisa are under police supervision;[94] the local police stations in Burao have lists of official NWGs, their members, control areas, and equipment.[95] The NWGs also use their ties to the police to involve them quickly in cases of violent encounters with criminals.[96] In these ways the distinctions between

Box 5.3 The Somaliland Police Force

When the newly formed Republic of Somaliland experienced a significant deterioration in security in the immediate post-conflict era, state authority was too weak and political power too contested to allow for the creation of a national police force. Consequently, the major clans and their militias sporadically formed nominal 'police units' in Boroma, Gabiley, Hargeisa, and elsewhere in order to ensure public safety (Bradbury, 2008, p. 97).[97] While this decentralized approach to security provision was approved in the National Peace Charter of 1993, regional administrations turned out to be unable or unwilling to raise sufficient revenue to finance their security forces. Hence, President Egal negotiated the establishment of a national and centrally controlled police force (Somaliland, 1994).

The establishment of the SLPF on 2 November 1993 not only laid the foundation for the creation of a national security apparatus, but also advanced demobilization and disarmament. The SLPF officers were largely recruited from the SNM militia (WSP, 1999, p. 70),[98] which represented the most powerful and potentially threatening group of fighters. Every new recruit was required to use his own gun in the service—an entry condition that contributed to the country's DDR efforts (Bulhan, 2004, p. 32); this situation still holds nearly two decades later.[99] This entry requirement is partly responsible for the wide array of small arms in the police arsenal (AK-47s, G3s, and M16s, among many others).

At the same time, the clan composition of the SLPF is relatively skewed. While they are recruited from every clan and region, police officers overwhelmingly hail from the Isaaq clan of the Hargeisa region.[100] The national police force under President Tuur counted 320 members in 1992, but its ranks swelled to some 4,000 by the mid-1990s, 4,300 by 2005,[101] and roughly 6,000 members in recent years.[102] About 1,800 police officers have graduated from the police academy in Mandera (UNDP, 2011).

The SLPF's effectiveness does not seem to have improved with its growth. According to UNDP, which has made a significant contribution to the expansion of the SLPF (Foineau, 2006), the police force is '[h]ampered by old and dysfunctional infrastructure, equipment and poor training and deployment' and faces 'challenges in winning respect from the citizenry' (UNDP, 2010). An estimated 30 to 40 per cent of the police force is illiterate.[103] Although salaries have increased over time from an average of USD 10 per month in the early 2000s (Fouineau, 2006) to some USD 60 by the end of that decade,[104] they have been insufficient to allow for the transfer of police officers to posts away from their families.

The uneven presence of the police throughout the territory and lack of enforcement capacity has meant that the population continues to conduct its own policing.[105] As a high-ranking official of the regional administration announced at a local meeting: 'We have the responsibility of protecting you, but we cannot do it without you—it is you [who] protect yourselves.'[106]

A policeman on patrol in Hargeisa, October 2007. © Sergey Vasilyev/WPN

criminal activity and publicly sanctioned protection business have become more clear-cut. The close cooperation between local watch groups and the police also helps explain the NWGs' low level of armament. While the police do not have the means to protect urban areas from theft, burglaries, and violent robberies at night, and therefore 'delegate' this service to private security groups, the NWGs, in turn, cannot enforce the law and rely on the police if they encounter criminal gangs or witness a serious crime such as murder. Police and NWGs thus complement each other in the provision of local security.

Overall, this arrangement is providing a viable, if still fragile, security architecture in the urban centres of Somaliland. With the SLPF now fairly well established in western and central Somaliland's cities, serious crimes occurring there are frequently punished and prison sentences enforced.[107] In fact, the contemporary SLPF is the most effective police force Somaliland has seen in the last two decades. Local NWGs fill in where SLPF capacities are exceeded, but, for the most part, the police also regulate and control their protection activities. Another indirect effect of this consolidation of state authority is that organized armed groups that contest the state have virtually disappeared from Burao.

Burao still struggles with common forms of criminal violence, whether committed by individuals or loosely organized gangs. But although impoverished youths remain a potential pool for the mobilization of militias, the kind of systematic violence committed by organized armed groups is absent. In contrast to most other parts of Somalia, Somaliland's state–private partnership has fostered a peaceful environment in most towns. Yet the arrangement remains dependent on political compromises underlying the Somaliland state. Indeed, the potential for deep social divides and the quick mobilization of large militias remains high, as discussed in the next section.

Communal disputes pose a challenge to peace and security in Somaliland.

COMMUNAL VIOLENCE

While political conflict has been a defining characteristic of Somaliland's violent process of state building, another important driver of conflict involves disputes over access to and ownership of land and real estate. According to a study undertaken by the Somaliland Academy for Peace and Development, land disputes have constituted an important source of violent conflict in Somaliland over the past two decades (APD, 2007, p. 3). Although these disputes generally occur at the local level, they can escalate into wider, inter-clan conflicts. As discussed above, communal and political violence were deeply interwoven during the civil wars of the 1990s. Although widespread collective violence has not occurred since then, violent confrontations in Kalshale and associated SSC activity in recent years have shown that communal disputes can still lead to political conflict. Consequently, communal disputes pose a challenge to peace and security in Somaliland, particularly as most of them have not yet been resolved decisively (APD, 2009, pp. 27–30).

Land disputes in a clan-based society

Just as Somaliland remains equipped with the 'hardware' (weapons) for warfare, it has also maintained its most powerful 'software' for the mobilization of armed groups—a social system based on clans. Quite in contrast to earlier Somali governments, the self-styled republic's administration has safeguarded rather than challenged the clan structure. In fact, the clan system and its traditional authorities came to form a key pillar of the young state. For one, the National Peace Charter of 1993 established a *beel* (clan or community) system of governance that recognizes kinship as the 'organizing principle' of Somali society. In principle, governance came to be exercised through a power-sharing coalition of Somaliland's main clans (Bradbury, Abokor, and Yusuf, 2003, pp. 460–61). Furthermore, traditional authorities came to be institutionalized and empowered in the form of the *Guurti,* the upper house of parliament. And two decades into Somaliland's existence the importance of the clan for society as well as for politics remains unbroken, as exemplified by the presidential elections of 26 June 2010, in which clans often voted nearly unanimously for 'their' candidate.[108]

Particularly volatile sources of communal violence in Somaliland are disputes over land and real estate that are rooted in recent history. Competing claims to water resources as well as unclear and illegal ownership records are one part of the story, the breakdown of traditional user rights and overlapping claims to land following the numerous civil conflicts, another. Overall, increased competition for economic resources in a resource-scarce environment has frequently resulted in communal conflicts. Examples include disputes in the environs of Gabiley (Awdal) and Erigavo (Sanaag), where members of the militarily victorious Isaaq clan laid claim to areas and assets that had formerly been controlled by other communities. The opposing claims have been violently contested over the course of the past two decades, during which whole sub-clans were mobilized to defend assets.

The abovementioned Kalshale conflict began as a land dispute, when members of the Garhajis–Habar Yonis sub-clan built a new water point in an area claimed by local Harti–Dhulbahante. About 50–100 individuals lost their lives in violent clashes between the sub-clans, while the SSC tried to recruit disaffected Dhulbahante against the intervening Somaliland army (OCVP, forthcoming e; UNSC, 2011, p. 130). Land disputes over water points and grazing lands, as well as cycles of revenge killings, have the potential to escalate. Thus, while the political integration of major (sub-) clans into the Somaliland state has prevented larger political conflict over the past 15 years, the basic ingredients for communal violence are still present and can reinforce clan identities and be used to mobilize affiliated militia. Further, critics of the *Guurti* allege that the body has lost its reputation for neutrality, hence leaving the polity with 'no strong cross-cutting institution in Somaliland that can hold a crisis';[109] such observations act as a reminder that the strength of the Somaliland state itself still depends on inter-clan harmony. Communal tensions therefore remain a serious potential threat to safety and security in Somaliland.

> Increased competition for economic resources has frequently resulted in communal conflicts.

Resolving communal conflict? From traditional mediation to state interventions

Particularly during the early years of the Somaliland Republic, when its central administration and law enforcement capacity were still very weak, the resolution of land conflicts was left to local and largely traditional authorities. Although Somaliland society's traditional and 'bottom-up' mechanisms of conflict resolution and reconciliation have frequently been praised, this system has not been able to bring about a durable settlement of many of these conflicts (APD, 2007, p. 3).[110]

Central state authorities have become increasingly involved in the resolution of these conflicts. In early 2011, for example, the Somaliland government announced the start of negotiations aimed at resolving a long-lasting land dispute between the Dir and Isaaq sub-clans in Ceel Bardaleh in western Somaliland. For this land dispute to be settled, President Ahmed Mahamoud 'Silanyo' appointed a mediation committee consisting of the ministers of interior, justice, foreign affairs, defence, and aviation (Waaheen Media Group, 2011). The involvement of state authorities has not yet brought about a solution to the conflict, partly because not a single important representative of the Dhulbahante clan participated in the meeting.[111]

In Burao, local authorities emphasize the mitigating impact of police interventions on cycles of revenge killings within the city as one of their major achievements.[112] According to Burao's police commander, the SLPF arrest two to three persons per month for a serious crime in the district, thereby preventing members of the victim's family from taking revenge.[113] However, the grip of the police remains limited. The OCVP finds that the majority of men would, in the event of being attacked, still turn to their community elders rather than the police (OCVP, forthcoming a, p. 40). In the end, police interventions to prevent revenge killings remain dependent on the collaboration of the local clans.[114] They continue to be somewhat less successful in rural areas, where the SLPF's presence is more limited.[115]

CONCLUSION

This chapter has explored the decrease in widespread political armed violence and improvements in security in Somaliland since the mid-1990s. It argues that, while DDR efforts and traditional mechanisms of conflict resolution have had a positive impact, they are not the most important factors. First, small arms, especially Kalashnikov-pattern assault rifles, remain ubiquitous in Somaliland society. Second, traditional mechanisms of conflict resolution, reconciliation, and governance have, in fact, been eclipsed by the institutions of the Somaliland state in responding to insecurity and violence.

Similarly, the effectiveness of Somaliland's collaborative security arrangement results from the state's increasing power. Only once the state had achieved an end to political conflict could it compel local groups to participate in a collaborative form of policing; that form has since brought important security gains to the population—most notably in urban areas such as Burao and Hargeisa. While local security is often provided by private NWGs, the dominant role of the police appears to have curbed the predatory behaviour documented in other contexts.

This chapter has thus emphasized the importance of the central administration's establishment of politico-military hegemony in reducing large-scale political violence and setting the scene for effective policing of criminal violence in Somaliland. This hegemony has been built on the creation of state institutions that have increasingly monopolized control over gunmen and heavy weapons, sidelined political competitors, and incorporated other political elites from most of the clan and sub-clan groups. Recently, the strength of the armed forces has also allowed the state to coerce some opponents and deter others. On the basis of political bargains, Somaliland has achieved significant security gains over the past 20 years. Yet its ability to maintain those gains appears to rest, to a large extent, on the continued integration of all relevant clan groups into the state.

LIST OF ABBREVIATIONS

DDG	Danish Demining Group
DDR	Disarmament, demobilization, and reintegration
DSC	District Safety Committee
NWG	Neighbourhood watch group
OCVP	Observatory of Conflict and Violence Prevention
SNA	Somali National Army
SNM	Somali National Movement
SLPF	Somaliland Police Force
SSC	Sool, Sanaag, and Cayn
SOS	Somali shilling
UNDP	United Nations Development Programme
UNITA	United Togdheer Association

ENDNOTES

1. See also ICG (2003, p. 6) and World Bank (2005, p. 19).
2. Bradbury (2008, p. 50); Compagnon (1993, p. 19); Helling (2010); Hoehne (2011); Huliaras (2002, p. 174); Spears (2004, p. 185).
3. *Deydey* is the common name for gangs that made a living through robbery and extortion (Bradbury, 2008, p. 88).
4. Author interview with member of SOOYAAL, the association for SNM war veterans, Hargeisa, 25 July 2008.
5. See NDC (1994); Brickhill (1994); Nyathi (1995).
6. An analysis of the DDR processes undertaken by President Egal—at the heart of which lay shrewd political manoeuvres—is beyond the scope of this study.
7. See also DDG (2007).
8. The most notable exception is the Sool, Sanaag, and Cayn rebel group in the area of Buuhoodle. In addition, a small number of private households appear to dispose of medium or heavy machine guns (Hughes and Lynge, 2010, p. 56).
9. This figure applies to the whole Togdheer region, of which Burao is the biggest city and regional capital (Hughes and Lynge, 2010, p. 54).
10. Author interviews with arms brokers, Burao, 21 June 2011, and correspondence with Jonah Leff, 22 November 2011. The price ranges for AK-47s quoted here echo those of the Hargeisa arms market (Somalia Report, 2011).
11. Author interview with an official of Somaliland's National Demobilization Commission, Hargeisa, 20 October 2011.
12. Author interview with a member of the Academy for Peace and Development, Hargeisa, 22 July 2008.
13. See, for example, Compagnon (1998, p. 79) and Cliffe (1999, p. 91).
14. See UCDP (2011) and SIPRI (2007; 2008; 2009). While the Uppsala Conflict Data Program does report on different actors and conflicts within one country, the data regarding northern Somalia or Somaliland is restricted to the armed confrontation between the SNM and the SNA until 1991, after which no armed conflict in Somaliland is reported (UCDP, 2011). The Stockholm International Peace Research Institute's list of major armed conflicts also fails to mention the massive outburst of political violence that took place in Somaliland at that time; see SIPRI (1995, p. 34; 1996, p. 29).
15. See WHO (n.d.).
16. Unless noted otherwise, information in this box is drawn from Leff (2011).
17. This price level was reported prior to an offensive of the African Union Mission in Somalia that began on 22 February 2011. In early March that year, the price rose to USD 0.40 per round (UNSC, 2011, p. 42).
18. Interview by Jonah Leff with confidential sources close to the arms trade in Mogadishu, Somalia, February 2011.
19. Interview by Jonah Leff with an arms dealer, Hargeisa, 16 October 2010.
20. Interview by Jonah Leff with the chief of Military Intelligence of Somaliland, Hargeisa, Somaliland, 11 May 2011.
21. Interview by Jonah Leff with an arms dealer, Hargeisa, 16 October 2010, and confidential interviews with sources close to the arms trade, October 2011.
22. Prior to this, firearm ownership, sale, and trade were regulated by the Public Order Law of 1963, enacted by the Mogadishu-based government.
23. Correspondence between Jonah Leff and a UNDP official, 18 October 2011.
24. Confidential interviews by Jonah Leff with sources close to the arms trade, October 2011.
25. See Geneva Declaration Secretariat (2011, p. 57).
26. See UCDP (2011) and SIPRI (1995, p. 34; 1996, p. 29; 2007, pp. 72–78; 2008, p. 80; 2009, pp. 46–52).
27. The population of Somaliland is estimated to be between 2 and 3.5 million people (UNOCHA Somalia, 2005; Somaliland Embassy, n.d.). Based on a population estimate of three million people, the annual average conflict-related death toll of 750 people leads to an annual conflict death rate of 25 per 100,000.
28. Press-reported casualties are lower than the estimate of 100 deaths per year since 2007. See Hoehne (2007, p. 1) and Somaliland Press (2010; 2011a; 2011b; n.d.).
29. Based on a population estimate of three million people, the annual average of 100 deaths means an annual conflict death rate of 3.33 per 100,000.
30. The Somaliland Police Force (SLPF) in Hargeisa registers on average two murders per year since 2003 for Togdheer region, whose capital is Burao. It is unclear whether this includes manslaughter. The Burao police record two to three serious crimes, including murder, each month in Burao (author interview with the police commander of Burao, Burao, 21 June 2011).
31. OCVP (forthcoming a, p. 16). Note that responses for 'became a little safer' and 'became a lot safer', as well as for 'became a little unsafe' and 'became very unsafe', were combined in each case. On average, each household was composed of 7.8 people: 4.1 male and 3.7 female (OCVP, forthcoming a, p. 7). In Table 5.1, all figures are rounded to the nearest per cent point.
32. Of the respondents, 498 were women and 288 were men (OCVP, forthcoming c, p. 18).
33. Of the respondents, 357 were women and 305 were men (OCVP, forthcoming b, pp. 21–22).
34. Of the respondents, 774 were women and 608 were men (OCVP, forthcoming d, p. 26).
35. OCVP (forthcoming a, p. 33).
36. OCVP (forthcoming c, p. 37).
37. The distribution of respondents from North and South Galkayo is not specified in the survey (OCVP, forthcoming c, p. 48).
38. OCVP (forthcoming d, p. 46).
39. Differing survey methodologies and a lack of time series data prevent a more thorough analysis. Despite these and other limitations, the surveys constitute the best data source currently available.

40 See OCVP (forthcoming a, p. 4).

41 While 98.0 per cent of the respondents said that safety had increased 'slightly', 0.8 per cent reported that they perceived a 'strong' increase in safety (OCVP, forthcoming a, p. 16).

42 At the same time, less than three per cent of the population surveyed proclaimed a worsening of the security situation (Hughes and Lynge, 2010, p. 13).

43 For the purpose of its victimization survey, the OCVP defined 'homicide' as 'murder or a death as a result of violence or crime' (OCVP, forthcoming b, p. 25).

44 Although the district of Las Anod that borders Puntland has been part of Somaliland since the violent takeover in 2007, it is something of an exception. In contrast to districts in western and central Somaliland, Las Anod experienced armed confrontations during the first decade of the 21st century, in particular between Somaliland, Puntland, and SSC militia forces, as discussed below.

45 The head of the SLPF in Burao claims that there are two to three serious crimes, including homicide, per month (author interview with head of SLPF Burao, Burao, 21 June 2011).

46 Cukier and Sidel (2006, p. 53) highlight the relative lethality of firearms compared to other weapons in attacks.

47 OCVP (forthcoming a, pp. 7, 19).

48 OCVP (forthcoming c, pp. 7, 24).

49 OCVP (forthcoming b, pp. 7, 25).

50 OCVP (forthcoming d, pp. 13, 31).

51 This category includes attacks with small arms and light weapons but also bombs, hand grenades, and other explosive devices. No attacks with explosives were reported in Burao, suggesting a different type of conflict than in Las Anod, Galkayo, and Mogadishu, where bombs and other explosive devices are regularly used in attacks; that is, in 5.2%, 10.0%, and 10.2% of cases, respectively.

52 OCVP (forthcoming a, p. 21).

53 OCVP (forthcoming c, p. 26).

54 OCVP (forthcoming b, p. 28).

55 OCVP (forthcoming d, p. 34).

56 Bradbury estimates that the civil war in Burao left about 4,000 dead, while the one in Berbera resulted in more than 1,000 deaths (Bradbury, 2008, pp. 87, 116).

57 According to Bradbury (2008, p. 116), the civil strife in Burao alone resulted in the displacement of some 180,000 people.

58 Author interview with a former SNM and National Demobilization Commission official, Hargeisa, 25 July 2008.

59 In the aftermath of the war against Barre, there were some 32 heavily armed clan militias in Somaliland; by the end of 1996, only three were left, all from the eastern regions of Sool and Sanaag (author interview with a former SNM and National Demobilization Commission official, Hargeisa, 25 July 2008).

60 Author interview with a member of parliament, Hargeisa, 3 July 2011.

61 The Garhajis are made up of the Isaaq sub-clans of the Idagalle and Habar Yonis.

62 Egal's portrayal of the opposition as being pro-unionist gained credence when their leader, Tuur, and some of his entourage accepted positions in the Mogadishu-based government of Gen. Abdi Farah Aidid.

63 See Wars in the World (2012). As an armed group, the SSC had already been formed in November 2007 by the Northern Somali Unionist Movement; see NSUM (n.d.).

64 The UN Monitoring Group report characterizes the SSC as an 'opportunistic and arguably mercenary militia force that has successfully appropriated legitimate local grievances and exploited radical diaspora sentiment for its own political and financial gain' (UNSC, 2011, p. 32).

65 See UNSC (2011, p. 130) and Hoehne (2007, p. 1).

66 Some leaders of the SSC have reportedly profited directly from the financial support the group received (UNSC, 2011, p. 131).

67 See OCVP (forthcoming a, p. 23; forthcoming c, p. 25).

68 *Catha edulis,* a green leaf rich in an amphetamine-like substance, is chewed as a stimulant and mood enhancer. *Qaat* has a long history of cultivation and customary use throughout the Horn of Africa and the Arabian peninsula, where millions of people use it on a daily basis.

69 Author interview with DSC members, Burao, 20 June 2011.

70 Author interviews with DSC members, Burao, 20 June 2011, and with the executive manager of the local NGO Havoyoko, Burao, 21 June 2011. See also OCVP (forthcoming a, p. 24).

71 See OCVP (forthcoming a, p. 23; forthcoming b, p. 30; forthcoming c, p. 29; forthcoming d, p. 36).

72 Regarding the percentage of attacks conducted by armed groups, Las Anod—where the SSC was reportedly active in 2010 (OCVP forthcoming c, pp. 10, 18)—takes a position between Burao, on the one hand, and districts in Puntland and south-central Somalia, on the other hand.

73 In Burao, only 3 per cent out of the 15 per cent of attacks by any type of armed group are conducted by what the OCVP refers to as 'organized armed groups'; 12 per cent of the attacks are perpetrated by 'clan groups' (OCVP, forthcoming a, p. 23). In all other districts, the categories of 'organized armed groups', 'foreign troops', and 'government forces' dominate the armed group category. See OCVP (forthcoming b, p. 30; forthcoming c, p. 29; forthcoming d, p. 36).

74 OCVP (forthcoming a, p. 19); author interview with members of the United Togdheer Association (UNITA), Burao, 20 June 2011. One exception is Las Anod, where women reported that land disputes were their primary concern (OCVP, forthcoming c, p. 21).

75 See OCVP (forthcoming a, pp. 21, 23; forthcoming b, pp. 28, 30; forthcoming c, pp. 26, 29; forthcoming d, pp. 34, 36).

76 Author interview with members of UNITA, Burao, 20 June 2011.

77 OCVP (forthcoming a, p. 32).
78 OCVP (forthcoming a, p. 39).
79 OCVP (forthcoming a, pp. 43, 45).
80 Author interview with members of UNITA, Burao, 20 June 2011.
81 See OCVP (forthcoming a, pp. 27–29).
82 See OCVP (forthcoming a, p. 27).
83 Author interview with an official of the Dahabshiil branch, Burao, 21 June 2011.
84 Author interview with the head of the Burao branch of Omar, Burao, 20 June 2011.
85 Author correspondence with Roda Ali, former research director at the OCVP, 5 December 2011.
86 Author interview with a journalist of Jamhuuriya, Hargeisa, 18 June 2011.
87 Author interview with members of an NWG, Hargeisa, 23 June 2011.
88 Author interview with members of an NWG, Hargeisa, 23 June 2011.
89 Author interview with members of an NWG, Hargeisa, 23 June 2011.
90 Author interview with members of an NWG, Hargeisa, 23 June 2011.
91 Author interviews with members of UNITA, Burao, 20 June 2011, and with the OCVP director, Hargeisa, 12 June 2011.
92 See, for example, author interview with members of a NWG, Hargeisa, 23 June 2011.
93 Author interviews with the OCVP director, Hargeisa, 12 June 2011, and with members of UNITA, Burao, 20 June 2011.
94 Author interviews with the police commander of Burao, Burao, 21 June 2011, with the governor of Togdheer region, Burao, 20 June 2011, and with the OCVP director, Hargeisa, 12 June 2011.
95 Author interview with the police commander of Burao, Burao, 21 June 2011.
96 This is reported by the District Safety Committee in Burao as well as the police commander of the city (author interviews with the DSC members, Burao, 21 June 2011, and with the police commander of Burao, Burao, 21 June 2011). Members of an NWG in Hargeisa also reported on their general arrangement with the local police (author interview with members of an NWG, Hargeisa, 23 June 2011).
97 Author interview with an SNM veteran, Hargeisa, 21 July 2011.
98 Author interview with a senior official of the SLPF, Hargeisa, 22 March 2009.
99 Author interview with the SLPF head of training, Hargeisa, 25 March 2009.
100 Author interview with a senior official of the SLPF, Hargeisa, 22 March 2009.
101 Author interview with a researcher of the Academy for Peace and Development, Hargeisa, 22 July 2008.
102 Author interview with UNDP Rule of Law and Security personnel, Hargeisa, 9 August 2008.
103 Author interview with a UNDP official, Hargeisa, 9 August 2008.
104 Author interview with the police commander of Burao, Burao, 21 June 2011.
105 Author interviews with officials of the SLPF, Hargeisa, 17 March 2009, and with a UNDP official, Hargeisa, 26 July 2011.
106 Author interview with a UNDP official, Hargeisa, 26 July 2008.
107 The prison system in Somaliland is maintained by a custodial guard of 1,540 men (Menkhaus, 2010, p. 362).
108 Author interview with a former journalist from Somaliland, Hargeisa, 4 July 2011.
109 Author interview with researcher of the Academy for Peace and Development, Hargeisa, 22 July 2008.
110 In the case of Erigavo it is estimated that some 90 per cent of all unrightfully occupied land was returned to the rightful owner by 1996 (author interview with a political analyst, Hargeisa, 4 August 2011).
111 Author interview with a member of parliament, Hargeisa, 3 July 2011.
112 Author interviews with the mayor of Burao, Burao, 20 June 2011, and with the police commander of Burao, Burao, 21 June 2011.
113 Author interview with the police commander of Burao, Burao, 21 June 2011.
114 Author interview with members of the DSC, Burao, 20 June 2011.
115 While revenge killings are the top concern of local communities in rural areas around Burao, they are much less important for urban populations (OCVP, forthcoming a, p. 19).

BIBLIOGRAPHY

Adam, Hussein. 1995. 'Somalia: A Terrible Beauty Being Born?' In William Zartman, ed. *Collapsed States: The Disintegration and Restoration of Legitimate Authority*. Boulder and London: Lynne Rienner Publishers, pp. 69–90.

APD (Academy for Peace and Development). 2007. *Land-based Conflict Project: Working Note*. Hargeisa: Academy for Peace and Development.

—. 2009. *Local Capacities for Peace: Addressing Land-Based Conflicts in Somaliland and Afghanistan*. Hargeisa: Academy for Peace and Development.

Baker, Bruce. 2002. 'When the Bakassi Boys Came: Eastern Nigeria Confronts Vigilantism.' *Journal of Contemporary African Studies*, Vol. 20, No. 2, pp. 223–44.

—. 2008. *Multi-choice Policing in Africa*. Stockholm: Nordiska Afrikainstitutet.

Bradbury, Mark. 1997. *Somaliland Country Report*. London: Catholic Institute for International Relations.

—. 2008. *Becoming Somaliland*. London: Progressio.
—, Adan Abokor, and Haroon Yusuf. 2003. 'Somaliland: Choosing Politics over Violence.' *Review of African Political Economy*, Vol. 30, No. 97, pp. 455–78.
Brickhill, Jeremy. 1994. 'Disarmament and Demobilisation in Somaliland (Northwestern Somalia).'
Bryden, Matt. 1999. *Somalia's New Order: Patterns of Political Reconstruction since State Collapse*. Halifax: Pearson International Institute of Peacekeeping.
— and Jeremy Brickhill. 2010. 'Disarming Somalia: Lessons in Stabilisation from a Collapsed State.' *Conflict, Security & Development*, Vol. 10, No. 2, pp. 239–62.
Bulhan, Hussein Abdillahi. 2004. *Survey on Small Arms in Somaliland*. Hargeisa: Center for Creative Solutions.
Cliffe, Lionel. 1999. 'Regional Dimensions of Conflict in the Horn of Africa.' *Third World Quarterly*, Vol. 20, No. 1, pp. 89–111.
Compagnon, Daniel. 1993. 'Somaliland: un ordre politique en gestation?' *Politique Africaine*, Vol. 50, pp. 21–31.
—. 1998. 'Somali Armed Units: The Interplay of Political Entrepreneurship and Clan-Based Factions.' In Christopher Clapham, ed. *African Guerrillas*. Oxford: James Currey, pp. 73–90.
Cukier, Wendy and Victor Sidel. 2006. *The Global Gun Epidemic: From Saturday Night Specials to AK-47s*. Westport, CT: Praeger Security International.
DDG (Danish Demining Group). 2007. *A Baseline Survey of Community Attitudes Toward Small Arms & Light Weapons (SALW) in North West Somalia (Somaliland)*. Copenhagen: DDG and Danish Refugee Council.
Flint, Julie. 1994. 'Struggling to Survive.' *Africa Report*, Vol. 39, No. 1, pp. 36–38.
Forberg, Ekkehard, and Ulf Terlinden. 1999. 'Small Arms in Somaliland: Their Role and Diffusion.' *BITS Research Report 99.1*. Berlin: Berlin Information-Centre for Transatlantic Security.
Fouineau, Julien. 2006. *Construction et viabilité des forces de sécurité dans un état sécessionniste: analyse des dynamiques et des faiblesses de la police au Somaliland (Somalie)*. Paris: Panthéon-Sorbonne University.
Geneva Declaration Secretariat. 2011. *Global Burden of Armed Violence: Lethal Encounters*. Cambridge: Cambridge University Press.
Gilkes, Patrick. 1993. 'Two Wasted Years: The Republic of Somaliland 1991–1993.' London: Save the Children Fund.
Hansen, Peter. 2009. *Governing Khat: Drugs and Democracy in Somaliland*. Working Paper 24. Copenhagen: Danish Institute for International Studies.
Helling, Dominik. 2010. 'Tillyan Footprints Beyond Europe: War-Making and State-Making in the Case of Somaliland.' *St Antony's International Review*, Vol. 6, No. 1, pp. 103–23.
Hoehne, Markus. 2007. 'Puntland and Somaliland Clashing in Northern Somalia: Who Cuts the Gordian Knot?' New York: Social Science Research Council. <http://hornofafrica.ssrc.org/Hoehne/>
—. 2011. 'Not Born a de facto State: The Complicated State Formation of Somaliland.' In Roba Sharamo and Berouk Mesfin, eds. *Regional Security in the Post-Cold War Horn of Africa*. Pretoria: Institute for Security Studies.
Hughes, Ed and Karina Lynge. 2010. *Community Safety and Small Arms in Somaliland*. Copenhagen: Danish Demining Group.
Huliaras, Asteris. 2002. 'The Viability of Somaliland: Internal Constraints and Regional Geopolitics.' *Journal of Contemporary African Studies*, Vol. 20, No. 2, pp. 157–82.
ICG (International Crisis Group). 2003. 'Somaliland: Democratisation and Its Discontents.' *ICG Africa Report*, No. 66. <http://www.crisisgroup.org/home/index.cfm?id=1682&l=1>
IRBC (Immigration and Refugee Board of Canada). 1995. *Somalia: Chronicle of Events September 1992–June 1994*. Issue Paper. July. Victoria: IRBC. <http://www.ecoi.net/file_upload/1684_1243950889_<http-www2-irb-cisr-gc-ca-en-research-publications-index-e-htm.pdf>
IRI (International Republican Institute). 2011. *Somaliland Opinion Survey: Hargeisa District*. 16 November. <http://www.iri.org/news-events-press-center/news/iri-releases-new-survey-somaliland-public-opinion>
Jhazbhay, Iqbal. 2008. 'Somaliland's Post-War Reconstruction: Rubble to Rebuilding.' *International Journal of African Renaissance Studies: Multi-, Inter- and Transdisciplinarity*, Vol. 3, No. 1, pp. 59–93.
Leff, Jonah. 2011. *Small Arms and Light Weapons in Somaliland*. Unpublished background paper. Geneva: Small Arms Survey.
Marks, Monique and Debby Bonnin. 2010. 'Generating Safety from Below: Community Safety Groups and the Policing Nexus in Durban.' *South African Review of Sociology*, Vol. 41, No. 1, pp. 56–77.
Mengisteab, Kidane. 2008. 'Reconciling Africa's Fragmented Institutions of Governance: Is Somaliland Charting a New Path?' Paper presented at the 7th Annual Conference on the Horn of Africa, Lund, Sweden. October 17–19, pp 179–92.
Menkhaus, Ken. 2007. 'Governance without Government in Somalia: Spoilers, State Building, and the Politics of Coping.' *International Security*, Vol. 31, No. 3, pp. 74–106.
—. 2010. 'Non-State Actors and the Role of Violence in Stateless Somalia.' In Klejda Mulaj, ed. *Violent Non-state Actors in World Politics*. New York: Columbia University Press, pp. 343–80.
NDC (National Demobilization Commission of Somaliland). 1994. *Disarmament and Demobilization in Somaliland (Northwest Somalia): Interim Emergency Programme*. Hargeisa: NDC.
NSUM (Northern Somali Unionist Movement). n.d. 'About Us.' Accessed 26 February 2012. <http://www.n-sum.org/?q=node/7>
Nyathi, P. 1995. 'Somaliland, Zimbabwe: Demobilisation and Development—The Tasks of Redesigning a Future without Conflict.' In A. Shepherd and Mark Bradbury, eds. *Rural Extension Bulletin: Development and Conflict*, No. 8. December, pp. 26–28.
OCVP (Observatory of Conflict and Violence Prevention). Forthcoming a. *Safety and Security Baseline Report: Burao*. Hargeisa: OCVP.
—. Forthcoming b. *Safety and Security Baseline Report: Galkayo* Hargeisa: OCVP.
—. Forthcoming c. *Safety and Security Baseline Report: Las Anod*. Hargeisa: OCVP.
—. Forthcoming d. *Safety and Security Baseline Report: Mogadishu*. Hargeisa: OCVP.

—. Forthcoming e. *Conflict Mapping: Kalshale Conflict.* Hargeisa: OCVP.
SIPRI (Stockholm International Peace Research Institute) 1995. *SIPRI Yearbook: Armaments, Disarmament, and International Security.* Oxford: Oxford University Press.
—. 1996. *SIPRI Yearbook: Armaments, Disarmament, and International Security.* Oxford: Oxford University Press.
—. 2007. *Armaments, Disarmament, and International Security.* Oxford: Oxford University Press.
—. 2008. *Armaments, Disarmament, and International Security.* Oxford: Oxford University Press.
—. 2009. *Armaments, Disarmament, and International Security.* Oxford: Oxford University Press.
Somalia Report. 2011. 'Somaliland's Booming Arms Business: Guns Freely Available in More Peaceful Region.' 2 November. <http://www.somaliareport.com/index.php/post/1920/Somalilands_Booming_Arms_Business>
Somaliland. 1994. Law on the Structure of the Somaliland Police Force, No. 54. 3 November. <http://www.somalilandlaw.com/police_law_.html>
Somaliland Embassy. n.d. 'Introduction.' Stockholm: Somaliland Embassy. <http://www.somalilandembassy.se/about_somaliland.html>
Somaliland Press. 2010. 'Somaliland Army Clashes with Militia, Five Dead.' 20 July. <http://somalilandpress.com/somaliland-army-clashes-with-militia-five-dead-17144>
—. 2011a. 'Army Boosting Forces after Deadly Unrests in Eastern Somaliland.' 9 February. <http://somalilandpress.com/army-boosting-forces-after-deadly-unrests-in-eastern-somaliland-20026>
—. 2011b. 'Somaliland: Kalshale Conflict Breaks out Again—Updated.' 19 February. <http://somalilandpress.com/somaliland-kalshale-conflict-breaks-out-again-updated-20288>
—. 2011c. 'SSC Militia Begin Peace Talks with Somaliland Government.' 28 June. <http://somalilandpress.com/somaliland-ssc-militia-begin-peace-talks-with-somaliland-government-22830>
—. 2011d. 'Somaliland: SSC Commander Reveals Split in His Group.' 20 August. <http://somalilandpress.com/somaliland-ssc-commander-reveals-split-in-his-group-17799>
—. n.d. 'Somaliland: Fighting Erupts in Kalshale between Troops and Clan Militia.' <http://somalilandpress.com/somaliland-fighting-erupts-in-kalshale-between-troops-and-clan-militia-20008>
Spears, Ian. 2004. 'Reflections on Somaliland and Africa's Territorial Order.' In Richard Ford, Hussein Adam, and Edna Ismail, eds. *War Destroys: Peace Nurtures—Somali Reconciliation and Development.* Asmara: Red Sea Press, pp. 179–92.
—. 2010. *Civil War in African States: The Search for Security.* Boulder: FirstForum Press.
UCDP (Uppsala Conflict Data Program). 2011. UCDP Battle-Related Deaths Dataset v. 5-2011, 1989–2010. <http://www.pcr.uu.se/research/ucdp/datasets/>
UNDP (United Nations Development Programme). 2010. 'Somaliland Progresses towards Police Reform.' <http://www.so.undp.org/index.php/1Somaliland-progresses-towards-police-reform.html>
—. 2011. *Rights-Based Partnership Policing Reform Efforts Press Ahead as 300 More Officers Graduate.* <http://www.so.undp.org/index.php/Somalia-Stories/Rights-Based-Partnership-Policing-Reform-Efforts-Press-Ahead-as-300-More-Officers-Graduate.html>
UNOCHA (United Nations Office for the Coordination of Humanitarian Affairs) Somalia. 2005. 'Overview of Humanitarian Environment in Somaliland.' <http://www.mbali.info/doc121.htm>
UNSC (United Nations Security Council). 2011. *Report of the Monitoring Group for Somalia and Eritrea pursuant to Security Council Resolution 1916 (2010).* S/2011/433 of 18 July. <http://www.un.org/ga/search/view_doc.asp?symbol=S/2011/433>
Waaheen Media Group. 2011. 'Odayaasha Labada Beelood ee isku dila Ceel Berdaale oo Madaxtooyada Hageysa kaga dhawaaqay inay Heshiinayaan iyo Madaxweyne Siilaanyo oo ku xoojiyay Guddi wasiiro ah.' 4 January. <http://waaheen.com/?p=7702_O3/Negotiations/20110105.3045>
Walls, Michael. 2009. 'The Emergence of a Somali State: Building Peace from Civil War in Somaliland.' *African Affairs*, No. 108, Vol. 432, pp. 371–89.
Wars in the World. 2012. 'Somalia: Created a New Federal State inside Somalia Called "The Unity and Salvation Authority of the SSC Regions of Somalia."' <http://www.warsintheworld.com/index.php/2012/01/13/somalia-created-a-new-federal-state-inside-somalia-called-the-unity-and-salvation-authority-of-the-ssc-regions-of-somalia>
WHO (World Health Organization). n.d. 'Health Statistics and Health Information Systems.' Accessed 27 October 2011. <http://www.who.int/healthinfo/global_burden_disease/estimates_country/en/index.html>
World Bank. 2005. *Conflict in Somalia: Drivers and Dynamics.* Washington, DC: World Bank.
WSP (War-Torn Societies Project). 1999. *A Self-Portrait of Somaliland: Rebuilding from the Ruins.* Hargeisa: Somaliland Centre for Peace and Development.

ACKNOWLEDGEMENTS

Principal authors
Dominik Balthasar and Janis Grzybowski

Contributor
Jonah Leff

Members of the pirate group Central Regional Coast Guard, one of whom carries a rocket-propelled grenade launcher, arrive on a beach near Hobyo, October 2008. © Veronique de Viguerie/Getty Images

Troubled Waters
SOMALI PIRACY

Pirate militiamen stand among fishing boats at a port in Hobyo, Somalia, October 2010.
Jehad Nga/*The New York Times*/Redux/laif

> *They kept us in a state of terror. Even when I could not see the torturing, I could hear the screams.*
> Dipendra Rathore, crewmember of a hijacked ship
> (Guardian, 2011)

In May 2010, 21-year-old Dipendra Rathore was training to become a naval officer onboard a Mumbai-owned chemical carrier when pirates hijacked the ship, 120 miles south of Oman. All 22 members of the crew were taken hostage. Eight months later, after mental and physical torture at the hands of their captors, they were released following the payment of a hefty ransom.

Experiences such as Dipendra Rathore's shed light on the harsh reality of hostage conditions. As discussed in the chapter on Somali piracy in this volume, these conditions appear to be deteriorating despite a decrease in overall hostage-taking at sea. Interviewed Somali pirates acknowledge their growing frustration—and increasing tendency to take that frustration out on hostages—as deployments of naval forces and private security companies on ships intensify and ransom negotiations are prolonged. As state and private actors continue to seek effective responses to stem pirate activity, however, the fate of the 199 hostages being held as of March 2012 remains uncertain (ICC CCS, n.d.).

This photo essay begins by considering the root causes of Somali piracy: strong criminal networks, profitability, and a lack of alternative economic opportunities. It then turns to the measures undertaken to respond to pirate activities—including local resistance, deployments of NATO vessels and private security companies, and incarceration.

Somali pirates have a significant presence at sea and on land, with team members fulfilling various roles, whether as part of an attack team on the water or as armed guards onshore **(Photos 1, 3)**. Ransom proceeds extend inland, benefitting not only pirate group members, but also local businesses and communities as well as armed groups that provide security **(Photo 5)**. Pirates report that 30 per cent of their profits are regularly paid to government officials **(Photo 2)**. Given the lack of economic and development prospects onshore, pirate groups can easily hire new recruits who are ready to risk their lives for a sum of money unobtainable elsewhere. Communities are left with few means with which to provide alternative livelihoods **(Photos 4, 6)**.

At sea, the area of pirate operations continues to expand, stretching from Somalia's coastline to the Gulf of Aden and large parts of the Indian Ocean. In 2008, a series of high-profile hijackings, including that of the Ukrainian cargo ship MV Faina, attracted international attention **(Photos 8, 9**; Leff, 2012; Lewis, 2009). Since then, many pirates have become increasingly violent during attacks and ransom negotiation periods. Although pirates have a financial interest in keeping their hostages alive, reports of torture and intimidation, and of the destruction of ships they are unable to capture, point to an escalation in the use of force at sea **(Photos 7, 12)**.

The international community has stepped up efforts to patrol waters under pirate threat and has been deploying three naval coalitions to the 'high-risk area' since 2008 **(Photos 10, 11)**. In Somaliland, an autonomous and stable region of Somalia, coast guards work to secure local waters **(Photo 13)**. The high number of attempted pirate attacks has also prompted states and shipping companies to rely more and more on private armed guards to secure their vessels **(Photo 15)**. Yet, while private security companies seem to have contributed to the recent reduction in successful hijackings, their use is also costly, poses complex legal challenges, and raises concerns over further escalations at sea.

The increasingly real possibility of capture and longer prison terms for pirates also contribute to this cycle of violence, with pirates claiming that higher risks prompt them to use greater force **(Photos 14, 16–17)**. Meanwhile, Somali prisons struggle to find space for growing numbers of captured pirates **(Photo 18)**. The economic incentives for piracy are clear, but the stakes for Somali pirates are high, and getting higher. Meanwhile, for those sailing within sight of the pirates, the risks have never been greater.

BIBLIOGRAPHY

ICC CCS (International Chamber of Commerce Commercial Crime Services). n.d. 'Piracy News & Figures.' <http://www.icc-ccs.org/piracy-reporting-centre/piracynewsafigures>

Leff, Jonah. 2012. *Reaching for the Gun: Arms Flows and Holdings in South Sudan.* HSBA Issue Brief No. 19. Geneva: Small Arms Survey. <http://www.smallarmssurveysudan.org/pdfs/HSBA-SIB-19-Arms-flows-and-holdings-South-Sudan.pdf>

Lewis, Mike. 2009. *Skirting the Law: Post-CPA Arms Flows to Sudan.* HSBA Working Paper No. 18. Geneva: Small Arms Survey. <http://www.smallarmssurveysudan.org/pdfs/HSBA-SWP-18-Sudan-Post-CPA-Arms-Flows.pdf>

Rathore, Dipendra. 2011. 'Experience: I Was Kidnapped by Somali Pirates.' *Guardian.* 11 June. <http://www.guardian.co.uk/lifeandstyle/2011/jun/11/kidnapped-by-somali-pirates-experience>

2 Farah Ismael Eid is serving 15 years for piracy. He says that his pirate team typically divided up their loot this way: 20 per cent for their bosses, 20 per cent for future missions, 30 per cent for the gunmen on the ship, and 30 per cent for government officials. 'Believe m a lot of our money has gone straight into the government's pockets.'
© Jehad Nga/*The New York Times*/Redux/laif

A truck passes through Galcayo on its way to Hobyo to supply pirates with boats, October 2009.
Veronique de Viguerie/Getty Images

4 Somali elders meet to discuss some of the challenges faced by the town of Hobyo, includ finding an alternative to piracy as a means of mak a living. © Roberto Schmidt/AFP Photo

5 Armed militiamen and a pirate walk on a rocky outcrop on the coast of Hobyo, as a hijacked Korean supertanker lies anchored on the horizon, August 2010. Both armed groups often work together and act as security forces for the region. © Roberto Schmidt/AFP Photo

6 A sign painted on a wall in Garowe, capital of Puntland and the home of several prominent and lesser-known pirates, reads 'No pirates allowed'. May 2009.
© Michael Kamber/*The New York Times*/Redux/laif

7 Suspected masked pirates guard the captain of a 24-member crew taken hostage when their Mozambican-flagged fishing vessel, Vega 5, was seized by Somali pirates, 25 April 2011. © Reuters

8 Armed Somali pirates watch over the MV Faina crew following a request by the US Navy to monitor the crew's health and welfare. The ship was released in February 2009 following ransom payment of USD 3.2 million. © AFP Photo/Getty Images

9 Somali pirates hijack the Ukrainian MV Faina on 25 September 2008. The cargo ship carried more than 30 Soviet-era tanks and other weapons destined for Southern Sudan.
© Reuters/US Naval Forces Central Command Public Affairs

A NATO armada, including ships from Germany, Turkey, and the UK, en route to the Indian Ocean to police the waters [off] the coast of Somalia. © Pio Luigi Cotrufo/AFP Photo/Getty

11 A magnetic board on the German ship Mecklenburg-Vorpommern shows the positions of various nations stationed in the of Aden. © Joerg Gläscher/laif

12 The MV Pacific Express is towed to the port of Mombasa after having been torched by suspected Somali pirates on 21 September 2011. Members of the crew had refused to open the cabin doors when the ship was hijacked. © AP Photo

Members of the Somaliland coast guard patrol the port of Berbera, tember 2011. © Sarah Hunter

14 Search and seizure team members from the guided-missile cruiser USS Vella Gulf close in to apprehend suspected pirates in the Gulf of Aden, 12 February 2009.
© Jason Zalasky/AFP Photo/Getty Images

15 Trainees take part in an anti-piracy drill during a maritime protection training programme, Haifa, Israel, June 2009.
© Baz Ratner/Reuters

6 One of 13 Iranian seamen whose fishing vessel was hijacked by suspected pirates prays on the ship's ck. The vessel was rescued by members of the US Navy, nuary 2012. © Tyler Hicks/*The New York Times*/Redux/laif

7 Suspected pirates on board the Iranian fishing vessel on which y were captured, January 2012. yler Hicks/*The New York Times*/Redux/laif

18 A prison warden tries to control inmates at Berbera prison, Somaliland. The increasing numb of pirate inmates is creating many challenges for priso in the region, March 2011. © Tony Karumba/AFP Photo

Escalation at Sea
SOMALI PIRACY AND PRIVATE SECURITY COMPANIES

INTRODUCTION

The frequency of pirate attacks on commercial ships worldwide has risen dramatically in the past six years, reaching record levels in 2010, much of it attributable to Somali groups operating in the Red Sea, Gulf of Aden, Arabian Sea, and Indian Ocean (IMB, 2011, p. 24; see Map 6.1). Acts of piracy are also becoming more costly, in both human and economic terms; higher ransom demands have resulted in longer negotiations and lengthier periods of captivity for the seafarers held hostage (Ince & Co., 2011, p. 2; UNSC, 2011b, p. 12).[1] International naval forces have increased their presence in affected waters, particularly since 2008 (Ghosh, 2010, pp. 14–15). While the navies have successfully increased maritime security in patrolled areas, pirates have begun to use captured vessels as 'mother ships' to transport provisions, weapons, and attack boats, allowing them to strike at ever greater distances from the coast (UNSC, 2011b, pp. 219–21).

Somali piracy's resilience to international action has prompted shipping companies to turn to maritime private security companies (PSCs) to provide security for their crews and vessels. This is a significant shift for an industry that long resisted placing weapons on ships due to inscrutable legal and insurance implications, concerns regarding crew safety, and fears of encouraging an escalation of violence at sea. Significantly, several governments and international organizations, including the International Maritime Organization (IMO), while falling short of encouraging the practice, have gradually recognized it as an option for protecting ships in dangerous areas.

This chapter takes a close look at the current stand-off between Somali pirates and PSCs, focusing on the associated small arms control challenges and rules of behaviour among all parties. The chapter also seeks to identify the types of small arms used by Somali pirates and PSCs, exploring whether the growing use of armed guards to protect ships increases security or leads to an escalation of violence at sea. Key findings include:

- While the number of attempted attacks by Somali pirates continued to increase in 2011, attacks were less successful than in 2010 and resulted in fewer hijackings. Operations by naval forces, the increasing use of PSCs to protect ships, and other self-protective measures applied by the shipping industry appear to be the main factors in the reduction of hijackings.
- Pirate groups increasingly resort to lethal violence during attacks and abuse their hostages. Pirates' growing frustration with the more robust deployment of naval forces and PSCs, longer periods of hostage release negotiations, and harsher prison terms for captured pirates seem to be the drivers behind this trend.
- Somali pirates continue to use primarily assault rifles, light machine guns, and rocket-propelled grenade launchers (RPGs). Allegations of the use of more destructive weapons remain largely anecdotal and unverified, but pirates' capacity to adapt tactics to changing circumstances, combined with weapons availability in Somalia, increase the risk of a pirate arms build-up.

- Due to the lack of harmonized regulations, there is no standard PSC 'weapon kit' and rules on the use of force vary greatly. Some countries allow maritime PSCs to carry only semi-automatic weapons; in practice, PSCs utilize a range of weapons, including sniper rifles, general-purpose machine guns, light machine guns, fully automatic assault rifles, bolt-action rifles, shotguns, and handguns.
- The presence of armed guards on ships poses complex legal and small arms control challenges related to the movement of armed guards in ports and territorial waters (see Box 6.1), as well as liability issues arising from guards' use of force and firearms.
- A number of states have sought to facilitate the provision of private armed security on ships, but the schemes they employ vary markedly. Some states offer PSCs the possibility to rent government-owned firearms.

After providing a contextualization of Somali piracy and the statistical trends observed in the past ten years, the chapter assesses Somali pirates' weapons and use of violence during attacks. It then documents the growing use of armed PSCs to protect commercial vessels, placing a particular focus on the weapons in their employ and the challenges this situation entails for small arms control. Interviews with representatives of PSCs and pirate groups, an analysis of International Maritime Bureau (IMB) data, and expert contributions are among the sources used in this chapter.

Box 6.1 Definitions

Following the approach adopted by the IMB, this chapter covers both acts of *piracy* and instances of *armed robbery at sea* (IMB, 2011, p. 3). Both terms refer to acts of violence, detention, or depredation committed against a ship, or the people or property onboard a ship.[2] Armed robbery against ships occurs in a state's internal, archipelagic, and territorial waters, the latter being the area within 12 nautical miles from a state's coast (IMO, 2009, p. 4; UN, 1982, art. 3).[3] Consonant with the UN Convention on the Law of the Sea, piracy refers to incidents that occur in waters beyond the territorial sea (UN, 2010).[4] These include the high seas, which generally start at 200 nautical miles from the coast, and the exclusive economic zones, which are the areas between the territorial and the high seas (UN, 1982, arts. 57, 58.2, 86). Unless stated otherwise, and because Somali pirates have carried out attacks in both territorial and high seas, the term piracy is used in this chapter to refer to both pirate attacks and incidents of armed robbery at sea.[5]

The term 'PSCs' in this chapter refers to all legally registered business entities that provide, on a contractual basis, security services at sea and in ports. Security services may entail the protection of persons and the guarding of objects (such as ships and their cargo), the maintenance and operation of weapons systems, the provision of advice or training, and associated surveillance and intelligence operations.[6]

SOMALI PIRACY IN CONTEXT

This section analyses recent developments in Somali piracy in the context of global pirate activity, highlighting trends in the frequency of attacks and their significance in human and economic terms. It then reviews available information on pirates' use of violence and firearms, and discusses the risks of an escalation of violence.

Pirate groups

The origins of modern Somali piracy are subject to ongoing debate. Interviewed pirates, and some analysts, date it back to the 1990s, when the local population—and fishermen in particular—exhibited a growing sense of anger against illegal fishing and waste dumping by foreign vessels in Somali waters; yet others argue that piracy has always been criminally motivated and clan-supported.[7] Although some of the first pirate attacks were aimed at foreign fishing ships, targeting quickly shifted to commercial boats with no direct link to the illegal use of Somali waters, illustrating the increasing criminalization of pirate groups, whose primary objective has become the ransoms secured through negotiation (Hansen, 2009; Shortland, 2012).

It is difficult to ascertain how many pirate groups operate in Somalia as they change over time and are sometimes forced to shift locations. Various reports identify five principal groups:

- the 'National Volunteer Coast Guard' based in the southern port of Kismayo;
- the 'Merca group' based in the port of Merca to the south of Mogadishu;
- a Haradheere-based group known variously as the 'Somali Marines', 'Defenders of Somali Territorial Waters', and 'Ocean Salvation Corps';
- a group based in Hobyo; and
- a group in Eyl in Puntland (Gettleman, 2011, p. 9; Harper, 2011; Murphy, 2009).[8]

Yet more recent accounts point to the emergence of large, highly organized groups, such as the 'Somali Marines', and numerous smaller, more informal units, which sometimes comprise several members of one family (Hansen, 2009; Harper, 2011; Gettleman, 2011, p. 9). Somali pirates are believed to number about 2,000—including 1,500 in the semi-autonomous region of Puntland—with the largest groups comprising as many as 500 members (OBP, 2011, p. 25; Hansen, 2009, p. 12).[9]

Members of the US Navy transport suspected pirates from the fishing vessel on which they were captured, January 2012.
© Tyler Hicks/The New York Times/Redux Pictures

194 SMALL ARMS SURVEY 2012

Map 6.1 'High-risk area', 2011 'actual' attacks, and reported pirate hubs*

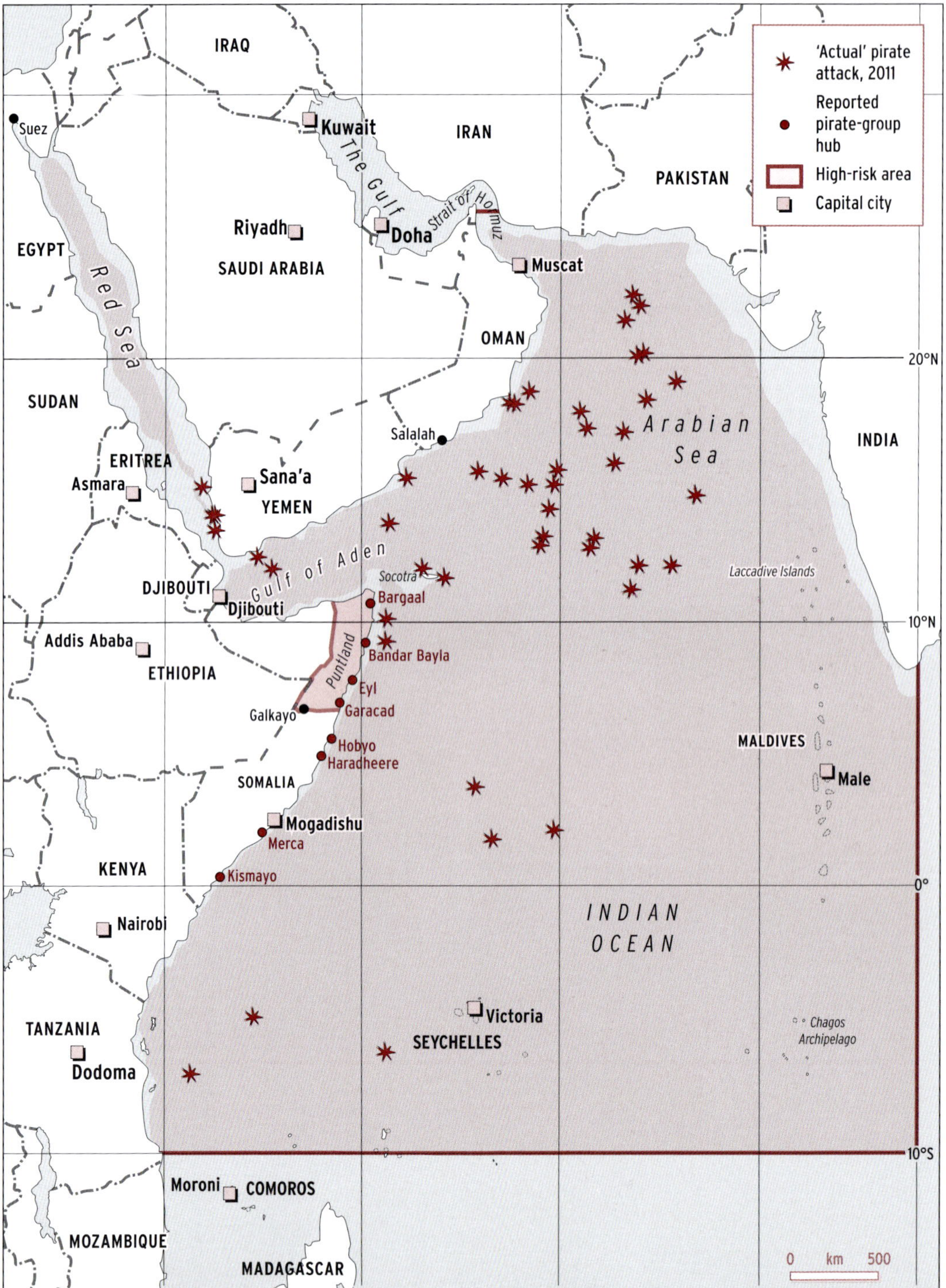

Notes: *The geographical limits of the 'high-risk area' are those defined in the Best Management Practices for Protection against Somalia Based Piracy, version 4; see UKMTO (2011). Actual attacks involve the successful boarding or hijacking of a ship by pirates.

Thanks to ransom payments for the release of ships and their crews, pirates have rapidly accrued the means to operate and recruit far beyond the local fishing communities, thereby evolving into increasingly organized criminal groups. Unlike most pirates active in other regions of the world, Somali pirates do not merely rob valuables they find on a ship; they engage in lengthy negotiations to secure ransoms against the release of the captured crew and vessel, which they often keep anchored near friendly coastal towns in Somalia until an agreement is reached (Murphy, 2009). Various sources estimate paid ransoms to have totalled USD 75–76 million in 2009, USD 80–112 million in 2010, and USD 135 million for the period from January to 7 December 2011. Average ransom payments increased steadily from an estimated USD 1.79–1.90 million in 2009 and USD 3.19–4.85 million in 2010, to USD 4.70–5.00 million in 2011 (UKHC, 2011, para. 111).[10] The proceeds of ransoms are reportedly distributed not only to the sea pirates and guards, but also to the initial 'investors' of the operation and local militia groups; they also tend to benefit local communities as the pirates spend and distribute their loot (UNSC, 2011b, p. 228; Shortland, 2012).

Pirate groups have become highly organized, with attack teams distinct from the guards who look after crews and prevent ships from being hijacked by other gangs (UNSC, 2011b, p. 209). Attack teams typically consist of at least two attack skiffs, or small boats, sometimes supported by a larger supply boat or mother ship loaded with weapons, equipment, and provisions (UNSC, 2010, p. 99; 2011b, pp. 219–21). In one well-documented incident, a mother ship supported two skiffs that each carried four men during the attack (Seychelles, 2010, paras. 13, 14). Testimonies from former hostages indicate that as many as 32 guards can be tasked with guarding a hijacked ship, with 16 to 18 always on the boat and 6 to 8 on active duty (UNSC, 2011b, p. 210).

The number of pirate attacks worldwide almost doubled between 2006 and 2011.

Piracy trends

The IMB Piracy Reporting Centre maintains the most comprehensive database of pirate attacks worldwide and publishes data on a quarterly and annual basis. Its database relies on the voluntary reporting of attacks by ship crews and owners, making the accuracy of IMB figures difficult to assess. The database records the initial report but does not follow up after an investigation is concluded; therefore, key elements emerging after the attack may be left out (Murphy, 2009, p. 60). Complicating the picture further is the fact that ship owners may choose to under-report attacks to avoid political retaliation from governments wishing their waters to be seen as safe (Murphy, 2009, p. 67). Companies may also under-report incidents to prevent an increase in insurance premiums, to avoid affecting employee morale, and to prevent the loss of time and money that would be caused by an official investigation (Phelps, 2011; Murphy, 2009, pp. 64–70). IMB data also tends to reflect primarily attacks against internationally registered commercial ships. Many local Somali fishermen and ship masters are unlikely to report to international institutions such as the IMB because of a lack of knowledge of the official procedures and insufficient communications equipment for reporting.[11]

Despite data limitations, comparisons between IMB statistics on hijackings by Somali pirates in 2010 and those of other sources—including the US Office of Naval Intelligence and the European Union Naval Force—reveal 'only minor discrepancies' (OBP, 2011, p. 7). Analysts usually consider reporting on Somali piracy the most reliable, given the international attention that Somali pirates have received in recent years.[12]

IMB data shows that the total number of pirate attacks worldwide has almost doubled, from 239 in 2006 to 439 in 2011 (see Figure 6.1). Somali pirates are largely responsible for the explosion of global piracy statistics; their share boomed from five per cent of total attacks in 2002 to 54 per cent of the world total in 2011.[13] This is a significant shift from 2002, when most reported incidents occurred in South-east Asia (IMB, 2002, p. 6).

Figure 6.1 The growth of Somali piracy, 2002-11

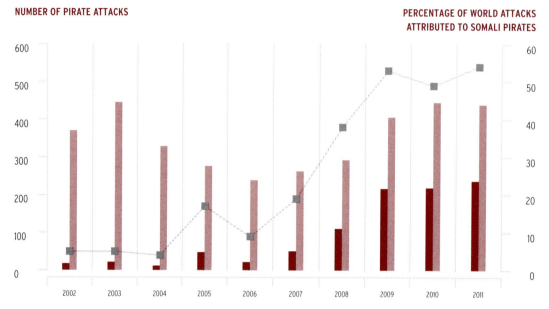

- Attacks by Somali pirates
- Pirate attacks, world
- Percentage of world attacks attributed to Somali pirates

Notes: 'Attacks' in this graph include both attempted attacks (in which a ship was merely fired upon or an unsuccessful attempt was made to board) and actual attacks (in which a ship was boarded or hijacked). Consistent with IMB methodology, attacks attributed to Somali pirates in all graphs in this chapter are those that occur in the waters of the Arabian Sea, the Gulf of Aden, the Indian Ocean, Oman, the Red Sea, and Somalia (IMB, 2011, pp. 5-6).

Source: IMB (2012)

While available indicators point to a remarkable boom in the frequency of Somali pirate attacks, several observers have questioned their overall significance in a wider context. Stephen M. Carmel, senior vice president of the shipping company Maersk Line, Limited, for instance, argues that piracy's effects on the shipping business were negligible, especially when put into perspective with the much higher costs of implementing new shipping regulations. He adds that 'there is a vast army of people in whose economic benefit it is to make everyone think piracy is bad and getting worse. [. . .] Piracy is a pain, but a manageable one that must be kept in context' (Carmel, 2011). Similarly, piracy analysts have argued that the costs of Somali piracy to the international economy—an estimated USD 6.6–6.9 billion in 2011—are 'minuscule' when compared with the total value of maritime commerce of USD 7.7 trillion for 2007 (IHS, 2009, p. 4; OBP, 2012, p. 1; Murphy, 2009, p. 51).

When placed in context, raw statistics on pirate attacks can provide a sense of scale and risk. On average, 30,000 commercial ships travel across the Red Sea and Gulf of Aden every year (OBP, 2011, p. 22). The odds of getting captured by Somali pirates are only about 0.1 per cent (Gettleman, 2011). Crime rates on land provide further points of comparison. As reported by Oceans Beyond Piracy, a US-based NGO that reports on the impact of piracy, a significant number of seafarers—697.5 per 100,000—are subjected to armed attacks on vessels—a rate that exceeds South Africa's 576 major assaults per 100,000 population. The rate of seafarers killed by Somali pirates is 1.3 per 100,000, however, much lower than the world average intentional homicide rate of 6 per 100,000 (OBP, 2011, p. 4; Geneva Declaration Secretariat, 2011, p. 51).

Pirate guns

Since pirates usually dump their weapons at sea when approached by naval forces, it is difficult to produce a clear picture of pirate gun holdings.[14] Available reporting suggests that, since the 1990s, Somali pirates have essentially used the same primary weapons, relying largely on pistols (Tokarev, Makarov), Kalashnikov rifles, light machine guns (PKM), and RPGs (UNSC, 2011b, pp. 205, 210, 215, 216). In a recent operation, the US Navy seized from captured pirates Chinese- and Yugoslav-made Kalashnikov rifle variants, all equipped with under-folding stocks—a practical feature for concealing weapons and using them in confined spaces such as pirate skiffs (Chivers, 2012).[15]

Reports of use of more advanced weapons systems have either been anecdotal or unconfirmed.[16] Pirates' reliance on unsophisticated weapon models may appear surprising given their access to ransom money and the presence of heavy weaponry in Somalia's black markets (ILLICIT SMALL ARMS). Other weaponry would not be particularly useful on pirates' small attack skiffs, which are both unstable and sensitive to recoil. Even AK-47 rifles and RPG-7s are essentially used for intimidation as long as they remain on the skiffs—as they lack the range and accuracy to pose any significant threat to most ships (Kain and Filon, 2011, p. 4). Yet heavy machine guns, mounted on more stable platforms such as mother ships, may be able to repel interventions by naval forces that use helicopters or rigid inflatable boats, making these weapons plausible candidates for future pirate firearms procurement.[17]

In addition, it appears that pirate groups do not necessarily procure weapons in an organized fashion; to some extent, they rely on their members' personal weapons. Pirates who contribute their own weapons and equipment to the group receive higher pay for their contribution (UNSC, 2010, p. 99). Interviews with Somali pirates shed light

Weapons seized by the US Navy following the capture of suspected pirates, January 2012. © Tyler Hicks/The New York Times/Redux Pictures

Figure 6.2 **Percentage of attacks during which pirates used firearms**

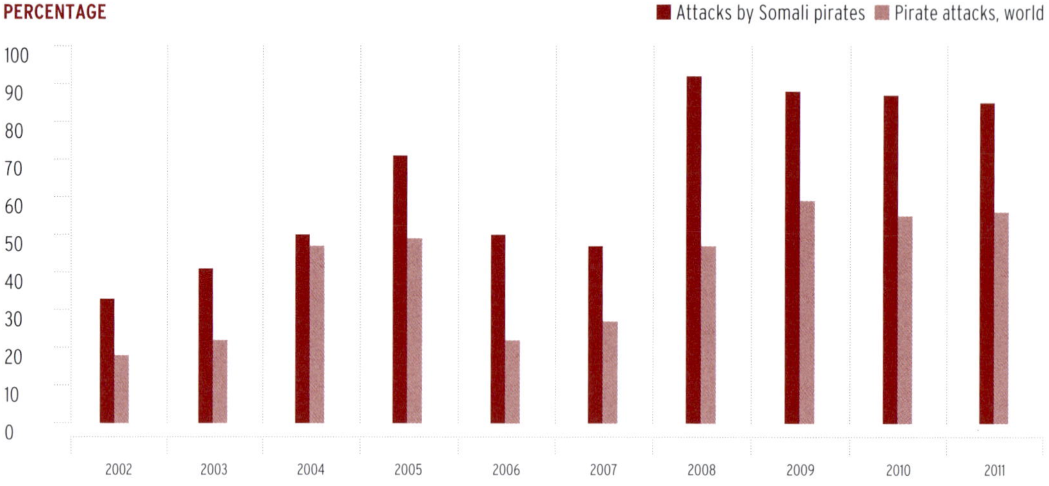

PERCENTAGE — Attacks by Somali pirates ■ Pirate attacks, world

[Bar chart showing percentages from 2002 to 2011. Somali pirate attacks involving firearms: ~35% (2002), ~42% (2003), ~51% (2004), ~72% (2005), ~51% (2006), ~48% (2007), ~94% (2008), ~90% (2009), ~88% (2010), ~85% (2011). World pirate attacks involving firearms: ~20% (2002), ~23% (2003), ~48% (2004), ~50% (2005), ~23% (2006), ~28% (2007), ~48% (2008), ~60% (2009), ~56% (2010), ~57% (2011).]

Note: This graph includes both attempted and actual attacks.
Source: IMB (2012)

on varying weapons procurement practices; some pirates explained that weapons were put in special stores belonging to the group after a mission was completed, while others said that they personally looked after their own weapons (Harper, 2011). Overall, pirate groups appear to be relatively poorly armed. Some attack teams may have fewer firearms than men; in 2010, international naval forces found only seven AK-47s, two RPGs, and ammunition after capturing a mother ship carrying 11 pirates (Seychelles, 2010, para. 19).

While their equipment may be limited, Somali pirates rarely operate unarmed. As Figure 6.2 illustrates, they used firearms in 85 per cent of attacks in 2011, whereas just 56 per cent of all pirate attacks around the world involved firearms. Somali pirates' systematic use of firearms appears to have taken hold in 2008, although that use has since stabilized or even declined slightly. Some sources observe not only the routine carrying of firearms, but also a shift from the use of weapons for intimidation to more frequent actual use.[18]

While it is difficult to establish with certainty what caused weapons use to increase in 2008, it is worth noting that several events in Somalia and at sea around that time changed the environment in which pirates evolved. In mid-2006, the Islamic Courts Union (ICU) seized control of most of South and Central Somalia and enforced a ban on piracy, leading pirate groups to relocate from the town of Haradheere in Central Somalia to Eyl in the semi-autonomous Puntland region (Bahadur, 2011, p. 37; Murphy, 2009, p. 105). After the ICU collapsed in December 2006, pirate groups were able to reorganize and expand in the comparatively more stable region of Puntland; a severe economic crisis facilitated this move as Puntland authorities were unable to pay the security forces needed to investigate piracy (Bahadur, 2011, pp. 38-42).

The year 2008 also marked the beginning of the international community's recognition of Somali piracy as a threat, following the much-publicized hijackings of three ships in particular: in September 2008, the target was the *Faina*, which carried weapons allegedly destined for South Sudan; in November it was the *MV Sirius Star*, a super tanker carrying USD 100 million worth of crude oil; and in April 2009, the *Maersk Alabama* became the first US cargo vessel hijacked in two centuries (Bahadur, 2011, p. 37). Increased media coverage ensued, as did the deployment of international navies to fight piracy in the area. The EU operation 'Atalanta' became active in December 2008, the US-led Coalition Task Force 151 was set up in January 2009, NATO's operation 'Ocean Shield' began in August 2009, and China, India, Iran, Japan, South Korea, and the Russian Federation sent warships independently (Ghosh, 2010, pp. 14–15). The UN Security Council formed the Contact Group on Piracy off the Coast of Somalia in January 2009, providing concerned states with a platform to coordinate military, political, and other efforts to tackle Somali piracy (Priddy and Casey-Maslen, 2012, pp. 11–12). Increased armed opposition at sea, and a greater ability to manoeuvre and operate inland, are factors that help explain Somali pirates' more systematic reliance on firearms from 2008.

Figure 6.3 **Number of seafarers killed per 100 actual attacks***

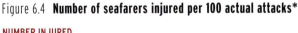

Note: * Actual attacks are those during which pirates boarded or hijacked a ship.
Source: IMB (2012)

Figure 6.4 **Number of seafarers injured per 100 actual attacks***

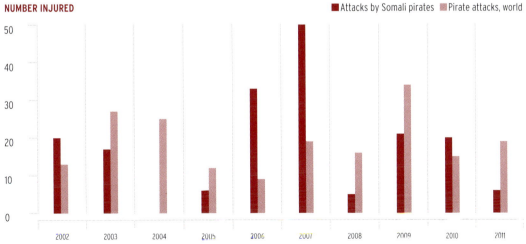

Note: * Actual attacks are those during which pirates boarded or hijacked a ship.
Source: IMB (2012)

More violence?

Recent statistics and analysis suggest that Somali pirates are becoming increasingly violent during attacks, especially if they succeed in boarding the targeted ship. IMB data—which records information provided during or shortly after attacks—shows that the number of seafarers killed has doubled from 8 per 100 'actual attacks'—in which pirates either boarded or hijacked the ship—in 2009 to 17 per 100 in 2011 (see Figure 6.3). In contrast, the rate of injuries inflicted on seafarers by Somali pirates fluctuated during that period, even decreasing in the past two years, from 20 per 100 actual attacks in 2010 to 6 per 100 in 2011 (see Figure 6.4).

At first glance, these trends may seem contradictory; in fact, they suggest that attacks by Somali pirates are becoming more lethal, possibly as a consequence of their increasingly systematic use of weapons. Since pirates are now operating in a larger, more dangerous area, they are required to act more swiftly—and with more determination—than in the past. Their targets are faster, larger ships that must be boarded before a naval ship can intervene—typically within 25–40 minutes.[19] As discussed below, the increased use of armed guards to protect commercial ships increases risks for pirates and is probably an additional factor pushing them to adopt more violent behaviour during attacks.

Less is known about violence perpetrated against hostages during the captivity period—after pirates successfully gain control of a ship—but available analysis suggests its incidence may be significant.[20] A recent report shows that close to 60 per cent of hostages taken by Somali pirates in 2010 were subjected to 'abuse' or used as human shields, or both (OBP, 2011, p. 3). Physical abuse included the 'deprivation of food and water, beating (often with the butt of a gun), shooting at hostages with water cannons, locking hostages in the ship's freezer, tying hostages up on deck exposed to scorching sun, and hanging hostages by their feet submerged in the sea' (p. 17). Pirates also engaged in psychological abuse of hostages, such as by firing weapons as an intimidation tactic, placing hostages in solitary confinement, calling family members while threatening hostages, parading hostages naked around the vessel, taking hostages ashore to see their supposed graves, making death threats, and orchestrating mock executions (OBP, 2011, pp. 17, 20).

Several interconnected factors help provide a better understanding of the dynamics behind such abuse. A private negotiator involved

A captain held hostage by Somali pirates for more than one year greets family members following his release, Karachi, Pakistan, June 2011.
© Shakil Adil/AP Photo

in hostage release negotiations explains that while pirates understand that an escalation of violence might trigger an even more drastic response by governments, the 'tensions and frustrations' that come with longer and more intense hostage release negotiations have made them more inclined to resort to violence.[21] The UN Monitoring Group also notes that as international naval forces deploy more systematically and push pirate groups to operate farther out in the Indian Ocean, the 'enhanced risks and costs to pirates associated with operating at greater distances from shore have helped to drive up ransom demands and prolong negotiations for the release of hijacked vessels' (UNSC, 2011b, p. 12). Indeed, in March 2011, the average duration of a hijacking reached 214 days, a doubling of early 2010 figures (Ince & Co., 2011, p. 2).[22] Interviewed pirates concur and identify the more robust international response, and harsher treatment of captured pirates, as among the drivers behind the increased violence (see Box 6.2).

Rules of behaviour

Pirate codes of conduct serve important organizational and disciplinary purposes.

The increasing use of violence among pirates has led analysts to warn that the 'supposed code of conduct for piracy is at risk of changing now that each side has reneged on the terms of hostage and ransom negotiations' (OBP, 2011, p. 12). The unspoken code required pirates to keep their hostages unharmed in exchange for the securing of generous ransoms.[23] While that tacit agreement may have contributed to restraining levels of violence for some time, it appears to have broken down, as evidenced by cases in which pirates refused to release all hostages although a ransom was paid and, conversely, in which naval forces attacked and killed pirates after hostages were freed (OBP, 2011, p. 12).

A closer look at known pirate rules of behaviour shows that while the treatment of hostages is an important component, documents referred to as codes of conduct also serve broader organizational and disciplinary purposes. Clear and precise hierarchies and organizational guidelines govern pirate group behaviour, with monetary fines and traditional clan codes of conduct serving as primary means of ensuring discipline among the ranks.[24] Documents found with some captured pirate groups, for instance, set out a number of rules, each subject to a monetary fine for non-compliance. Mistreating the crew of a hijacked vessel can carry a fine of up to USD 5,000 and dismissal, and order refusal is punishable by a fine of up to USD 10,000.[25] Another document retrieved by the UN Monitoring Group on Somalia and Eritrea specifies a USD 1,500 fine for members who steal from the captured ship, while providing a USD 2,000 merit-based reward to any group member who performs well (UNSC, 2010, pp. 97–98).

These rules suggest that pirates' codes of conduct are important instruments for group cohesion and efficiency. Interviews carried out with pirates in July and August 2011 suggest that Somali pirate groups continue to rely on these rules of behaviour, either in written form or transmitted orally by team leaders, and that the rules remain largely unchanged (see Box 6.2). Anecdotal reports confirm that some pirate groups have indeed taken steps to protect hostages. In one case, a female crew member of a captured ship was promised she would not be harmed, a pledge the pirates respected, and all of the hostages' private belongings were returned to them shortly before the end of the hijacking (UNSC, 2011b, p. 210, ns. 14, 18).

While the contents of pirate codes of conduct appear not to have changed dramatically over time, there is evidence of a shift towards more violent tactics. Some pirate documents now seem concerned with tactical questions, such as how pirates should respond to naval forces (see Box 6.2). Recent incidents confirm a shift in pirates' attack tactics. In late 2011, pirate groups carried out attacks with as many as 6 to 12 skiffs—in contrast to earlier attacks, which typically involved 2 or 3 skiffs (Seychelles, 2010, paras. 13, 14; Stratfor, 2012; Thomas, 2011). In August 2011, Somali

Box 6.2 Violence trends and rules of behaviour: the perceptions of five Somali pirates[26]

Several reports document the existence of Somali pirate codes of conduct (Murphy, 2011, p. 121). Originally, at least, some of these codes reportedly contained provisions prohibiting pirates from mistreating hostages, thus contributing to low levels of violence on captured ships. In light of the increased incidence of violence inflicted to hostages, some observers have suggested that pirates have abandoned or renounced their codes.

In-depth interviews with five pirates from different groups based in coastal areas of south-central Somalia (Hobyo and Haradheere) and the semi-autonomous north-eastern region of Puntland (Eyl and Garacad) provide a more nuanced assessment.[27] All five pirates said they used to be fishermen and were between their early twenties and early forties. Some worked as 'foot soldiers'–part of the attack teams that take to sea in small boats in search of ships to hijack. One of them said he was the leader of a pirate gang, one man was a spokesman for several groups, and another said he sometimes organized operations but also took part in attacks. The existence of several different roles within some of the groups suggests that at least some of them are well organized and structured.

All the pirates confirmed that their groups had codes of conduct with sanctions for those who broke the rules. Four of the five pirates said their groups had written codes of conduct. All the pirates had a detailed knowledge of the rules, which were read out to them by their leader before they went out on a mission. The nature of the rules differed somewhat across the different groups but did not seem to change much over time. For some groups, the most important rules concerned the capture of ships and the treatment of hostages. For others, the rules concentrated on how to deal with attacks from foreign navies. One pirate said the key rule was that the pirates must not fight among themselves. Some of the rules dealing with the treatment of hostages stated that they must not be killed or tortured, but other pirates suggested the rules for hostages from countries whose navies were more 'violent', such as South Korea's, involved a more brutal approach from the pirates. Some rules dealt with 'taxation' (ransom) issues, including how the money should be divided. The pirates said there were clear rewards and punishments for members of the group who obeyed or broke the rules. One interviewee referred to a special 'pirate court' where transgressors were tried and sentenced.

All groups offered rewards for obedient pirates who closely followed the rules. The most common form of reward was increased respect and admiration–which is highly prized in Somali culture. Material rewards were also mentioned; one interviewee said good pirates were rewarded with 'whatever they were interested in at the time'. There were heavy sanctions for those who broke the rules; transgressors were expelled from the group, fined, or even imprisoned.

Despite the existence of such codes, all five pirates recognized they had become more violent over time. Pirates explained that they become violent towards hostages when ransom negotiations are advancing slowly; they reported taking out their frustration by beating and humiliating their hostages. They also blamed international naval patrols, the increased armed protection of cargo ships, the killing of pirates by foreign forces, and the long prison sentences handed out by foreign courts to Somalis found guilty of piracy. One interviewee spoke of captured pirates being tortured. Overall, pirates explained that increased violence was a reaction to what they perceive as the increasingly robust, organized, and militarized international response to Somali piracy. Most of the interviewees said no pirate group was more violent than any other. Only one said the more established gangs were the most violent because the money they had gained from ransoms enabled them to buy more sophisticated weapons, including anti-aircraft guns and PKM general purpose machine guns.

Source: Harper (2011)

---

pirates hijacked an Indian chemical tanker that was anchored at the Omani port of Salalah after the private security guards protecting it had disembarked (Reuters, 2011a; 2011b). The incident was a bold and surprising move as Somali pirates usually targeted ships in the high seas and not in ports considered safe, such as Oman's. One pirate group even turned inland for hostage taking, kidnapping three humanitarian workers of the Danish Demining Group in Galkaayo, Puntland, in October 2011 (CFC, 2011; *Somalia Report*, 2011b). Taken together, these events suggest that pirates can and will adapt their tactics in order to maintain a steady flow of ransoms.

ARMED GUARDS AT SEA

Faced with increasing numbers of pirate attacks and rising ransom demands in spite of the intervention of naval forces, the IMO, governments, and shipping companies have increasingly come to recognize the use of PSCs as an option for protecting commercial vessels in the areas that are exposed to Somali piracy. This section documents the growing scale of the use of armed guards in this 'high-risk area' and the types of weapons employed by PSCs.

A controversial shift

The use of private armed guards on commercial ships is a relatively new development. It challenges a long-established division of labour between private companies that were responsible for commercial shipping, on the one hand, and states' coast guard and naval forces that guaranteed security at sea, on the other (Murphy, 2009; Noakes, 2011). The shipping industry has been particularly reluctant to endorse the use of armed guards. The Baltic and International Maritime Council, an international shipping association that represents ship owners controlling around 65 per cent of the world's tonnage, is among the few actors that still criticize the move (BIMCO, 2011; Swedish Club, 2011a). The main concerns relating to the use of private armed guards on ships include:[28]

Ransom money is dropped by parachute near the Ukrainian cargo ship MV Faina, February 2009.
© Michael R. McCormick/US Navy

- the risk of an escalation of violence and a change of tactics and weapons by pirates;
- the time and resources required to obtain clearance for the passage of armed guards in ports and territorial waters;
- legal liability issues when the ship is damaged or the crew or security team are injured;
- the accreditation and 'policing' of PSCs;
- the challenges that the presence of armed teams poses to the ship masters' authority as enshrined in the International Convention for the Safety of Life at Sea; and
- the possibility that other preventive measures may be as effective.

Despite these concerns, most key maritime players have, one by one, recognized the use of PSCs as a legitimate option for dealing with piracy in the high-risk area. The IMO issued a series of circulars between May and September 2011, providing guidance and recommendations for states and shipping companies on the use of private armed teams (IMO, 2011a–2011f). That same year, a coalition of international counter-piracy actors and shipping industry associations issued the *Best Management Practices for Protection against Somalia Based Piracy* (BMP)—now in its fourth version—acknowledging that the use of armed guards is a 'matter for individual ship operators to decide following their own voyage risk assessment and approval of respective Flag States' and making reference to the guidance

included in the IMO circulars (UKMTO et al., 2011, pp. 39–40). At the same time, several governments that had previously firmly opposed the practice either changed their legislation to allow the use of armed PSC personnel or were in the process of reviewing their policies.[29] In November 2011, the US Department of State declared that 'the shipping industry's use of [BMP] and the increasing use of Privately Contracted Armed Security Personnel are among these measures, which have proven to be the most effective deterrents against pirate attacks' (USDOS, 2011). Significantly, in late 2011, insurance companies also expressed growing support for placing armed guards on ships (Saul and Barker, 2011).

While most stakeholders fall short of an outright endorsement of the use of armed guards, the growing number and geographical spread of pirate attacks have left them little choice but to accept the practice. As noted by the Security Association for the Maritime Industry (SAMI), the growth of the world's trade combined with the shrinking of Western navies is giving the private security industry an indisputable role (Cook, 2011). Analysts also argue that as the use of private guards expands to become the norm, hijacking victims on unarmed ships may be able to sue their employers on the basis that not hiring PSCs constitutes negligence and a failure to provide seafarers with a safe working place (Friedman and Smith, 2011). The risk of being sued may lead more companies to rely on armed guards and provide additional momentum to the private maritime security industry.

> The number of successful hijackings by Somali pirates decreased in 2011.

Reports document a rapid increase in the use of armed guards in 2011. From May to November 2011, the proportion of ships carrying private armed guards in the high-risk area reportedly increased from ten to as high as 25 per cent (BBC News, 2011a; Bloomberg, 2011b).[30] Industry representatives predicted that the proportion would continue to increase in 2012 (Bloomberg, 2011a). Some 1,000 armed guards were believed to be deployed to protect ships against Somali pirates in 2011 (Saul and Barker, 2011). As of October 2011, 70 SAMI members were carrying out 550 to 600 escorts per month, providing protection to about 25 to 28 per cent of all transits across the Indian Ocean region, 90 per cent of which were armed (Cook, 2011; *Somalia Report*, 2011a). In 2011, the annual cost of providing armed private security on ships was estimated at USD 530.6 million, amounting to more than three times the ransoms total paid to Somali pirates during that year (OBP, 2012, pp. 1, 19).

Despite a continued increase in attempted pirate attacks, the new protective measures have shown encouraging results. The success rate of attacks—that is, the proportion of Somali pirate attacks resulting in a successful hijacking—decreased to 12 per cent in 2011 compared with 22 per cent in 2010 and a ten-year high of 38 per cent in 2008 (see Figure 6.5). Maritime security industry representatives claim that armed security teams deterred 90 per cent of all unsuccessful acts of piracy (Cook, 2011). Although 2011 saw more attacks by Somali pirates overall, pirate activity declined in the latter part of the year; Somali pirates carried out 31 attacks and hijacked 4 vessels in the last quarter of 2011, compared to 90 attacks and 19 hijackings over the same period in 2010 (BBC News, 2011d; IMB, 2011, p. 24). Analysts also indicate that pirates' area of operation shrank slightly in 2011, noting, however, that matters could hardly have become worse than they were in 2010 (Startfor, 2012). While other factors—such as the increased deployment of naval forces since 2008 and the more systematic use of BMP, including the hardening of ships and use of secured rooms or 'citadels' where crews can seek shelter during attacks—may also have had an impact, informed observers usually agree that the more systematic deployment of private armed guards in 2011 played an important role in reducing the effectiveness of pirate attacks.[31]

It must be stressed, however, that neither the decrease in the rate of successful hijackings nor the deployment of armed guards has succeeded in preventing the increase in ransoms paid to pirates in 2011, as discussed above. According to one source, Somali pirates had already collected more ransom money in the first seven months of 2011

Figure 6.5 Percentage of attacks by Somali pirates resulting in a hijacking

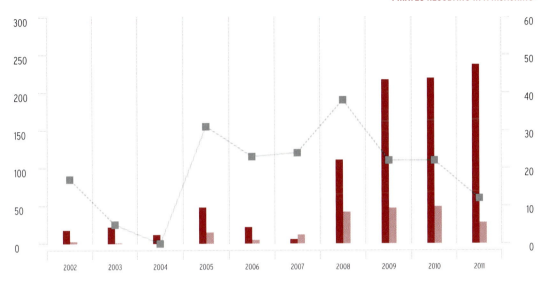

Notes: This graph includes both attempted and actual attacks.
Source: IMB (2012)

than in the whole of 2010 (UKHC, 2011, para. 111). This trend suggests that pirates may have felt less financial pressure to seize ships in late 2011.[32]

PSC firearms

Private maritime security firms essentially offer two types of generally defensive armed services:[33] teams of armed guards that remain onboard the protected vessel itself, and separate escort vessels that accompany and protect up to four commercial ships.[34] Armed teams on commercial ships usually comprise three to four men, and according to reports usually cost about USD 5,000 per team per day, although figures range from USD 1,500 to USD 21,000 per team per day (Carmel, 2011; Friedman and Smith, 2011; Saul, 2011). Escort vessels carry about six to eight armed security personnel and cost USD 30,000–55,000 per vessel for a three- or four-day journey (Bloomberg, 2011b; Ghosh, 2010, p. 29).[35]

There is no standard 'weapon kit' used by the private maritime security companies. Industry sources reveal that the ratio of firearms per PSC personnel varies between 0.75 and 2.00.[36] There appear to be two main practices with respect to the types of firearms used by PSCs. Some companies use a single type of firearm—most commonly assault rifles or shotguns.[37] These weapons provide a range not exceeding 300–400 metres at sea;[38] they appear to be used primarily in self-defence and to intimidate pirates. Some private armed guards have even been provided with World War II-era German carbines (Tammik, 2011). In fact, some PSCs have no choice but to use whatever equipment is legally available, even if inappropriate for guarding ships. Key transit points such as Oman and South Africa only allow semi-automatic weapons onboard ships that transit through their waters, forcing companies to choose from the few such weapons they can procure on the legal markets—often shotguns and hunting rifles.[39] Companies that use any available type of firearm rely primarily on the deterrent effect offered by the mere presence of armed guards on ships, regardless of the equipment at their disposal, hoping it will persuade pirates to backtrack and look instead for unarmed targets. Indeed, a factoid regularly used by supporters of the use of PSCs states that 'no ship with an armed security team embarked has been boarded and hijacked' (Cook, 2011; Thomas, 2011). While it may be accurate, this claim cannot be verified.[40]

Other firms rely on a combination of more specialized weapons that are effective from close range to more than 1,200 m, including pistols and shotguns (20-m range), light machine guns (400–600 m), general-purpose machine guns (1,000–1,200 m), and sniper rifles (1,000–1,200 m) (Thomas, 2011).[41] These companies use a strategy that rests on a more graduated use of force.[42] They explain that for PSC firearms to play a deterrent effect, they must be visible and more powerful than the pirates' own guns. This requires 'specialist' weapons that either have significant firepower or are accurate enough to disable pirate skiffs without injuring their men and to fire warning shots from great distances (Kain and Filon, 2011, p. 5).[43] This practice raises important issues, however; to be in accordance with applicable law, the use of force in self-defence must not exceed what is strictly necessary and must be proportionate to the threat (Priddy and Casey-Maslen, 2012, p. 2).[44] Sniper rifles and general-purpose machine guns, in particular, provide PSCs with ranges and accuracy far greater than the weapons currently deployed by the pirates, and for this reason their use could be characterized as disproportionate.[45]

There is no international standard on what type of firearms PSCs may use.

The IMO began providing general guidance on the matter in a September 2011 circular, indicating that PSC weaponry needed to provide an 'accurate and graduated level of deterrence, at a distance' (Swedish Club, 2011d; IMO, 2011d, annex, p. 6). While this guidance appears to support the second practice described above—the graduated use of different types of firearms at various ranges—it lacks specificity with respect to what constitutes an appropriate range for firing warning shots, for instance, and could be broadly interpreted. A lack of clear regulations on the types and quantities of weapons used by PSCs may lead to a great disparity in the approaches used, with less responsible firms using excessive and inappropriate equipment for the task. As reported by the UN Monitoring Group, for instance, Clear Ocean, a company based in the United Arab Emirates, planned to acquire heavy weaponry to undertake a contract for Somalia's Transitional Federal Government, including a '20mm Regimental Ship Gun, an AK630 Gatling machine gun, [and] NSV caliber 12.7mm' (UNSC, 2011b, p. 261). As a shipping industry executive noted:

> *there is no international standard on what types of weapons, on the training and vetting of shooters, or even any requirement [that] they are different than the normal crew. Nor is there any international standard on what types of weapons are considered appropriate [. . .] limits on weapons and actually no useful guidance on training. That is all up to us* (Carmel, 2011).

Weapons procurement issues

Shipping companies wishing to place armed guards on their vessels must consider at least three different sets of laws. Most important, the laws of the ship's flag state—the state where the ship is officially registered—but also the laws of the state where the ship owners or managers are incorporated, as well as the regulations of the coastal states or ports where the vessel will transit or stop (Swedish Club, 2011a). IMO guidelines state that the shipping industry must respect flag, coastal, and port state firearm regulations when using private armed personnel onboard ships;

they also require that procedures be in place for the storage and inventorying of firearms on the boat and during transfer (IMO, 2011d, sec. 3.4). The IMO also calls on flag states to have policies in place to determine whether the use of armed PSC personnel is authorized and, if so, under what conditions (IMO, 2011e). Finally, the IMO urges governments of coastal states bordering the Indian Ocean, the Gulf of Aden, and the Red Sea to develop policies and procedures to 'facilitate the movement of [private armed guards] and of [their] firearms' (IMO, 2011f, annex, p. 1).

As described above, several governments have either made arrangements to allow the use of armed guards on registered vessels or were reviewing their policies in this area in late 2011.[46] Some of the countries in and around the high-risk area reportedly allow PSCs to store weapons in their ports between escorts and sign them on and off the protected ships, allowing for a degree of transparency and accountability in PSC arms procurement.[47] Although individual states and the IMO have made efforts to facilitate the work of PSCs, industry sources explain that regulations are often inconsistent. Ports in Oman and South Africa, for instance, only allow semi-automatic firearms in their territorial waters, complicating the work of PSCs that use automatic weapons.[48] German-flagged ships cannot carry guards armed with semi-automatic rifles (Dabelstein & Passehl, 2011, p. 2). Very few ports allow PSCs to store weapons for extended periods of time after a contract is finished, creating logistical challenges for companies that are waiting for the next client.[49]

In addition, some countries' regulations may change unexpectedly. The Suez Canal authority temporarily prohibited merchant vessels from transiting the canal with firearms on board in 2010, obliging armed ships to hand over their weapons to Egyptian authorities, who drove the weapons to other end of the canal, where they returned them to the ships (Bennett, 2010). For PSCs that acquired their weapons from the United States, this procedure violated the International Traffic in Arms Regulations and prompted the US Coast Guard to make special legal arrangements to allow the temporary handover of US-exported weapons to the Suez Canal authority (Bennett, 2010; USCG, 2010).

A boat in the Red Sea believed to serve as a floating platform for the embarking and disembarking of weapons onto ships protected by PSCs, October 2011. © DS

This complex web of legal requirements has led to surprising situations and violations of established rules. Reports emerged in early 2011 that PSCs were dumping weapons at sea to avoid violating arms transfer regulations when arriving at ports of call or a final destination (Hope, 2011; Saul, 2011). The UN Monitoring Group found that one PSC violated the arms embargo on Eritrea in December 2010 (UNSC, 2011b, para. 182). The *Sea Scorpion*, a ship operated by Protection Vessels International, effectively served as a 'floating platform for storing and transferring weapons, equipment, and personnel between operations' (UNSC, 2011b, p. 310). Inclement weather and a need for fuel had led the ship into Eritrean waters, in violation of the UN arms embargo; the event instigated a stand-off with Eritrean authorities. Sources indicate that other companies were using a similar 'floating arms platform' model in late 2011.[50] PSC personnel have also faced trouble when flying to their ports of embarkation with their firearms. Five employees of the US firm Greyside, for instance, were detained at Nampula airport in Mozambique in September 2011 for illegally possessing weapons—including a FN 5.56 mm rifle and ammunition—that they had procured in Kenya (AllAfrica, 2011; BBC News, 2011b).

> PSCs can rent government-owned firearms for a fee in some ports.

Some governments whose ports are strategically located on the main shipping routes crossing the high-risk area have set up special arrangements to allow PSCs to embark and disembark weapons. Some countries, such as Djibouti, sell annual permits for USD 150,000 or more that allow PSCs to operate from their ports with weapons (UNSC, 2011b, para. 179, n. 154).[51] Djibouti has also put in place a gun-rental scheme whereby ships carrying PSC personnel may rent and embark government-owned weapons at its port in exchange for daily fees. A presidential decree appointed Djibouti Maritime Security Services (DMSS) as the only private entity tasked with authorizing and controlling the activities of PSCs operating from the country, including the temporary transit, rental, and storage of weapons (Republic of Djibouti, 2009, arts. 2, 3). As of June 2011, only about 200 semi-automatic firearms—including Browning semi-automatic rifles, .30-06 Benellis, .308 Winchesters, and Saiga M3s—imported for this purpose from Malta, were available for rent, and all firearms had been rented out.[52] Representatives of several private security firms, as well as a government source, stated that DMSS also rents out fully automatic weaponry.[53] As of mid-2011, DMSS's price list included AK-47s, AR10s, Browning BARs, Steyrs, and Dragunov Tiger rifles available at the rate of USD 30 per day, RPKM light machine guns at USD 50 per day, and ammunition at the rate of USD 5 per round used.[54] DMSS also operates a fast supply vessel that allows it to go to sea to retrieve weapons from returning ships before they move on to other ports that do not allow armed guards to enter (UNSC, 2011b, p. 305).

On the other side of the high-risk area (see Map 6.1), Sri Lankan authorities have set up a similar system out of the port of Galle. Through private Sri Lankan companies that act as intermediaries, PSCs can rent weapons, ammunition, and equipment belonging to the Sri Lankan armed forces, including Type 56 automatic assault rifles, 84S semi-automatic rifles, and 12-gauge repeater shotguns.[55] The rental cost was of USD 210 per weapon per day as of November 2011. Rented weapons come with lockable safe boxes for storage and a set amount of ammunition (120 rounds per Type 56, five rounds per 84S and shotgun); clients may also request additional rounds.[56] Ammunition costs USD 0.50 per spent round. All rented equipment is to be returned to Galle within one month of issue.[57] An important difference with the Djibouti scheme is that ships renting out Sri Lankan weapons must also embark a retired or off-duty Sri Lankan Navy or Army officer who will stay onboard in his private capacity to monitor the use of the weapons.[58] PSCs must pay a USD 850 daily fee for the presence of the officer and cover the costs of food, accommodation, insurance, and travel back to Galle.[59]

Other governments offer the services of their own security forces to escort commercial ships. The Yemeni Navy reportedly collaborated with a British PSC to offer escorts through the Gulf of Aden, even proposing full military escorts along its coast at the rate of USD 50,000 for a three-day journey (Ghosh, 2010, p. 29).[60] Sri Lanka does not only rent out government-owned weapons, but it also permits the hiring of three-man teams of retired or off-duty Sri Lankan officers to provide security on ships.[61] Reports released in late 2011 speculate that the shipping industry itself was considering adopting its own arrangements. A group of container lines was reportedly discussing the possibility of using a common pool of armed security guards that would shuttle across the high-risk area on different companies' ships (Wallis, 2011). The move appeared to result from the increasing security costs borne by the industry; one shipping company declared having to increase its security surcharge by 20 to 50 per cent in 2011, and spending USD 200 million in 2011 on security, or double the amount spent in 2010 (Wallis, 2011).

Rules on the use of force and firearms

As with weapons procurement, a complex web of laws and non-binding guidelines regulates PSC use of force and firearms. Human rights principles require that use of force be avoided whenever possible and that, in the case of an act of self-defence, it be proportional to the threat (Priddy and Casey-Maslen, 2012, p. 2). PSCs are bound by the law of the state in which the ship is registered and by that of the states whose waters it transits. National regulations tend to vary greatly across countries and do not always provide sufficient guidance on PSC use of force (OBP, n.d.; UKHC, 2011, para. 37). The IMO attempts to address these discrepancies and gaps by specifying that 'it is essential that [PSC personnel] have a complete understanding of the rules for the use of force as agreed between shipowner, [PSC], and Master and fully comply with them' (IMO, 2011d, sec. 3.5). PSCs are to provide a 'detailed graduated response plan to a pirate attack' and prevent their personnel from using firearms against persons except in self-defence or defence of others against the imminent threat of death or serious injury (sec. 3.5). These guidelines are not legally binding, however.

> Some PSCs fire disabling shots aimed at a pirate boat's propulsion system.

The shipping industry has also sought to provide a degree of standardization, including the BMP, a standard contract for PSCs developed by the Baltic and International Maritime Council, and guidelines developed by national or international shipping associations (see OBP, n.d., pp. 2–6). Maritime PSC associations such as the International Association of Maritime Security Professionals and SAMI were, at the time of writing, also in the process of putting in place industry standards—including rules on the use of force—and accreditation mechanisms.[62] A number of maritime PSCs are signatories of the International Code of Conduct for Private Security Providers, a document developed mainly for land-based PSCs (OBP, n.d.).[63]

A review of actual PSC rules on the use of force made available to the Small Arms Survey suggests that current guidelines and obligations remain subject to broad interpretation and would benefit from more specificity. For some companies, the use of firearms is only justified when pirates shoot first; for others, team members may shoot if they conclude that the crew or security team is at risk of death or serious bodily harm.[64] Some companies adopt a graduated response approach, beginning with the firing of warning shots at more than 1,200 m, followed by disabling shots aimed at the skiff's propulsion system, and ending with the use of lethal force in self-defence at close distances.[65] The lack of consistent rules governing the use of force and firearms poses challenges for the oversight of PSCs and may also result in miscommunication with pirate groups and other ships.[66]

A related issue is the extent to which the presence of armed guards threatens established command and control procedures on ships. The International Convention for the Safety of Life at Sea clearly gives the ship master responsibility for the crew and authority on the vessel (IMO, 1974, art. 34.1; Ince & Co., 2011, p. 2; Swedish Club, 2011a). If an armed guard faces an immediate lethal threat, he or she is not necessarily obliged to consult the master since the right to self-defence would arguably outweigh the master's authority (Ince & Co., 2011, p. 2).[67] Some insurers recommend that the private security team seek the master's consent as long as it is 'reasonable' or 'feasible' to do so (Swedish Club, 2011b). This approach would require that the ship master be fully briefed and trained in rules on the use of force.

Legal issues also arise with respect to armed escort ships carrying firearms in territorial waters. Article 19(2)(b) of the UN Convention on the Law of the Sea states, for instance, that the 'passage of a foreign ship shall be considered prejudicial [. . .] if in the territorial sea it engages in [. . .] any exercise or practice with weapons' (UN, 1982). Escort vessels may also face legal challenges in arguing that the use of force to protect another vessel is an act of self-defence.[68]

There is little information available on the actual use of force by PSCs, largely due to the fact that current IMB and other reporting focuses on violence perpetrated by pirate groups and not by private guards or naval forces.[69] Although a lack of detail precludes attribution to PSCs or naval forces, a UN report suggests that 200–300 pirates went missing and that at least 62 were killed at sea during the first five months of 2011 (OBP, 2011, p. 25; UNSC, 2011a, para. 17). The lack of knowledge about the circumstances of these pirates' deaths, together with reports of PSCs mistakenly firing on fishing boats, shows that the effective monitoring of the use of force and firearms by PSCs is a ways off (Ince & Co., 2011, p. 3).

Rapid PSC deployment has outpaced regulation.

Based on the limited available data in this area, this study indicates that the increased use of PSCs risks leading to an escalation of violence at sea—and possibly on land. Private security representatives argue that in the vast majority of cases, showing weapons or firing warning shots continues to be sufficient to deter approaching pirates (Thomas, 2011). But this may only be the case because most ships—about 75 per cent according to the above-mentioned estimates—remain unarmed and pirates generally avoid fights with armed guards if they can simply wait for the next unarmed ship.[70] As the proportion of armed ships continues to increase and as private guards become the norm rather than the exception, pirates could become more inclined to use force against PSCs. Another blowback risk is that pirates may step up attacks on local fishermen's ships, or dhows, for loot or use as mother ships, further undermining local development.[71] Above-mentioned examples of pirates' capacity to adapt to changing circumstances and to take retaliatory measures suggest that the arming of ships, once it becomes the norm rather than the exception, could lead to a more systematic use of force by pirates and armed guards, and a further escalation of violence.

CONCLUSION

The last two years have seen dramatic changes in the provision of security at sea. Once the exclusive domain of the world's navies and coast guards, maritime security now involves a complex web of private and public players, sometimes intermingled in new, unusual partnerships. Somali piracy has been the key factor contributing to this development. By adapting their tactics and stretching their geographical reach, Somali pirates have demonstrated the limits of state security provision at sea, leaving the shipping industry and government regulators few alternatives but to adopt self-protective measures and accept the use of private armed guards. From being a negligible player, maritime PSCs have grown to serve as protectors of roughly one-quarter of the ships travelling in the high-risk area exposed to Somali piracy, and their importance appears set to increase in the near future.

Whether this new paradigm increases overall security on the seas remains an open question. PSCs appear to have reduced the success rate of pirate attacks. The relative decline in pirate attacks of late 2011 provides further reasons for hope. The PSC presence has not influenced ransoms paid, however, which increased again in 2011. Moreover, rapid PSC deployment has outpaced regulation, with issues such as the types, quantities, procurement, and use of firearms requiring focused attention. Available evidence also suggests that in response to increased armed opposition at sea, pirates have exposed seafarers to more lethal violence during attacks and greater abuse during captivity. Overall, pirates have adapted their tactics in response to international maritime efforts to curb their activities.

Should pirates one day run out of unarmed ships to attack, they may shift to more violent and innovative methods in order to keep the ransom money flowing, as they have in the past when confronted with similar challenges. As of the end of 2011, new tactical developments included increasing the number of attack skiffs, striking ships close to or within ports, and kidnapping foreigners on land. In the absence of serious efforts to engage Somali pirates non-violently and to address their deeper motivations, the use of private armed guards on ships may blow back on the ostensible protectors and protected.

LIST OF ABBREVIATIONS

BMP	Best Management Practices for Protection against Somalia Based Piracy
DMSS	Djibouti Maritime Security Services
ICU	Islamic Courts Union
IMB	International Maritime Bureau
IMO	International Maritime Organization
PSC	Private security company
RPG	Rocket-propelled grenade launcher
SAMI	Security Association for the Maritime Industry
UNODC GPML	United Nations Office on Drugs and Crime Global Programme against Money Laundering, Proceeds of Crime and the Financing of Terrorism

ENDNOTES

1   Author correspondence with hostage release negotiator 10, 12 October 2011.
2   Adapted from UN (1982, art. 101) and IMO (2009, p. 4).
3   On the definition of armed robbery and its relationship to piracy, see Geiss and Petrig (2011, pp. 72–75).
4   Coastal states have the primary responsibility for law enforcement and combating armed robbery in their territorial waters, while all states have universal jurisdiction in countering piracy in international waters (UN, 2010). Since adopting Resolution 1816 of June 2008, as prolonged by Resolutions 1846, 1897, 1959, and 2020, the Security Council has authorized states and regional organizations to tackle piracy in the Somali territorial sea, subject to permission from the Transitional Federal Government (Geiss and Petrig, 2011, pp. 70–80; Roach, 2010, p. 400).
5   For more extensive discussions of piracy and armed robbery at sea in international law, see Geiss and Petrig (2011), Murphy (2009, pp. 11–16), Priddy and Casey-Maslen (2012), and Roach (2010).
6   Adapted from Small Arms Survey (2011, p. 102).
7   See Bahadur (2011); Hansen (2009); and Harper (2011). Author correspondence with Martin Murphy, senior fellow, Atlantic Council, 4 December 2011. While evidence of illegal fishing and toxic dumping in Somalia's territorial waters has proved difficult to verify, allegations have been serious enough to warrant the UN Security Council to ask the Secretary-General to produce a report on the subject; see UNSC (2011c).
8   While pirate bases shift regularly, as of late 2011 they appeared to operate from ports along the north-central Somali coast, between Mogadishu and Puntland, from Haradheere in the south to Bandar Bayla in the north, and in Bargaal and Kismayo (Stratfor, 2012).
9   Author correspondence with the United Nations Office on Drugs and Crime Global Programme against Money Laundering, Proceeds of Crime and the Financing of Terrorism (UNODC GPML), 2 February 2012. Other estimates range from 1,500 to 3,000 Somali pirates (UKHC, 2011, n. 10). Naval forces reportedly caught as many as 1,500 pirates between early 2010 and November 2011; most have been released, however, given the difficulties in finding a country willing to try them (Bloomberg, 2011b). More than 1,000 pirates were behind bars in 21 countries as of September 2011 (*Somalia Report*, 2011a).

10   Author correspondence with UNODC GPML, 2 February 2012.

11   The IMB accords the same weight to reports of incidents involving local fishing vessels and dhows as to those affecting international ships. Author correspondence with Pottengal Mukundan, director, IMB, 18 January 2012.

12   Author interviews with several maritime crime and security analysts, London, 24–25 August 2011.

13   West Africa's Gulf of Guinea is another fast-growing piracy hotbed, especially along the coasts of Benin, Nigeria, Guinea, and Togo (Phelps, 2011). According to the IMB, Benin faced 20 pirate attacks in 2011, compared with none in 2010 (IMB, 2011, p. 5). The IMB also notes that while the period of hostage captivity is shorter in the Gulf in Guinea, attacks are also considerably more violent than those perpetrated by Somali pirates (IMB, 2011, p. 24).

14   Naval forces have also dumped the weapons of captured pirates at sea and released the captors if no country was willing to try them. Author correspondence with informed source 10 and with a Western naval officer, both on 11 December 2011.

15   The seized Kalashnikov ammunition was mainly manufactured by Wolf, a US firm that sells Russian-made ammunition. The captured weapons also included a Singapore-built SAR 80 NATO-calibre assault rifle that appears to have originated from former Somali state stockpiles (Chivers, 2012).

16   Following a reported RPG attack against the *MV Brillante Virtuoso* in July 2011, speculation emerged on maritime blogs that pirates might have gained access to the RPG-29, a powerful grenade launcher designed to defeat explosive reactive armours (Jones, 2011; Mwangura, 2011). The allegations were never confirmed, however, and weapons specialists tend to refer to them as unsubstantiated. Author correspondence with informed source 10, 11 December 2011.

17   Sources indicate that a pirate group mounted a 12.7 mm heavy machine gun on a mother ship, the *MV Polar*, in late 2010 (author telephone interview with private security representative 1, 20 April 2011; interviews with informed sources, London, August 2011; correspondence with informed source 10, 11 December 2011). A UN Monitoring Group also published photos of a pirate skiff equipped with a 'universal mount for heavy or light machine guns' (UNSC, 2011b, p. 213).

18   Author correspondence with UNODC GPML, 2 February 2012, and with Martin Murphy, senior fellow, Atlantic Council, 29 November 2011.

19   Author correspondence with Martin Murphy, senior fellow, Atlantic Council, 29 November 2011, and with a Western naval officer, 11 December 2011.

20   This lack of public information on the treatment of hostages results from 'sensitivities shown towards victims, military classification restrictions, liability concerns, and fears of retribution' (OBP, 2011, p. 3).

21   Author correspondence with hostage release negotiator 10, 12 October 2011.

22   One source argues that in addition to the greater distances now involved for Somali pirates, the involvement of inexperienced private negotiators and the tendency to 'pay more to release hostages quickly' has also contributed to spiking ransom demands. Author correspondence with private security representative 8, 12 December 2011.

23   Author interview with Pottengal Mukundan, director, IMB, 25 August 2011.

24   Author correspondence with UNODC GPML, 2 February 2012.

25   Author correspondence with UNODC GPML, 2 February 2012.

26   The author of this box, Mary Harper, is the BBC's Africa Editor and the author of a book on Somalia (Harper, 2012).

27   Mary Harper carried out the interviews on the telephone with the assistance of a Somali interpreter. She possesses significant field experience in Somalia and worked with trusted intermediaries in the field to set up the discussions.

28   See Ghosh (2010, p. 22); IMO (1974, art. 34); Kain and Filon (2011, p. 2); Mair (2011, pp. 15–16); Noakes (2011); and Swedish Club (2011c).

29   These countries include Cyprus, Finland, Germany, Greece, Hong Kong, India, Italy, the Netherlands, Norway, Spain, the UK, and the United States. BBC News (2011c); Ince & Co. (2011, p. 2); OBP (n.d., p. 8); Sanyal (2011); Swedish Club (2011a); Thuburn (2011).

30   See also UKHC (2011, para. 26).

31   Author correspondence with Pottengal Mukundan, director, IMB, 23 November 2011.

32   Author correspondence with private security company representative 8, 12 December 2011.

33   Reports suggest that in September 2011 the US firm Greyside planned to undertake an offensive operation aimed at freeing a ship captured by pirates, but this practice appears to be uncommon (BBC News, 2011b).

34   Private security companies do not have the right granted to states under the UN Convention on the Law of the Sea to board and seize suspected pirate ships or to arrest pirates (Priddy and Casey-Maslen, 2012, p. 2).

35   Author correspondence with private security companies, September 2011, and telephone interview with private security representative 3, 22 June 2011.

36   Author correspondence with four private security companies, September 2011.

37   Author correspondence with private security companies, September 2011; interview with private security representative 6, Geneva, 19 September 2011; and telephone interview with private security representative 3, 22 June 2011.

38   Author correspondence with private security company representative 8, 12 December 2011.

39   Author correspondence with private security company representative 8, 12 December 2011.

40   IMB data, for instance, does not specify whether attacked ships had private guards on board (IMB, 2011).

41   Author correspondence with private security companies, September 2011; telephone interview with private security representative 3, 22 June 2011; and correspondence with private security company representative 8, 12 December 2011.

42 Author correspondence with private security companies, September 2011, and interview with private security representative 6, Geneva, 19 September 2011.
43 Author interview with private security representative 6, Geneva, 19 September 2011.
44 Author correspondence with Anna Petrig, researcher, Max Planck Institute, 13 February 2012.
45 Author correspondence with private security company representative 8, 12 December 2011.
46 See also ICS and FCSA (2011).
47 Author correspondence with private security representative 8, January 2012.
48 Author interview with private security representative 7, Washington, DC, 19 October 2011.
49 Author interview with private security representative 7, Washington, DC, 19 October 2011.
50 Author interview with private security representative 7, Washington, DC, 19 October 2011, and with private security representative 5, 10–11 November 2011.
51 Author interview with private security representative 7, Washington, DC, 19 October 2011, and correspondence with private security representative 5, 10–11 November 2011.
52 Author telephone interview with informed source 2, 14 June 2011.
53 Author correspondence with private security representative 8, 11 July 2011; interview with private security representative 7, Washington, DC, 19 October 2011; and interview with government source 9, Washington, DC, 19 October 2011.
54 Author correspondence with private security representative 8, 11 July 2011.
55 Author correspondence with private security representative 5, 10–11 November 2011, and interview with private security representative 7, Washington, DC, 19 October 2011.
56 Author correspondence with private security representative 5, 10–11 November 2011.
57 Author correspondence with private security representative 5, 10–11 November 2011.
58 Author correspondence with private security representative 8, 12 December 2011.
59 Author correspondence with private security representative 5, 10–11 November 2011, and interview with private security representative 7, Washington, DC, 19 October 2011.
60 Author correspondence with private security representative 8, 9 January 2012.
61 Author correspondence with private security representative 5, 10–11 November 2011, and interview with private security representative 7, Washington, DC, 19 October 2011.
62 See Cook (2011); IAMSP (2011); Saul (2011); and Small Arms Survey (2011, pp. 124–26).
63 See Small Arms Survey (2011, pp. 125–26).
64 Confidential responses from numerous private security companies, September 2011; private security company rules on the use of force 1 and 2.
65 Confidential responses from numerous private security companies, September 2011; private security company rules on the use of force 2; interview with private security representative 6, Geneva, 19 September 2011.
66 Author correspondence with a private security company, September 2011.
67 Author correspondence with Anna Petrig, researcher, Max Planck Institute, 13 February 2012.
68 Author interview with private security representative 6, Geneva, 19 September 2011.
69 A lack of data on use of force by private guards also applies to land-based private security firms, as reported by the Small Arms Survey in 2011 (Small Arms Survey, 2011, pp. 122–23).
70 Author telephone interview with private security representative 3, 22 June 2011.
71 Author correspondence with an informed source, 11 December 2011.

BIBLIOGRAPHY

AllAfrica. 2011. 'Mozambique: Security Company Lies about Nampula Arrests.' 21 September. <http://allafrica.com/stories/printable/201109220065.html>
Bahadur, Jay. 2011. *Deadly Waters: Inside the Hidden World of Somalia's Pirates*. London: Profile Books.
BBC News. 2011a. 'Piracy: IMO Guidelines on Armed Guards on Ships.' 21 May.
—. 2011b. 'Mozambique Holds "Pirate Hunters."' 16 September.
—. 2011c. 'Somali piracy: Armed Guards to Protect UK Ships.' 30 October.
—. 2011d. 'Somali Attacks Sharply Down in November.' 5 December.
Bennett, John C. W. 2010. 'State Dep't & USCG Respond to Suez Canal Firearms Prohibition.'
    <http://mpsint.com/2010/09/08/state-dep%E2%80%99t-uscg-respond-to-suez-canal-firearms-prohibition/>
BIMCO (Baltic and International Maritime Council). 2011. 'About BIMCO.' <https://www.bimco.org/About/About_BIMCO.aspx>
Bloomberg. 2011a. 'Somali Pirates Thwarted as Attacks Worldwide Rise to Record.' 18 October.
—. 2011b. 'Somalia Piracy Spurs Private Navy to Start within Five Months.' 7 November.

Carmel, Stephen. 2011. *Pirates vs. Congress: How Pirates Are a Better Bargain.* Remarks by Stephen M. Carmel, senior vice president of Maersk Line, Limited, made at the Commander Second Fleet Intelligence Symposium of the US Navy. 3 August.

CFC (Civil-Military Fusion Center). 2011. *Anti-Piracy Review.* Week 45. Norfolk: NATO Allied Command Operations. 8 November.
<https://www.cimicweb.org/Documents/CFC%20Anti-Piracy%20Review/CFC_Anti-PiracyReview_08%20November.pdf>

Chivers, Chris. 2012. 'Somali Pirate Gun Locker: An Oddball Assault Rifle, at Sea.' *At War* Blog. *The New York Times.* 25 January.
<http://atwar.blogs.nytimes.com/2012/01/25/somali-pirate-gun-locker-an-oddball-assault-rifle-at-sea/>

Cook, Peter. 2011. 'Security Association for the Maritime Industry (SAMI).' Presentation at the International Conference on Piracy at Sea, Malmö, 20 October.

Dabelstein & Passehl. 2011. *Criminality Liability Risks When Using Armed Private Security Teams.* Hamburg: Dabelstein & Passehl. September.

Friedman, Darren and Lauren Smith. 2011. 'Fighting Fire with Fire: The Debate over Arming Merchant Vessels.' *Maritime Executive.* 16 November.

Geiss, Robin and Anna Petrig. 2011. *Piracy and Armed Robbery at Sea: The Legal Framework for Counter-Piracy Operations in Somalia and the Gulf of Aden.* New York: Oxford University Press.

Geneva Declaration Secretariat. 2011. *Global Burden of Armed Violence 2011: Lethal Encounters.* Cambridge: Cambridge University Press.

Gettleman, Jeffrey. 2011. 'Taken by Pirates.' *The New York Times.* 5 October.

Ghosh, P. K. 2010. *Somalian Piracy: An Alternative Perspective.* Occasional Paper No. 16. New Dehli: Observer Research Foundation. September.

Hansen, Stig Jarle. 2009. *Piracy in the Greater Gulf of Aden: Myths, Misconception, and Remedies.* NIBR Report 2009:29. Oslo: Norwegian Institute for Urban and Regional Research.

Harper, Mary. 2011. *Trends of Violence and Rules of Behaviour in Piracy: The Perceptions of Somali Pirates.* Unpublished background paper. Geneva: Small Arms Survey.

—. 2012. *Getting Somalia Wrong: Faith, War and Hope in a Shattered State.* London: Zed Books.

Hope, Bradley. 2011. 'Firearms: An Odd Casualty of Piracy.' *National* (Abu Dhabi). 6 February.

IAMSP (International Association of Maritime Security Providers). 2011. Website. <http://iamsponline.org/>

ICS and ECSA (International Chamber of Shipping and European Community Shipowners Associations). 2011. 'Flag State Rules and Requirements on Arms and Private Armed Guards on Board Vessels: Combined ICS/ECSA Table.' July.
<http://www.marisec.org/Piracy%20Flag%20State%20Laws%20July%2011.pdf>

IHS (Information Handling Services). 2009. *Valuation of the Liner Shipping Industry.* Englewood: IHS. December.
<http://www.worldshipping.org/pdf/Liner_Industry_Valuation_Study.pdf>

IMB (International Maritime Bureau). 2002. *Piracy and Armed Robbery against Ships.* London: IMB, International Chamber of Commerce.

—. 2011. *Piracy and Armed Robbery against Ships.* London: IMB, International Chamber of Commerce

—. 2012. *Piracy and Armed Robbery against Ships.* Archive of annual and quarterly reports, 2002–11. London: IMB, International Chamber of Commerce.

IMO (International Maritime Organization). 1974. International Convention for the Safety of Life at Sea.

—. 2009. Code of Practice for the Investigation of Crimes of Piracy and Armed Robbery against Ships. Resolution A.1025(26).

—. 2011a. *Guidelines to Assist in the Investigation of the Crimes of Piracy and Armed Robbery Against Ships.* MSC.1/Circ.1404. 23 May.

—. 2011b. *Interim Guidance to Shipowners, Ship Operators, and Shipmasters on the Use of Privately Contracted Armed Security Personnel on Board Ships in the High Risk Area.* MSC.1/Circ.1405. 23 May.

—. 2011c. *Interim Recommendations for Flag States Regarding the Use of Privately Contracted Armed Security Personnel on Board Ships in the High Risk Area.* MSC.1/Circ.1406. 23 May.

—. 2011d. *Revised Interim Guidance to Shipowners, Ship Operators, and Shipmasters on the Use of Privately Contracted Armed Security Personnel on Board Ships in the High Risk Area.* MSC.1/Circ.1405/Rev.1. 16 September.

—. 2011e. *Revised Interim Recommendations for Flag States Regarding the Use of Privately Contracted Armed Security Personnel on Board Ships in the High Risk Area.* MSC.1/Circ.1406/Rev.1. 16 September.

—. 2011f. *Interim Recommendations for Port and Coastal States Regarding the Use of Privately Contracted Armed Security Personnel on Board Ships in the High Risk Area.* MSC.1/Circ.1408. 16 September.

Ince & Co. 2011. *Piracy: Issues Arising for the Use of Armed Guards.* London: Ince & Co.

Jones, Steven. 2011. 'Tooled up.' *Maritime Security Review.* 12 July.

Kain, Andrew and Ric Filon. 2011. 'Are Weapons the Answer to Counter Ship Piracy.' *Maritime Executive.* 6 June.
<http://www.maritime-executive.com/article/are-weapons-the-answer-to-counter-ship-piracy-pt-1>

Mair, Stefan (ed.). 2011. *Piracy and Maritime Security: Regional Characteristics and Political, Military, Legal, and Economic Implications.* Research Paper 3. Berlin: German Institute for International and Security Affairs (SWP). March.
<http://www.swp-berlin.org/fileadmin/contents/products/research_papers/2011_RP03_mrs_ks.pdf>

Murphy, Martin. 2009. *Small Boats, Weak States, Dirty Money: Piracy and Maritime Terrorism in the Modern World.* London: Hurst and Company.

—. 2011. *Somalia: The New Barbary?* London: Hurst and Company.

Mwangura, Andrew. 2011. 'Weekly Piracy Report.' *Somalia Report.* 15 July.

Noakes, Giles. 2011. 'Privately Contracted Armed Security Personnel (PCASP) on Board Merchant Vessels.' Presentation at the International Conference on Piracy at Sea, Malmö, 18 October.

OBP (Oceans Beyond Piracy). 2011. *The Human Cost of Somali Piracy.* Boulder: One Earth Future Foundation. 6 June.

—. 2012. *The Economic Cost of Somali Piracy 2011.* Working Paper. Boulder: One Earth Future Foundation. February.

—. n.d. 'Introduction to Private Maritime Security Companies (PMSCs).' <http://oceansbeyondpiracy.org/sites/default/files/pmsc_map_final.pdf>

Phelps, Steve. 2011. *Small Arms Usage by Maritime Criminals in West Africa*. Unpublished background paper. Geneva: Small Arms Survey.

Priddy, Alice and Stuart Casey-Maslen. 2012. *Counter-Piracy Efforts and Operations: Law and Policy Issues*. Background paper for the Wilton Park Roundtable (draft). Geneva: Geneva Academy of International Humanitarian Law and Human Rights. 13 January.

Republic of Djibouti. 2009. Décret No. 2009-030/PRE Instituant un Contrôle des Services en Matière de Sécurité et de Protection des navires et des Equipages. *Journal Officiel de la République de Djibouti*. 12 February.

Reuters. 2011a. 'In Brazen Attack, Somali Pirates Hijack Ship from Omani Port.' 20 August.

—. 2011b. 'Pirates Attack Second Tankers in Two Days near Omani Port.' 22 August.

Roach, J. Ashley. 2010. 'Countering Piracy off Somalia: International Law and International Institutions.' *American Journal of International Law*, Vol. 104, No. 3, pp. 397–416.

Sanyal, Santanu. 2011. 'Private Armed Guards Only a Quick Fix against Piracy.' *Hindu Business Line*. 25 September.
<http://www.thehindubusinessline.com/industry-and-economy/logistics/article2484922.ece?homepage=true&ref=wl_home>

Saul, Jonathan. 2011. 'Facing Piracy, Ship Security Firms Set Ethics Code.' Reuters. 9 May.

— and Anthony Barker. 2011. 'Marine Insurers Backing Armed Guards as Piracy Threat Grows.' *Insurance Journal*. 20 September.

Seychelles. 2010. *Seychelles Republic vs. Ali et al.* Judgement. Victoria: Supreme Court of the Seychelles. 3 November.

Shortland, Anja. 2012. *Treasure Mapped: Using Satellite Imagery to Track the Developmental Effects of Somali Piracy*. Africa Programme Paper: AFP PP 2012/01. London: Chatham House. January.

Small Arms Survey. 2011. *Small Arms Survey 2011: States of Security*. Cambridge: Cambridge University Press.

Somalia Report. 2011a. 'ICOPAS 2011 Conference Presentations Now Online.' 26 September.

—. 2011b. 'Locals Fight Pirates Holding DDG Aid Workers.' 1 November.

Stratfor. 2012. *Somali Piracy: 2011 Annual Update*. Austin: Stratfor. 13 January.

Swedish Club. 2011a. *Piracy & Armed Guards: General Overview*. Member Alert. Goteborg: Swedish Club. 21 April.

—. 2011b. *Piracy & Armed Guards: Specific Contractual Provisions*. Member Alert. Goteborg: Swedish Club. 21 April.

—. 2011c. *Piracy: FAQs*. Member Alert. Goteborg: Swedish Club. September.

—. 2011d. *Update: Piracy & Armed Guards*. Member Alert. Goteborg: Swedish Club. 3 October.

Tammik, Ott. 2011. 'Anti-Piracy Guards Cheated by Advanfort, Left Defenseless in Hazardous Waters.' Estonian Public Broadcasting. 17 November.
<http://news.err.ee/6082c763-d9b3-4458-abce-5dc727697cb1>

Thomas, Gavin. 2011. 'Cardiff Security Company Sea Marshals Takes on Pirates.' BBC News. 7 November.

Thuburn, Dario. 2011. 'Pirates Hijack Italian Ship with 18 Crew off Oman.' Agence France-Presse. 27 December.

UKHC (United Kingdom House of Commons). 2011. *Piracy off the Coast of Somalia*. Tenth Report. London: Foreign Affairs Committee. 20 December.
<http://www.publications.parliament.uk/pa/cm201012/cmselect/cmfaff/1318/131802.htm>

UKMTO (United Kingdom Maritime Trade Operations) et al. 2011. *BMP 4: Best Management Practices for Protection against Somalia Based Piracy*. Version 4. Edinburgh: Witherby Publishing Group. August.

UN (United Nations). 1982. Convention on the Law of the Sea.

—. 2010. *Legal Framework for the Repression of Piracy under UNCLOS*. New York: Division for Ocean Affairs and the Law of the Sea. September.
<http://www.un.org/depts/los/piracy/piracy_legal_framework.htm>

UNSC (United Nations Security Council). 2010. *Report of the Monitoring Group on Somalia*. S/2010/91 of 10 March.

—. 2011a. *Report of the Special Adviser to the Secretary-General on Legal Issues Related to Piracy off the Coast of Somalia*. S/2011/30 of 25 January.

—. 2011b. *Report of the Monitoring Group on Somalia and Eritrea*. S/2011/433 of 18 July.

—. 2011c. *Report of the Secretary-General on the Protection of Somali Natural Resources and Waters*. S/2011/661 of 25 October.

USCG (United States Coast Guard). 2010. *Port Security Advisory (4-09)(Rev 4)*. International Port Security Program. 3 September.

USDOS (United States Department of State). 2011. 'Contact Group on Piracy off the Coast of Somalia Meets in New York.' Media Note. Washington, DC: Office of the Spokesperson, USDOS. 16 November.

Wallis, Keith. 2011. 'Shippers Rock Boat with Security Fees.' *China Business Watch*. 31 October.

ACKNOWLEDGEMENTS

Principal author
Nicolas Florquin

Contributors
Mary Harper, Matthias Nowak, Steve Phelps

A solitary footprint marks mankind's first steps on the moon, 20 July 1969. © NASA

Precedent in the Making
THE UN MEETING OF GOVERNMENTAL EXPERTS

INTRODUCTION

How to stop a criminal from removing the identifying marks on a polymer-frame handgun? This was the kind of question asked, and sometimes answered, at the Open-ended Meeting of Governmental Experts (MGE),[1] convened at UN headquarters in New York from 9 to 13 May 2011. For the first time at a UN small arms meeting, the discussions were expert-led and relatively interactive as delegations focused on the practical details of weapons marking, record-keeping, and tracing, specifically as dealt with in the International Tracing Instrument (ITI) (UNGA, 2005).

The MGE produced an official report (UNGA, 2011a) and a more substantive Chair's Summary (New Zealand, 2011c). Yet, as of early 2012, it had not produced much in the way of concrete follow-up. The ideas, proposals, and lessons learned that states shared at the meeting, although reflected in the Chair's Summary, face an uncertain future. Nor have UN member states decided to convene any future MGEs. Still, the potential impact of the 2011 meeting appears significant.

Drawing on the Chair's Summary and the author's own observations from the meeting, this chapter presents details of the MGE discussions with a view to identifying some of the key impediments to full ITI implementation, as well as the various means of overcoming them. It does not reach any conclusions concerning progress UN member states have made in their implementation of the ITI. Its aim, rather, is to examine the 'challenges and opportunities' inherent in such implementation, specifically as discussed at the MGE.

The chapter's main findings include the following:

- A key recommendation emerging from the MGE was for the establishment of a Technical Committee that would draft recommendations for marking in light of new developments in weapons manufacture and design.
- Although the subject was broached at the MGE, differences between the marking of light weapons and that of small arms remain to be explored in the UN framework.
- MGE delegations highlighted a series of challenges associated with the conversion of paper-based record-keeping systems into electronic form, including a lack of qualified personnel and software problems.
- Meeting participants cited a lack of information in tracing requests, along with the inaccurate identification of weapons and weapons markings, as the leading causes of tracing failures. Weapons produced under licence in a second country were often misidentified because of the incorrect identification of the manufacturer or country of manufacture.
- The MGE discussions revealed that when their national and international lines of communication were good, national points of contact were often instrumental in resolving even the most complex weapons cases.
- The MGE highlighted the role of technology, both in complicating implementation of certain ITI provisions (as with the import marking of polymer-frame weapons) and in overcoming critical implementation challenges (such as through the use of digital photography for weapons identification).

- UN member states have yet to develop specific means of following up on the ideas, proposals, and lessons learned that are shared at MGEs.

The chapter begins with a brief overview of the history leading to the convening of the first MGE in May 2011. It then focuses on the meeting discussions, topic by topic, with particular emphasis on the question of implementation challenges. The chapter conclusion provides a brief assessment of the meeting and situates it in the broader framework of the UN small arms process, noting some unfinished business from the 2011 MGE.

THE MGE: A SHORT HISTORY

The possibility of a meeting focused on implementation has long been part of UN Programme of Action (PoA)[2] discussions. At the PoA's First Review Conference in 2006, many states expressed dissatisfaction with the first two Biennial Meetings of States (BMSs), held in 2003 and 2005. Both had involved mostly non-specific discussions of the PoA and its implementation; neither had produced an agreed substantive outcome. Despite relatively broad, though not unanimous, dissatisfaction with the two BMSs, the Review Conference reached no agreement on a new format or focus for future meetings.[3]

Change came in 2008. At BMS3, UN member states discussed a limited set of PoA-related subjects. The meeting also produced a substantive outcome document that summarized key points from the discussions and outlined follow-up measures in each of the thematic areas.[4] The same format was followed in 2010 for BMS4, which also focused on a limited number of discussion topics and produced a substantive outcome. The shift towards an expert-led discussion was not, however, complete. BMS3 and BMS4 blended the politically minded discussions that had dominated UN small arms meetings to that point with a more focused consider-

Weapons recovered from crime scenes are displayed at a crime lab in Ciudad Juárez, Mexico, March 2009. © Tomas Bravo/Reuters

ation of the details of PoA and ITI implementation. An Informal Meeting on Transfer Controls,[5] hosted by the Government of Canada in August 2007 in Geneva, had demonstrated the merits of bringing together states, inter-governmental organizations, and civil society for interactive, in-depth discussions of international small arms control issues. Canada framed the meeting as a possible stepping stone to an 'inter-sessional process' that would complement the BMS approach.[6]

A proposal for 'periodic meetings of governmental experts' as part of 'a forward-looking implementation agenda for the Programme of Action' was made during the 'Other issues' session of BMS3 and reflected in the meeting's outcome document (UNGA, 2008a, para. 29b). Several months later, the UN General Assembly nailed down the idea with its decision:

to convene an open-ended meeting of governmental experts for a period of one week, no later than in 2011, to address key implementation challenges and opportunities relating to particular issues and themes, including international cooperation and assistance (UNGA, 2008b, para. 13).

MGEs and other aspects of PoA follow-up were on the agenda of BMS4. Although there was agreement on a six-year meeting cycle for the PoA, comprising two BMSs and one review conference, there was no agreement to include regular MGEs in the cycle. Instead, the BMS4 outcome merely acknowledged that MGEs 'had a potential role to play in [the PoA] implementation architecture' if adequately prepared and 'action-oriented' (UNGA, 2010a, paras. 32, 44). UN member states left it to the 2012 PoA Review Conference to address the question of convening additional MGEs beyond that scheduled for 2011 (para. 44).[7] With respect to the 2011 MGE, states emphasized the need to limit the number of issues under discussion, presumably in order to foster a 'pragmatic, action-oriented' exchange (paras. 32, 47).

Some months after BMS4, the UN General Assembly adopted Resolution 65/64, which provided further detail on the objectives and format of the 2011 MGE. It recapped earlier language emphasizing the meeting's focus on the practical details of PoA implementation, in particular 'key implementation challenges and opportunities' (UNGA, 2010b, paras. 6–7).[8] In this regard, it encouraged states 'to contribute relevant national expertise' to the meeting (basically by sending experts) (para. 9). It also stressed the importance of civil society contributions to PoA implementation, specifically for purposes of preparing for the MGE (para. 10). In relation to international cooperation and assistance, which Resolution 63/72 had already identified as an MGE theme, Resolution 65/64 encouraged states 'to consider ways to enhance cooperation and assistance and to assess their effectiveness' (para. 15). Finally, the resolution set the dates for the meeting: 9 to 13 May 2011 (para. 6).

The chair-designate encouraged states to send relevant experts to the MGE.

The chair of the 2011 MGE, Ambassador Jim McLay of New Zealand, was designated at the time of BMS4, in June 2010. He immediately undertook consultations with UN member states regarding such questions as meeting format and themes (New Zealand, 2010a). Many delegations expressed support for a format that would 'discourage set-piece national statements in favour of focused interactive dialogue' (New Zealand, 2010b, p. 2). The possibility of convening parallel sessions 'to facilitate interactive, technical discussions amongst experts' was considered, but ultimately rejected due to a lack of meeting space and also because many smaller delegations would have had difficulty covering parallel meetings (New Zealand, 2010d, p. 3).

The subject of civil society participation in the MGE also came up. Whereas UN small arms meetings had hitherto allowed civil society representatives to make statements only during a separate dedicated session, the Canadian Informal Meeting of August 2007 had set aside at least one half-hour for interventions from non-state participants following the initial interventions of states in each session (Canada, 2007, para. 6). Political opposition precluded such an arrangement for the MGE although, continuing a practice begun at BMS3, representatives of civil society, along with representatives of inter-governmental organizations and states, made presentations at the beginning of the thematic sessions to introduce the subject at hand. These were complemented by national or regional case studies related to the topic.[9]

The chair-designate emphasized, in general terms, the importance of 'the interactive sharing of information and experiences among experts' (New Zealand, 2011a, p. 2). More specifically, he encouraged states to send relevant experts to the MGE (p. 3). The UN Development Programme established a voluntary sponsorship programme 'to facilitate attendance by relevant experts from developing states' (p. 3).

The question of meeting themes was also the subject of much discussion and debate in advance of the 2011 MGE. General Assembly Resolution 63/72 had put international cooperation and assistance on the agenda. Other possibilities included:

> *tracing, trade across borders, illicit brokering and stockpile management. In addition, some states have suggested a focus on key aspects of national implementation infrastructure, such as national legislation, national reporting or national coordinating bodies* (New Zealand, 2010c, p. 2).

At the end of the day, the field was narrowed down to marking, record-keeping, and tracing, principally as addressed in the ITI, but also in the PoA and the UN Firearms Protocol, to the extent these instruments added normative value to the ITI.[10] Sessions on those three themes were complemented by others on national frameworks (national implementation of the ITI in general terms), regional cooperation, and international assistance and capacity-building. In keeping with the mandate, as first articulated in General Assembly Resolution 63/72, international cooperation and assistance was made a cross-cutting theme, relevant to each of the substantive topics. In order to assist states in their preparations, Ambassador McLay distributed a set of thematic discussion papers in advance of the meeting (New Zealand, 2011b).

THE MGE

Most of the delegations that took the floor during the MGE offered information on their national practices in the area of marking, record-keeping, and tracing and related legislative and enforcement efforts. Sometimes they made specific reference to the ITI (or the PoA). More often, they did not.

Overall, the information states provided on their implementation of the ITI at the MGE did not add significantly to the existing store of knowledge, which is based on national reporting.[11] It seldom contained the level of detail that would be needed to determine the extent of national implementation of the ITI. For example, the states that took the floor on marking methods mostly articulated the objectives they sought to fulfil in this area, such as making the erasure of markings difficult. Only occasionally did they provide details as to the methods they used (such as stamping and engraving).

In any case, as indicated above, the purpose of the MGE was not to elicit information from states that would allow for an assessment of their implementation of the ITI. Rather, the meeting was designed to facilitate the sharing of detailed information and experiences that might eventually enhance implementation. In the event, often as part of their account of national implementation, many MGE delegates did have something to say about 'implementation challenges and opportunities'. With varying degrees of candour and specificity, states described the obstacles they had encountered in implementing the ITI or, more simply, in establishing effective systems for small arms marking, record-keeping, and tracing, including lessons learned and successes in coping with implementation challenges.

While this chapter provides some indication of the information states offered on national implementation in each of the thematic areas, it focuses on the 'implementation challenges and opportunities' states highlighted at the MGE. Using both the Chair's Summary (New Zealand, 2011c) and the author's own observations from the meeting, it seeks to provide a record of some of the current sticking points in ITI implementation as recounted by MGE delegates.

> Many MGE delegates did say something about 'implementation challenges and opportunities'.

MARKING

National interventions during the MGE session on marking covered both the methods and content of marking. While states provided relatively little information on marking at the time of manufacture, they offered more on post-manufacture—and especially import—marking. In accordance with paragraph 8d of the ITI, several delegations indicated they had ensured, or were in the process of ensuring, that all small arms held by government armed and security forces were marked. Some states reported that they marked weapons that were found or seized on national territory but not destroyed. Several others said they were strengthening existing legislation or adopting new legislation to fill gaps relating to weapons marking. A few countries provided information on the enforcement of these laws, especially those relating to the falsification, removal, or defacement of weapons markings.

New developments in weapons manufacture and design. Some states with significant small arms production called attention to recent developments in weapons manufacture and design that made certain aspects of ITI implementation more difficult. They noted, for example, that the increased popularity of modular weapons designs, which provide for the routine changing of major components, could result in the marking of different serial numbers on distinct parts of the same weapon, increasing the risk of misidentification.

Whereas the ITI prescribes the application of a 'unique marking [. . .] to an essential or structural component of the weapon [. . .] such as the frame and/or receiver', it also encourages the marking of 'other parts of the weapon such as the barrel and/or slide or cylinder' (UNGA, 2005, para. 10). Depending on the type of firearm, more than one of these components could be marked with the same serial number (for handguns: frame, barrel, and slide). If one or more parts are subsequently changed, however, the identifying numbers will be different.[12]

Another recent trend in firearm manufacture that gave rise to considerable discussion at the MGE was the increasing use of polymer frames, especially in guns destined for the civilian market, given their important advantages in cost, weight, and performance. In contrast to the marking of metal-frame weapons, which typically leaves an imprint on the metal underlying the mark, it is difficult to mark polymer-frame weapons durably, as the ITI stipulates (UNGA, 2005, para. 7). As several states pointed out, metal strips containing serial numbers can help to overcome this obstacle, but these can be removed by a criminal.[13] Delegations also noted that since the use of polymer frames for military firearms was limited, the tracing of conflict weapons would not be greatly affected by this problem.

Nevertheless, states called on governments and industry to discuss and develop practical solutions for the durable marking of other, mostly civilian, polymer-frame weapons. In fact, the key recommendation emerging from this discussion of new manufacturing trends was for the establishment of a Technical Committee, comprising representatives of governments and industry; this group would draft recommendations for weapons marking in light of new developments, such as polymer casing and modular design.

Polymer frames, increasingly used in firearms destined for the civilian market, present challenges in marking weapons durably.
© Robin Ballantyne/Omega Research Foundation

Import marking. Import marking under the ITI, although not mandatory, is strongly encouraged (UNGA, 2005, para. 8b). At the MGE, several states emphasized its importance for tracing. If a small arm or light weapon lacks an import mark, efforts to trace it have to rely on a record-keeping trail, which may reach back many years, and possibly several decades, to the date of manufacture.[14] Nevertheless, as several participants underlined during the MGE discussions, certain factors can make import marking difficult.

The key problem is that post-manufacture marking methods, such as stamping, that are sufficiently 'durable' (UNGA, 2005, para. 7) to thwart many attempts to remove them may harm the weapon (or at least invalidate manufacturer warranties) because of the force applied during marking.[15] As discussed at the MGE, there are two ways of dealing with this problem. The first is for the manufacturer to make the import mark prior to import. This is possible in cases of direct international sale. If, however, the weapons are acquired some time after manufacture, through a dealer, for example, or if the manufacturer refuses to include import marks in the production run, perhaps because of the additional expense, then the importing entity must make the import marks itself. Several participants pointed out that, in such cases, laser engraving poses no danger to the physical integrity of the weapon but is less resistant to attempts at sanitization (alteration or erasure of markings).

Some states that took the floor during the MGE said that the markings on imported small arms were carefully recorded or that import was refused if serial numbers were not already present on the weapon. While both practices are important in ensuring the traceability of the weapon, neither replaces the application of an import mark identifying the country of last legal import. As explained above, this step can determine the success or failure of a trace.

Falsification, alteration, and erasure of markings. Much of the MGE marking discussion centred, explicitly or implicitly, on the problem of criminal attempts to falsify or sanitize markings. As indicated above, in conjunction

A pistol with its serial number scraped off, Rio de Janeiro, Brazil, May 2004. The gun was seized during the arrest of a 26-year-old drug dealer accused of killing several policemen. © Alaor Filho/Agência Estado/AE

with other factors, this difficulty influences the choice of marking methods. It also shapes weapons tracing strategies. Participants emphasized law enforcement tools such as the use of covert markings, applied by some manufacturers in addition to regular (visible) markings, along with proof markings, which, although unique, are often untouched by traffickers. They also mentioned strategies such as the development of new techniques for the recovery of sanitized markings and the use of evidentiary rules in the prosecution of weapons-related offences (shifting the burden of proof for suspects in possession of firearms with sanitized markings). Participants underlined the importance of criminalizing the removal or distortion of weapons markings.

Trade in illicit parts. In its 2003 report, the Group of Governmental Experts on Tracing raises the problem of traffickers reconstituting an unmarked weapon from unmarked components (UNGA, 2003, para. 62h). In response, the Open-ended Working Group that negotiated the ITI included a provision specifying that a 'unique marking [serial number] should be applied to an essential or structural component of the weapon', meaning the frame or receiver in the case of a firearm (UNGA, 2005, para. 10). The importance of this provision in combating the trade in illicit parts (and the reconstitution of an unmarked weapon) was highlighted at the MGE.

Temporary export and re-import. At the MGE, one state explained that a judicial ruling mandated the import marking of weapons (for example, hunting rifles) that had been temporarily exported and then *re-imported* into the country, notwithstanding ITI language exempting temporary imports from import marking. As reflected in the MGE Chair's Summary, it is important that national control frameworks cover all aspects of related transactions (temporary export as well as import) when translating ITI commitments into domestic law (New Zealand, 2011c, p. 4).

Craft production. Craft production, which, by definition, is not authorized by the state that has jurisdiction over the activity, poses a challenge to national efforts to ensure compliance with ITI marking standards. Meeting discussions emphasized the importance of bringing this activity under regulatory control—and of informing craft producers of applicable laws and penalties, and training them in weapons marking.

Officials from the Criminal Investigation Division register weapons at the Registro Balistico in Tegucigalpa, Honduras, August 2005. © Ginnette Riquelme/AP Photo

Marking small arms v. light weapons. At the MGE, states mentioned the fact that small arms, on the one hand, and light weapons, on the other, are marked differently, because of their different physical characteristics, but they did not elaborate or engage in follow-up discussion. In fact, several distinguishing features of the two weapons categories have implications for marking. These include the greater surface area of light weapons; the greater fragility of many light weapons components (such as electronic control systems); and the integration of ammunition with the launcher in some light weapons systems. To date, discussions of the ITI marking commitments have focused on firearms (small arms and a narrow range of light weapons); there has been little consideration of the marking of light weapons generally.

RECORD-KEEPING

Given constitutional differences among states, and particularly the presence or absence of a federal structure, national practices in the area of record-keeping often vary significantly. Record-keeping systems may be centralized or decentralized. Decentralization can take different forms, such as the separation of record-keeping systems among sub-national units of government, between government and the private sector (manufacturers or dealers), or between the police and the military.

Yet, whatever form they take, record-keeping systems need to fulfil certain minimum functions. Prompt access to accurate records allows a country to respond to tracing requests from other states 'in a timely and reliable manner' (UNGA, 2005, para. 11). At the national level, accurate records are needed for the prosecution of weapons-related offences. Insufficient or inaccurate record-keeping thwarts the achievement of these objectives.

Legislative framework. As in other areas discussed at the MGE, several states underlined the importance of an adequate legislative framework for record-keeping, applicable to all relevant actors, both governmental and non-governmental. They stressed that national laws needed to establish an obligation to keep records and provide for sanctions for non-compliance, as underpinned by ITI marking provisions. Several participants also emphasized the importance of the ITI provision requiring manufacturers and dealers that cease activity to forward their weapons records to the state (UNGA, 2005, para. 13).

Maintenance of weapons registers and data. The challenge of maintaining effective record-keeping systems elicited comment from several national delegations. They underscored the need to recruit qualified and sufficiently numerous personnel, pointing out that targeted, sustained training of these officials facilitated the accurate identification of weapons and weapons markings and, consequently, an accurate record. They cited measures that states could implement to ensure the continued reliability of record-keeping systems, namely regular spot checks of data accuracy and consistency, together with computer surveillance software that searches electronic systems for incompatible records. States also stressed the importance of safeguarding against unauthorized access to and use of record-keeping systems.

Computerization. Several states that took the floor at the MGE described projects, ongoing or completed, to convert paper-based record-keeping systems into electronic form. Some delegations also requested technical assistance in order to help them undertake such a conversion. The challenges that states highlighted in this area included a lack of qualified personnel and software problems, such as in the electronic conversion of non-alphanumeric scripts into alphanumeric form. One state recounted that such difficulties had prevented it from completing a conversion process. States reported on several strategies that had proven successful in managing such conversions, including:

- adequate training of personnel (in particular, to ensure they understood what information was needed for a record);
- the provision of necessary equipment;
- defining minimum content for the creation of an electronic record;
- the development of software to convert non-alphanumeric markings into alphanumeric form; and
- strong project control, with clear definitions of software, personnel, and security requirements.

Integration of multiple systems. Several participants highlighted particular challenges to effective record-keeping, such as a lack of uniformity and appropriate linkages across multiple registers. Some states indicated that they were integrating separate police and military systems. Others said they had centralized or were centralizing civilian firearm records, although legal restrictions precluded this in some countries.

Record retention. The MGE discussions revealed that not all states were complying with ITI norms on record retention. Very few delegations made explicit reference to the ITI when indicating how long they kept records of small arms and light weapons, although several states did cite figures consistent with the ITI minimum of 30 years for manufacturing records and 20 years for all other records, including import and export records (UNGA, 2005, para. 12). Many states that took the floor on this issue said they kept weapons records indefinitely, as encouraged in the ITI (para. 12), given the utility to tracing and reductions in the cost of long-term electronic data storage.

Yet one state gave a figure of ten years, citing the outdated UN Firearms Protocol standard (UNGA, 2001a, art. 7). Another indicated that it destroyed corresponding records one year after the final disposal of a weapon. This, another delegate pointed out, could facilitate the diversion of a weapon that had not actually been destroyed, the elimination of the record rendering the weapon untraceable.

Record-keeping in post-conflict settings. Several MGE participants noted the need to build capacity for effective record-keeping in post-conflict situations and other contexts in which states are seeking to increase their control over the circulation of small arms and light weapons.

COOPERATION IN TRACING

The discussion of cooperation in tracing at the MGE saw delegations recount national experiences in the conduct of weapons tracing, highlight its potential in a range of contexts, and call attention to particular problems that impeded successful tracing.

States provided little information on the outcomes of specific tracing operations, but in several cases they offered an overview of their experiences. Some states reported a relatively high rate of success in their tracing efforts, while others indicated that they received no response at all to some of their requests. Non-response is, in fact, a breach of ITI commitments to 'acknowledge receipt [of a tracing request] within a reasonable time' and subsequently explain any delay or restriction in the contents of a response, or refusal to respond (UNGA, 2005, paras. 19, 22–23).

Despite such limits to tracing cooperation (and ITI implementation), delegations that took the floor during the session broadly emphasized the importance of weapons tracing in crime and conflict settings. Participants argued that, as a law enforcement tool, tracing could be used not only to prosecute individuals guilty of weapons offences, but also to identify illicit trafficking networks and neighbourhoods prone to gun crime, and to focus police resources on these

A Massachusetts State Police Crime Lab forensic chemist holds up a gun produced as evidence during a murder trial in Woburn, Massachusetts, June 2008.
© Bill Greene/Pool/Reuters

problems. Yet some pointed out that tracing was only one instrument in a broader law enforcement arsenal that included, for example, ballistics information systems.

A number of states also emphasized the value of tracing small arms and light weapons during and after armed conflict in an effort to curb proliferation and enhance security; several delegations cited weapons traces conducted by UN expert panels in support of investigations of arms embargo compliance. MGE participants also highlighted the potential utility of tracing to the control of international arms transfers, noting that tracing results could be used to evaluate the effectiveness of national import controls in preventing arms smuggling. Some also observed that export licensing authorities could use tracing data to identify destinations and recipients that present a significant risk of diversion before authorizing arms shipments to them.

States mentioned a series of challenges for tracing during the MGE session, as discussed below.

Insufficient information. Along with the problem of weapons misidentification (see below), MGE participants consistently cited a lack of information in tracing requests as a key reason for tracing failures. When discussing this issue, most states emphasized the failure to provide full information on weapon type and model, as well as weapons markings. Some participants highlighted the need for more information on the case motivating the tracing request.

> Insufficient information and weapons misidentification were leading causes of tracing failures.

Misidentification of weapons and markings. Several participants highlighted the inaccurate identification of weapons and weapons markings as the leading cause of tracing failures. They cited poor weapons design or model recognition and the misinterpretation of different types of markings as common failings. Some remarked that the development of weapons families that shared similar design features had further increased the risk of misidentification. Yet participants also identified a range of solutions to the problem of misidentification, including continuous training to maintain police identification skills; the use of digital photography; and the use of electronic databases, such as the INTERPOL Firearms Reference Table, to enhance firearm identification.

Licensed production. Several states indicated that, in their experience, weapons produced under licence in another country were often misidentified because of the incorrect identification of the manufacturer or country of manufacture. They said that in some cases the problem lay with the party requesting the trace (due to a misinterpretation of weapon type or model, or of weapons markings); in others, particularly cases of unlicensed manufacture abroad, the markings were fraudulent or absent (such as when the country of manufacture was not indicated). MGE delegations noted that proof marks, located on the frame or barrel of a firearm in participating countries, could be used to overcome the lack of information on the country of origin.

Delays. Several delegations complained of delays in receiving responses to tracing requests they had submitted to other states. Some noted that such delays could, for example, force the state requesting tracing information to release a suspect for lack of evidence once the time limit for their provisional detention had been reached. Delegations stressed that national-level cooperation among relevant government agencies, and between government and industry, was important in minimizing the delays that could occur in responding to tracing requests. In this regard, participants also highlighted the importance of direct lines of communication between relevant officials in different countries.

Neglecting weapons offences. States considered whether it generally made sense to drop a weapons charge in favour of a criminal charge that was easier to prove, such as drug possession or trafficking, partly to avoid conducting a time-consuming, potentially unsuccessful trace. While some participants asserted that this was the general tendency, a number of countries said that they did not normally abandon weapons prosecutions, especially as the penalties for such offences were often quite severe. Their preference was, whenever possible, to bring the most serious charge.

Confidentiality. Several MGE participants noted the importance of transmitting tracing-related information, including on intermediate and final weapon purchasers, through secure channels. Some delegates reported that their states had passed legislation to that effect. Others mentioned that such exchanges usually involved law enforcement personnel, although in recent years INTERPOL had granted certain UN peacekeeping missions and other UN bodies[16] access to its police information systems, including secure channels of communication for the dispatch and receipt of tracing requests.

Participants called attention to the fact that some states, especially common law jurisdictions, allowed for the disclosure of tracing information during judicial proceedings. Delegates noted, however, that although confidentiality rules could make weapons-related prosecutions more difficult, they did not make them impossible; they pointed out that in most cases, prosecutors worked over the long term to complete the investigation.

Long lifespans. At the MGE, several states noted that the long lifespan and complex chain of ownership of many small arms and light weapons, especially those that had crossed several borders, made tracing difficult. In this context, they singled out poor record-keeping and the frequent absence of markings noting the country of last legal import. Delegates said that newer weapons were easier to trace, not only because they had normally seen fewer changes of ownership, but also because there was a better chance that records still existed and, moreover, could be easily accessed in electronic form. They observed that older weapons, especially those without import markings, were often untraceable and that, even if the manufacturer of the weapon still held the original record, there was a high risk of a break in the record-keeping chain (records reflecting changes in ownership) following the point of manufacture. Some countries reported the wholesale loss of records from earlier periods in their history. Others noted that apparently complex traces were sometimes straightforward and that national records occasionally provided information on the weapon's most recent history, obviating the need for tracing assistance from the country of manufacture or of last legal import.

> Older weapons, especially those without import markings, were often untraceable.

NATIONAL FRAMEWORKS

The MGE discussion of national implementation frameworks focused on ITI provisions that address broad aspects of implementation, such as points of contact, as well as the interface between national implementation and bilateral, regional, and international action. Legislation was a key theme of the session. Several participants outlined plans to develop or adopt new legislation, or to strengthen existing laws. Yet participants highlighted the need to evaluate implementation gaps and needs before developing national legislation and structures. A number of states noted the importance of linking national frameworks for marking, record-keeping, and tracing to national programming in related areas, such as national development.

National points of contact. Much of the national frameworks discussion was devoted to the topic of national points of contact, including their role in tracing and in broader aspects of ITI implementation, such as information exchange. Several states indicated that they had not yet designated a point of contact or had initially delayed doing so.

Ugandan police markings applied to a Chinese Type 56 assault rifle as part of Uganda's initiative to mark all small arms and light weapons in defence and security force inventories. © Conflict Armament Research Ltd., 2012

Some delegates said that these delays stemmed from uncertainty surrounding the relationship between the point of contact for the PoA and that for the ITI; others spoke of disagreement about which national agencies—police, defence, or foreign affairs—should fulfil this function.

Several states noted that the ITI's reference to 'one or more national points of contact' (UNGA, 2005, para. 25) pointed to a division of functions, particularly between tracing operations and other aspects of ITI implementation, such as the exchange of information on national marking practices (para. 31b) or assistance needs (paras. 27–29). Numerous participants asserted that the tracing point of contact needed to be police-based given long-standing police experience in the protection of confidential information and international tracing practice, including cooperation among national police forces through INTERPOL's National Central Bureau system. Several states said that they had designated a single point of contact both for the PoA and for broader aspects of ITI implementation, in particular information exchange.

MGE participants noted several challenges in ensuring the effective functioning of national points of contact. In particular, they argued that the tracing point of contact should have ready access to all of a country's record-keeping systems (such as those for military, police, and civilian weapons). One state said that its ITI point of contact convened regular inter-ministerial meetings in order to ensure the coordination of marking, record-keeping, and tracing policy within the country. At the international level, participants stressed the important role of the UN Programme of Action–Implementation Support System in communicating point of contact information to all UN member states.[17] The discussion revealed that when their national and international lines of communication were good, points of contact were often instrumental in resolving even the most complex weapons cases; wherever these conditions held, critical information could be exchanged in a matter of days.

National reporting. Several MGE participants expressed concern over the low levels of national reporting on ITI implementation and the resulting shortfall in communication among states. Some countries noted that the administrative burden associated with reporting was alleviated by the ITI's incorporation of a biennial reporting schedule. Others stated that the reporting task, in particular the collection of information from different government agencies, was eased through the use of national coordination agencies.[18]

Implementation mechanisms and policy instruments. In considering challenges in the area of national frameworks, a number of states cited difficulties in ensuring the full implementation of existing laws, including their effective enforcement. Several states noted that a lack of coordination within government could hinder ITI implementation; they spoke of a need for a 'whole of government' approach that employed implementation mechanisms, as well as policy instruments, to structure participation and coherent action across government.

> Several states encouraged a 'whole of government' approach to ITI implementation.

Among the implementation mechanisms they used for improved national coordination, states cited national firearms (or small arms) commissions, national firearms platforms, and national management committees. They indicated that these mechanisms helped ensure continuity in the face of personnel changes, as well as adequate cooperation and expertise among relevant staff. Delegations emphasized the importance of broad participation in such institutions, not only of the government agencies involved in ITI and PoA implementation, but also of industry and other civil society representatives. With respect to policy instruments, several states underscored the utility of national action plans in coordinating ITI implementation across all sectors of government.

The MGE discussion highlighted a broad range of applications for these mechanisms and policy instruments, including the review of implementation; the identification of implementation needs and gaps; information exchange and policy coordination across government; and the development or revision of national small arms policy.

Additional challenges. Among other challenges states cited in relation to national implementation frameworks were language barriers preventing full uptake of relevant technology (such as user manuals in a foreign language). During this and other MGE sessions, several countries cited the ITI's politically binding nature as an obstacle to its full and effective implementation. A number of states asserted that a legally binding framework would better support national implementation efforts, including inter-agency coordination, and would enhance linkages between the ITI (and the PoA) and other international processes that dealt with arms trafficking.

REGIONAL COOPERATION

Both the International Tracing Instrument and the Programme of Action acknowledge the importance of regional cooperation to their implementation.[19] During the corresponding session at the MGE, participants outlined some of the activities conducted by regional organizations, or within a national framework, to support work on marking, record-keeping, and tracing. These included the development of model legislation, regional implementation standards, and best practice guidelines; training and other capacity-building activities; and the provision of marking machines.[20] More broadly, states emphasized continuity, complementarity, and cost-effectiveness as guiding principles for regional-level work.

Regarding implementation challenges, MGE participants underlined the need for regional organizations to remain responsive to the needs of member states; they called attention to the risk of large organizations losing 'proximity' (and relevance) to these countries (New Zealand, 2011c, p. 13). States saw another key challenge in ensuring inter-state cooperation where regional cooperation was limited; they suggested bilateral and tri-lateral relationships as useful alternatives in such cases. Some states also highlighted the importance of cooperation between regional and sub-regional organizations. They identified meetings, workshops, and other forms of interaction as ways to facilitate the exchange of information and experience and to strengthen relationships between these organizations. Participants also mentioned challenges such as the duplication of efforts among organizations in certain regions and differences in legislation, capacity, and interest that made common action between states in a region more difficult.[21]

INTERNATIONAL ASSISTANCE AND CAPACITY-BUILDING

Given its pivotal role in ITI (and PoA) implementation, international assistance and capacity-building was a cross-cutting theme at the MGE. Many of the assistance and capacity-building needs that states articulated

Seized munitions are showcased to the media at a military compound in Bara, Pakistan, January 2012. © Khuram Parvez/Reuters

during the MGE were relevant to two or more substantive areas. These included:

- equipment (such as marking machines and record-keeping software);
- training (such as on the use and maintenance of equipment, weapons identification, and data entry);
- sharing of technical expertise (such as in combating the falsification or sanitization of markings);
- legislation (such as the strengthening of existing legislation and assistance in the adoption of new legislation);
- building institutional capacity (such as for effective tracing); and
- support for the development of national action plans as well as associated national legislation.

During the session on record-keeping, states formulated a range of assistance needs specific to that topic, including:

- technical assistance for the conversion of paper-based records into electronic form;
- building capacity for record-keeping in post-conflict settings as part of broader weapons collection programmes; and
- addressing the problem of under-staffed and under-resourced firearm registries.

With respect to building national capacity for effective implementation, MGE participants underlined the importance of several existing tools:

- mechanisms that help to match needs and resources (such as the New York-based Group of Interested States and the Programme of Action–Implementation Support System);

- model legislation, guidelines, and standards;
- multilateral funding mechanisms for ITI and PoA implementation (such as the UN Trust Fund for Global and Regional Disarmament Activities); and
- small arms research, seminars, and workshops.

Participants reported that assistance took several forms—financial, material, and technical—and occurred within bilateral, regional, and international frameworks. They also noted the importance of South–South, as well as North–South, cooperation. Several delegations emphasized the role of assistance efforts in building capacity in both recipient and donor states, citing the enhancement of inter-agency cooperation in the latter. Some states criticized the imposition of conditions on assistance, stressing the importance of equal access to assistance by all states that require it.

In keeping with the MGE mandate 'to consider ways to enhance cooperation and assistance *and to assess their effectiveness*' (UNGA, 2010b, para. 15, emphasis added), delegations also identified factors that facilitate (or impair) the provision, uptake, and long-term effectiveness of assistance. States mentioned such elements as:

- national ownership of assistance and capacity-building initiatives in recipient states, including sustained political support for implementation;
- the capacity of recipient states to assess their needs;
- the ability of recipient states to draw on national resources, including human resources, as a complement to international assistance programmes and projects; and
- the adaptation of assistance efforts to the specific needs and contexts of recipient countries ('no "one size fits all" approach'; New Zealand, 2011c, p. 15).

Several delegations stressed that the long-term effectiveness of assistance initiatives depended on the provision of comprehensive and ongoing support. They argued that it was not sufficient to provide marking equipment, for example, but that relevant personnel needed to be trained in its use and maintenance. Moreover, they pointed out that a machine that marked weapons would have little impact without associated equipment, such as computers and record-keeping software to record information on marked weapons. In general terms, participants said it was important to ensure the sustainability of any transfer of knowledge and technology. They also cited broader challenges such as the avoidance of overlap and duplication in the provision of assistance, specifically through improved transparency and coordination.

CONCLUSION

This review of the MGE discussions reveals that the meeting was, as intended, largely 'pragmatic [and] action-oriented' in nature (UNGA, 2010a, para. 32). In every session, states identified a range of factors that were impeding or slowing ITI implementation, as well as practical solutions to such problems. The chair contributed to this success, posing questions on the various themes, distilling key points from national interventions, and, in many cases, following up with specific questions to delegates. Ambassador McLay also encouraged participants to respond to points raised or questions posed by other delegations. In contrast to other UN small arms meetings, this one was not only expert-led, but also quite interactive.

That said, meeting expertise and interactivity had somewhat tenuous footing. Although many states had experts on their delegations, a significant number remained silent during the meeting. Some delegations were represented solely by New York-based diplomats. At the end of the day, a relatively small number of experts, typically from industrialized countries, made a disproportionately large contribution—both to the content of the discussions and to their interactive nature. Nevertheless, for the first time at a UN small arms meeting, the term 'implementation challenges and opportunities' was more than a mere slogan.

The 2011 MGE revealed considerable breadth and depth in weapons marking, record-keeping, and tracing practice throughout the world. It was not the role of the meeting to assess the extent to which that activity was tied to the ITI, but the MGE can be expected to have some influence in raising awareness of the Instrument's existence and spurring strengthened implementation. There is some early evidence that the MGE did just that.

The number of national points of contact notified to the UN Office for Disarmament Affairs—one key marker of ITI implementation—saw a huge boost from the meeting, rising from 18 in mid-January 2011 to 67 by 12 May, the second-to-last day of the MGE (McDonald, 2011, pp. 49–50; UNODA, 2011).[22] Moreover, INTERPOL figures show an increase in the number of tracing requests that the organization is copied on: from an average of 25 per month during the two-year period preceding the MGE, to an average of 36 per month thereafter (representing thousands of firearms, through January 2012).[23] It also appears likely that the MGE discussions, including those conducted among participants in the margins of the meeting, will catalyse follow-up action in some cases.[24] One of the 2011 MGE's most important legacies could be the development of contacts among the experts who attended the meeting and their subsequent interaction.

As of early 2012, the implications of the 2011 MGE for the UN small arms process were unclear. The UN membership had yet to agree to convene any further MGEs, leaving this question to the PoA's Second Review Conference, scheduled for August–September 2012 (UNGA, 2011b, para. 14). Although the UN's general ('omnibus') resolution on small arms endorsed the formal (largely non-substantive) MGE report[25] and took 'note with appreciation of the Chair's summary of discussions' (para. 5), it did not follow up on the many recommendations that emerged from the meeting, some of which, such as the establishment of a Technical Committee for weapons marking, require multilateral action.

The MGE highlighted the role of technology, both in making implementation of certain ITI provisions more difficult (as with the import marking of polymer-frame weapons) and in overcoming key implementation challenges (such as through the use of digital photography for weapons identification). While these findings and others are set out in the Chair's Summary (New Zealand, 2011c), it is not yet clear whether or how this text will translate into concrete follow-up. There is also a need to distil, presumably in UN document form, the various elements of the meeting that contributed to its success, including the expert-led nature of the discussions, their interactive character, and the chair's role in facilitating such processes. Among other things, such a document might help address the—as yet unanswered—question of how to distinguish the mandates of BMSs, review conferences, and MGEs.[26]

The place of MGEs in the PoA meeting cycle is not yet assured. Specific means of following up on the ideas, proposals, and lessons learned shared at such meetings still have to be developed. Yet, if the aim of UN small arms meetings is to foster the strengthened implementation of the PoA and ITI, the logical first step is to examine the 'challenges and opportunities' inherent in implementation. The 2011 MGE shows what can be done in this respect, but concrete follow-up remains uncertain given, among other things, the current lack of institutional footing for MGEs generally. Precedent in the making, but not yet made.

LIST OF ABBREVIATIONS

BMS	Biennial Meeting of States
INTERPOL	International Criminal Police Organization
ITI	International Instrument to Enable States to Identify and Trace, in a Timely and Reliable Manner, Illicit Small Arms and Light Weapons ('International Tracing Instrument')
MGE	Open-ended Meeting of Governmental Experts
PoA	Programme of Action to Prevent, Combat and Eradicate the Illicit Trade in Small Arms and Light Weapons in All Its Aspects

ENDNOTES

1   The full title of the event was the Open-ended Meeting of Governmental Experts on the Implementation of the Programme of Action to Prevent, Combat and Eradicate the Illicit Trade in Small Arms and Light Weapons in All Its Aspects.
2   The PoA is the Programme of Action to Prevent, Combat and Eradicate the Illicit Trade in Small Arms and Light Weapons in All Its Aspects; see UNGA (2001b).
3   See McDonald, Hasan, and Stevenson (2007, p. 125). In fact, like BMS1 and BMS2, the First Review Conference produced no substantive outcome of any kind.
4   See Bevan, McDonald, and Parker (2009, pp. 136–43).
5   The full title is the Informal Meeting on Transfer Control Principles for Small Arms and Light Weapons.
6   See Canada (2007, 'Conclusion'). The Chair's Summary can be requested at <ida@international.gc.ca>.
7   See also UNGA (2010b, para. 20).
8   See also UNGA (2010b, para. 8).
9   For more on these presentations, see New Zealand (2011c) and UN (n.d.a).
10 On this issue, see the opening (normative) paragraph in sections II to VII of the MGE Chair's Summary (New Zealand, 2011c). See also UNGA (2001a).
11 See Parker (2011, pp. 46–69).
12 Although not discussed at the MGE, one solution to this problem is to identify a 'control component' (for a firearm: the frame or receiver) and use only the markings on that component to identify the weapon. At the same time, it is important to track component changes (especially of the frame or receiver) through accurate and up-to-date record-keeping.
13 The methods used to recover markings on metal-frame weapons that criminals seek to erase cannot be employed on polymer frames. For some polymer-frame firearms, covert markings inserted at the time of manufacture can instead be used to defeat attempts at sanitization (alteration or erasure). Author correspondence with Firearms & Explosives Programmes, INTERPOL, 14 February 2012.
14 See Bevan (2009, pp. 118–19).
15 See Persi Paoli (2010).
16 The UN bodies include sanctions committees, special political missions, and special tribunals.
17 See UN (n.d.b).
18 See UNGA (2001b, para. II.4).
19 Regarding the ITI, see UNGA (2005, para. 26). Regarding the PoA, see UNGA (2001b, para. III.11).
20 See New Zealand (2011c, pp. 12–13).
21 See New Zealand (2011c, p. 13).
22 See also UNGA (2005, para. 31). Note that as of 15 February 2012, the PoA–Implementation Support System listed ITI point of contact information for 74 UN member states.
23 These figures represent tracing requests (973 total), not numbers of firearms traced (several thousand), and may include a limited number of repeat requests. Note that only tracing requests were counted, not requests for additional information, responses to tracing requests, or reports of firearm seizure not involving a tracing request. These figures reflect only tracing requests sent through INTERPOL's I-24/7 communication system, on which the INTERPOL General Secretariat was copied; they do not reflect bilateral requests between countries on the system. Author correspondence with Firearms & Explosives Programmes, INTERPOL, 9 February 2012.
24 For example, within one month of the meeting, MGE discussions had led to plans for a nationwide training initiative for police in Papua New Guinea, involving firearms identification, record-keeping, and tracing. Author correspondence with the New Zealand Permanent Mission to the United Nations in New York, 20 July 2011.
25 See UNGA (2011a).
26 See UNGA (2010a, paras. 34, 48).

BIBLIOGRAPHY

Bevan, James. 2009. 'Revealing Provenance: Weapons Tracing during and after Conflict.' In Small Arms Survey. *Small Arms Survey 2009: Shadows of War*. Cambridge: Cambridge University Press, pp. 106–33. <http://www.smallarmssurvey.org/publications/by-type/yearbook/small-arms-survey-2009.html>

—, Glenn McDonald, and Sarah Parker. 2009. 'Two Steps Forward: UN Measures Update.' In Small Arms Survey. *Small Arms Survey 2009: Shadows of War*. Cambridge: Cambridge University Press, pp. 134–57. <http://www.smallarmssurvey.org/publications/by-type/yearbook/small-arms-survey-2009.html>

Canada. 2007. *Chair's Summary: Informal Meeting on Transfer Control Principles for Small Arms and Light Weapons*.

McDonald, Glenn. 2011. 'Fact or Fiction? The UN Small Arms Process.' In Small Arms Survey. *Small Arms Survey 2011: States of Security*. Cambridge: Cambridge University Press, pp. 42–67.

—, Sahar Hasan, and Chris Stevenson. 2007. 'Back to Basics: Transfer Controls in Global Perspective.' In Small Arms Survey. *Small Arms Survey 2007: Guns and the City*. Cambridge: Cambridge University Press, pp. 116–43. <http://www.smallarmssurvey.org/publications/by-type/yearbook/small-arms-survey-2007.html>

New Zealand. 2010a. Letter dated 12 July from Jim McLay, Permanent Representative, New Zealand Permanent Mission to the United Nations.

—. 2010b. Letter dated 4 August from Jim McLay, Permanent Representative, New Zealand Permanent Mission to the United Nations.

—. 2010c. Letter dated 21 October from Jim McLay, Permanent Representative, New Zealand Permanent Mission to the United Nations.

—. 2010d. Letter dated 13 December from Jim McLay, Permanent Representative, New Zealand Permanent Mission to the United Nations.

—. 2011a. Letter dated 14 March from the New Zealand Permanent Mission to the United Nations. <http://www.poa-iss.org/MGE/>

—. 2011b. *Programme of Action on Small Arms and Light Weapons: Thematic Discussion Papers Submitted by the Chair of the Open-ended Meeting of Governmental Experts*. A/CONF.192/MGE/2011/CRP.1 of 29 April.

—. 2011c. *Summary by the Chair of Discussions at the Open-ended Meeting of Governmental Experts on the Implementation of the Programme of Action to Prevent, Combat and Eradicate the Illicit Trade in Small Arms and Light Weapons in All Its Aspects, 9 to 13 May 2011, New York*. A/66/157 of 19 July (Annexe). <http://www.poa-iss.org/MGE/>

Parker, Sarah. 2011. *Analysis of National Reports: Implementation of the UN Programme of Action on Small Arms and the International Tracing Instrument in 2009–10*. Occasional Paper No. 28. Geneva: Small Arms Survey. May. <http://www.smallarmssurvey.org/fileadmin/docs/B-Occasional-papers/SAS-OP28-Analysis-of-National-Reports.pdf>

Persi Paoli, Giacomo. 2010. *The Method behind the Mark: A Review of Firearm Marking Technologies*. Issue Brief No. 1. Geneva: Small Arms Survey. December. <http://www.smallarmssurvey.org/fileadmin/docs/G-Issue-briefs/SAS_IB1_Method-behind-the-mark.pdf>

UN (United Nations). n.d.a. 'Learn More about the MGE Topics.' <http://www.poa-iss.org/MGE/Topics.html>

—. n.d.b. 'International Tracing Instrument.' <http://www.poa-iss.org/Poa/NationalContacts.aspx>

UNGA (United Nations General Assembly). 2001a. Protocol against the Illicit Manufacturing of and Trafficking in Firearms, Their Parts and Components and Ammunition, Supplementing the United Nations Convention against Transnational Organized Crime ('UN Firearms Protocol'). Adopted 31 May. Entered into force 3 July 2005. A/RES/55/255 of 8 June. <http://www.unodc.org/pdf/crime/a_res_55/255e.pdf>

——. 2001b. Programme of Action to Prevent, Combat and Eradicate the Illicit Trade in Small Arms and Light Weapons in All Its Aspects ('UN Programme of Action'). A/CONF.192/15 of 20 July. <http://www.poa-iss.org/PoA/PoA.aspx>

——. 2003. *Report of the Group of Governmental Experts Established Pursuant to General Assembly Resolution 56/24 V of 24 December 2001, Entitled 'The Illicit Trade in Small Arms and Light Weapons in All Its Aspects'*. A/58/138 of 11 July.

——. 2005. International Instrument to Enable States to Identify and Trace, in a Timely and Reliable Manner, Illicit Small Arms and Light Weapons ('International Tracing Instrument'). A/60/88 of 27 June (Annexe). <http://www.poa-iss.org/InternationalTracing/InternationalTracing.aspx>

——. 2008a. *Outcome of the Third Biennial Meeting of States to Consider the Implementation of the Programme of Action to Prevent, Combat and Eradicate the Illicit Trade in Small Arms and Light Weapons in All Its Aspects*. A/CONF.192/BMS/2008/3 of 20 August (s. IV). <http://www.poa-iss.org/DocsUpcomingEvents/ENN0846796.pdf>

——. 2008b. Resolution 63/72. Adopted 2 December. A/RES/63/72 of 12 January 2009. <http://www.un.org/documents/resga.htm>

——. 2010a. *Outcome of the Fourth Biennial Meeting of States to Consider the Implementation of the Programme of Action to Prevent, Combat and Eradicate the Illicit Trade in Small Arms and Light Weapons in All Its Aspects*. A/CONF.192/BMS/2010/3 of 30 June (s. V). <http://www.poa-iss.org/BMS4/Outcome/BMS4-Outcome-E.pdf>

——. 2010b. Resolution 65/64. Adopted 8 December. A/RES/65/64 of 13 January 2011. <http://www.un.org/documents/resga.htm>

——. 2011a. *Report of the Open-ended Meeting of Governmental Experts on the Implementation of the Programme of Action to Prevent, Combat and Eradicate the Illicit Trade in Small Arms and Light Weapons in All Its Aspects*. A/CONF.192/MGE/2011/1 of 6 June. <http://www.poa-iss.org/MGE/>

——. 2011b. Resolution 66/47. Adopted 2 December. A/RES/66/47 of 12 January 2012. <http://www.un.org/documents/resga.htm>

UNODA (United Nations Office for Disarmament Affairs). 2011. 'ITI National Points of Contact.' Handout. 12 May.

ACKNOWLEDGEMENTS

Principal author

Glenn McDonald

Parts of Beretta 92S semi-automatic pistols are stored at the manufacturing plant before the final touches are made, Italy, December 2008. © Andreas Solaro/AFP Photo

Piece by Piece
AUTHORIZED TRANSFERS OF PARTS AND ACCESSORIES

8

INTRODUCTION

The authorized international trade in small arms and light weapons is diverse and dynamic, affecting every region of the world and all levels of society. Recreational hunters and other private individuals buy millions of imported rifles, shotguns, and rounds of ammunition each year. Millions of additional foreign-sourced weapons are procured by military and law enforcement agencies worldwide. Most of these weapons are used in accordance with national and international laws, but a small percentage is misused, poorly managed, or diverted, often with disastrous consequences. Yet, despite the profound implications of this trade, much of it remains opaque. Publicly available sources of data on international transfers of small arms and light weapons cover only a fraction of the total trade, and much of the data that is available is vague and incomplete. As a result, each year thousands of transfers of small arms and light weapons go unreported, and thousands more are inadequately documented. This lack of transparency hinders efforts to monitor arms transfers to problematic recipients and to identify the accumulation of excessively large or destabilizing stockpiles of weapons.

In 2009, the Small Arms Survey launched a four-year project aimed at enhancing our understanding of the authorized trade in small arms and light weapons, their parts, accessories, and ammunition. This chapter summarizes the findings from the fourth and final phase of the project, whose focus is on parts and accessories. Using these findings and those presented in previous phases of the project, the chapter provides a new global estimate for the annual value of the international authorized small arms trade (see Box 8.1). The new estimate is significantly higher than the previous estimate of USD 4 billion, reflecting both an absolute increase in the value of transfers of certain items and a more complete accounting of these and other transfers. Key findings from this chapter include the following:

- Authorized international transfers of small arms, light weapons, their parts, accessories, and ammunition are estimated to be worth at least USD 8.5 billion annually.
- The annual value of authorized international transfers of parts of small arms and light weapons is estimated to be worth at least USD 1,428 million, USD 146 million of which is not documented in publicly available sources.
- The trade in parts for military firearms and light weapons is dominated by weapons-producing countries. The 56 countries that produce military firearms and light weapons imported 97 per cent of parts by value, while the 117 countries that have no known domestic production capacity imported only 3 per cent.
- The value of the authorized international trade in weapon sights is estimated at more than USD 350 million. Available data suggests that sights account for most of the trade in major accessories for small arms and light weapons, but data gaps preclude a definitive assessment.
- The civilian market in weapon sights in Chile, Paraguay, Peru, and Uruguay is dominated by Chinese producers and exporters.

- In 2009 the top exporters of small arms and light weapons (those with annual exports of at least USD 100 million), according to available customs data, were (in descending order) the United States, Italy, Germany, Brazil, Austria, Japan, Switzerland, the Russian Federation, France, South Korea, Belgium, and Spain (see Box 8.4).

- In 2009 the top importers of small arms and light weapons (those with annual imports of at least USD 100 million), according to available customs data, were (in descending order) the United States, the United Kingdom, Saudi Arabia, Australia, Canada, Germany, and France (see Box 8.4).

The chapter begins with a brief summary of key terms and definitions, which is followed by an overview of the methodology used to generate the revised estimate for the value of international transfers. The chapter then looks at international transfers of parts and accessories for small arms and light weapons. The trade in parts is explored through an analysis of supply chains and import patterns. The assessment of accessories is divided into two sections. The first section provides a basic overview of five categories of major accessories, indicating how they work, who uses them, and how they are used. The second section sheds light on the trade in accessories through case studies, one on the civilian market for weapon sights in four South American countries and a second on procurement of accessories by the armed forces of six countries. The chapter concludes with a brief recap of major themes from the four-year study, including the need for more transparency in the small arms trade.

Box 8.1 The four-year study on international transfers of small arms and light weapons

In 2009, the Small Arms Survey launched an unprecedented multi-year study of authorized international transfers of small arms, light weapons, their parts, accessories, and ammunition. The goal of the study was to use new and potentially rich sources of data to reassess the Survey's previous estimate of USD 4 billion for the annual global trade, which was first published in 2001. Over the course of the study, the Survey compiled tens of thousands of records on national procurement and international transfers of small arms and light weapons, including previously unreleased data obtained directly from governments (see Box 8.2). The resulting data review is the largest and most detailed of its kind. To fill in the gaps that remained despite these efforts, the Survey developed new estimation techniques, including that described below.

The study was undertaken in four phases. The first phase consisted of a comprehensive overview of multiple data sources on transfers of small arms, including heavy machine guns and anti-materiel rifles up to 14.5 mm in calibre. During this phase, the Survey compiled more than 10,000 records from more than a dozen sources; these records were individually compared and assessed (Dreyfus et al., 2009, pp. 26–30).

Transfers of ammunition for small arms and light weapons were assessed during the second phase of the project. To overcome a near-total absence of usable data on transfers of light weapons ammunition, the Survey contacted more than 70 governments, several of which provided previously unreleased data. The Survey then used this data to generate an estimate for the rest of the world (Herron et al., 2010, pp. 17–20). Similar methods were used to derive an estimate for the value of international transfers of light weapons, including guided missiles, during the project's third phase (Herron et al., 2011, pp. 19–22).

In the fourth and final phase, presented in detail in this chapter, the Survey assessed international transfers of parts and some accessories for small arms and light weapons. It also reviewed previous findings, updating them as necessary. A brief summary of the methodologies used during the study is provided in this chapter, with more detailed information available on the Small Arms Survey's website.[1]

Based on the findings from the four-year study, the Survey estimates the annual value of authorized international transfers of small arms, light weapons, their parts, accessories, and ammunition to be at least USD 8.5 billion (see Figure 8.1). Note that the previous estimate for small arms and light weapons was revised to reflect a recent methodological refinement.

Figure 8.1 Annual estimated value of international transfers of small arms, light weapons, parts, accessories, and ammunition (in USD million)

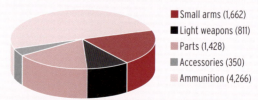

- Small arms (1,662)
- Light weapons (811)
- Parts (1,428)
- Accessories (350)
- Ammunition (4,266)

Note: Reflecting a recent refinement of the methodology, the estimated value for transfers of small arms and light weapons differs from that published in Dreyfus et al. (2009) and Herron et al. (2011).

> **Box 8.2 Assistance from governments**
>
> The Small Arms Survey would like to thank the following governments for the assistance they provided over the course of the four-year study: Bosnia and Herzegovina, Canada, Colombia, France, Germany, Ireland, Italy, Liechtenstein, the Netherlands, Norway, Poland, Portugal, Slovakia, Sweden, Thailand, Ukraine, the United Kingdom, and the United States. Without the data and expertise provided by officials from these governments, which included previously unreleased data on the procurement and transfer of thousands of small arms and light weapons, this study would not have been possible.

TERMS AND DEFINITIONS

Weapons

For the purposes of this chapter, the term 'small arms' refers to the following items:

- pistols and revolvers;
- sporting rifles and sporting shotguns; and
- military firearms, meaning light machine guns, heavy machine guns with a calibre of 14.5 mm or less, sub-machine guns, assault rifles, non-automatic military rifles, military shotguns, and anti-materiel rifles with a calibre of 14.5 mm or less.

The term 'light weapons' is used to refer to the following items:

- mortar systems up to and including 120 mm;
- handheld (stand-alone), under-barrel, and automatic grenade launchers;
- recoilless guns;
- portable rocket launchers, including rockets in single-shot disposable launch tubes; and
- portable missiles and launchers, namely anti-tank guided weapons (ATGWs) and man-portable air defence systems (MANPADS).

Heavy machine guns and anti-materiel rifles—which the UN has defined as light weapons (UNGA, 1997)—are categorized here as 'small arms' because data on transfers of these items is often (inextricably) aggregated with data on transfers of other firearms. In line with previous Survey definitions, mortars up to and including 120 mm calibre are also considered light weapons in this study.[2]

Common parts of small arms

Figure 8.2 depicts a typical assault rifle. The buttstock rests against the shoulder and is used to aim the rifle. One hand holds the pistol grip, and a finger rests on the trigger mechanism, which is protected by the trigger guard. Another hand holds the hand guard, which covers part of the barrel. The magazine feeds ammunition into the receiver (also known as the frame), which contains the working parts of the small arm and is the mechanism that actually fires a cartridge. Sights rest on the top and are used for aiming. Rails are often attached to the hand guard and are used to attach accessories to the assault rifle.

Other types of small arms contain similar parts. The one feature all small arms have in common is a receiver. Designs vary, but receivers house the gun's moving parts and usually contain springs, levers, and pistons. Pistols contain the magazine in the pistol grip, and their short barrel does not require a hand guard. Many rifles and shotguns designed for hunting and sport do not have a pistol grip or magazine.

Parts of light weapons

What follows is a partial overview of the parts of various types of light weapon.[3] Mortars, which are primarily muzzle-loaded, are of simple construction and usually consist of a tube, base plate, and bipod. Mortar bombs are fired when they strike a firing pin at the bottom of the tube. Rocket launchers and recoilless rifles, which fire unguided projectiles, consist of a launch tube that is connected to a firing mechanism. MANPADS and ATGWs are complex, technology-intensive systems. Both are based around a missile, which usually contains sensors, a central guidance unit, a warhead and rocket motor, and propellant. The missile is usually propelled from a launch tube. In the case of MANPADS, a gripstock and battery unit are usually attached under the launch tube, and both are necessary to fire the weapon. ATGWs are more diverse, but many contain a tripod and an aiming and fire-control unit (in addition to the missile and launch tube).

An employee of the Colombian weapons manufacturer INDUMIL (Industria Militar) displays the different parts of an assault rifle, May 2006. © Mauricio Dueñas/AFP Photo

Figure 8.2 **Parts of an assault rifle**

Accessories for small arms and light weapons

An 'accessory' is defined here as an item that physically attaches to the weapon and increases its effectiveness or usefulness but, generally speaking, is not essential for the basic, intended use of the weapon. This definition captures a wide array of items, ranging from extended magazine releases for pistols to thermal night sights for anti-tank guided weapons. This chapter focuses on the following major accessories:

- sights (telescopic, reflex, thermal, image-intensifying, and holographic);
- aiming lasers and illuminators that attach directly to the weapon;[3]
- night vision devices that attach directly to the weapon;
- laser rangefinders that attach directly to the weapon; and
- fire-control systems that attach directly to the weapon.

Two elements of this definition call for some elaboration. The first concerns the requirement that the item be physically attached to the weapon. As explained in more detail below, many items are not attached to small arms and light weapons but enhance their usefulness or effectiveness nonetheless. Examples are numerous and include handheld and helmet- and vehicle-mounted variants of the items listed above. Including items that are not physically attached to the weapon would so dramatically expand the list of items categorized as accessories that it would render the term 'accessory' meaningless, especially on a networked battlefield. The greater the battlefield connectedness, the more crucial the physical attachment requirement becomes. Without this requirement, an unmanned aerial vehicle that collects data on a potential target and relays that information to a sniper in range of the target could fit the definition of an 'accessory'. Categorizing the unmanned aerial vehicle, the operating base, and the other networked platforms as 'accessories' for the sniper rifle would be impractical, however.

The second element that requires some clarification is the definition's exclusion of items that are essential for the basic, intended use of the weapon. While most of the major accessories listed above conform to this definition, there are some notable exceptions, including telescopic sights for long-range sniper rifles and some fire-control systems.

The physical limitations of the human eye preclude effective sniping beyond a certain range without a telescopic sight. Similarly, the 25 mm airburst munitions fired from the XM25 would be of little use without the fuse setter in the weapon's fire-control system. As the number of small arms and light weapons that fire 'smart' munitions increases, the line between a 'part' and an 'accessory' will become increasingly blurred, but for now most of the items listed above fit the definition of 'accessories' adopted for this study (see Box 8.3).

Box 8.3 When parts become accessories: an introduction to modular weapons

This chapter defines 'accessories' as items that physically attach to a weapon to increase its effectiveness or usefulness but, generally speaking, are not essential for the basic, intended use of the weapon. In contrast, 'parts' are defined as items that are integral to the weapon and necessary for its basic functioning (such as a barrel). This distinction reflects the characteristics of the majority of current small arms involving a combination of the basic weapon and several 'add-ons'. Nevertheless, this distinction is becoming increasingly blurred with the entry into the market of weapons whose main parts, such as barrels and receivers, can be easily changed by soldiers to adapt them to the specific operational context. These interchangeable components remain 'parts' in nature but become 'accessories' in use; they are necessary for the basic functioning of a weapon, but they can be switched in just a few moments to alter its characteristics and increase its performance in a given context.

The need for a more flexible type of weapon that could be easily reconfigured to meet different operational needs and could accommodate a range of sophisticated accessories led to the evolution of infantry rifles into 'modular weapons'. The idea behind the concept of modularity is simple: each rifle has a core section (usually the upper receiver) around which all other parts can be switched directly by the soldier to obtain different configurations, depending on the need. These reconfiguration operations can be done without the use of any tool and are simple and quick. The Beretta ARX160 illustrates the advantages of a modular weapon. This assault rifle enables the user to switch between three different types of barrels: the special forces (12-inch barrel), carbine (16-inch barrel), and designated marksmen or light sniper (16-inch heavy barrel). In addition, the soldier can choose between the standard configuration with a 5.56 x 45 mm NATO calibre or easily and 'tool-lessly' swap the bolt head, lower receiver, and barrel, to reconfigure the rifle to use 5.45 x 39 mm, 7.62 x 39 mm, or 6.8 mm SPC rounds (Beretta, n.d.).

The diffusion of modular weapons could have significant implications for the international small arms trade by altering current patterns of procurement. To date, national holdings typically include several types of weapon, purchased from different producers and representing national preferences in each small arm and light weapon category. With the spread of modularity, procurement will probably select the modular weapon that, through its configurations, best meets national needs. This 'single-source' approach to small arms and light weapons procurement would have the following main consequences:

- from a state perspective: reduce acquisition and maintenance costs;
- from an industry perspective: reduce production costs and provide a strong incentive for research and development; and
- from a market perspective: reduce the number of procurement contracts, increase the quantities per individual contract, and, possibly, reduce the number of suppliers.

In addition, the market for weapon components is likely to become larger, with procurement of spare parts for maintenance purposes as well as of main components for reconfiguration purposes.

Modular weapons pose several challenges from an arms control perspective. Record-keeping, and consequently tracing, could become more difficult unless relevant control measures, such as the International Tracing Instrument, are adapted to reflect the new trend towards modular weapons design. It will be especially important to identify a 'control component' of a modular weapon, for example the upper receiver, so that, whatever changes occur in the weapon's configuration, it can be tracked during its life cycle using the markings (including serial number) on the control component and other basic information about it, including the manufacturer, type, and model. This discussion has barely begun, however (UN, 2011).

Author: Giacomo Persi Paoli

ESTIMATING THE INTERNATIONAL SMALL ARMS TRADE

A well-known and indeed chronic problem in studying global transfers of small arms and light weapons is the lack of comprehensive data. Information is lacking on virtually all segments of the international trade, including the weapons themselves, their parts, accessories, and ammunition. Sporting rifles, sporting shotguns, pistols, revolvers, and small-calibre ammunition are the only categories for which there is customs data that is both widely reported and sufficiently disaggregated by weapon type. The UN Commodity Trade Database (Comtrade) remains the most extensive source of data on international transfers of such items, even though not all states report to the instrument.

The trade in light weapons and their ammunition, as well as parts and accessories for military firearms and light weapons, is more difficult to estimate since UN Comtrade combines transfer data on these items with that of other goods. Moreover, transfers of these items are only partially reported in other public data sources, such as the UN Register of Conventional Arms, national reports, and procurement documentation. Through outreach to approximately 70 governments, the Small Arms Survey was able to obtain hundreds of detailed records on transfers of thousands of weapons. This data provides new insight into the small arms trade, including transfers of items that are often poorly (publicly) documented, even by transparent countries. Nonetheless, these records were not sufficient to fill in all of the data gaps and, consequently, models and estimation techniques have become the methodological backbone of the endeavour to generate a global estimate of the annual trade in small arms and light weapons, their parts, accessories, and ammunition.

> The authorized trade study uses an interpolation model to fill in the missing data points.

The estimation technique

A key technique for estimating the value of global transfers in small arms and light weapons is the use of interpolation models. This method was used to estimate the value of international imports of small arms and light weapons ammunition as well as of light weapons (Herron et al., 2010; 2011). Imports, as opposed to exports, were studied because more data is available; at the global level, more countries report on their imports than on their exports. For this chapter, interpolation models were used to estimate the value of the trade in parts of military firearms and light weapons as well as to revise previous estimates of the transfers in small arms (Dreyfus et al., 2009, pp. 7–59).

The authorized trade study uses an interpolation model to collect detailed and fairly complete import data on a sufficiently large sample of countries, then 'interpolates' from this sample to fill in the missing data points for all non-sample countries worldwide, and finally adds all of the data points to generate a global figure. The underlying assumption of this method is that, generally speaking, similar countries have similar levels of imports and that any variation between the imports of different countries is due to specific explanatory factors.

Key explanatory factors are used to create different country groups. In the case of military firearms, light weapons, and light weapons ammunition, the groups are based on military spending per soldier and the size of each country's armed forces.[5] Once the average imports of sample countries in a certain group had been calculated, that average was applied to group members outside of the sample. In some cases, modifiers were applied to non-sample countries to reflect domestic production of small arms and light weapons or involvement in armed conflict. When applied to all countries, this process yielded estimated dollar values for the undocumented international trade in military firearms, light weapons, and light weapons ammunition.

The number of available sample countries depends principally on two factors: the quality of the data provided by individual countries and the number of years for which import data is available. Data over a longer period is generally preferred to balance fluctuations in military procurement and the resulting variations in imports from year to year. In some cases, however, there was a lack of hard data, and the number of sample countries was comparatively small: 11 for the trade in light weapons ammunition, 26 for non-guided light weapons, and 25 for ATGWs (Herron et al., 2010, pp. 18–19; 2011, p. 20).

While the Survey had previously estimated the annual value of authorized transfers of small arms, light weapons, and their ammunition, the assumptions and data on which the estimates were based were revisited and, in the case of small arms and light weapons, revised, leading to a new estimate (see Annexe 8.1). The documented trade in small arms has an average annual value of USD 1.560 billion, while the undocumented trade is estimated at USD 102 million, yielding a total value of USD 1.662 billion. Annual transfers of ammunition for small arms and light weapons are estimated at USD 4.266 billion,[6] while the trade in light weapons is believed to total USD 811 million (see Table 8.3).[7]

Estimating the trade in parts of small arms and light weapons

All countries were divided into four principal and mutually exclusive groups.

There are significant variations in data on transfers of different types of parts. Customs data on transfers of parts of pistols, revolvers, sporting rifles, and sporting shotguns is sufficiently plentiful and disaggregated to allow for a calculation of their overall value without the use of an estimation model. The model is needed, however, for parts of military firearms and of light weapons. As disaggregated data on the transfers of such parts are neither reported to the UN Register, nor provided in most national reports, customs data was the only suitable data source.

Customs data for imports, rather than exports, was used because some major exporters, such as China and the Russian Federation, do not report comprehensively on their exports.[8] While the export trade is dominated by a few states, imports of parts of military firearms and light weapons are distributed among a large number of countries and are reported by many of them, rendering non-reporting by a few states less of a problem than in the case of major exporters. While the available customs data is not very precise, in that the categories of parts for military firearms and for light weapons are partially conflated with larger conventional weapons, reporting is relatively comprehensive: 83 countries have reported their imports for the period from 2005 to 2009.[9] Chapter calculations rely on a long time period because many small countries do not appear to import parts for military firearms and for light weapons every year; a significant proportion of these imports would not appear in a dataset covering a shorter time period.

This section explains how the data from the 83 countries was used to estimate the undocumented imports of parts of military firearms and of light weapons of 90 countries for which no data was available.[10] It also explains how data on transfers of parts for light weapons was disaggregated from data on larger-calibre conventional weapons.

The first challenge was to identify the main factors that best explain variation in imports of light weapons parts. To this end, a dataset containing all available data on imports in parts as well as numerous potential explanatory variables was set up.[11] Through a regression analysis, three factors were found to be relatively important and statistically significant: gross domestic product (GDP), membership in the Organisation for Economic Co-operation and Development (OECD), and whether the production of light weapons in a country was mainly state-owned (see below). The regression analysis did not yield results strong enough to warrant the use of the statistical relationships to compute import values for non-sample countries directly.[12] However, the three most important variables could be used to construct country groups as part of a simple interpolation model. Thus, all countries were divided into four principal and mutually exclusive groups: OECD members with and without predominantly state-owned production and non- OECD states with and without predominantly state-owned production (see Table 8.1).

Table 8.1 Principal country groups in the interpolation model

	Predominantly state-owned production	No predominantly state-owned production
OECD member	**Group 1:** OECD members with predominantly state-owned production (such as Poland)	**Group 2:** OECD members without predominantly state-owned production (such as Germany)
Non-OECD states	**Group 3:** Non-OECD states with predominantly state-owned production (such as the Russian Federation)	**Group 4:** Non-OECD states without predominantly state-owned production (such as Singapore)

The four groups were further subdivided into categories based on GDP in order to reflect large differences in GDP within the principal groups. In each sub-group, the average ratio between GDP and imports was calculated. Next, the imports of every non-sample country were estimated using the country's known GDP value and the average ratio for its GDP-based sub-group. The resulting import estimates for all countries were then compared to mirror data from exporting states in order to distinguish between documented imports, which are captured by mirror data, and undocumented imports, which are not.[13] The sum of all import values constitutes a preliminary global estimate for the imports of light weapons parts.

One problem remained after this step. As mentioned before, UN Comtrade combines data on parts of military firearms and of light weapons with data on parts of larger conventional weapons, such as artillery (Harmonized System (HS) code 930591). In order to distinguish between these sets of parts, the share of imports of complete light weapon systems, on the one hand, and of imports of larger artillery systems, on the other hand, was calculated (see Table 8.2). In 2009, 75 per cent of the total imports in these categories concerned military firearms and light weapons combined. The other 25 per cent consisted predominantly of larger artillery systems and other major conventional weapon systems. Consequently, a proportion of 25 per cent was subtracted from the overall value of all imports of military weapons parts listed under HS code 930591. The resulting figure of USD 969 million is taken to be the average annual value of imports in parts of military firearms and of light weapons. Together with the aggregate value for imports of parts of pistols, revolvers, sporting rifles, and sporting shotguns, it represents the overall value of the trade in parts of small arms and light weapons.

Table 8.2 Share of imports, by weapon category and type, 2009

Weapon category (HS code)	Weapon type	Percentage of imports[14]
Military firearms (930190)	Small arms	41%
Grenade launchers, rocket launchers, etc. (930120)	Light weapons	33%
Self-propelled artillery (930111)	Larger conventional weapons	11%
Mortars, non-self-propelled artillery (930119)	Larger conventional weapons but mixed with some light weapons	14%

Source: UN Comtrade (n.d.)

As mentioned above, data on transfers of parts of pistols, revolvers, sporting rifles, and sporting shotguns is sufficiently robust to conclude that the undocumented trade is probably small for these categories. There was thus no need to estimate the undocumented trade. Documented annual transfers of parts of pistols, revolvers, sporting rifles, and sporting shotguns are worth USD 459 million, based on UN Comtrade data. Thus, the value of the annual trade in all parts of small arms and light weapons is USD 1.428 billion. Of this value, transfers worth USD 1.282 billion are documented transfers, whereas undocumented transfers account for USD 146 million.

Estimating the trade in accessories (weapon sights) for small arms and light weapons

This process yielded a global estimate of approximately USD 350 million for imports of weapon sights annually.

The methodology used to estimate the value of authorized international transfers of accessories is very different from the interpolation model described above. Data on international transfers of accessories is extremely sparse. Customs data made available through UN Comtrade is overly aggregated; all of the relevant HS codes contain data on transfers of unrelated items. The same is true of regional reporting mechanisms and national reports on arms transfers. The UN Register of Conventional Arms—an important source of data on transfers of small arms and light weapons—contains almost no data on transfers of accessories. Publicly available data on national military procurement of accessories is a bit more plentiful, but most contract award notices and other data sources reviewed for this study were too vague[15] or incomplete to serve as substitutes for trade data. Hoping to fill these data gaps, the Small Arms Survey contacted more than 30 governments, four of which responded with data on procurement of accessories for their armed forces. This data—provided by Colombia, Portugal, Sweden, and the UK—sheds important light on the procurement of accessories but does not constitute a sufficiently large sample size for use in an interpolation model. Furthermore, the data on government procurement does not reflect imports for civilian end users, which represent a significant percentage of international transfers in certain types of accessories, such as weapon sights.

While these data gaps preclude the calculation of a comprehensive estimate for the international trade in all accessories, there is sufficient data on transfers of weapon sights to capture that portion of the trade. Transfers of weapon sights are reported under several HS categories in UN Comtrade. Sights exported with rifles and other small arms and light weapons are reported with the weapon and therefore most are captured in the abovementioned estimates for the weapons themselves.[16] Most sights exported separately are reported under a different commodity code, namely HS code 901310. Data submitted to UN Comtrade under this HS code is plentiful. From 2007 to 2010, nearly 130 countries submitted data to UN Comtrade under this code.

Since the data under HS code 901310 includes items other than sights,[17] converting it into an estimate for the global trade in weapon sights required several additional steps. The first step was to gather disaggregated data on imports of weapon sights from as many sample countries as possible. To this end, using various sources, the authors obtained data on the following countries: Chile, India, Paraguay, Peru, Taiwan, Uruguay, and the United States. For 2007–10, the combined annual value of imports of weapon sights transferred to these seven countries was approximately USD 124 million.

The data from the seven sample countries was then used to calculate the value of weapon sights as a percentage of the total value of imports reported under HS code 901310 for each of these countries. Next, the ratios of sights to other items for each country were used to generate an average ratio for all seven sample countries, which was then applied to (overly aggregated) data from 125 countries and territories submitted to UN Comtrade for the period 2007–10. This process yielded a global estimate of approximately USD 350 million for imports of weapon sights annually.[18]

Table 8.3 The average annual value of transfers of small arms and light weapons, their ammunition, parts, and accessories (USD million)

	Annual average value of documented transfers	Annual average value of undocumented transfers	Overall average annual value
Small arms	1,560	102	1,662
Light weapons	256	555	811
Parts of small arms and light weapons	1,282	146	1,428
Accessories of small arms and light weapons (weapon sights)[19]	350	n/a	350
Ammunition	1,903	2,363	4,266
All small arms and light weapons, their parts, accessories, and ammunition	5,351	3,166	8,517

Note: All figures are rounded to the nearest USD million. Differences between the column totals and the sum of the individual figures in the columns are due to rounding.

The annual global value of small arms transfers

As a result of this multi-year project, the global annual value of the trade in small arms and light weapons, their parts, accessories, and ammunition has been found to be at least USD 8.517 billion (see Table 8.3)—a number significantly higher than all previous estimates. This number does not include the trade in parts of guided missiles and components of light weapon ammunition and accessories other than weapon sights, none of which could be accounted for due to the near-total lack of data. Assuming the value of the missing elements is high (see below), it is conceivable that the total trade value could reach, or perhaps even exceed, USD 10 billion.

TRANSFERS OF PARTS FOR SMALL ARMS AND LIGHT WEAPONS

The analysis of authorized international transfers of parts in this chapter can only provide a snapshot. There is insufficient data on parts of several important types of equipment—particularly man-portable guided missiles and other light weapons ammunition. As parts of these weapons are not included in the estimates presented in this section, they will underestimate the value of the trade, probably significantly, especially in view of the high value of production of man-portable missiles and other types of light weapons ammunition and the value of associated parts, which is most probably high (Herron et al., 2010; 2011). In a previous Small Arms Survey chapter dealing with global transfers of ammunition, parts of small-calibre ammunition and shotgun shells were accounted for separately and included in exports of finished ammunition (Herron et al., 2010, pp. 23–27).

In general, data on the trade in parts is available only from UN Comtrade. While this source provides much useful information, it was not designed to be a transparency mechanism for the arms trade, as discussed above. Indeed, UN Comtrade aggregates some types of parts with other equipment to the extent that the data cannot be used. Other data sources that have been used in previous phases of the authorized transfers project, such as national reports or the UN Register of Conventional Arms, do not contain useful information on parts. Parts are either completely absent from the data or aggregated to the extent that no useful information can be gleaned from the sources.

In addition, the authors have had to set conceptual boundaries on what could be studied. The production chain can be traced much further back than the import of the parts discussed in this chapter. For example, this study considers the trade in shotgun barrels, but not in the steel used to make them. In addition, parts of accessories (such as lenses used in optical sights) were not examined, nor were intangibles, such as blueprints and production or export licences. Again, the paucity of available information means that estimating the global value of such transfers is currently impossible.

An overview of the parts trade

As will be shown, the most important use of internationally traded parts is in the production of small arms and light weapons. Imports of parts for repair and maintenance—by parties that already own the finished weapons—is of secondary importance.

Supply chains

Over the past few decades, one of the most striking developments in economics has been the creation by manufacturing companies of global supply chains. This model of production is a radical departure from the traditional plants of old, in which all aspects of production occurred under one roof. Now, in globalized firms, a final factory assembles the finished products from parts that have been produced elsewhere. These parts were themselves assembled from smaller components, which were supplied by other factories, often located in other countries. Globalized production is organized in the form of a supply chain, with firms progressively supplying more complex parts up the chain until they are assembled at the top into a finished product.

Research conducted for this chapter indicates that, in many countries, the production of small arms and light weapons is carried out through such globalized production chains. Countries with small arms production industries import much larger quantities of parts than countries that do not produce small arms. Other countries, such as Norway, export more parts than finished weapons. Their industries specialize in the production of parts rather than finished weapons.

Parts of small arms and light weapons are transferred through the production chain between several types of organizations and under various contractual conditions:

- **Outsourcing** occurs when parts are purchased from a separate company, rather than being produced by the manufacturer of the finished product. In some cases, a large number of firms may compete and produce similar parts, which may be sold to manufacturers of finished products through an intermediary, such as a dealer. In others, one company may be highly reliant on another as a sole supplier or purchaser (Williamson, 1981).
- Under **licensed production** agreements, one organization grants another a licence (usually for a fee) to produce a particular weapon (or other product). Licensed production is normally accompanied by the transfer of intellectual property (such as designs). In addition, the licensor may also provide production machinery and parts that will be used by the licensee to produce the finished weapons. Initially, the licensee may be dependent upon imports of parts from the licensor, but, over time, they may be able to switch to local suppliers or to producing the parts themselves (Gimelli Sulashvili, 2007).
- In **co-production**, two or more companies (often located in different states) agree to develop and produce a weapon system jointly. An example is the MILAN anti-tank weapon, which was originally developed by Euromissile, a consortium of the French Aerospatiale Group and Germany's Daimler-Benz Aerospace (Gander and Cutshaw, 1999, p. 355).

- **Offset and countertrade** arrangements occur when a supplier agrees to buy products from the country purchasing the finished weapon (Brauer and Dunne, 2004). For example, a ministry of defence might procure missiles from a producer in another country, and a condition of the deal might be that the missile producer buys parts from companies based in the same country as the ministry of defence.
- Transfers of parts also occur within elements of **multinational corporations**. Parts produced in a plant in one country may be sent across the border for assembly in another.

The industry is not uniformly globalized. This research finds that OECD members import many more parts than non-members. Membership in the OECD serves as a proxy for countries whose industries have embraced a globalized production chain.[20] In contrast, non-members, including some with large production industries, import relatively few parts from abroad. One example is Brazil, which has one of the world's largest pistol-exporting industries but imports very few parts of pistols. Brazilian firms are either engaged in producing parts themselves or prefer to purchase parts from domestic suppliers. Countries and firms may choose not to source parts internationally for several reasons. Ministries of defence that place security of supply as the highest priority may decide that they do not wish to become dependent on foreign suppliers for crucial parts of weapons used by their armed forces. State-owned firms may prioritize local employment over cost savings (Dimitrov and Hall, 2012).

Furthermore, there is a strategic concern that distinguishes small arms and light weapons from other industrial production. Exports of a majority of parts used in small arms and light weapons manufacture are controlled goods that require an export licence and, potentially, re-export controls. Such transfer controls can restrict the direct or indirect transfer of parts, for example from the United States to arms-producing firms in China. Such retransfer controls also influence trade among Western countries. The United States has strict retransfer controls over exports of parts that require importing firms to request permission before the parts are transferred to another country (for example, as part of a finished weapon). In some cases, these controls have led European producers to source parts from domestic or other European suppliers in order to avoid US retransfer controls.[21]

International transfers of most parts are controlled through export and import licences.

Producers of parts for small arms and light weapons may specialize in those industries or, alternatively, they may specialize in manufacturing a wide variety of parts and other products. For example, Lothar Walther is a specialist manufacturer of gun barrels with a worldwide network of dealers (Lothar Walther, n.d.). The Dandong Xunlei Technology Company of China has a much wider range of products but still targets the small arms market. The company makes parts for small arms, such as buttstocks, hand guards, and rails, along with other accessories, such as telescopic sights, flashlights (to be attached to the weapon), and laser pointers (DXT, n.d.). A company far less focused on small arms production is the US-based Connecticut Spring & Stamping Corporation. It makes a wide variety of springs and stamped metal products for the medical, aerospace, defence, automotive, and small arms industries (CSS, n.d.). While both Lothar Walther and the Dandong Xunlei Technology Company are part of the small arms industry, the designation is not entirely appropriate for the Connecticut Spring & Stamping Corporation. It is a general manufacturer that includes parts for small arms in its diverse range of products.

Commercially available off-the-shelf—or COTS[22]—products are items that are sold in large quantities in the commercial marketplace and can be purchased by an arms-producing company in exactly the same form as is available to the public. For example, commercially available electronics components, such as semi-conductors or printed circuit boards, are often used in military equipment. National laws and regulations differ, but such parts can often be exported freely without a licence.

Repair and maintenance

A further use for internationally transferred parts is for maintenance, repair, and upgrade. Parts are replaced when weapons are routinely maintained. An example is replacement barrels for machine guns. Additional repairs may occur when a weapon is unexpectedly damaged. Weapons may also be upgraded when old parts are replaced with new, frequently better, ones. Research carried out for this chapter indicates that repair, maintenance, and upgrade are not significant components of the overall trade in parts, as imports are affected by production industries rather than the size of armed forces. The analysis shows that, in 2005–09, 56 countries—all of which engage in domestic production—imported 97 per cent of the financial value of all parts for military firearms and light weapons.

Although there is no statistical relationship between the size of a state's armed forces and the import of parts of military firearms and light weapons, there is one with the production of those weapons. This may be because finished weapons that have been imported are often returned to their manufacturer abroad for repair and upgrade.[23] In such cases, the import of parts associated with repair and maintenance would be carried out in the country with the production industry and not necessarily the country with large armed forces. It is important to note, however, that light weapons parts, in particular, are often supplied with the finished weapon when it is initially exported and so may be recorded as part of that transaction.[24] The value of the parts may thus be included in the value of the finished weapon, thereby 'hiding' some parts transfers within other data categories. This effect, though, is likely to be small.

With respect to pistols, revolvers, sporting rifles, and sporting shotguns, civilian purchasers of small arms may simply replace a damaged weapon (which may have cost only a few hundred dollars in the first place) rather than attempt to repair it.[25] In this way, the demand for small arms parts, like that for light weapons, is centred on the production of new small arms rather than the maintenance of old ones. For example, for Sturm, Ruger & Co., a US manufacturer of pistols, revolvers, and rifles, parts and accessories accounted for about half of one per cent of total small arms sales in 2010 (USD 11.5 million out of USD 251.7 million). The company's average for 1993–2010 is 0.32 per cent.[26]

The international trade in parts of pistols, revolvers, sporting rifles, and sporting shotguns

A great deal of data on the international trade in pistols, revolvers, sporting rifles, and sporting shotguns can be found in the UN Comtrade database. The data on parts of revolvers and pistols includes the following: back sights,

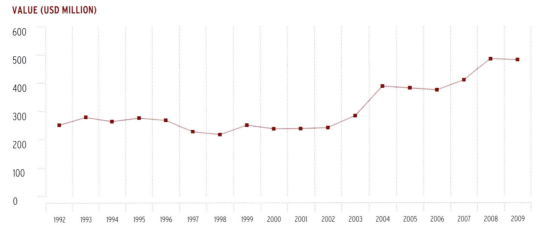

Figure 8.3 **Value of exports of parts of pistols, revolvers, sporting rifles, and sporting shotguns (in USD million)**

Note: All values in constant 2005 USD.

breeches, butt plates, butts, buttstocks, butt swivels, castings, cocking pieces, cylinders for revolvers, extractors, and forgings. Parts of sporting rifles and sporting shotguns include: back sights, breeches, butt plates, butts, buttstocks, butt swivels, castings, cocking pieces, ejectors, extractors and extracting equipment, forgings, front sights, rifled barrels, and shotgun barrels.[27]

The period 1992–2009 witnessed an 88 per cent increase in the value of the documented trade in these parts (after adjusting for inflation). As is shown in Figure 8.3, the trade fluctuated within a relatively narrow range over the 12 years from 1992 to 2003, after which there was a marked increase.

As is shown in Figure 8.4, the overall rise in exports is explained by overall increases by Belgium, Germany, Israel, and Italy. The United States and Japan were important exporters, but over the eight-year period US exports declined by 1.4 per cent and Japan's increased by a modest 13.3 per cent. Figure 8.5 shows that the six largest increases in 2002–09 were by Israel, China, South Korea, Turkey, the Czech Republic, and Mexico. Aside form Israel, these countries' increases began at a low level and thus did not significantly affect the overall value of the trade.

Figure 8.4 **Value of parts of pistols, revolvers, sporting rifles, and sporting shotguns exported by six most significant exporters, 2002-09**

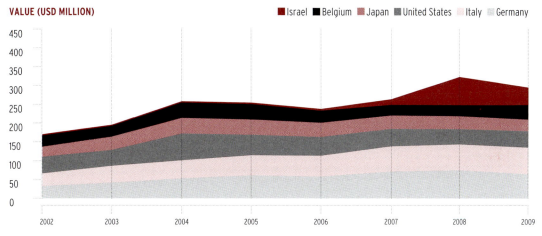

Note: All values in constant 2005 USD. The selected countries accounted for at least one per cent of global exports.

Figure 8.5 **Six largest percentage increases in exports of parts of pistols, revolvers, sporting rifles, and sporting shotguns, 2002-09**

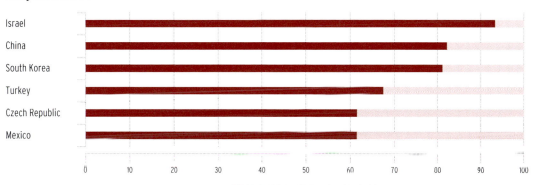

Figure 8.6 **Value of parts of pistols, revolvers, sporting rifles, and sporting shotguns imported by six most significant importers, 2002–09**

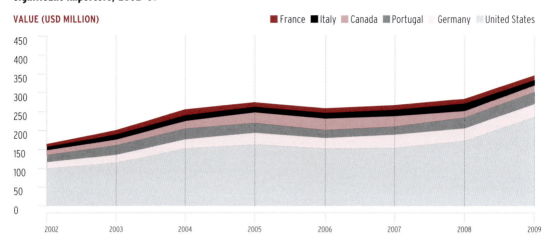

Figure 8.7 **Six largest percentage increases in imports of parts of pistols, revolvers, sporting rifles, and sporting shotguns, 2002–09**

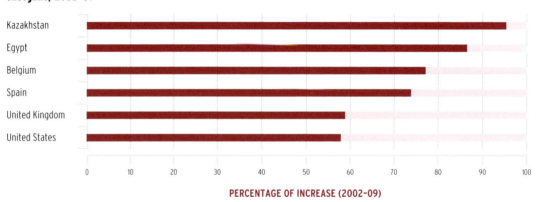

Imports are the other half of the market analysis. As Figure 8.6 shows, the United States dominated the global import market; the rise in imports by the country was largely responsible for the general increase in the global trade, although increases in imports until 2009 by other countries—particularly Germany, Portugal, and Italy—contributed to the overall increase. Several of these countries are also important exporters of small arms to the United States (Dreyfus et al., 2009).

Figure 8.7 presents the six countries with the largest relative increase in exports. Aside from the United States, their absolute increases were not enough to significantly affect the overall value of the trade.

The transfers of parts of pistols, revolvers, sporting rifles, and sporting shotguns can further be disaggregated into parts of pistols and revolvers; shotgun barrels; and parts of shotguns and rifles. Each of these market segments is analysed below. This disaggregation also reflects the available data on the production of small arms in the United States; a further research task could be to analyse this source of information, as well as UN Comtrade, in order to gain a better understanding of manufacturing of small arms by US firms.

Transfers of parts of pistols and revolvers

Documented exports of parts of pistols and revolvers were worth USD 195 million in 2009. Four countries accounted for half of that amount: Italy (USD 43 million), Germany (USD 26 million), Israel (USD 25 million), and Austria (USD 21 million). Three further countries had exports of more than USD 10 million: the United States (USD 15 million), India (USD 11 million), and South Korea (USD 11 million). The largest market by far for these exporters is the United States.

The importance of the United States is reflected in documented imports of parts of pistols and revolvers. The country imports slightly more than half of these parts—at a value of USD 108 million in 2009 (its main suppliers being the first four countries listed above). Germany has imports worth USD 14 million (mainly from the Netherlands, Italy, Spain, and Switzerland). No other country has imports worth more than USD 10 million.

The transfers of parts to the United States appear to be associated with globalized production. For example, the Austrian producer Glock and the Italian manufacturer Beretta both make pistols that are widely sold in the United States, where both also have plants.[28] Glock exports parts from Europe for assembly at their US-based plants (Sweeney, 2008, pp. 98–99); these transfers may explain the reported exports of parts from Austria to the United States.

Beretta has similarly exported parts from Italy to its US plant. Its production of the M9 pistol is an example of how the import of parts for manufacture can change over time. In 1985, it won a major USD 75 million contract to supply the US armed forces with a new standard-issue pistol (Gabelnick, Haug, and Lumpe, 2006, p. 21). Initially, Beretta's new US plant used parts manufactured in Italy (under the auspices of US inspectors) and assembled the pistols in the United States. But in 1989, it started producing parts for the M9 in its US factory and, as of the following year, the pistols were fully produced there (Thompson, 2011, pp. 21–22). After the large Department of Defense orders were completed with the delivery of more than 600,000 pistols, Beretta's US plant continued to produce a variety of pistols for the law enforcement and civilian markets (p. 23). The continued large-scale export of parts of pistols from Italy to the United States suggests that the US-based manufacture of Beretta pistols for civilian and police markets still employs parts sourced from Italy, probably along with parts made in the United States.

The United States imports half of all transfers of parts of pistols and revolvers.

Transfers of shotgun barrels

The documented trade in barrels for sporting shotguns was worth about USD 47 million in 2009. Some USD 37 million of this trade was accounted for by the top five exporters: Belgium (USD 10 million), Mexico (USD 9 million), Japan (USD 8 million), Italy (USD 7 million), and the United States (USD 4 million). Again, the United States was, by far, the largest importer of shotgun barrels from the (other) top exporters. Other countries with exports of more than USD 1 million were: Germany (USD 4 million), Israel (USD 3 million), and Turkey (USD 2 million). The largest recipient of shotgun barrels manufactured in Turkey was Italy. This relationship can perhaps be explained in part by the ownership by Beretta Holding (which controls Fabbrica d'Armi Pietro Beretta) of Stoeger Silah Sanayi, a shotgun-manufacturing firm based near Istanbul since 2005 (Stoeger, n.d.).

As reflected in the export figures, the United States dominates the import market. In 2009, it accounted for USD 33 million of all imports, or 70 per cent of the global market. Other countries with imports worth more than USD 1 million were: France (USD 3 million), Mexico (USD 2 million), and Italy, the UK, and Greece (with around USD 1 million each).

US Customs and Border Protection (CBP) provides rare insight into the supply chain involving shotgun barrels exported from Mexico to the United States. In a ruling concerning O. F. Mossberg & Sons, a shotgun manufacturer based in the US state of Connecticut, CBP reports that:

Mossberg exports lengths of U.S. origin steel rod and barrel extensions to Mexico for processing into unfinished shotgun barrels (consisting of a barrel tube and a barrel extension). Upon importation into the U.S., the shotgun barrels are shipped directly to a Mossberg facility where they are subjected to certain finishing operations, which include attachment of the sights, polishing, and applying the finish (bluing). The shotgun barrels are then assembled with U.S. origin components (i.e., stock, butt, forearm, receiver, magazine, trigger and bolt assemblies, etc.), into finished shotguns. The shotguns are tested to ensure proper functioning and precision fit, and then are partially disassembled to facilitate packaging and shipping (CBP, 2001).

Transfers of parts of sporting rifles and sporting shotguns

The value of the documented trade in parts of sporting rifles and sporting shotguns was worth USD 293 million in 2009 (this figure excludes shotgun barrels, which are covered above). The top ten exporters accounted for some USD 230 million. They were: Germany (USD 42 million), Belgium (USD 33 million), the United States (USD 28 million), Italy (USD 28 million), Japan (USD 26 million), Israel (USD 19 million), China (USD 19 million), Turkey (USD 17 million), Mexico (USD 10 million), and Austria (USD 8 million).

The United States dominates the import markets, but to a lesser extent than with shotgun barrels or parts of pistols and revolvers. Its 2009 imports were worth USD 118 million, or 40 per cent of the market. The ten largest importers accounted for USD 233 million (80 per cent of all imports). After the United States, they were: Portugal (USD 30 million), Germany (USD 23 million), Canada (USD 14 million), Italy (USD 10 million), Austria (USD 9 million), France (USD 9 million), the United Kingdom (USD 8 million), Spain (USD 6 million), and Thailand (USD 4 million).

Portugal's position as the second-largest importer of shotgun and rifle parts in the world can be explained by the assembly of Browning Arms Company sporting rifles and sporting shotguns at a plant at Viana, using parts made in Belgium by FN Herstal and exported to Portugal (Browning, n.d.a; n.d.b). Belgium is the largest supplier of parts to Portugal.

CBP provides additional insight into the highly globalized manufacture of sporting rifles. In a ruling, it states that:

We are informed that the barreled actions are completed rifles without rifle stocks. The barreled actions are manufactured in Belgium by E. Dumoulin and Co. from a series of components consisting of trigger mechanisms from England or Belgium, trigger guards from Spain, screws from Belgium and rifle barrels from the United States, as well as incomplete bolt assemblies, bolt stop assemblies and receivers from China (CBP, 1996).

The international trade in parts of military firearms and light weapons

Data on the international trade in parts of military firearms and light weapons can be found in the UN Comtrade database.[29] The data covers parts of military firearms and of some light weapons (including grenade launchers, mortars, and rocket launchers). Relevant parts include: back sights, barrels, breeches, butt plates, butts, buttstocks, butt swivels, carriages, castings, cocking pieces, ejectors, extractors, forgings, front sights, hammers, levers, liners, locks, magazines, Morris tubes, mountings, percussion hammers, piling swivels, plates, protective cases and covers, recoil mechanisms, safety catches, slings, sound moderators (silencers), stampings, triggers, tripods, and turrets. UN Comtrade data does not disaggregate parts of guided missiles or commercially available off-the-shelf components (see above). The data also comprises parts for weapons larger than what the Small Arms Survey defines as a light weapon, in particular artillery and mortars larger than 120 mm. In calculating the figures presented below, an attempt has been made, using the methodology described above, to remove such large equipment, but doing so adds a further element of imprecision.

AUTHORIZED TRANSFERS 259

Figure 8.8 **Value of documented exports in parts of military firearms and light weapons**

VALUE (USD MILLION)

[Chart showing values from 1992 to 2009, ranging from approximately 400 in 1992 to peaks near 1,300 in 2008]

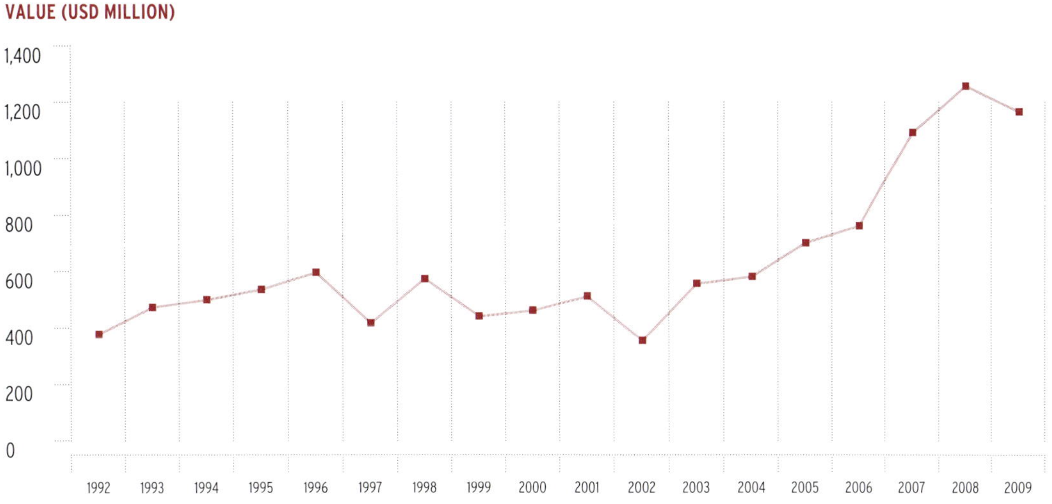

Note: All values in constant 2005 USD.

Between 1992 and 2004, the global value of documented transfers of parts of military firearms and light weapons (adjusted for inflation) remained largely stable. After 2005, though, there was a dramatic threefold increase until 2009 (see Figure 8.8).

Figure 8.9 shows the main components of this increase between 2005 and 2009. After a rise in 2007, US exports slipped back; they were 17 per cent lower in 2009 than in 2005. The most dramatic absolute increases were by the UK and Norway. Canada and Sweden also increased their exports of parts. For Canada, Norway, and the UK, the United States was, by far, the most important recipient of parts. For Sweden, India was the most important export market (see below). South Korea had a large relative increase but its absolute exports were much smaller than those cited above. The significant variations in exports are shown in Figure 8.10; aside from France, the countries with the largest absolute increases also had the largest relative increases.

Figure 8.9 **Value of parts of military firearms and light weapons exported by six most significant exporters, 2005-09**

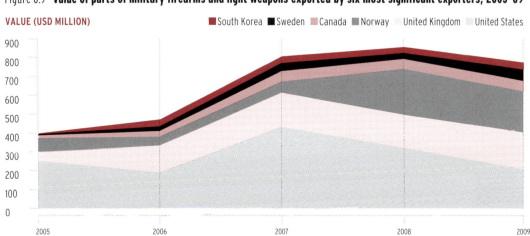

Figure 8.10 **Six largest percentage increases in exports of parts of military firearms and light weapons, 2005–09**

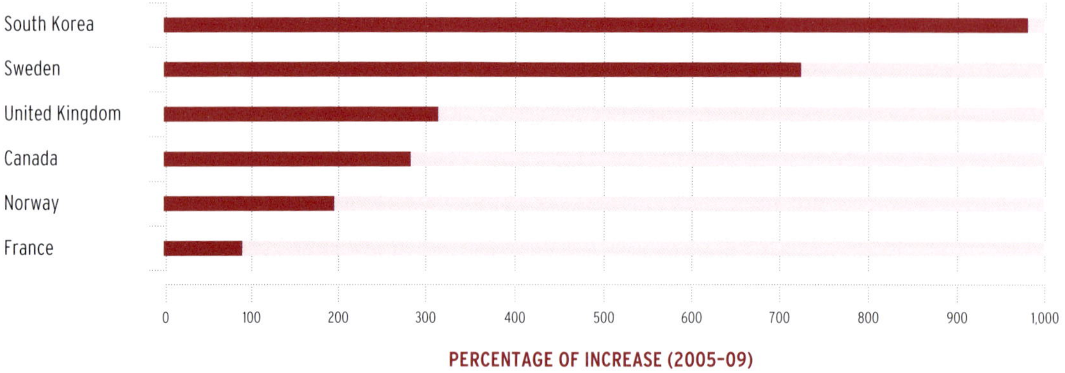

[Bar chart showing percentage of increase (2005–09):
- South Korea: ~980
- Sweden: ~720
- United Kingdom: ~320
- Canada: ~280
- Norway: ~200
- France: ~90]

PERCENTAGE OF INCREASE (2005–09)

Figure 8.11 indicates the destinations of this increased export trade. Increased US imports are responsible for the overall increase over the five-year period. Imports by South Korea and Japan peaked in 2007 and were responsible for the overall peak of global transfers in that year. Figure 8.12 summarizes the six largest increases in imports in 2005–09; India, Thailand, Norway, and Canada had large relative changes but their absolute imports were not large enough to significantly affect the overall total.

Exports of parts from Norway to the United States include the Kongsberg remote turret. In 2007, the US Army awarded Norwegian company Kongsberg Defence Systems a USD 1.4 billion contract to supply remote turrets. These high-technology systems allow a machine gun to be mounted with sensors on the roof of an armoured vehicle; the weapon is controlled and fired by personnel safely ensconced within (Cox, 2007). Following a CBP decision, the remote turret was defined as a part (as are other turrets) (CBP, 2010).

India also offers an interesting illustration of the trade in parts. It produces the Carl Gustaf recoilless rifle under licence from Saab of Sweden.[30] Given that the largest destination of parts from Sweden was India, it is likely that some of the parts exported from Sweden to India are for the production of Carl Gustafs. India has also, since 1985, produced under licence the Franco-German MILAN anti-tank guided missile system. The production of this missile

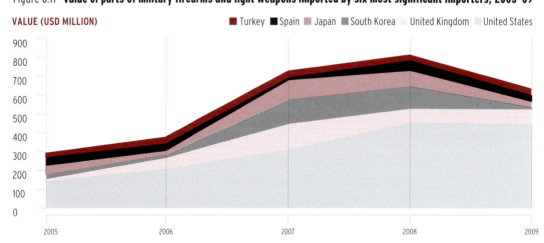

Figure 8.11 **Value of parts of military firearms and light weapons imported by six most significant importers, 2005–09**

Figure 8.12 **Six largest percentage increases in exports of parts of military firearms and light weapons, 2005-09**

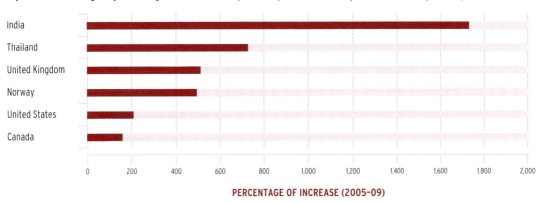

system illustrates how the trade in parts can change. Bharat Dynamics of India started production in 1985 (Mohanty, 2004, pp. 19–21). Initially, production was based on the assembly of imported parts; by 2004, the local content of the updated MILAN2 was reportedly 'expected to reach 75%' (Khan, 2004, p. 251). In all, some 30,000 MILAN missiles were manufactured in India in the period up to 2009, and production continues to this day, with an increasing emphasis on domestic production (*Vayu*, 2010).

ACCESSORIES FOR SMALL ARMS AND LIGHT WEAPONS

Despite their widespread military and civilian use and increasingly important role on the battlefield, accessories for small arms and light weapons have received little attention from researchers and policy-makers. This section attempts to improve our understanding of accessories for small arms and light weapons by providing an overview of the types and models of major accessories and their roles, capabilities, and characteristics. The contours of the authorized international trade are also captured through an analysis of data on recent international transfers to Chile, Colombia, India, Paraguay, Peru, Portugal, Sweden, the UK, the United States, and Uruguay.

As explained above, the Small Arms Survey defines accessories as items that physically attach to small arms and light weapons and increase their effectiveness or utility but, generally speaking, are not essential for the basic, intended use of the weapon. This section focuses on the following items:

- weapon sights;
- aiming lasers and illuminators;
- night vision devices;
- laser rangefinders; and
- fire-control systems, including ballistics calculators.

Accessories are sold with the weapons on which they are used, as part of upgrade packages,[31] and as stand-alone orders.[32] Many are attached to rails mounted on the weapon, including the widely used Picatinny rail. While most of the items categorized above are not essential for the basic use of the weapon to which they are attached, as indicated earlier, there is a small but growing list of exceptions, including telescopic sights for long-range sniper rifles and fire-control systems for weapons that fire airburst munitions. These items are nevertheless considered accessories for the purposes of this chapter.

Many of the technologies used in accessories for small arms and light weapons have a wide range of military and commercial applications. Roles for thermal imagers range from long-range sniping to detecting leaks in thermal insulation and spotting cases of bird flu (Saletan, 2009; Dove, 2010). Similarly, laser rangefinders, which are used by militaries to calculate superelevation angles for grenade launchers and programme airburst munitions, are also used by interior designers to measure floor space and by golfers to determine the distance to the next green (Baily, 2009; Bosch, n.d.).

These technologies take several different forms on the battlefield. Image-intensifier tubes used in night vision weapon sights are also used in night vision goggles, helmet-mounted monoculars, handheld binoculars, and other surveillance systems. Thermal imagers are used in weapon sights, unmanned aerial vehicles, goggles, handheld cameras, and vehicles (Gething, 2008). Laser rangefinders, fire-control systems, and the other accessories listed above are also deployed on various platforms.

Small arms and light weapons accessories and the modern battlefield

Accessories for small arms and light weapons are an important part of the modern battlefield. The need to engage elusive enemy forces operating in heavily populated areas quickly and accurately while minimizing civilian casualties requires precision, situational awareness, and robust command and control, all of which can be enhanced by the accessories listed above. However, these benefits must be weighed against the budgetary and logistical strain caused by the ever-expanding list of items issued to dismounted infantry. Balancing these often-competing demands is the key challenge facing military procurement officers (Gelfand, 2011).

The various soldier modernization programmes being pursued by militaries worldwide provide some insight into trends in military technology and the goals of those attempting to harness their tactical and operational potential. Common to most of these programmes is the vision of a networked battle space in which the dismounted soldier is one of several platforms contributing to—and benefiting from—a real-time exchange of data. Accessories for small arms and

A US Army soldier makes his way through underbrush with his Land Warrior System at Fort Benning, August 2001.
© Ric Field/AP Photo

light weapons are key components of this vision, which is epitomized by the Pointer system developed by the British firms QinetiQ and Qioptiq. The Pointer system collects data from gunshot detectors, rangefinders, and other sensors and transmits that data to all networked (Pointer) weapon sights, instantaneously if desired. The data is displayed in the sight as intuitive symbols that direct the shooter to the target while providing information on the range and position of the target and nearby friendly forces (QinetiQ, n.d.; Brown, 2011).

This technology could be game-changing. In theory, at least, a single shot fired by an enemy sniper in an area where Pointer is deployed would result in the near-instantaneous identification and targeting of that sniper by every networked soldier within range. Weapon sights fielded in recent years also have the capacity to capture and transmit video imagery through wireless links, allowing the commander to see what the shooter sees (Pengelley, 2008). This capability has the potential to reduce friendly fire and civilian casualties, both of which are high-priority goals of most modern militaries.

Yet the tactical and operational benefits from accessories must be balanced against real-world constraints, such as finite procurement budgets and physically overburdened soldiers. The latest accessories are expensive. For example, a ThOR 3 colour-imaging scope costs more than USD 13,000 per unit (Firearm Blog, 2011)—more than some countries spend annually on all equipment for individual soldiers.

Weight and power consumption are other factors that shape demand for weapons and equipment, including accessories. UK soldiers conducting patrols in Iraq carried an average of 54–64 kg of equipment (Gelfand, 2011). Reducing this burden (or at least not adding to it) has become a central preoccupation of military procurement officers in the UK and elsewhere.[33] In response, defence firms have reduced the weight of their products. The defence firm ITT, for example, markets night vision weapon sights that are 'the size of a middle finger' (Gelfand, 2011). But even as the weight of individual items has decreased, the number of items issued to soldiers has increased, offsetting many of the gains from industry efforts to reduce the weight of accessories and other items (see photo). During an interview with Jane's Information Group, a Canadian official pithily summarized this dilemma: 'Twenty years ago our soldiers were compelled to carry 100 lb of heavy equipment, now they carry 100 lb of very light equipment' (Pengelley, 2009b).

The tension between operational needs and budgetary and logistical constraints is likely to continue to shape the development and procurement of military technology, including accessories for small arms and light weapons, for the foreseeable future.

Weapon sights, aiming lasers, and clip-on night vision devices

Weapon sights[34] are among the most widely exported accessories for small arms and light weapons. There is significant variation regarding which types of sights are deployed and how widely, however. Use of telescopic sights, for example, ranges from universal (weapons with built-in sights)[35] or near universal (long-range sniper rifles) to negligible (pistols and single-shot disposable rocket launchers). Many models are used on several different types of weapons. For example, versions of the AN/PAS-13 thermal sights are used on the US M4 assault rifle, the M136 recoilless rifle, the M249 light machine gun, the M240B machine gun, the M24 sniper rifle, the MK19 automatic grenade launcher, and 'surface-to-air missile launchers' (Brown and Wasserbly, 2009).

For the purposes of this chapter, sights are divided into seven (sometimes overlapping) categories: iron sights, telescopic sights, reflex sights, image-intensifying sights, thermal sights, holographic sights, and laser sights. Since many aiming lasers often take the form of sights, they are discussed with laser sights. A brief description of each category follows.

A Spanish soldier is equipped with an H&K G36 rifle fitted with accessories, procured as part of the 'Future Soldier' programme. © Spanish Ministry of Defence

Iron sights are the oldest and most widely used type of sight. Iron sights come standard on most small arms and many light weapons and are the only type of sight used on some weapons.[36] Most iron sights for small arms consist of two main components: a front sight, which is typically positioned at the end of the muzzle barrel, and a rear sight, which is commonly positioned on or over the receiver. Iron sights are generally divided into two categories: open sights and aperture sights. With open sights, the front sight is typically a post or bead, and the rear sight is a notch. The shooter lines up the front sight (post or bead) inside of the rear sight (notch) and below the target (see Figure 8.13).

A German Bundeswehr Army soldier uses the laser pointer on his weapon during a joint patrol with Afghan National Army soldiers north of Kabul, September 2008.
© Fabrizio Bensch/Reuters

Figure 8.13 **Various open sights, along with one aperture sight (H)**

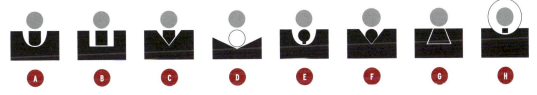

Key:
A) U-notch and post, B) Patridge, C) V-notch and post, D) Express, E) U-notch and bead, F) V-notch and bead, G) trapezoid, H) ghost ring.
The grey dot represents the target.

Source: Wikipedia (n.d.)

With aperture sights, the rear sight consists of a ring instead of a notch. There is a wide variety of iron sights, which range from a simple groove milled into the receiver of some pistols to the elaborate adjustable aperture sights used by competition target shooters (see photo).[37]

Telescopic sights, which are often referred to as 'scopes', are basically weapon-mounted telescopes with a reticle (crosshair). Telescopic sights aid in targeting and improve accuracy by magnifying the image of—and projecting a reticle onto—the target. They also require less eye coordination than iron sights. Drawbacks of telescopic sights include the need to position the eye a specific distance from the sight and a limited field of view, meaning that the operator sees less of the surroundings while looking into the sight. Generally speaking, telescopic sights are also less durable than iron sights and can be significantly more expensive.

Telescopic sights have been widely used by hunters, military snipers, and others engaged in longer-range shooting for decades. Their use has grown in recent years as militaries attempt to improve the accuracy and range of standard-issue rifles. Low-magnification sights are now widely issued to soldiers in the militaries of many countries.

> Image intensifiers for military end users have been produced in at least 26 countries.

Reflex sights display an illuminated reticle that is superimposed on the image of the target in the sight window.[38] Because the reticle often takes the form of a red dot, reflex sights are often referred to as 'red dot' sights. Reflex sights are popular because the shooter's eye can be positioned at more angles and at greater distances from reflex sights than from telescopic sights. Consequently, target acquisition is much faster with reflex sights than with telescopic and iron sights. The shooter can also look through the sight with both eyes open, allowing for a fuller field of view and therefore better situational awareness than with telescopic sights (White, 2010). For these reasons, reflex sights 'have become a staple requirement for [military] small-arms programmes around the world' (White, 2010); they are particularly useful in close-quarter combat and other scenarios in which targets must be engaged quickly (Ring Sights, 1998).

Holographic sights feature a hologram of a reticle that is recorded at the time of manufacture and then displayed on the sight window by a laser. The major advantage of holographic sights is that the reticle is fully visible regardless of the angle and distance at which the sight is viewed. The image remains visible even if the sight itself is partially obscured by mud, snow, or rain (Jones and Ness, 2011, p. 617; L3 Communications, n.d.). The major disadvantages of holographic sights are their comparatively high cost and power consumption.

Image-intensifying sights use image-intensifier tubes to gather existing (ambient) light, such as starlight, moonlight, and certain infrared light. The light is then amplified and converted into an image that is displayed in the sight. Since the first military night vision devices were fielded in the late 1930s, several additional generations of image-intensifier tubes have been developed. Generational improvements include brighter and sharper images, improved performance in low-light conditions and light-polluted areas (such as cities), and longer target detection ranges.

Image intensifiers for small arms and light weapons are available as stand-alone sights and clip-on units used with day sights. An example of the latter is the Clip-on Sniper Night Sight used with the US military's Semi-Automatic Sniper System (IQPC, 2010). Sights or clip-on units featuring image intensifiers for military end users have been produced in at least 26 countries (Jones and Ness, 2011, pp. 633–63).[39]

Thermal sights differ from image-intensifying sights in that they detect infrared radiation emitted by the target rather than light reflected off the target (Electrophysics, n.d.). The vast majority of thermal sights feature uncooled detectors, which are less sensitive than cooled detectors but are also lighter, quieter, and less expensive (Gething, 2008).[40] Since thermal sights do not rely on ambient light, they are much more effective than image intensifiers in the low-light environments encountered in underdeveloped and sparsely populated countries such as Afghanistan. They can also 'see' through dust, fog, sand, and other obscurants (USMC, 2004, p. 143). The major drawback of thermal sights is their cost. According to US Army budget documentation, the per-unit cost of the AN/PAS family of thermal sights procured in 2010 ranged from approximately USD 7,900 for the light sight designed for use with assault rifles to USD 9,500 for 'heavy' sights used with sniper rifles, heavy machine guns, and automatic grenade launchers (US Army, 2011; DRS Technologies, 2010).[41] These costs are likely to decline, however, which will make them increasingly attractive vis-à-vis image-intensifying sights.[42]

Laser sights and other aiming lights[43] project a beam of visible or infrared light at the target. The beam is typically aligned with the barrel of the weapon (boresighted) and therefore laser sights are often used instead of iron sights and telescopic sights when rapid target acquisition is required (White, 2010; Jones and Ness, 2011, p. 622). When the shooter is operating as part of a team, laser sights are also used to identify and hand off targets. Some aiming lights are used for illumination (Laser Devices, 2010, p. 8). Aiming lights, including laser sights, are used by armed forces, law enforcement agencies, and civilians and vary significantly in price and performance. Low-powered laser sights for pistols, which typically have a range of a few dozen metres, can be acquired for as little as USD 20 online (Opticsplanet.com, n.d.a). On the other end of the spectrum is Insight Technology's CNVD-T2 Clip-On Thermal Imaging Sight with Laser Pointer, which sells for USD 25,000 (Opticsplanet.com, n.d.b).

Figure 8.14 **M4 MWS with accessories**

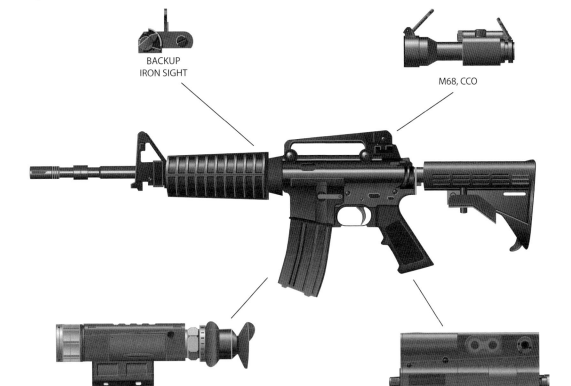

Laser sights and other aiming lights have several limitations. Visible lasers can reveal the presence and the position of the shooter, alerting the target and exposing the shooter to counter-fire. The same is true of infrared lasers if the enemy is equipped with night vision devices. Many aiming lights are also less effective in bright light and against backgrounds in which there is no surface, such as the sky (Tai et al., 1996).

Figure 8.14 includes several examples of the accessories described above and indicates where they are mounted on a typical assault rifle. These accessories include a reflex sight (M68 Close Combat Optic), a back-up iron sight, a thermal sight (AN/PAS-13), and a range-finder (AN/PSQ-23).

Other accessories

The term 'rangefinder' is used to refer to a variety of instruments ranging from the 'stepped slot, indicating the apparent length of an average tank at [specific] ranges' in the foresight of the RPG-75 to the US military's AN/PSQ-23 Small Tactical Rifle Mounted Micro-Laser Range Finder (Jones and Ness, 2007, p. 451). This chapter focuses on **laser rangefinders**, which measure the time it takes for a laser projected at a target to 'bounce back' to the rangefinder. A microprocessor in the unit calculates the distance to the target by measuring the length of time between when the beam is projected and when it bounces back (Shideler and Sigler, 2008, p. 49).

As mentioned above, laser rangefinders have numerous military and civilian purposes. Features found in more expensive hunting rangefinders include filtering technology that improves readings in obscurants such as rain, haze, and dust; a scanning mode that allows for the tracking of moving objects; and the ability to exclude false readings from objects in the path of the target, such as tree branches (Shideler and Sigler, 2008, p. 49). Some are integrated into riflescopes, but most appear to be sold as binoculars. Many military rangefinders are integrated into fire-control systems that perform multiple functions (see below) and can significantly increase the range and accuracy of the weapons to which they are attached. Use of a laser rangefinder and thermal optic reportedly increases the range of the Shoulder-launched Multipurpose Assault Weapon II rocket by 200–500 m and can increase first-round hit rates to more than 80 per cent (Gething, 2010).

Fire-control systems vary significantly in terms of technological sophistication and complexity.

A **fire-control system** (FCS) is a device that assists in acquiring and tracking targets, computing targeting data, and controlling the rate and direction of fire.[44] While commonly associated with mortars, fire-control systems are used with many small arms and light weapons and take various forms. Some are mounted on vehicles, others are handheld—often consisting of a personal digital assistant or tablet computer (Pengelley, 2011)—and still others are attached to dismounted weapons. Some fire-control systems are not 'accessories' as defined in this chapter since they are built into the weapon at the time of manufacture and are essential for one or more intended uses.

Fire-control systems vary significantly in terms of technological sophistication and complexity. The COCOS COmmando COntrol System for 60 mm mortars, for example, does only one thing. Its microprocessor uses data on the range of the target and the position of the barrel to determine when the angle of the barrel matches the range data (Jones and Ness, 2011, p. 678). Other systems are more technologically sophisticated and perform a variety of functions. Rheinmetall's Vingmate FCS for heavy machine guns and automatic grenade launchers features a ballistics computer, day camera, magnetic compass, global positioning system (GPS), and laser rangefinder. With the Vingmate, the operator of an automatic grenade launcher can: (1) detect and recognize man-size targets at distances greater than 1,100 m, (2) calculate the distance of targets located up to 4,500 m away, (3) compute the proper superelevation angle for the grenade launcher, (4) programme the fuses of airburst munitions, (5) send streaming video and other data to command posts and other networked units, and (6) receive data, including the locations of friendly forces (Pengelley, 2009a; Rheinmetall Defence, n.d.b).

For the militaries that can afford them, fire-control systems offer many benefits. The systems eliminate the need to 'walk' rounds into the target, a major liability of light weapons such as automatic grenade launchers (Pengelley, 2011). To hit a target with a conventional grenade launcher, the gunner often has to fire a round, observe where it lands, adjust the position of the barrel, fire another round, and repeat this process until a round hits the target. This process wastes ammunition, increases the likelihood of collateral damage, allows the target to take cover, and exposes the gunner to counter-fire. Systems such as the Vingmate increase the likelihood of a first-round hit, thereby addressing many of these problems.

Advanced fire-control systems can be costly. According to one industry official the cost of a basic FCS for the MK19 automatic grenade launcher is approximately US 50,000,[45] significantly more than the grenade launcher itself. Fire-control systems for mortars appear to be even more expensive. The estimated unit cost of the FCS for the M150 dismounted mortars budgeted for procurement by the US Army in 2010 was USD 107,000 per unit (US Army, 2011).

The international trade in accessories for small arms and light weapons

As discussed above, data on international transfers of accessories for small arms and light weapons is less plentiful and detailed than data on transfers of the weapons themselves. Detailed, disaggregated data on accessories is generally not published through any of the multilateral reporting mechanisms, and few national reports include such data. Consequently, comparatively little is known about the international trade in small arms accessories.

As a first step towards correcting this deficiency, the Small Arms Survey obtained detailed data on imports of sights and other accessories for small arms and light weapons for ten countries.[46] Data on six of these countries—Colombia, India, Portugal, Sweden, the UK, and the United States—reflects military procurement. Data on the four remaining countries—Chile, Paraguay, Peru, and Uruguay—was obtained from Datamyne, a US provider of trade data collected from national customs agencies, and primarily reflects imports of weapon sights for civilian end users.

The following section summarizes and assesses this data through two case studies. The first case study draws on detailed customs data to assess the international trade in weapon sights used by civilians.[47] The procurement of sights and other accessories by militaries in six countries is the focus of the second case study. The case studies shed important light on the types of accessories most frequently acquired, the major suppliers and countries of origin of these accessories, and recent procurement trends.

Case study: the civilian market for weapon sights in South America

Data on transfers of weapon sights in Chile, Paraguay, Peru, and Uruguay reveals a civilian import market that is dominated by inexpensive sights manufactured and, to a lesser extent, sold by companies located in China. Most of these items appear to be basic telescopic sights for use with hunting and air rifles but other types of sights, including red dot, laser, and night vision sights, are also listed.

China is identified as the country of origin for nearly 90 per cent of the roughly 133,000 imported sights.[48] Spain was the second-largest producing state, accounting for approximately 7,000 units, or roughly 5 per cent, of all imports. Table 8.4 lists the ten largest countries of origin for the imported sights.

China also tops the list of exporting countries, but its share of exports is notably lower. According to the data, approximately two-thirds of all documented imports came directly from exporters based in China. Most of the

Table 8.4 Weapon sights transferred to Chile, Paraguay, Peru, and Uruguay: top ten countries of origin

Country/territory of origin	Quantity of imported sights[49]	Percentage of total imports
China	116,857	88%
Spain	6,898	5%
Hong Kong	3,660	3%
United States	2,199	2%
Unidentified	1,077	1%
Uruguay	492	<1%
Italy	394	<1%
South Korea	286	<1%
Japan	167	<1%
Philippines	135	<1%

Source: Datamyne (n.d.)

remaining sights were purchased from exporters located in the United States and Spain, which captured 15 per cent and 7 per cent of the import market, respectively. Even in these cases, however, China retained a presence in the production and distribution chain; more than half of the sights exported from the United States and Spain were produced in China. Table 8.5 lists the top exporting states and their respective share of imports in three South American countries.

Table 8.5 Weapon sights transferred to Chile, Paraguay, and Uruguay: top ten countries of export[50]

Country/territory of export	Quantity of imported sights[51]	Percentage of total imports
China	83,422	64%
United States	19,021	15%
Spain	9,017	7%
Iquique Free Zone	5,051	4%
Uruguay	4,287	3%
Hong Kong	3,562	3%
Chile	1,950	2%
Montevideo Free Zone	1,201	<1%
Italy	1,199	<1%
Argentina	775	<1%

Source: Datamyne (n.d.)

Table 8.6 Weapon sights transferred to Chile, Paraguay, Peru, and Uruguay: unit values

Unit value (USD)	Quantity	% of total quantity	Countries of origin (top five)	Countries of export (top five)
>1,000	6*	<1%	United States, Austria	United States, Austria
500–1,000	20*	<1%	Germany, Austria, United States	Germany, Austria, United States
100–500	464	<1%	United States (53%), Japan (33%), Sweden (5%), South Korea (4%), China (3%)	United States (92%), Sweden (5%), Italy (2%), Germany (<1%), Spain (<1%)
10–100	28,650	22%	China (75%), Spain (13%), United States (6%), Uruguay (2%), Unspecified (1%)	United States (42%), China (23%), Spain (17%), Unspecified (5%), Uruguay (5%),
<10	103,467	78%	China (92%), Hong Kong (3%), Spain (3%), United States (<1%), Unspecified (<1%)	China (74%), United States (6%), Iquique Free Zone (5%), Spain (4%), Hong Kong (3%)

Note: * This quantity excludes one sight whose importer is identified as the Peruvian military.
Source: Datamyne (n.d.)

Also noteworthy is the large percentage of very inexpensive imported sights and the clear stratification of producers and exporters of inexpensive and expensive sights. The data indicates that the vast majority of imported sights had unit values of less than USD 100, and most had values of less than USD 10 per unit. Nearly all of the least expensive items (such as items with a unit value of less than USD 10) were produced in China. US exporters competed more successfully with their Chinese counterparts for the market in sights valued at USD 10–100, but most producers of these sights were also located in China. The commodity descriptions submitted with the customs documentation indicate that most of the least expensive items were intended for use with sporting, hunting, and air rifles.

The trade in more costly sights was small and dominated by companies located in Austria, Germany, and the United States. Imports of sights with a unit value of USD 1,000 or more were negligible, consisting of just six units, four of which were exported to Chile from the United States. These sights include a PS22 night vision scope along with high-end telescopic sights produced by well-known US and European manufacturers. Similarly, only 26 sights with unit values exceeding USD 500 were imported; all of them came from Austria, Germany, and the United States. Sights valued at between USD 100 and USD 500 were slightly more numerous but still constituted less than one per cent of all imported sights. Most were produced in the United States or Japan, and nearly all were exported from the United States. Of the 464 sights in this price band, only 15 were reportedly produced in China, and none were exported from China. Table 8.6 lists the major producers and exporters of sights in each price band, along with the quantities of sights imported per band.

Whether the civilian market for weapon sights in South America is representative of the global market is unknown, but US import data suggests that at least some of its features, including the influential role of Chinese producers and exporters, are more broadly applicable. According to US customs data, three-quarters of imported rifle sights[52] were imported from China (USCB, n.d.). The data also indicates that the unit value of these sights is significantly lower than sights from other countries. Comparably detailed customs data from other countries would allow for a more definitive assessment of the international civilian market for weapon sights and other accessories.

Case study: military procurement of accessories

Data on the acquisition of accessories, including sights, by a small but diverse group of governments highlights several noteworthy features of military procurement of accessories for small arms and light weapons. The data covers procurement, including imports, by the militaries of Colombia, India, Portugal, Sweden, the UK (see Table 8.7), and the United States.

The data suggests that the market for military sights is different from the civilian market in several notable ways. Whereas most of the sights exported to civilian end users in South America came from China, nearly all of the imported sights procured by the six militaries studied were purchased from Canada, Germany, Israel, South Africa, Sweden, or the United States. Common to both markets, however, is the important presence of US producers and exporters, which are listed as the contractors for nearly all accessories procured by the US Army and a large percentage of the sights imported by the other countries.

The data also reflects the rapid expansion of the military market for thermal weapon sights, one of the most significant trends in the development and procurement of weapon sights.[53] In the late 1990s, thermal sights for small arms were comparatively rare, primarily because of their cost, weight, and excessive power consumption (Brown and Wasserbly, 2009). Today, thousands are procured each year by militaries worldwide. This trend is evident in the data obtained for this study and other reports on recent military procurement. Data from budget documents published by the US Army reveals planned procurement of nearly 110,000 thermal sights in US fiscal years 2008–2010, the combined estimated value of which was nearly USD 1 billion (US Army, 2009; 2010; 2011). Deliveries of thermal sights to the US Army by BAE Systems alone topped 100,000 units as of August 2011 (BAE Systems, 2011).

Table 8.7 **Accessories for small arms and light weapons delivered to the UK Ministry of Defence in 2010**

Equipment	Supplier	Country of manufacture[54]	Quantity delivered
Maxikite	Qioptiq	UK	892
Lightweight thermal imager (VIPIR2)	Qioptiq	UK	1,119
Lightweight thermal imager (VIPIR2+)	Qioptiq	UK	18
Thermal weapon sight	DRS Optronics	US	1,066
Sniper thermal imaging capability	Qioptiq	UK	300
AN-PEQ2A	Thomas Jacks Ltd.	US	876
Small arms thermal imager	Insight Technology	US	130
FIST thermal sight	THALES	UK	252
FIST lightweight day sight	THALES	Canada	1,299
FIST close quarter battlesight	THALES	UK	2,778
FIST underslung grenade launcher system	THALES	UK	264
FIST CWS MK2	THALES	UK	200
FIST Maxikite	THALES	UK	180

Source: UKMoD (2011), obtained by Small Arms Survey through a Freedom of Information Act request

While probably the biggest consumer of thermal sights, the US military is not alone. Nearly 2,900 of the approximately 8,200 sights delivered to the British armed forces in 2010 were thermal imagers (see Table 8.7). Other countries that have recently procured—or are planning to procure—thermal sights for small arms include Australia, Bulgaria, France, Germany, India, Israel, Romania, Singapore, and South Korea.[55]

This trend is reflected in interviews with defence industry representatives. According to one representative, the use of thermal sights and image intensifiers on the widely exported Carl Gustaf recoilless rifle is growing, in part because of the preference among militaries to engage in combat operations at night. There is also widespread interest in thermal imagers for the BILL 2 anti-tank guided weapon and the RBS-70 man-portable air defence system.[56]

The data also sheds some light on the military market for other accessories, the procurement of which appears to be small compared to sights. Budget data from the US Army reveals planned annual procurement of fewer than 10,000 rangefinders and fewer than 100 M150 mortar fire-control systems, as compared to more than 100,000 weapon sights. The United States procured more aiming lasers and lights than fire control systems and rangefinders, but the total value of that procurement was a fraction of the budget for sights (US Army, 2009; 2010; 2011). According to data collected for this study, the number of sights procured by Colombia, Portugal, and the UK also far exceeded procurement of other accessories.[57] In addition, interviews with industry

Box 8.4 Trends in the small arms trade

Every year UN Comtrade receives data on arms transfers from more than 150 countries. It is thus the richest single data source and the best means by which trends over time and comparisons between countries can be made. Nevertheless, a lack of reporting by some countries and aggregation of different types of equipment mean the figures presented in this box and in Table 8.8 only reflect a portion of the trade.

An analysis of UN Comtrade data reveals that the total value of reported exports of small arms, light weapons, their parts, and ammunition in 2009 was USD 4.6 billion. This is an absolute increase over the USD 4.3 billion reported for 2008 (Herron et al., 2011, p. 11); these two years continue the trend of a steadily increasing financial value of global small arms transfers (Dreyfus et al., 2009, p. 11). The difference between the global figure presented in this box and the USD 8.5 billion estimate of the value of the trade presented elsewhere in this chapter is explained by the four-year project's use of additional data sources and estimation of the undocumented trade to calculate the higher estimate. However, that methodology is unsuitable for making rankings and assessing trends. For that, a single data source—UN Comtrade—is used.

The analysis shows that 37 countries exported more than USD 10 million. Classified as major or top exporters, these 37 collectively accounted for transfers worth USD 4.5 billion, or some 98 per cent of the global total. Twelve countries exported more than USD 100 million; in descending order, they are the United States, Italy, Germany, Brazil, Austria, Japan, Switzerland, the Russian Federation, France, South Korea, Belgium, and Spain, all classed as top exporters. Together, the 12 countries exported small arms worth USD 3.5 billion, or 76 per cent of the global total. Since 2008, France and Japan have risen to the top category, as they exported more than USD 100 million in 2009. Four countries—Canada, Israel, Norway, and Turkey—exported less than the threshold in 2009 and thus dropped down to the major first tier category (USD 50-99 million). Twenty-five countries exported between USD 10 million and USD 99 million and are therefore classed as major exporters. Together, they exported arms worth USD 1 billion, or 22 per cent of global exports. Four countries—Australia, Bulgaria, Hungary, and the Philippines—were newly classified as major exporters since they exported less than USD 10 million in 2008.

The seven top importers (acquiring more than USD 100 million) in 2009 were (in descending order): the United States, the United Kingdom, Saudi Arabia, Australia, Canada, Germany, and France. Overall, these top importers accounted for 59 per cent of global transfers. The United States was the largest importer by far; its 2009 imports were worth USD 1.8 billion or 38 per cent of the total. There were 47 major importers (USD 10-99 million), which collectively imported 35 per cent of global transfers.

Table 8.8 Exporter rankings for 2009			
Ranking	Country	Exports in 2009 (USD million)	Change from 2008 to 2009
Top first tier (>USD 500 million)	United States	706	
	Italy	507	
Top second tier (USD 100-500 million)	Germany	452	
	Brazil	382	
	Austria	249	
	Japan	249	Moved up from major first tier
	Switzerland	188	
	Russian Federation	168	
	France	161	Moved up from major second tier
	South Korea	159	
	Belgium	142	
	Spain	101	
Major first tier (USD 50-99 million)	Israel	98	Moved down from top second tier
	Turkey	94	Moved down from top second tier
	Czech Republic	93	
	Canada	87	Moved down from top second tier
	United Kingdom	68	
	Finland	68	
	Croatia	63	Moved up from major second tier
	Norway	61	Moved down from top second tier
Major second tier (USD 10-49 million)	China	46	
	Portugal	45	
	Sweden	38	Moved down from major first tier
	Serbia	35	
	Mexico	31	
	Poland	27	
	Taiwan	21	
	Romania	16	
	India	16	

Hungary	14	Exported less than USD 10 million in 2008
Philippines	14	Exported less than USD 10 million in 2008
Argentina	13	
Bulgaria	13	Exported less than USD 10 million in 2008
Cyprus	12	
Singapore	12	
Australia	11	Exported less than USD 10 million in 2008
Netherlands	10	

representatives suggest that procurement of many accessories is modest compared to sights. For example, interest in weapon-mounted laser rangefinders for Carl Gustaf recoilless rifles is growing but the potential global market is estimated at only 1,000 units annually.[58] The same is true for fire control systems for automatic grenade launchers, whose total *potential* global market probably numbers in the thousands of units annually, including units procured from domestic sources, according to one well-placed industry representative.[59]

Finally, the data highlights the rising cost of accessories vis-à-vis the weapons on which they are mounted. As mentioned above, US budget data indicates that the estimated unit value of the M150/M151 mortar FCS was more than twice the cost of the M120 mortar system itself (US Army, 2011). The difference between the unit cost of small arms and some accessories is even starker. The estimated unit cost of the small arms thermal sights budgeted for procurement by the US Army in 2010 was six times the cost of the M4 rifle with which many of the sights were to be used. However, some weapons still cost as much or more per unit than even the most expensive accessories. The M110 semi-automatic sniper system, for example, is comparable in cost to high-end thermal sights for sniper rifles (US Army, 2011). But such examples are increasingly the exception rather than the rule as accessories become more technologically sophisticated, while the weapons themselves remain largely unchanged. If this trend continues, it is possible that, at least in some countries, budgets for accessories will eventually dwarf budgets for the weapons themselves, with significant implications for the international trade in small arms and light weapons.

CONCLUSION

The four-year study of authorized international transfers of small arms, light weapons, their parts, accessories, and ammunition has yielded valuable insights. Among the most notable findings is the highly concentrated nature of the global trade. A handful of countries accounted for most of the documented transfers of small arms and light weapons during the ten-year period covered in the study. Twenty exporting states accounted for more than 80 per cent of transfers of small arms and light weapons from 2000 to 2006, according to customs data (Dreyfus et al., 2009, p. 8). The dominant role of the United States in much of the small arms trade is also highlighted; from 2000 to 2009, it was the top importer of small-calibre ammunition, sporting shotguns, pistols, revolvers, and parts for small arms and light weapons. While data limitations preclude a definitive accounting of the US role in the trade in light weapons and the other items the project has studied, available data suggests that it was significant.

Another remarkable trend is the absolute growth in the value of international transfers since 2002. This growth was first highlighted in the 2009 *Survey* chapter on the trade in small arms. Subsequent chapters have also found evidence of increases in the trade in small-calibre cartridges and shotgun shells (2010), some light weapons (2011), and now parts. The causes of this increase in world trade need further attention, but there are two probable factors. The first is increased spending by US civilians on small arms and ammunition. The US market accounts for such a large proportion of global sales that an uptick in US demand can have a major effect on exports around the world (Dreyfus et al., 2009, pp. 40, 42; Herron et al., 2010, p. 21). Second, governments made large-scale purchases of military firearms and light weapons for forces involved in fighting in Iraq and Afghanistan; these were intended for both armed forces participating as international forces and for Iraqi and Afghan security forces.[60]

The study also highlights the huge disparity in data on—and public understanding of—transfers to and from the most transparent countries versus transfers between the least transparent countries. As documented throughout the four-year study, transfers to and from Europe and North America are the most thoroughly documented and best understood. This reflects the greater trade transparency of major exporting countries in these regions (TRANSPARENCY). The Survey has also documented transfers in much of South America, filling in some gaps with data obtained through Datamyne and direct requests to government officials. Other key countries for which data was collected include India, Israel, and South Korea—all significant producers of small arms and light weapons.

For these countries, the four-year study yielded a fairly complete picture of transfers of small arms, small-calibre ammunition, parts for non-military firearms, and some light weapons, including MANPADS. The trade in other items remains opaque, however. Few countries publish detailed, disaggregated data on transfers of light weapons parts and ammunition or on most accessories for small arms and light weapons, notwithstanding their widespread use by armed groups and their potentially game-changing effects on combat operations and the capabilities of non-state groups.[61]

Despite the Survey's exhaustive data review, the trade in Africa, Asia, and the Middle East also remains opaque. Few countries in these regions routinely report on their transfers of small arms and light weapons to the UN Register or publish detailed national reports on transfers or military procurement of these items. UN Comtrade provides some information on transfers of small-calibre ammunition and certain small arms in these regions, but it offers few details on the trade in machine guns, assault rifles, sniper rifles, light weapons and their parts and ammunition, and most accessories. Other sources utilized for this study—including military procurement notices, direct requests to governments, and commercial aggregators of shipping data (Datamyne)—capture some of these transfers, but huge gaps remain. Of particular concern is the lack of information on possible exports to unstable or abusive regimes from China and the Russian Federation, neither of which releases data on exports of pistols, military firearms, light weapons, or light weapons ammunition (except MANPADS). Arms transfers from Iran and North Korea and re-exports from states with large surplus stockpiles, such as Angola, are also poorly understood.

Filling these data gaps and building on the findings of this study will improve public understanding of the sources and means through which authorized arms transfers fuel the illicit trade. From Afghanistan to Mexico, arms flows can threaten efforts to stabilize countries, enhance security, and combat organized crime (ILLICIT SMALL ARMS and DRUG

VIOLENCE). Yet much of the authorized trade—and in particular those aspects that appear most closely related to the illicit trade—remain shrouded in secrecy. The more we know, the more we need to know.

LIST OF ABBREVIATIONS

ATGW	Anti-tank guided weapon
CBP	United States Customs and Border Protection
FCS	Fire-control system
GDP	Gross domestic product
HS	Harmonized System
MANPADS	Man-portable air defence system
OECD	Organisation for Economic Co-operation and Development
UN Comtrade	United Nations Commodity Trade Database

ANNEXE

Online annexe at <http://www.smallarmssurvey.org/publications/by-type/yearbook/small-arms-survey-2012.html>

Annexe 8.1 Methodology

This annexe provides a detailed summary of the methodology used in this chapter.

ENDNOTES

1   See Purcena et al. (2009) and Herron, Marsh, and Schroeder (2010; 2011) and Annexe 8.1.
2   See Berman and Leff (2008, pp. 8–11). For a definition of 'authorized transfers' and an explanation of the various types of transfers captured in the data used in this study, see Dreyfus et al. (2009, pp. 9–10).
3   Detailed information on numerous light weapons can be found in Jones and Ness (2011).
4   Also included are items attached to other accessories that are fastened to the weapon.
5   See Dreyfus et al. (2009); Herron et al. (2010, pp. 18–20; 2011, pp. 20–21).
6   See Herron et al. (2010, p. 18).
7   See Herron et al. (2011, p. 21).
8   See Lazarevic (2010, pp. 36, 39, 64–65, 116–17).
9   The trade category used is parts of military weapons under HS code 930591.
10 The dataset comprised 201 countries, which is more than the number of UN member states because some autonomous territories report their imports and exports separately to UN Comtrade (such as Greenland, New Caledonia, and French Polynesia); however, the import values of 28 of the countries in the dataset—most of them very small countries—could not be estimated due to a lack of GDP data.
11 The tested variables include: the extent of light weapons production, military expenditure, size of the armed forces, military expenditures per member of the armed forces, GDP, military expenditure as a proportion of GDP, total population, GDP per capita, firearms per capita, total number of guns, OECD membership, NATO membership, EU membership, and the existence of state-owned production.
12 See Annexe 8.1.
13 For details on the use of mirror data, see Khakee (2004).
14 The figures for the different weapon categories do not add up to 100 per cent due to rounding.
15 Many of the contract award notices identify the procured items simply as 'night vision devices' without identifying the model or intended platform. Since night vision technology is used on a variety of platforms, many of which are outside the scope of this chapter, data from these notices is of little use for the purposes of this study.

16 Data submitted to UN Comtrade is not sufficiently detailed to distinguish the value of the sight from the value of the weapon with which it is exported.

17 Other items reported under HS code 901310 include telescopes, periscopes, and optical devices used for industrial purposes.

18 Datamyne (n.d.); GTIS (n.d.); UN Comtrade (n.d.); USCB (n.d.). The estimates reflect sights imported separately from the weapon with which they are to be used. The value of sights imported with weapons is recorded in the same customs category as the weapon itself.

19 This HS category reflects data on weapon sights shipped separately from the weapon for which they were intended. Weapon sights imported with the weapon are included in the same commodity category as the weapon itself.

20 Interestingly, observers find that 'greater openness to trade and Foreign Direct Investment (FDI) lowers small arms imports per capita' (de Soysa, Jackson, and Ormhaug, 2009). Their finding underlines the points made in this chapter regarding the fact that the major importers and exporters involved in production chains differ from those trading in finished weapons.

21 Author communication with an industry representative, 27 October 2011.

22 The acronym COTS can also refer to 'components off the shelf', which has a similar meaning.

23 The Small Arms Survey includes international movement of arms for repair and maintenance in transfers studied as part of the four-year project to assess the trade (Dreyfus et al., 2009, p. 9).

24 Author interview with an industry representative, 4–5 October 2011.

25 Author interview with an industry representative, 4–5 October 2011.

26 Author communication with analyst Jurgen Brauer, who compiled statistics on Sturm, Ruger & Co., 16 December 2011.

27 For more information on UN Comtrade, see Dreyfus et al. (2009, pp. 10–11).

28 Glock has a plant in Smyrna, Georgia, while Beretta has one in Accokeek, Maryland.

29 Parts are described as being 'military' by the state reporting the transfer. This designation is based upon the nature of the finished weapon (such as being fully automatic) rather than the identity of the intended recipient.

30 For more information, see Herron et al. (2011) and Gander and Cutshaw (1999, p. 382).

31 An example is the MK19 Capability Upgrade Package from Rheinmetall (Rheinmetall Defence, n.d.a).

32 Author telephone interview with a defence industry representative, 26 September 2011.

33 The onus is on the vendor to demonstrate that the operational benefits of a given piece of equipment exceed the opportunity cost of the equipment it displaces. As articulated by a British official interviewed by Jane's Information Group, '1 kg of extra capability must exceed the value of one litre of water or two rifle magazines' (Pengelley, 2009b).

34 Weapon sights are often referred to as 'optics' or 'scopes'.

35 An example is the optical sight in Singapore Technology Kinetic's SAR 21 rifle (Jones and Ness, 2007, p. 209).

36 The RPG-2 rocket-propelled grenade launcher is an example. According to Jane's Information Group, '[t]here is no provision for the fitting of an optical sight' on the RPG-2 (Jones and Ness, 2007, p. 476).

37 Since iron sights come standard on most small arms and are widely viewed as essential for the basic, intended use of these weapons, iron sights are considered 'parts' for the purposes of this study. Exceptions include emergency sights for use when primary sights are damaged or lost and specialized sights for target shooting.

38 The US Defense Department defines a reflex sight as an 'optical or computing sight that reflects a reticle image (or images) onto a combining glass for superimposition on the target' (USDoD, 2009).

39 These countries are Austria, Belgium, Canada, China, Croatia, the Czech Republic, Finland, France, Germany, Greece, India, Israel, Italy, the Netherlands, Norway, Pakistan, Poland, Romania, the Russian Federation, Serbia, South Africa, South Korea, Spain, Turkey, the United Kingdom, and the United States.

40 A small number of sights for light weapons, including the BILL 2 ATGW, feature cooled imagers (Jones and Ness, 2011, p. 666).

41 The AN/PAS-13(D)3 heavy sight is also mounted on squad leaders' assault rifles (DRS Technologies, 2010).

42 Author interview with a defence industry representative, Washington, DC, 11–12 October 2011.

43 The terms 'laser sight', 'aiming laser', and 'aiming light' are used interchangeably.

44 This definition is based on the US military's definition of 'integrated fire control system', which is a 'system that performs the functions of target acquisition, tracking, data computation, and engagement control, primarily using electronic means and assisted by electromechanical devices' (USDoD, 2009).

45 Author telephone interview with a defence industry representative, 23 June 2011.

46 This dataset differs from the data used to generate the annual estimate for the value of transfers of weapon sights in that the data on military procurement was not used in the global estimate because of concerns about double-counting.

47 While the intended end user is often unclear, the commodity descriptions suggest that most of the sights are intended for hunting or sporting rifles and air guns.

48 Note that data on items with an average per-unit cost of less than USD 1 was excluded based on the assumption that they may be parts for sights (incorrectly categorized) rather than complete sights.
49 While the relevant HS commodity category under which these transfers were recorded is for complete sights, it is possible that some shipments of parts were declared in this category.
50 Data on transfers to Peru is excluded from the table since it does not identify the country of export.
51 While the relevant HS commodity category under which these transfers were recorded is for complete sights, it is possible that some shipments of parts were declared in this category.
52 This data reflects imports of rifle sights that are sold separately from small arms.
53 As articulated by analysts Brown and Wasserbly, 'the main theme evident in sights developments and acquisitions over the last five years is the democratization of access to thermal sights' (Brown and Wasserbly, 2009).
54 Brown and Wasserbly (2009); Gething (2008; 2011); Jones and Ness (2011); White (2009).
55 The data provided by the UK Ministry of Defence includes the following caveat: 'The country of origin is based on the supplier we purchased from, as although some items are imported the delivery to MOD is through a UK supplier' (UKMoD, 2011). The information was supplied in response to a Freedom of Information Act request by the Small Arms Survey.
56 Author telephone interview with a defence industry representative, 26 September 2011.
57 The one exception is the Swedish military.
58 Author telephone interview with an industry representative, 26 September 2011, and author interviews with industry representatives, Washington, DC, 11–12 October 2011.
59 Author correspondence with an industry representative, 7 November 2011.
60 Dreyfus et al. (2009, pp. 17, 32, 37, 54); Herron et al. (2010, pp. 22, 31; 2011, pp. 9, 23–24, 27–28, 31–33, 35).
61 For an example of the significant tactical advantages provided by latest-generation accessories, see the section titled 'Small arms and light weapons accessories and the modern battlefield', above. For a discussion of the potential consequences of acquisition and use of latest-generation light weapons by armed groups, see Bonomo (2007, pp. 71–75).

BIBLIOGRAPHY

BAE Systems. 2011. 'BAE Systems Hits Delivery Milestone for Thermal Weapon Sight.' 23 August.

Baily, Mike. 2009. 'Bushnell's Latest Golf Course Laser Rangefinder—the Pro 1600 Slope Edition—Hits the Mark.' *WorldGolf*. April.
<http://www.worldgolf.com/golf-equipment/bushnell-pro-1600-slope-edition-laser-rangefinder-10063.htm>

Beretta. n.d. 'ARX160 and GLX160: Unrivaled Soldier's Defender.' <http://www.berettadefence.com/index.aspx?m=53&did=145>

Berman, Eric G. and Jonah Leff. 2008. 'Light Weapons: Products, Producers, and Proliferation.' In Small Arms Survey. *Small Arms Survey 2008: Risk and Resilience*. Cambridge: Cambridge University Press, pp. 6–41.

Bonomo, James, et al. 2007. *Stealing the Sword: Limiting Terrorist Use of Advanced Conventional Weapons*. Santa Monica: RAND Corporation.

Bosch. n.d. 'Laser Rangefinder DLE 50 Professional.' <http://www.bosch-pt.com/productspecials/professional/dle50/uk/en/start/index.htm>

Brauer, Jurgen and Paul Dunne, eds. 2004. *Arms Trade and Economic Development: Theory, Policy, and Cases in Arms Trade Offsets*. London: Routledge.

Brown, Nick. 2011. 'First Sighting of Pointer Networked Target Indication System.' *Jane's International Defence Review*. 22 February.

— and Daniel Wasserbly. 2009. 'Piercing the Fog of Battle: Forces Set Sights on Thermal Technology.' *Jane's International Defence Review*. 14 April.

Browning. n.d.a. 'BAR Rifles, Firearms.' Accessed 4 November 2011. <http://www.browning.com/products/catalog/family.asp?webflag_=002B>

—. n.d.b. 'Question: Where are Browning Firearms Manufactured?' Accessed 4 November 2011.
<http://www.browning.com/customerservice/qna/detail.asp?id=90>

CBP (United States Customs and Border Protection). 1996. 'Country of Origin Marking for Barelled Actions from Belgium.' HQ 558849. 2 February.
<http://rulings.cbp.gov/index.asp?ru=559392&qu=HQ+559392&vw=detail>

—. 2001. 'Country of Origin Marking of Unfinished Shotgun Barrels.' HQ 561919. 31 May.
<http://rulings.cbp.gov/index.asp?ru=561919&qu=HQ+561919&vw=detail>

—. 2010. 'Internal Advice: Classification of Kongsberg Remote Weapons System (RWS).' HQ H080595. 5 August.
<http://rulings.cbp.gov/index.asp?ru=h080595&qu=H080595&vw=detail>

Cox, Matthew. 2007. 'Army Buys 6,500 Remote Turret Controls.' *Army Times*. 23 August.
<http://www.armytimes.com/news/2007/08/army_crows_070823w/>

CSS (Connecticut Spring & Stamping). n.d. Company website. Accessed 4 November 2011. <http://www.ctspring.com/>

Datamyne. n.d. Datamyne Database. Accessed September 2011. <http://www.datamyne.com/index.html>

Dimitrov, Dimitar and Peter Hall. 2012. 'Small Arms and Light Weapons Production as Part of a National and Global Defence Industry.' In Owen Greene and Nicholas Marsh, eds. *Small Arms, Crime and Conflict: Global Governance and the Threat of Armed Violence*. London: Routledge.

Dove, Bill. 2010. 'Applications for Infrared Thermal Imagers.' *Air Conditioning, Heating & Refrigeration News*. 14 June.

Dreyfus, Pablo, et al. 2009. 'Sifting the Sources: Authorized Small Arms Transfers.' In Small Arms Survey. *Small Arms Survey 2009: Shadows of War*. Cambridge: Cambridge University Press, pp. 7–59.

DRS Technologies. 2010. 'Thermal Weapon Sight (TWS) II.' <http://www.drs.com/Products/RSTA/PDF/TWSII.pdf>

DXT (Dandong Xunlei Technology Co., Ltd.). n.d. Company website. Accessed 4 November 2011. <http://ddlaser.en.alibaba.com/>

Electrophysics. n.d. 'How Night Vision Works.' <http://www.electrophysics.com/nl/HNVW/index.html>

Firearm Blog. 2011. 'ATN ThOR 3 Thermal Weapon Sight: $13,000 of Fun.' 22 July.
    <http://www.thefirearmblog.com/blog/2011/07/22/atn-thor-3-thermal-weapon-sight-13000-of-fun/>

Gabelnick, Tamar, Maria Haug, and Lora Lumpe. 2006. *A Guide to the US Small Arms Market, Industry, and Exports, 1998–2004*. Occasional Paper 19. Geneva: Small Arms Survey.

Gander, Terry and Charles Cutshaw, eds. 1999. *Jane's Infantry Weapons: 1999–2000*. Coulsdon: Jane's Information Group.

Gelfand, Lauren. 2011. 'Streetwise: Fighting in Built-up Areas.' *Jane's Defence Weekly*. 28 April.

Gething, Michael. 2008. 'Lighting the Way: Armed Forces Seek Greater Thermal Imaging Capabilities.' *Jane's International Defence Review*. 11 September.

—. 2010. 'The Perfect Shot: Military and Industry Partners Look to Develop Effective Target Suppression Measures.' *Jane's Defence Weekly*. 7 June.

—. 2011. 'India.' *Jane's World Armies*. 15 July.

Gimelli Sulashvili, Barbara. 2007. 'Multiplying the Sources: Licensed and Unlicensed Military Production.' In Small Arms Survey. *Small Arms Survey 2007: Guns and the City*. Cambridge: Cambridge University Press, pp. 7–37.

GTIS (Global Trade Information Services). n.d. 'Global Trade Atlas.' Accessed September 2011. <http://www.gtis.com/english/>

Herron, Patrick, Nic Marsh, and Matthew Schroeder. 2010. 'Annexe 1.3: Methodology.' June.
    <http://www.smallarmssurvey.org/fileadmin/docs/A-Yearbook/2010/en/Small-Arms-Survey-2010-Chapter-01-Annexes-1-3-Methodology.pdf>

—. 2011. 'Annexe 1.3: Methodology.' June.
    <http://www.smallarmssurvey.org/fileadmin/docs/A-Yearbook/2011/en/Small-Arms-Survey-2011-Chapter-01-Annexe-1.3-EN.pdf>

Herron, Patrick, et al. 2010. 'Emerging from Obscurity: The Global Ammunition Trade.' In Small Arms Survey. *Small Arms Survey 2010: Gangs, Groups, and Guns*. Cambridge: Cambridge University Press, pp. 6–39.

—. 2011. 'Larger but Less Known: Authorized Light Weapons Transfers.' In Small Arms Survey. *Small Arms Survey 2011: States of Security*. Cambridge: Cambridge University Press, pp. 9–41.

IQPC (International Quality & Productivity Center). 2010. 'IF Technologies: Examples of Equipment that Use Image Fusion Technology.' November.

Jones, Richard and Leland Ness. 2007. *Jane's Infantry Weapons 2007–2008*. Coulsdon: Jane's Information Group.

—. 2011. *Jane's Infantry Weapons 2010–2011*. Coulsdon: Jane's Information Group.

Khakee, Anna. 2004. 'Back to the Sources: International Small Arms Transfers.' In Small Arms Survey. *Small Arms Survey 2004: Rights at Risk*. Oxford: Oxford University Press.

Khan, J. A. 2004. *Air Power and Challenges to IAF*. New Delhi: APH Publishing.

L3 Communications. n.d. *EOTech HOLOgraphic Weapon Sights: 2010 Complete Catalogue for Holographic Weapon Sights, and Night Vision Products*.

Laser Devices. 2010. *Laser Devices, Inc.: 2010 Product Catalogue*.

Lazarevic, Jasna. 2010. *Transparency Counts: Assessing State Reporting on Small Arms Transfers, 2001–2008*. Geneva: Small Arms Survey. June.

Lothar Walther. n.d. Company website. Accessed 4 November 2011. <http://www.lothar-walther.de/3.php>

Mohanty, Deba. 2004. *Changing Times? India's Defence Industry in the 21st Century*. Bonn: Bonn International Center for Conversion.

Opticsplanet.com. n.d.a. 'Firefield Mini Red Laser Sight.' Accessed 1 November 2011. <http://www.opticsplanet.net/firefield-mini-red-laser-sight.html>

—. n.d.b. 'Insight Technology CNVD-T2 Clip-on Thermal Imaging Sight with Laser Pointer.' Accessed 1 November 2011.
    <http://www.opticsplanet.net/insight-technology-cnvd-t2-clip-on-thermal-device-with-laser-pointer-cqt-001-a12.html>

Pengelley, Rupert. 2008. 'Moving towards a Digitised Future: France Steals a March with FELIN.' *Jane's International Defence Review*. 8 May.

—. 2009a. 'Vingmate Receives Baptism of Fire on Operations.' *Jane's International Defence Review*. 6 March.

—. 2009b. 'Soldiers Look to Infantry C4I Aids for Tempo without Torment.' *Jane's International Defence Review*. 9 July.

—. 2011. 'Mortar Fire Control in the Modern Battlespace Mix.' *Jane's International Defence Review*. 14 February.

Purcena, Júlio Cesar, et al. 2009. 'Annexe 1.3: Methodology.'
    <http://www.smallarmssurvey.org/fileadmin/docs/A-Yearbook/2009/en/Small-Arms-Survey-2009-Chapter-01-Annexe-3-EN.pdf>

QinetiQ. n.d. 'Pointer Advanced Targeting Technology.'
   <http://www.qinetiq.com/what/capabilities/land/Documents/Pointer-Advanced-Targeting-Technology.pdf>

Rheinmetall Defence. n.d.a. 'MK19 Capability Upgrade.' Product spec sheet.

—. n.d.b. 'Vingmate: Advanced Sight and Fire Control System for Heavy Machine Guns and Grenade Launchers.' Product spec sheet.

Ring Sights. 1998. 'Sights for the Modern Military Rifle.' 20 February. <http://www.ringsights.com/Tech%20Papers/13733.PDF>

Saletan, William. 2009. 'Heat Check.' *Slate Magazine*. 28 April.

Shideler, Dan and Derrek Sigler. 2008. *The Gun Digest Book of Tactical Gear*. Iola, WI: Krause Publications.

de Soysa, Indra, Thomas Jackson, and Christin Ormhaug. 2009. 'Does Globalization Profit the Small Arms Bazaar?' *International Interactions*, Vol. 35, No. 1, pp. 86–105.

Stoeger. n.d. Company website. Accessed 4 November 2011. <http://www.vursan.com.tr/index_en.htm>

Sweeney, Patrick. 2008. *The Gun Digest Book of the Glock*, 2nd edn. Iola, WI: Gun Digest Books.

Tai, Anthony, et al. 1996. 'United States Patent: Compact Holographic Sight.' Patent No. 5,483,362. 9 January.
   <http://www.google.com/patents/about/5483362_Compact_holographic_sight.html?id=M0MdAAAAEBAJ>

Thompson, Leroy. 2011. *The Beretta M9 Pistol*. Oxford: Osprey Publishing.

UKMoD (United Kingdom Ministry of Defence). 2011. '20110920_FOI_Schroeder_Enclosure_Data-U.' 22 September.

UN (United Nations). 2011. 'Why Traces Fail: Challenges Involved in Issuing and Responding to Tracing Requests.' Summary Report of Side Event for the Open-ended Meeting of Governmental Experts, New York. 10 May. http://www.poa-iss.org/mge/sideevents.html

UN Comtrade (United Nations Commodity Trade Database). n.d. 'United Nations Commodity Trade Statistics Database.' <http://comtrade.un.org/db/>

UNGA (United Nations General Assembly). 1997. *Report of the Panel of Governmental Experts on Small Arms*. A/52/298 of 27 August. New York: UNGA.

US Army (United States Army). 2009. *Committee Staff Procurement Backup Book*. May.
   <http://www.asafm.army.mil/offices/BU/BudgetMat.aspx?OfficeCode=1200>

—. 2010. *Committee Staff Procurement Backup Book*. February. <http://www.asafm.army.mil/offices/BU/BudgetMat.aspx?OfficeCode=1200>

—. 2011. *Committee Staff Procurement Backup Book*. February. <http://www.asafm.army.mil/offices/BU/BudgetMat.aspx?OfficeCode=1200>

USCB (United States Census Bureau). n.d. 'USA Trade Online.' Accessed September 2011. <https://www.usatradeonline.gov/>

USDoD (United States Department of Defense). 2009. *Department of Defense Dictionary of Military and Associated Terms*. Amended 31 October.

USMC (United States Marine Corps). 2004. *Marine Corps Concepts and Programs 2004*.
   <http://www.marines.mil/unit/pandr/Documents/Concepts/2004/TOC1.HTM>

Vayu. 2010. 'The Missile Powerhouse: Visiting MBDA in France.' Iss. 3, pp. 92–97.

White, Andrew. 2009. 'Romania Unleashes Assault Rifle Requirement.' *Jane's International Defence Review*. 17 December.

—. 2010. 'CQB Technologies Help Forces Grapple with Short-range Front Line.' *Jane's International Defence Review*. 11 February.

Wikipedia. n.d. 'Iron Sight.' <http://en.wikipedia.org/wiki/Iron_sight>

Williamson, Oliver. 1981. 'The Economics of Organization: The Transaction Cost Approach.' *American Journal of Sociology*, Vol. 87, No 3, pp. 548–77.

ACKNOWLEDGEMENTS

Principal authors

Janis Grzybowski, Nicholas Marsh, Matt Schroeder

Contributor

Giacomo Persi Paoli

Bullets hang from a machine gun on display at the Defence Systems and Equipment International Exhibition, London, September 2005. © Kirsty Wigglesworth/AP Photo

Point by Point
TRENDS IN TRANSPARENCY

9

INTRODUCTION

In a 1991 study intended to encourage the development of the nascent UN Register of Conventional Arms, UN Secretary-General Boutros Boutros-Ghali reasons that 'transparency can contribute to the building of confidence and security, the reduction of suspicions, mistrust and fear, and the timely identification of trends in arms transfers' (UNGA, 1991, p. 3). Since that study was published, the commitment to transparency in small arms and light weapons transfers has only grown. In 2008, transparency was recognized as 'a core element in preventing conflict and securing peace and stability' (UNSC, 2008, para. 37). In late 2012 the Group of Governmental Experts working on the scope of the UN Register of Conventional Arms will convene to conduct its triennial review of the instrument. At that meeting, experts will—among other things—take stock of states' reporting practices over time.

The Small Arms Trade Transparency Barometer, published annually by the Small Arms Survey since 2004, uses a standardized set of guidelines to analyse the transparency of small arms exporters. Applying criteria drawn from actual state reporting practices, the Barometer assesses changes in states' transparency over time.[1] This chapter presents the 2012 edition of the Barometer, which covers reports on export activities conducted in 2010 by the 52 countries the Survey has classified as 'major exporters'—those exporting at least USD 10 million in small arms, light weapons, their parts, accessories, and ammunition in at least one calendar year since 2001.

This chapter also reviews ten years of reporting on the small arms trade by those same exporting states. Without assessing the accuracy of the data states provide, the chapter examines changes in reporting practices—as evidenced in national arms export reports and submissions to instruments such as UN Comtrade and the UN Register—with respect to the Barometer's seven parameters and 43 criteria. It unpacks reporting and identifies areas where transparency has improved—and where it has not.[2]

The main findings of the chapter include:

- The 2012 edition of the Small Arms Trade Transparency Barometer identifies Switzerland, the United Kingdom, and Romania as the most transparent of the major small arms and light weapons exporters.
- The 2012 Barometer identifies Iran, North Korea, and the United Arab Emirates as the least transparent major exporters. They all score zero points.
- State transparency on small arms and light weapons transfers improved by more than 40 per cent between 2001 and 2010, but the average score for all states combined remains below half of all available points.
- Switzerland achieved the highest Transparency Barometer score over the ten-year period, gaining 21.00 out of 25.00 points for reporting on 2007–10 activities. It is the only country to have produced a dedicated national report on small arms and light weapons exports.

- States have made significant improvements under the parameters of *comprehensiveness* and *clarity* since 2001. Reporting on *licences granted* and *licences refused* remains the weakest among the seven parameters.
- The single most important way states can improve their transparency on small arms and light weapons transfers is through the timely publication of comprehensive national arms export reports.

The first section of the chapter is divided into two parts: a scene-setter, which describes the methodology used in the chapter and notes trends in reporting, and the 2012 edition of the Small Arms Trade Transparency Barometer, which analyses the level of transparency for reporting on the 2010 activities of 52 states.

The chapter's second section examines multi-year changes in national reporting practices in relation to the Transparency Barometer's seven parameters. In addition to highlighting which country has achieved top scores and why, the section also asks why states fail to earn full points. Most importantly, the chapter identifies how states can improve the level of transparency in their reporting.

THE TRANSPARENCY BAROMETER

Setting the scene

The Transparency Barometer evaluates 52 major small arms exporters.

The Small Arms Trade Transparency Barometer was introduced in the 2004 edition of the *Small Arms Survey* (Khakee, 2004, pp. 115–18). This volume presents the 9th edition of the Barometer. In the 2009 edition, the Small Arms Survey revised the Barometer to provide a clearer and more comprehensive assessment of reporting. The scoring guidelines were redesigned to reflect national best practices and encourage the use of important reporting instruments such as the UN Register. The overall points distribution was maintained, but greater emphasis was placed on consistent, timely, and more frequent reporting.[3] All the countries rated were retroactively scored to better establish comparisons and observe trends in transparency for the past ten years. Results of the retroactive scoring are found in this chapter as well as in Lazarevic (2010). This chapter examines a longer time span—data for 2001–10 rather than 2001–08—and includes the analysis for more states (52 instead of 48) than Lazarevic (2010). It focuses on the seven parameters and individual criteria, whereas the previous publication provides a country-specific analysis.

The Barometer evaluates all reporting states that have exported[4]—or are believed to have exported—at least USD 10 million[5] worth of small arms, light weapons, their parts, accessories, and ammunition in at least one calendar year since 2001.[6] This edition of the Barometer evaluates 52 countries, of which 9 reached the USD 10 million threshold once since 2001,[7] 4 reached it twice,[8] and 21 reached it every year. They include 19 'top exporters'—countries that have exported at least USD 100 million worth of materiel in a calendar year. The number of states under review has increased with every edition; Greece, Lithuania, and Luxembourg joined this year, and the total number of countries evaluated might increase further in future editions.

The Barometer encompasses seven parameters: *timeliness, access and consistency, clarity, comprehensiveness, deliveries, licences granted, and licences refused*. Each parameter has a set of criteria that states must fulfil in order to receive points. The more overall points a state receives, the higher its ranking in the Barometer. Table 9.1 lists the seven parameters and the distribution of the 43 associated criteria and sub-criteria used to evaluate transparency. Scores are awarded based on a 25.00-point scale. A country is accorded full, partial, or zero points on each criterion.[9] All 43 criteria correspond to actual reporting practices.

Table 9.1 Number of Transparency Barometer criteria for each of the seven parameters

Parameters		Number of criteria	Maximum points
I	Timeliness	3	1.50
II	Access and consistency	4	2.00
III	Clarity	11*	5.00
IV	Comprehensiveness	13	6.50
V	Deliveries	4	4.00
VI	Licences granted	4	4.00
VII	Licences refused	4	2.00
Total		43	25.00

Note: * One of the eight clarity criteria is composed of four sub-criteria.

Transparency in reporting exports is assessed using a series of publicly available reporting instruments that provide official information on small arms transfers (see Figure 9.1).[10] These include the United Nations Commodity Trade Statistics Database (UN Comtrade), the UN Register of Conventional Arms (UN Register), and national arms export reports, including the EU Report for EU member states.

UN Comtrade

UN Comtrade provides important insight into the value of the arms trade, including in small arms and light weapons. States usually submit their customs data in the 12 months following the year of the transfer activities. Over the ten-year period under review, at least 42 of the 52 surveyed countries reported to UN Comtrade, and 38 countries submitted data for every year,[11] making it the most widely used reporting instrument[12] (see Figure 9.3). Bulgaria, North Korea, South Africa, and Ukraine are the only countries under review that did not submit any customs data to UN Comtrade during this period.

Figure 9.1 Number of national submissions, 2001–10

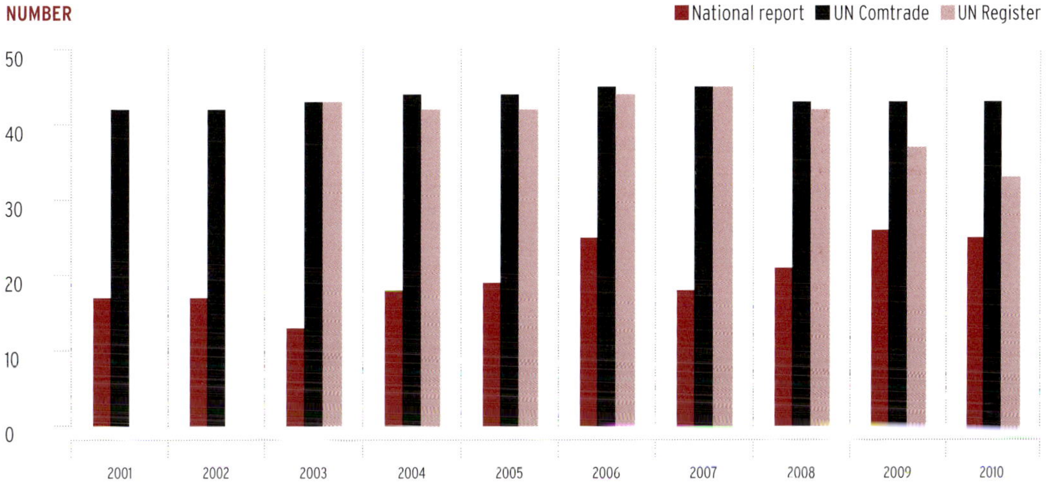

A closer look at the UN Comtrade categories reveals progress regarding national reporting in some of the small arms and light weapons categories. In reporting on 2010 activities, more than three-quarters of the 52 countries provided data on sporting and hunting guns and rifles (11 per cent more than for 2008 activities), on ammunition that is smaller than 12.7 mm, and on parts and accessories for small arms (8 per cent more than for 2008 activities). Likewise, slightly more than half provided data on pistols and revolvers (4 per cent more than for 2008 activities). One-third of the states provided information on military firearms (21 per cent more than for 2008 activities) and guided and unguided light weapons (6 per cent more than for 2008 activities).

The UN Register

Participation in the UN Register's[13] conventional arms categories is slightly lower than contributions to UN Comtrade. Of the 52 major exporters, 22 have submitted data for transfers in each year since 2003. Iran, North Korea, Saudi Arabia, and the United Arab Emirates have never submitted a report to the UN Register.[14] Reporting under the 'voluntary background information' category for international small arms and light weapons transfers peaked in 2007, with 26 countries providing information on small arms exports (see Box 9.1). In 2010, the number of countries reporting dropped to 23.

Thirty-three countries reported to the UN Register on their transfer activities in 2010, four fewer than for the previous year (see Figure 9.3). However, efforts to improve the Register's online interface have led to delays in access to reporting, with the consequence

---

Box 9.1 The UN Register's virtual eighth category: background information

States may report 'background information' on international small arms and light weapons transfers under the UN Register (UNODA, 2009). While this reporting is optional, states are increasingly utilizing this heading (see Table 9.2). States are unrestricted in the type of information that they can provide on their international small arms and light weapons transfers, and they have used it to document weapon categories transferred, detailed information on quantities and destination countries, and individual accounts of transfers, re-exports, brokers, and end users, among other details.

With the introduction of the standardized reporting form in 2006,[15] the number of states submitting background information on their small arms exports or nil reports increased from 5 to 18 and rose to at least 23 for the four subsequent years. Between 2003 and 2010, 42 countries provided information on exports of small arms and light weapons at least once, while 21 countries submitted a nil report, which signals that they did not import or export any materials. As Table 9.2 indicates, the Netherlands and the United Kingdom submitted background information for every year. The following 13 countries submitted information on exports only once: Andorra, Austria, Belgium, Cyprus, Finland, Latvia, Lithuania, Luxembourg, Mexico, Montenegro, the Philippines, Spain, and Trinidad and Tobago. In total, 37 of the 52 countries under review have submitted information on small arms exports under the virtual eighth UN Register category. The highlighted cells in Table 9.2 indicate which states are included in the 2012 Barometer.

In December 2009 the UN General Assembly passed a resolution asking states to submit their views on 'whether the absence of small arms and light weapons as a main category in the Register has limited its relevance and directly affected decisions on participation' (UNGA, 2009, para. 6). By September 2011, Colombia, Israel, Japan, Mauritius, Mexico, the Netherlands, Singapore, and Switzerland had submitted their views. Most of these states explicitly expressed their support for the inclusion of a proper small arms and light weapons category in the UN Register. Singapore was the sole country to urge that the category remain optional, arguing that a new mandatory category might decrease states' participation (UNGA, 2010a; 2010b; 2011a).

These views, along with data on the frequency and universality of reporting, will be considered at upcoming meetings of the Group of Governmental Experts on the continuing operations and further development of the UN Register. The next meeting is scheduled for November 2012 (UNGA, 2011b).

Table 9.2 States submitting information on exports or nil reports to the UN Register as background information on international small arms and light weapons transfers carried out in 2003–10

Country	2003	2004	2005	2006	2007	2008	2009	2010
Albania				X			X	
Andorra								X
Antigua and Barbuda					Nil		Nil	
Argentina					Nil	Nil	Nil	Nil
Australia					X		X	X
Austria								X
Bangladesh				Nil				Nil
Belgium						X		
Bosnia and Herzegovina				X	X		X	X
Bulgaria						X	X	X
Canada				X	X	X	X	X
Colombia					X		X	
Croatia					X	X		X
Cyprus				X	Nil	Nil		
Czech Republic				X	X	X		X
Denmark				X	X	X	X	
El Salvador					Nil			
Finland		X						
France		X	X	X	X	X		
Fiji					Nil			
Germany			X	X	X	X	X	X
Georgia				Nil				
Ghana					Nil			
Greece				X	X			X
Guyana								Nil
Haiti				Nil				
Hungary				X	X	X	X	X
Italy					X	X	X	
Jamaica				Nil				
Japan				Nil				
Latvia	X			Nil				
Lebanon							Nil	Nil	Nil

Country	2003	2004	2005	2006	2007	2008	2009	2010
Liechtenstein					Nil		X	X
Lithuania					Nil	X		
Luxembourg						X		
Mali					Nil			
Malta						Nil	Nil	Nil	Nil
Mexico					Nil			X
Moldova					Nil	Nil		
Mongolia								Nil
Montenegro						X		
Netherlands	X	X	X	Nil	X	X	X	X
New Zealand					X	X	X	
Norway					X	X	X	
Panama					Nil			
Philippines					Nil		X	
Poland	X	X	X	X	X	X	X	
Portugal					X	X	Nil	
Romania						X	X	X	X
Saint Lucia					Nil			
Senegal					Nil			
Serbia							X	X
Slovakia					X	X	X	X	X
South Korea					X	X	X	X	X
Spain							X	
Swaziland					Nil	Nil	Nil	
Sweden	X			X	X	X	X	
Switzerland							X	X	X
Togo					Nil	Nil		
Trinidad and Tobago					Nil			X
Turkey					X	X	X	X
Ukraine						X	X	X	X
United Kingdom	X	X	X	X	X	X	X	X
Total on exports	5	5	5	18	26	24	23	23
TOTAL including Nil	5	5	5	37	36	29	28	29

Note: Highlighting indicates that states are included in the 2012 Transparency Barometer.

that some states' contributions for 2010 activities have not been made available for analysis. The 2012 Barometer therefore relies on data provided in the UN Secretary-General's 2011 report on the UN Register (UNGA, 2011a). While the Secretary-General's report is typically followed by the publication of addenda that contain additional states' reports, as of 22 February 2012 no addenda had been published. This explains why the level of reporting for 2010 activities is lower than expected.

National arms export reports

National arms export reports vary significantly across countries. While certain states do not publish their reports on time,[16] some fail to report altogether. In the past ten years, 29 countries (more than half of all reviewed) published a national arms export report at least once. Some states also publish reports that cover activities for multiple years. Eight countries published an 'up-to-date' (current) report each year.[17] Of the 31 European countries under review, Cyprus, Greece, Lithuania, Luxembourg, the Russian Federation, and Turkey have not published a national arms export report.[18] This stands in contrast to reporting practices in other regions. Of the 21 non-European countries reviewed, only Australia, Canada, South Africa, and the United States have published national arms export reports.

Such reports are an important tool for assessing transparency, as they can contain more detailed information on small arms transfers than the UN instruments. For this reason, not publishing a national arms export report limits the number of points a state can receive in the Barometer even if it reports to both UN Comtrade and the UN Register. A total of 10.00 of the possible 25.00 points (that is, 40 per cent) can only be achieved if the relevant data is made available in a national arms export report; even complete reports to UN Comtrade and the UN Register do not provide this information.

> Switzerland, the United Kingdom, and Romania are the three most transparent small arms exporters.

The 2012 Transparency Barometer

The 2012 edition of the Transparency Barometer assesses the reporting practices of the 49 countries covered in the 2011 Barometer, plus three countries that are appearing for the first time: Greece, Lithuania, and Luxembourg (see Map 9.1). These three states appear to have exported USD 22 million, USD 11 million, and USD 16.7 million worth of relevant materiel in 2010, respectively.

This year's Barometer identifies Switzerland, the United Kingdom, and Romania as the three most transparent countries. Italy and Denmark broke into the top ten, replacing Norway and the United States, which now rank 11th and 12th, respectively. The least transparent countries are Iran, North Korea, and the United Arab Emirates, all scoring zero points (see Table 9.3). Of a maximum of 25.00 possible points, the average score is 11.22, a drop of almost 2 per cent (0.18 points) since 2011.[19] The average score of the ten most transparent countries remains the same as last year (18.00 points). Just over half of the countries reviewed received fewer than 12.50 points, suggesting that, despite some progress among some states, there remains much scope for improved reporting.

The countries that slipped the most were the United Arab Emirates, Poland, and the United States, whose scores declined by 8.50 points (100 per cent), 2.25 points (15 per cent), and 1.25 point (8 per cent), respectively. The United Arab Emirates score dropped to zero because it did not use any of the instruments to report on its small arms transfer activities for two consecutive years. Although Poland provided information on military holdings and procurement through national production, it failed to report on its transfers to the UN Register. Poland also omitted information on end users for its deliveries and its reports on quantities exported were less comprehensive than in the previous year (Small Arms Survey, 2012, para. 9.6). The United States did provide more detailed information on exports of larger-

290 SMALL ARMS SURVEY 2012

Map 9.1 **Countries reviewed by the Transparency Barometer**

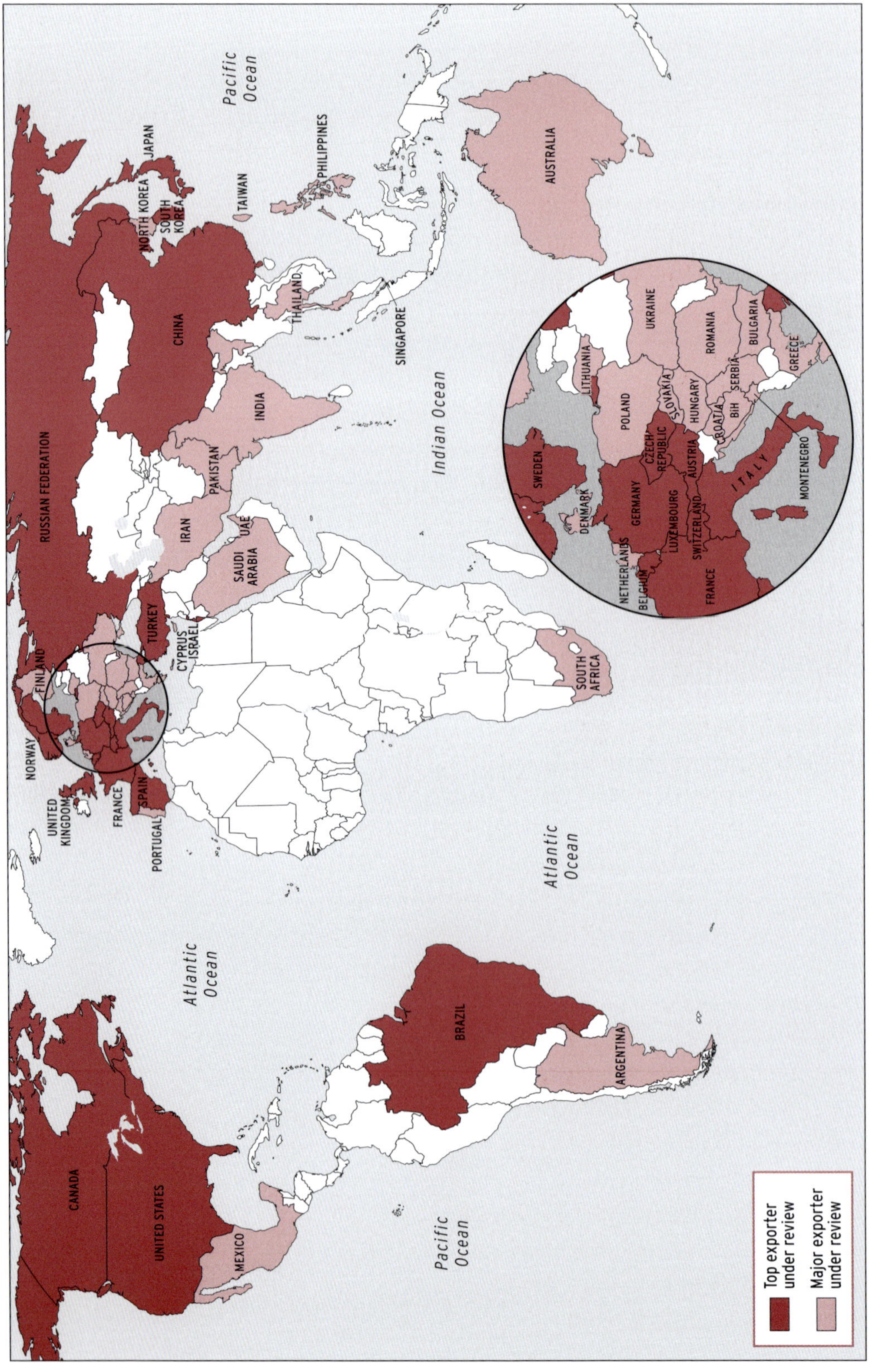

- Top exporter under review
- Major exporter under review

Guns are displayed at a stall during a defence exhibition in Karachi, 2006. © Zahid Hussein/Reuters

calibre ammunition, but it did not report on 'intangible transfers' and was less comprehensive regarding quantities and values for licences granted. Its information on quantities exported was also less comprehensive than last year. The United States, the world's largest exporter of small arms and light weapons in dollar value, now ranks 12th of the 52 countries in overall transparency.

Several countries improved their scores this year, led by Greece[20] and Croatia. Greece's score increased by almost 30 per cent due to better reporting to the UN Register on different weapon types, such as man-portable air defence systems, rifles, and pistols and revolvers. Its reporting on re-exports and transits, as well as licence denials and types of end users, to the UN Register and the EU Report also improved. Croatia increased its score by 20 per cent because it provided new information on licences granted and also reported on light weapons and larger ammunition under UN Comtrade (it also lost a point for no longer providing information on re-exports under the UN Register).

Another positive development captured by this year's Barometer is that Austria and Mexico, in reporting to the UN Register on activities carried out in 2010, provided background information on their international small arms and light weapons exports for the first time. Croatia, Hungary, and Poland continued to publish a national arms exports report after having produced their first reports the previous year.

Table 9.3 Small Arms Trade Transparency Barometer 2012, covering major exporters*

	Total (25.00 max)	Export report**/ EU Annual Report***	UN Comtrade**	UN Register**	Timeliness (1.50 max)	Access and consistency (2.00 max)	Clarity (5.00 max)	Comprehensiveness (6.50 max)	Deliveries (4.00 max)	Licences granted (4.00 max)	Licences refused (2.00 max)
Switzerland	21.00	X	X	X	1.50	1.50	4.00	5.25	3.00	4.00	1.75
United Kingdom	19.75	X / EU Report	X	X	1.50	2.00	3.75	5.25	3.50	2.50	1.25
Romania	19.00	X / EU Report	-	X	1.50	2.00	2.50	4.50	3.00	3.50	2.00
Serbia	18.75	X (09)	X	X	1.50	1.00	3.25	5.00	3.50	2.50	2.00
Germany	18.50	X / EU Report	X	X	1.50	1.50	3.75	4.25	2.50	3.50	1.50
Netherlands	18.50	X / EU Report	X	X	1.50	2.00	4.25	4.75	2.50	2.50	1.00
Belgium	17.00	X / EU Report	X	X	1.50	2.00	3.00	3.00	3.00	2.50	2.00
Denmark	16.50	X / EU Report	X	X (09)	1.50	1.50	4.75	3.25	2.50	2.00	1.00
Italy	16.00	X / EU Report	X	X (09)	1.50	1.50	3.25	5.00	2.50	2.00	0.25
Spain	15.75	X / EU Report	X	X (09)	1.50	2.00	2.25	4.00	3.50	1.50	1.00
Slovakia	15.50	X / EU Report	X	X	1.50	1.50	2.50	3.50	2.50	2.00	2.00
Norway	15.25	X	X	X	1.50	1.50	3.75	3.00	3.00	2.50	0.00
Sweden	15.25	X / EU Report	X	X (09)	1.50	2.00	3.50	4.00	2.50	1.50	0.25
United States	15.00	X	X	X	1.50	1.50	2.75	4.25	3.00	2.00	0.00
Croatia	14.75	X	X	X	1.50	1.00	3.00	3.25	3.00	3.00	0.00
Montenegro	14.50	X (09)	X	-	1.50	0.50	3.00	5.00	2.50	2.00	0.00
Finland	14.25	X / EU Report	X	X	1.50	1.50	3.25	3.25	2.50	2.00	0.25
Czech Republic	14.00	X / EU Report	X	X	1.50	1.50	2.50	3.25	3.00	1.50	0.75
France	14.00	X / EU Report	X	-	1.50	1.50	4.00	2.75	2.50	1.50	0.25
Austria	13.75	X (09) / EU Report	X	X	1.50	1.50	2.25	3.75	3.00	1.50	0.25
Poland	12.75	X / EU Report	X	-	1.50	1.00	2.00	3.75	3.00	1.50	0.00
Hungary	12.50	X / EU Report	X	X	1.50	1.00	3.00	2.75	2.50	1.50	0.25
Canada	12.25	X (07-09)	X	X	1.50	1.00	2.75	4.00	3.00	0.00	0.00
Greece	12.00	EU Report	X	X	1.50	0.50	2.00	3.25	3.00	1.50	0.25
Portugal	11.75	X (08) / EU Report	X	X	1.50	1.50	2.00	2.25	2.50	2.00	0.00

Country	Score										
Bulgaria	11.25	X / EU Report	-	X	1.50	1.50	2.00	2.25	2.50	1.50	0.00
Luxembourg	10.75	EU Report	X	-	1.50	0.50	1.75	3.00	2.50	1.50	0.00
Australia	10.00	-	X	X	1.50	1.00	1.50	3.00	3.00	0.00	0.00
Lithuania	10.00	EU Report	X	X	1.50	1.00	1.75	1.75	2.50	1.50	0.00
Israel	9.75	-	X	X (09)	1.50	0.50	1.75	3.50	2.50	0.00	0.00
South Korea	9.75	-	X	X	1.50	1.00	1.50	3.25	2.50	0.00	0.00
Thailand	9.75	-	X	X (09)	1.50	0.50	1.50	3.25	3.00	0.00	0.00
Pakistan	9.50	-	X	X	1.50	0.50	1.50	3.50	2.50	0.00	0.00
Mexico	9.00	-	X	X	1.50	1.00	1.50	2.50	2.50	0.00	0.00
Bosnia and Herzegovina	8.75	X (09)	-	X	1.50	0.50	1.50	1.00	1.50	1.50	1.25
Turkey	8.75	-	X	X	1.50	0.50	1.50	2.75	2.50	0.00	0.00
Argentina	8.50	-	X	X	1.50	1.00	1.50	2.00	2.50	0.00	0.00
Japan	8.50	-	X	X	1.50	1.00	1.00	2.00	3.00	0.00	0.00
India	8.25	-	X	X	1.50	1.00	1.50	1.75	2.50	0.00	0.00
Philippines	8.25	-	X	-	1.50	0.50	1.50	2.25	2.50	0.00	0.00
Ukraine	8.00	X	-	X	1.50	1.50	1.00	2.00	2.00	0.00	0.00
Brazil	7.50	-	X	X	1.50	1.00	1.00	1.50	2.50	0.00	0.00
Cyprus+	7.50	-	X	X	1.50	0.50	1.00	2.00	2.50	0.00	0.00
Taiwan	7.50	-	X	-	1.50	0.50	1.50	1.50	2.50	0.00	0.00
China	7.00	-	X	X (09)	1.50	0.50	1.00	1.50	2.50	0.00	0.00
Singapore	6.50	-	X	X (09)	1.50	0.50	1.00	1.50	2.00	0.00	0.00
Russian Federation	5.50	-	X	X	1.50	1.00	0.50	0.50	2.00	0.00	0.00
Saudi Arabia	2.75	-	X (09)	-	1.00	0.00	0.50	0.75	0.50	0.00	0.00
South Africa	2.00	X	-	X (09)	1.50	0.50	0.00	0.00	0.00	0.00	0.00
Iran	0.00	-	-	-	0.00	0.00	0.00	0.00	0.00	0.00	0.00
North Korea	0.00	-	-	-	0.00	0.00	0.00	0.00	0.00	0.00	0.00
United Arab Emirates	0.00	-	-	-	0.00	0.00	0.00	0.00	0.00	0.00	0.00

Note: The online version of the Transparency Barometer incorporates updates and corrections, and fills in reporting gaps, all of which affect states' scores as well as their rankings for current and previous years. For these reasons, the online editions—rather than the printed versions—should be considered definitive.[21]

Notes:

* Major exporters are countries that export–or are believed to export–at least USD 10 million worth of small arms, light weapons, their parts, accessories, or ammunition in a given calendar year. The 2012 Barometer includes all countries that qualified as a major exporter at least once during the 2001–10 period.

** X indicates that a state submitted a report on activities in 2010. Reports provided for earlier years are indicated in parentheses.

*** The Barometer assesses information provided in the EU's 13th Annual Report (CoEU, 2011b), reflecting military exports by EU member states in 2010.

Scoring system

The scoring system for the 2012 Barometer is identical to that used in 2011, providing comprehensive, nuanced, and consistent thresholds for the various categories. The Barometer's seven categories assess: timeliness as well as access and consistency in reporting (categories i–ii), clarity and comprehensiveness (iii–iv), and the level of detail provided on actual deliveries, licences granted, and licences refused (v–vii). For more detailed information on the scoring guidelines, see Small Arms Survey (2012, para. 17.9).

Explanatory notes

Note A: The Barometer is based on each country's most recent arms export report, made publicly available between 1 January 2010 and 31 December 2011.

Note B: The Barometer takes into account information that states have submitted to UN Comtrade for their 2010 exports through 17 January 2012, and national reporting to the UN Register through 31 December 2011. However, efforts to improve the Register's online interface have led to delays in access to reporting, with the consequence that some states' contributions for 2010 activities have not been made available for analysis. The 2012 Barometer therefore relies on data provided in the UN Secretary-General's 2011 report on the UN Register (UNGA, 2011a). While the Secretary-General's report is typically followed by the publication of addenda that contain additional states' reports, as of 22 February 2012 no addenda had been published. This explains why the level of reporting for 2010 activities is lower than expected.

Note C: The fact that the Transparency Barometer is based on three sources–national arms export reports (including reporting to the EU Report), reporting to the UN Register, and UN customs data–works to the advantage of states that publish data in all three outlets. All information provided to the three sources is reflected in the scoring. The same information is not credited twice, however.

Country-specific notes

⁺ Cyprus declared no exports of defence equipment to the 13th EU Report.

Source: Small Arms Survey (2012)

UNPACKING THE SEVEN PARAMETERS: NATIONAL PRACTICES

Transparency in reporting on small arms transfers has increased over the past ten years. Figure 9.2 illustrates the average points received by all covered states each year. States received an average of 7.98 points for their reports on 2001 activities; the average peaked at 11.40 for 2009 activities, declining slightly to 11.22 for 2010 activities. This represents an increase of 40 per cent over the period, but the average points earned by states still remains below half the total available points (that is, below 12.50 out of 25.00).

Figure 9.2 **Average total points received by all 52 countries for reports on 2001–10 activities**

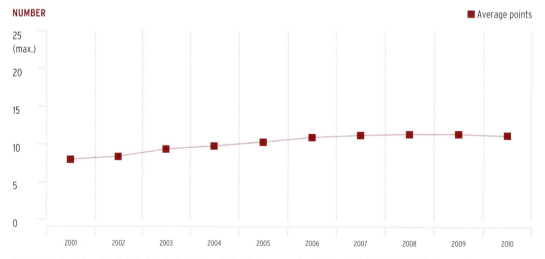

Note: The average for reporting on 2001–05 activities is based on 51 states because Serbia and Montenegro were not separate independent states. For 2006–10 activities the average is calculated for 52 states.

Figure 9.3 Average level of transparency of 52 states for reports on 2001–10 activities

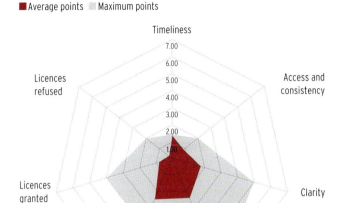

The progress is not generally uniform across all the countries under review and for all the Barometer's parameters. To reveal where progress has been made and where it has lagged, this section unpacks reporting by parameter, criterion, and sub-criterion. Figure 9.3 shows the average points earned by states for each of the seven main parameters. The grey area illustrates the maximum point allotment for each parameter, and the red area illustrates the average points received for all states for their reports on activities in 2001–10. It shows that states came close to full transparency for *timeliness* but fell far shorter with respect to the *licences refused* and *licences granted* parameters.

The following sections offer more details on both high and low scores as regards highlighted parameters and criteria.

Timeliness

The timely release of information allows for public scrutiny of recent exports and export licensing decisions as well as early warning about developments that could threaten peace and security. Timeliness is also a valuable confidence-building measure among states. This parameter awards a maximum of 1.5 points for the prompt reporting of export

Table 9.4 Countries scoring zero points for reports on 2001–10 activities

Year of activities	Countries scoring zero points
2001	Bosnia and Herzegovina, Bulgaria, Iran, North Korea, Pakistan, South Africa, Ukraine, United Arab Emirates
2002	Bosnia and Herzegovina, Bulgaria, North Korea, Pakistan, Ukraine, United Arab Emirates
2003	North Korea, United Arab Emirates
2004	North Korea, United Arab Emirates
2005	North Korea
2006	Iran, North Korea
2007	Iran, North Korea
2008	Iran, North Korea
2009	Iran, North Korea
2010	Iran, North Korea, United Arab Emirates

information through any one of three means: by publishing an up-to-date national arms export report (or, for European countries, contributing to the EU Report), by reporting to UN Comtrade, or by submitting a report to the UN Register. Submissions are considered 'on time' if they are published by 31 December of the year following the reported activities.[22] This parameter does not consider the content or the quality of the information.

In contrast to the other parameters, the average state score for *timeliness* is quite high; over the past ten years, the states under review scored an average of 1.40 (93 per cent) of the 1.50 available points. This is primarily because the number of states that did not provide any information on their arms transfers range from eight to one over the period (see Table 9.4). In most other cases, states received the total maximum points available for *timeliness* (see Table 9.5).

Access and consistency

Regular and accessible small arms and light weapons transfers data is critical for evaluating transparency. The parameter *access and consistency* (maximum 2.00 points) encourages states to report regularly to all reporting instruments and to make their reports available to a wide audience. It is divided into four criteria that assess the accessibility of the information states provide, the frequency and regularity of submissions, and the use of multiple reporting instruments.

A container of ammunition is loaded onto a US navy ship, Naval Magazine Indian Island, June 2011.
© Larry Steagall/Kitsap Sun/AP Photo

Over the past ten years, states have received an average of 0.98 points, or about half the available points for this parameter. The number of states earning the maximum available points gradually increased from zero to six over the period (see Table 9.5). The number of states receiving zero points under the *access and consistency* parameter varied from nine (for 2002 reporting) to three for reporting on 2004, 2006, and 2007 activities.

Table 9.5 Countries achieving maximum points under any of the seven parameters for reporting on 2001–10 activities

Year of activity	Timeliness (1.50 max)	Access and consistency (2.00 max)	Clarity (5.00 max)	Comprehensiveness (6.50 max)	Deliveries (4.00 max)	Licences granted (4.00 max)	Licences refused (2.00 max)
2001	43 countries	–	–	–	–	–	Denmark
2002	43 countries	–	–	–	–	–	–
2003	49 countries	Netherlands, Sweden, UK	–	–	Poland	–	–
2004	49 countries	Belgium, Netherlands, Sweden, UK	–	–	–	France	–
2005	49 countries	Belgium, Netherlands, Sweden, UK	–	–	Canada, Poland	–	–
2006	49 countries	Belgium, Netherlands, Sweden, UK	–	–	Poland	–	Montenegro, Romania
2007	50 countries	Belgium, Netherlands, Sweden, UK	–	–	Spain	Switzerland	–
2008	49 countries	Belgium, Netherlands, Sweden, UK	–	–	Poland, Spain	Switzerland	Germany, Romania, Serbia
2009	49 countries	Belgium, Netherlands, Romania, Sweden, UK	–	–	Poland, Spain	Switzerland	Romania, Serbia
2010	48 countries	Belgium, Netherlands, Romania, Spain, Sweden, UK	–	–	–	Switzerland	Belgium, Romania, Serbia, Slovakia

A state can earn half the available points under this parameter by publishing a national arms export report online. The national arms export report fulfils the criterion of 'availability of interim information on transfer activities' and publishing online fulfils the 'online accessibility' criterion. With the exception of Bosnia and Herzegovina and South Africa, all the countries that publish a national arms export report make it available on their government websites and receive the full points for this criterion.

Interim reports help hold governments accountable for their export licensing procedures and help civil society react to licensing decisions. Of the 29 countries that have published a national arms export report, six produce additional interim reports.[23] The Netherlands and Sweden publish monthly reports, Romania and the United Kingdom produce quarterly reports, and Spain and Belgium's Flanders and Brussels regions produce biannual reports. These interim reports can contain more detailed information than the annual arms export reports. States accrue points if they publish interim information; they earn further points if these reports provide additional relevant information that is required by the Barometer criteria.

The criterion on the regular use of a reporting tool can be fulfilled if a state reports annually to UN Comtrade or the UN Register or if it publishes a yearly national arms export report for three consecutive years. Should a country fail to publish a report or submit data to one of the other two instruments, it will not be awarded points for this criterion for three consecutive years, unless it reports regularly to another tool. For example, Saudi Arabia does not publish a national arms export report, nor does it report to the UN Register; it is thus scored on its reporting to UN Comtrade. Saudi Arabia did not submit data on its 2008 activities to UN Comtrade. It was therefore scored with its reporting on 2007 activities, and has, as a result, lost one half point under *access and consistency*. This half point loss will remain for three consecutive years, even if Saudi Arabia were to resume reporting to UN Comtrade in the meantime. This example illustrates that irregular reporting has an impact on a country's score across multiple years.

Six states produce monthly, quarterly, or biannual national arms export reports.

Clarity

Transparency on small arms transfers can make an important contribution to regional and international confidence building, but only if small arms transfers data is reported in a clear and understandable way. The parameter *clarity* is divided into 11 criteria, one of which is divided into four sub-criteria. Its main purpose is to reflect the extent to which information on small arms and light weapons transfers, including their ammunition, can be distinguished from transfers of other conventional arms transfers. It asks whether information on temporary exports is provided and whether the transfers are supplied by private industry or by the government. It also evaluates the information that countries provide on relevant legislation, such as measures to prevent diversion and international, regional, and sub-regional commitments to small arms transfers and brokering controls (see Box 9.2). Finally, it grants points for data on aggregated totals of deliveries, licences granted and refused, and brokers. A maximum of 5.00 points can be awarded.

Over the past ten years, states received an average of 1.81 points for *clarity*, slightly more than one-third of the available points. No state received the maximum available points. Denmark received the highest score for this parameter (4.75 points) once, for reporting on 2010 activities.

The *clarity* parameter provides a quantifiable measure of the extent to which state reporting distinguishes transfers of small arms and light weapons from those of other conventional weapons; it also assesses reporting on aggregate totals of deliveries (see Figure 9.4). Almost all states satisfy both aspects of the parameter. The exceptions are Iran and North Korea, which have not provided any information on their transfer activities since 2006 and since 2001, respectively. South Africa publishes a national arms export report using a classification for exported goods that is

Figure 9.4 Average points earned under 11 criteria of the clarity parameter, based on reporting on 2001, 2006, and 2010 activities

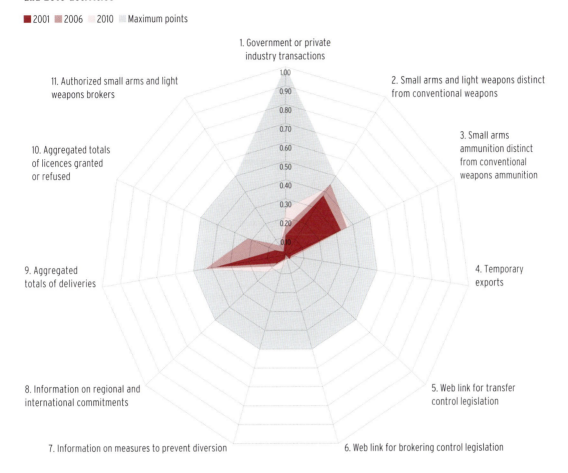

not defined in the national report. For this reason, South Africa has also failed to receive points for the classification of small arms and light weapons since reporting on its 2002 activities (Small Arms Survey, 2012, paras. 12.1–12.3).

Of the 52 countries under review, 14 fully reported whether their transfers were government or private industry transactions.[24] The national arms export reports of Finland, the United Kingdom, and the United States provide this information for all ten years of reporting.[25] The US Department of State, for example, provides a direct commercial sales report on US industry transactions. Meanwhile, the US Department of Defense prepares a foreign military sales report that lists its own transactions (Small Arms Survey, 2012, paras. 16.1a–b, 16.9 a–b). Other states have started providing this kind of information recently; for example, Norway began with its report on 2005 activities and Switzerland with its report on 2007 activities. While some states made the information available in the past ten years, they did not do so consistently. These include France, the Netherlands, Portugal, and Romania. Since reporting on 2005 activities, Germany[26] and Sweden ceased to provide full information on both government and private industry transactions and therefore only receive partial points. Many other countries under review received partial points as well because they mention either private or government transactions but do not provide further details. Similarly, Serbia, in its report on 2009 activities, states that 85 per cent of its arms industry was privately owned and the remainder was mixed or state-owned;

however, it does not indicate the proportion of small arms exports that were sourced from private, rather than public, manufacturers. Serbia therefore received partial points for acknowledging private and government-owned industries (Small Arms Survey, 2012, para. 11.3).

Reporting on temporary exports is not widespread. Denmark and the United Kingdom[27] reported in such a way as to receive full points for all ten years. Denmark indicates if the goods have been exported temporarily in the 'Comments on the transfer' column in the UN Register background information category (Small Arms Survey, 2012, paras. 4.2–4.6). It provides details on weapons type, quantity, and destination country. Bosnia and Herzegovina, Croatia, Germany, Hungary, Italy, the Netherlands, and Norway have received full points for individual years. The Czech Republic indicates that licences for 2009 activities were given for temporary transactions but does not provide further details, and so has received partial points (para. 3.1).

Comprehensiveness

A thorough understanding of the global and regional small arms trade requires complete and consistent reporting from states. Distinguishing small arms transfers from other conventional arms transfers is also important for monitoring excessive and destabilizing transfers and accumulations of small arms in regions at risk of conflict. The parameter *comprehensiveness* examines the level of detail provided on small arms and light weapons types.[28] It also evaluates reporting of different types of transfers, such

Box 9.2 Reporting on brokering activities

Brokers and their activities are an element in the chain of legal arms transfers, but they also play an important role in the illicit small arms trade. Several global and regional initiatives have been launched for the control of brokering activities in recent years. A UN Group of Governmental Experts, for example, recommended in August 2007 that states dedicate a specific section to brokering in their reporting under the UN Programme of Action on small arms and light weapons; they also urged states to consider reporting on the prevention of illicit brokering at the Biennial Meetings of States (UNGA, 2007, paras. xx, xxii).[29]

EU member states have been invited to exchange information on brokering licences granted and refused under the EU Report since reporting on 2008 activities (that is, since the *11th EU Annual Report*). So far, 15 EU member states have reported on brokering transactions in the EU Report (CoEU, 2009; 2011a; 2011b). South-east European states, in cooperation with the South Eastern and Eastern European Clearinghouse for the Control of Small Arms and Light Weapons, developed a regional database on registered brokers in order to improve information sharing and control over arms brokering activities–the first such database in the world (Bromley, 2011, p. 39).

The Transparency Barometer has two criteria relating to brokering activities. First, the Barometer assesses whether a country provides a reference to a website that offers free, full-text access to its brokering control legislation and whether the country explains how this legislation is implemented. Second, it determines whether the country provides information on authorized small arms brokers. The former criteria can be fulfilled in a national arms export report and the latter either in a national arms export report or through reporting to the UN Register.

Few countries provide information on their brokering control legislation in their national arms export reports. France, Finland,[30] and Norway are three countries that started to provide this information while reporting on 2001 and 2002 activities. Denmark, Germany, Spain, and Switzerland began to provide the required information on their brokering control legislation later. In its report on 2005 activities, Norway notes a web page for its export control legislation; it also outlines the responsibilities of a broker and how the country implements its brokering control legislation. Relevant guidelines were attached in the annexe (Small Arms Survey, 2012, para. 8.1). Sweden has managed to receive full points for its reporting on brokers. In its reports documenting 2005–10 activities, it provides a full list of registered brokers (para. 13.1).

The sensitivity of the issue and commercial confidentiality are the main reasons why most states do not provide information on registered small arms brokers. If they provide any sort of information, they generally list the names of companies that received a licence to trade in military materiel. These lists rarely distinguish which companies were licensed as brokers. Croatia is the only country that has received partial points for this criterion, since it reported to the UN Register on its 2008 activities. Under the column 'Comments on the transfer', Croatia provides the name of the company that has undertaken the export (Small Arms Survey, 2012, para. 2.1).

as permanent re-exports and transit or transhipment activities. States can earn one half point for each of the parameter's 13 criteria, for a maximum of 6.50 points.

States have received an average of 2.45 points (38 per cent) for *comprehensiveness* over the past ten years. No state received the maximum available points. Two countries managed to achieve 5.25 points—the United Kingdom twice, for reporting on 2006 and 2010 activities, and Switzerland four times, for reporting on 2007–10 activities.

To provide a sense of the changes in reporting on this parameter since 2001, Figure 9.5 illustrates the average score of all the countries under review for each of the 13 criteria of *comprehensiveness* for reporting on 2001, 2006, and 2010 activities. Data for 2006 is included because of improvements to state reporting following the introduction of the voluntary background information on small arms and light weapons transfers under the UN Register.

A general improvement in reporting standards can be observed for many of the criteria. In particular, reporting on guided light weapons (such as man-portable air defence systems and anti-tank guided weapons), unguided light weapons (such as mortars, rocket-propelled grenades, and grenade launchers), pistols and revolvers, and military firearms has improved over the period. Since reporting on 2001 activities, states have increasingly provided information on small arms exports in their national arms export reports and given data on different weapon types. The

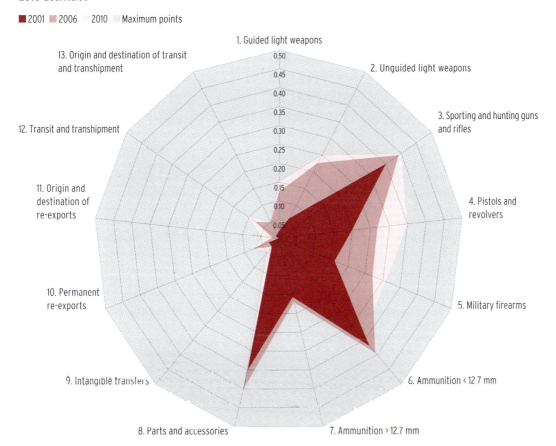

Figure 9.5 **Average points earned under 13 comprehensiveness criteria, based on reporting on 2001, 2006, and 2010 activities**

introduction of voluntary background information on small arms and light weapons transfers in the UN Register has unquestionably contributed to enhanced reporting in this area.

Figure 9.5 also shows where reporting is weaker, such as for sharing information on 'intangible transfers'. Intangible transfers include the provision of technical plans, blueprints, know-how, schematics, and software related to small arms and light weapons, their ammunition, and their parts and accessories. Seven countries received maximum points for reporting on intangible transfers.[31] In its 2007 national arms export report, the United Kingdom notes the transfer of technology and technical assistance licences; the report's annexe specifies that licences were issued for technology for sniper rifles and other small arms (Small Arms Survey, 2012, para. 15.1, p. 26, annexes). In its 2009 national arms report, Montenegro states that no licences were granted for technical assistance, whether for transfer of know-how, development, or production (para. 6.2, p. 28). Austria, Belgium, Germany, Greece, and Serbia received partial points for reporting on intangible transfers, primarily because the reporting was not consistent and included examples and mixed categories of arms and equipment. In Belgium's case, one of the country's regional reports highlights intangible transfers, but the other regions failed to provide such information (paras. 1.1a–b, 1.5 a–d).[32]

> Overall reporting remains poor on permanent re-exports and transit and transhipment activities.

Despite improvement among some states in recent years, overall reporting remains poor on permanent re-exports and transit and transhipment activities, including information sharing on countries of origin and destination for re-exports and materiel in transit. Most states do not share any information on permanent re-exports and transit and transhipment activities.

Deliveries

Providing information on actual deliveries can build confidence among states and help identify and potentially monitor destabilizing accumulations of arms in countries and regions at risk of conflict. It also facilitates verification of transfers between exporter and importer states (Bromley and Holtom, 2011, p. 34). The parameter *deliveries* is divided into four criteria, each worth 1.00 point for a total of 4.00 points (see Figure 9.6). It awards points for information shared on actual deliveries and the destination countries; end users; and the types, values, and quantities of delivered weapons.

Over the past ten years, states have received an average of 2.56 points on this parameter, or almost two-thirds of the possible points. Poland received the total points available for reporting on 2003, 2005, 2006, 2008, and 2009 activities and Canada and Spain received full points for reporting on 2005 and 2007–09 activities, respectively. The number of states that received zero points varied from nine (for 2001 reporting) to three for reporting on 2005 and 2007–09 activities.

States steadily increased the quality of reporting for deliveries, with the exception of reporting on quantities in 2010 (see Figure 9.6). All criteria can be fully or partially fulfilled via reporting to UN Comtrade or the UN Register,[33] or via publication of a national arms export report or submission to the EU Report. The exception is information on specific end users, which cannot be captured in reporting to UN Comtrade.

Details on end users are useful for informing the licensing officials of other states and the public on the specifics of small arms transfers. Every importing state issues end-user certificates; exporting states are supposed to review the original end-user certificate before granting an authorization to export small arms or light weapons. This means that both parties should know the purported end users for each transaction. Nevertheless, few countries provide information on the end user in their reporting. For reporting on 2001, 2002, and 2003 activities, for example, no country received points for the end-user criterion. Since then, 17 countries have received partial points and three countries—Canada, Poland, and Spain—earned full points for providing information on end users. Poland received

Figure 9.6 **Average points earned under the four deliveries criteria, based on reporting on 2001, 2006, and 2010 activities**

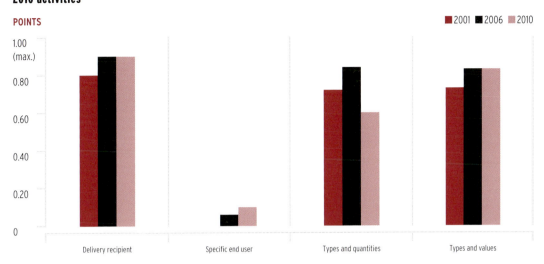

full points for detailed information provided under the UN Register in the column 'Comments on the transfer'. It provided information on end users of small arms and light weapons, such as the Ministry of Defence of recipient countries, collectors, museums, NATO (including in Afghanistan), and other international forces (Small Arms Survey, 2012, paras. 9.1–9.5).

Under the EU Report, countries can also provide information on end users (that is, exports to UN-mandated or other international missions). While some EU countries identify end users, they do not necessarily do so systematically, or they provide only selected examples. A case in point is the Czech Republic's report on 2009 activities, which includes a footnote indicating that small arms exported to Cambodia went to police forces and the United Nations Assistance to the Khmer Rouge Trials Mission. Because the information was not provided systematically for all small arms exports, the country earned partial points for this disclosure (Small Arms Survey, 2012, para. 3.1).

Licences granted

Information on licences granted sheds light on how national export criteria are interpreted. It also affords civil society and the media the opportunity to alert stakeholders to sensitive and potentially excessive and destabilizing transfers before actual exports occur. The parameter *licences granted* is divided into four criteria, each of which has a value of 1.00 point, for a total of 4.00 points. It awards points for information shared on licence recipients; end users; and the types, values, and quantities of weapons for which licences were granted.

Over the past ten years states received an average of 0.81 points, or one-fifth of the total points available. No country received the maximum points for reporting on 2001–03 and 2005–06 activities. France received the full points for reporting on 2004 activities and Switzerland for reporting on 2007–10 activities.

The very low average of points received and the high number of states receiving zero points (an average of 30 countries over the past decade) under the parameter is linked to the fact that points cannot be earned by reporting to UN Comtrade or the UN Register.[34] If a country reports only to the EU Report, it cannot earn full points.[35] States must thus publish a national arms export report to be able to earn full points.

Figure 9.7 illustrates what type of information states provide about the recipient country. Information on quantities and values of weapon types for which licences were granted are not provided often; information on the end user is rarely shared.

Switzerland published its first national arms export report specifically dedicated to small arms and light weapons transfers activities in 2008. To date, no other government has published a separate small arms export report. This practice has contributed to Switzerland being the most transparent country since reporting for 2007 activities. In its reports, Switzerland consistently provides detailed data on licenses granted, identifying delivery recipients of granted licenses as well as weapons types, quantities, and values. A separate table provides a list with all destination countries and the quantity of weapons delivered. The table identifies the percentage of licences granted for various end users, such as brokers, armouries, private persons, the police, armed forces, and other government agencies (Small Arms Survey, 2012, paras. 14.1–14.3).

Licences refused

Information on licence denials provides insight on how national export criteria are interpreted and helps neighbouring states and entire regions form a common position on their export policies. By sharing information on licence denials, states can be warned, for example, about exporting to a country in close proximity to a conflict zone or where the risk of diversion is high. The parameter

Firearms manufactured by SAN Swiss Arms AG in Neuhausen, Switzerland, June 2001. © Michele Limina /AP PHOTO/KEYSTONE

Figure 9.7 **Average points earned under the four criteria of licences granted, based on reporting on 2001, 2006, and 2010 activities**

POINTS ■ 2001 ■ 2006 ■ 2010

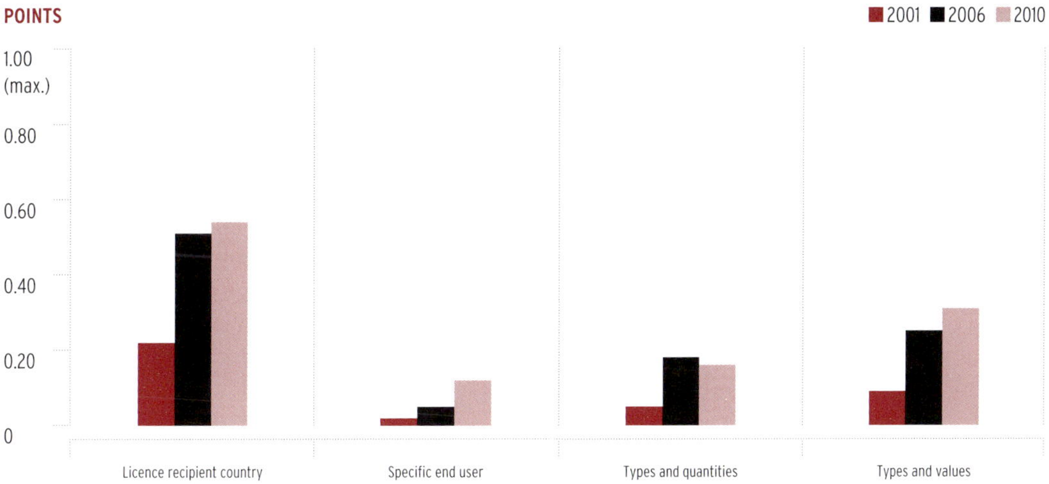

licences refused is divided into four criteria, each of which entails one half point, for a total of 2.00 points. It considers whether the country identifies destination countries that were refused licences; provides an explanation for such refusals; and offers information on the types, values, and quantity of weapons for which licences were refused.

Over the past ten years states received an average of less than ten per cent of the possible points (0.20 points) (see Figure 9.8). No country received the maximum points for reporting on 2002–05 and 2007 activities. Denmark received full points for reporting on 2001 activities. Montenegro and Romania received the full points for reporting on 2006 activities. Germany, Romania, and Serbia received full points for reporting on 2008 activities; for reporting on 2009 activities, Serbia and Romania received full points (see Box 9.3). They were joined by Belgium and Slovakia for reporting on 2010 activities (see Table 9.5). Since they submitted nil reports, Belgium, Denmark, Montenegro,

Figure 9.8 **Average points earned under the four criteria of licenses refused, based on reporting on 2001, 2006, and 2010 activities**

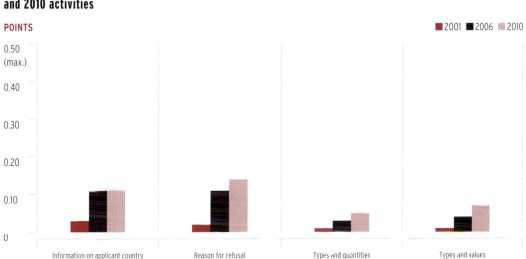

Romania, and Slovakia received the full points for reporting on licence refusals; these submissions indicate that these five countries did not reject any licences (Small Arms Survey, 2012, paras. 1.7a–d; 4.1; 6.1; 10.1; 18.1). Serbia provided complete information on its licence denials—including export destination and the end-user country—in the national arms export reports for 2007 and 2008 activities. It states that the licences were denied due to incomplete documentation and provides details on the type, value, and quantity of the materiel for which a licence was refused (paras. 11.1, 11.2).

Box 9.3 Efforts to increase transparency in South-east Europe: Serbia and Romania

Since the early 1990s a growing number of European states have produced national arms export reports. The common practice of parliamentary scrutiny and the obligation of EU member states to exchange data on their export licence approvals, actual exports, and denials for licences have encouraged non-EU states to publish national arms export reports as well. Prospective EU candidacy or EU membership are powerful incentives to provide better and more detailed documentation of small arms and light weapons transfers. These factors are driving South-east Europe's increasing transparency in their arms export activities.

In 2009, Albania, Bosnia and Herzegovina, Macedonia, Montenegro, and Serbia launched the initiative to publish a regional report on arms exports for South-eastern Europe. The report, modelled on the EU Report, was published for three consecutive years; the contents are also available in an online database (Bromley, 2010; SEESAC, n.d.). The data for the regional report is based on the participating countries' national arms export reports; however, those reports include more detailed information than the recently established regional report.[36] Some national reports, such as those of Romania and Serbia, also tend to be more detailed than many of the national reports produced by EU member states. Both states have steadily increased their level of transparency since reporting on 2006 transfers (see Figure 9.9); they are among the most transparent countries with respect to reporting on 2006–10 activities.

Since its independence in 2006, Serbia has achieved its highest scoring results, ranking among the most transparent countries for reporting on 2006–10 activities. Serbia produced its first annual report on arms exports in 2007 (Small Arms Survey, 2012, para. 11.4); furthermore, the country started to provide background information on its international small arms and light weapons transfers to the UN Register in 2010. Serbia's first national arms export report includes extensive information on arms export licences granted, actual arms exports, end users, and licences denied. The information is broken down by destination country and control list category, with additional descriptions of the goods and weapon types. By posting its national arms export report on the government website, Serbia increased its score for reporting on 2008 activities by one half point. The following year the country provided more details on ammunition larger than 12.7 mm and permanent re-exports and transit and transhipment. These improvements resulted in an increase of 1.5 points, raising Serbia to fourth most transparent country for reporting on 2009 and 2010 activities.

Romania entered the EU on 1 January 2007. The same year, it took a number of ambitious steps towards transparency with regard to its arms transfers information. First, it retroactively published national reports for 2003–05 activities. These reports already provide detailed information on licence denials. Subsequently, Romania introduced quarterly reports and provided specifics on licences granted, increasing its total score for 2007 and 2008 activities. Romania's 2009 report includes new information on transit and transhipment activities, increasing the score by another half point. It also shares more comprehensive information on end users, distinguishing licences granted for governments, industry, and commercial markets. For 2010 activities, Romania provides more details on quantities in licences granted; in its national arms export report, it reports on ammunition larger than 12.7mm (Small Arms Survey, 2012, para.10.3). With these improvements, Romania increased its total score by one full point. It now ranks third in the 2012 Barometer, after Switzerland and the United Kingdom. Romania's arms export report provides a level of detail on small arms and light weapons ahead of several EU member states' reporting. On a regional level, it is the most transparent country in South-east Europe.

Individual efforts by South-east European countries and the regional reports reflect the willingness and capacity in the region to comply with the EU model of reporting and export controls. The regional efforts to increase transparency serve as a good example of how states can improve the contents of national reports through long-established reporting practices. Many EU and non-EU member states should be encouraged to follow this example.

Figure 9.9 **Romania and Serbia's total points compared to the average and maximum points, based on reporting on 2001–10 activities**

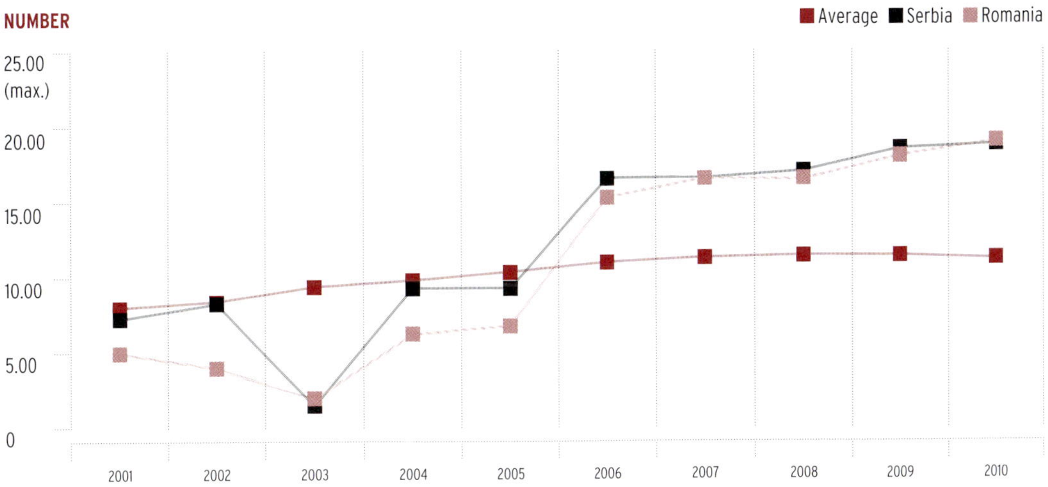

[Figure: Line chart showing Average, Serbia, and Romania total points from 2001 to 2010, Y-axis NUMBER 0 to 25.00 (max.)]

Note: Serbia's political status over the period was as follows: 2001–02: part of the Federal Republic of Yugoslavia; 2003–05: part of Serbia and Montenegro; and 2006–10: independent state.

An average of 38 countries scored zero points over the past ten years for *licences refused*, yielding a very low average for all states on this parameter. This is primarily because points can only be earned by publishing a national arms export report or reporting to the EU Report. In fact, only European countries managed to receive points for this parameter.

The maximum that EU members can receive is one quarter point for reporting on licence refusals to the EU Report, if they do not provide information on brokering licence refusals.[37] Unfortunately, due to format changes in the EU Report for reporting on 2009 activities, many countries did not receive these quarter points as the information on licence refusals was no longer national, but instead aggregated at the EU level (Herron et al., 2011, p. 19). This resulted in a decrease from 19 countries receiving points under *licences refused* for reporting on 2008 activities to ten countries for reporting on 2009 activities. For reporting on 2010 activities, the information on licence refusals was once again provided at the national level.

CONCLUSION

Over the past ten years, major exporting states have become increasingly transparent in reporting on their small arms and light weapons transfers. The average score of all 52 states surveyed has increased by at least 40 per cent over the period. The trend towards greater transparency spans the entire sample, including both high-scoring and low-scoring countries.

Some of this improvement can be traced to specific reporting practices by states. More states are producing national arms export reports and utilizing the UN Register's optional category for small arms and light weapons transfers. These reporting mechanisms provide states with more opportunities to supply comprehensive information. As a result, some states' scores in the Transparency Barometer's *comprehensiveness* parameter have improved.

The importance of national arms export reports can also be seen in some of the parameters in which most states remain opaque, such as reporting on *licences granted* and *licensed refused*. There is no mechanism for reporting on these parameters other than through a national arms export report. While the number of states producing an export report has increased by nearly 50 per cent over the past decade, fewer than half of all states surveyed in this chapter submitted a report in 2011.

This is one reason why, despite the general increase in transparency over time, the average score of all states remains below 50 per cent of the maximum possible score. Reporting practices remain generally poor with reference to *licences granted and refused* and the *clarity* and *comprehensiveness* parameters. To enhance clarity, many states still have to standardize the process of including information on temporary exports and brokering control legislation, on measures taken to prevent and detect international diversion, and on licensed brokers. The inclusion of information on ammunition larger than 12.7 mm, intangible transfers, and permanent re-export, transit, and transhipment activities would increase comprehensiveness.

One inescapable conclusion from this ten-year review of transparency is that there is major room for improvement among most states, and that the best way to achieve it is through national arms export reports. These are now the norm for European countries—25 of 31 European countries surveyed publish them—but they remain the exception in other regions, where only 4 of 21 states surveyed provide them. In fact, all of the top 25 most transparent states in the 2012 Barometer, 23 of them European, issued national arms export reports in addition to using the other reporting instruments. These countries are leading the way in transparency in the small arms trade. But membership in the 'club' is open.

LIST OF ABBREVIATIONS

NISAT	Norwegian Initiative on Small Arms Transfers
UN Comtrade	United Nations Commodity Trade Statistics Database
UN Register	United Nations Register of Conventional Arms

ENDNOTES

1. For more information on the Transparency Barometer, see Khakee (2004) and Lazarevic (2010).
2. The chapter complements a recent country-specific analysis of transparency published by the Small Arms Survey. See Lazarevic (2010).
3. For more information on the previous scoring methods, see Lazarevic (2010).
4. Exports can refer to the sale of newly produced goods, transfers, temporary exports, re-exports, and the sale of stockpiles or surplus materiel.
5. The Small Arms Survey relies on the Norwegian Initiative on Small Arms Transfers (NISAT) Database on Authorized Small Arms Transfers to determine which states meet the minimum export threshold for inclusion. The NISAT database draws exclusively on the United Nations Commodity Trade Statistics Database (UN Comtrade). For more details about the methodology, see Marsh (2005) and Small Arms Survey (2012, paras. 7.1–7.8).
6. For a more detailed description of which countries were added to the Transparency Barometer, and when, see Lazarevic (2010, p. 164).
7. These states were Denmark, Greece, Lithuania, Luxembourg, the Philippines, Saudi Arabia, Thailand, Ukraine, and the United Arab Emirates.
8. These states were Hungary, India, Iran, and Slovakia.
9. For more information on the scoring guidelines, see Small Arms Survey (n.d.a).
10. For more information on the reporting instruments, see Lazarevic (2010, pp. 16–24).
11. The exception is Serbia, which has only reported to UN Comtrade since its independence in 2006.
12. This finding may be somewhat biased as reporting to UN Comtrade is the basis for selecting the countries analysed in the Transparency Barometer.
13. For more information on the seven UN Register categories, see UN (n.d.).
14. Taiwan has not submitted a report to the UN Register either, as it is prevented from doing so.

15 The optional standardized format for reporting on small arms and light weapons transfers was adopted in 2006 after an Expert Group reviewed the UN Register. It provides six categories for small arms and six categories for light weapons, representing a clearer breakdown than UN Comtrade codes.

16 For the purpose of the Transparency Barometer, national arms export reports are 'on time' if they are published by 31 December of the year following the period analysed. For example, the 2012 Transparency Barometer covers national arms export reports on 2010 data that were published by 31 December 2011.

17 These were the Czech Republic, Italy, the Netherlands, Norway, Sweden, Switzerland, the United Kingdom, and the United States.

18 EU member states are required under the EU Council's Common Position Article 8(3) to publish an arms export report if they export technology or equipment on the EU Common Military List. Of the EU member states analysed in the Transparency Barometer, Cyprus, Greece, Lithuania, and Luxembourg are the only EU members that have not yet published a national arms export report. Latvia and Malta, which have not produced such a report either, are not included in the Barometer (Weber and Bromley, 2011).

19 All previous editions of the Transparency Barometer are available online; see Small Arms Survey (n.d.b).

20 Although Greece was included in the Barometer for the first time for 2010 activities, it has been retroactively scored and included in all previous editions, as have all newly included states.

21 The online version of the Barometer is available at Small Arms Survey (n.d.b).

22 For example, the 2012 Transparency Barometer considers national arms export reports on 2010 data that were published by 31 December 2011. Reports to UN Comtrade and to the UN Register are also considered if they were submitted by 31 December of the year following the period analysed.

23 These interim reports are also included in the analysis of transparency.

24 These states are the Czech Republic, Denmark, Finland, France, Germany, Hungary, the Netherlands, Norway, Portugal, Romania, Sweden, Switzerland, the United Kingdom, and the United States.

25 The UK did not earn full points for reporting on 2001 activities; since Switzerland started producing a national small arms export report in 2007, it has also provided information if the transfer is supplied by the government or private entities.

26 For its documentation of 2010 activities, however, Germany improved its reporting and received full points.

27 The UK did not earn full points for reporting on 2004 data.

28 For example, a detailed report might identify materiel as (un-)guided light weapons, sporting and hunting guns, pistols and revolvers, military firearms, small arms ammunition, ammunition larger than 12.7 mm, or parts and accessories.

29 For more information on initiatives to control brokering activities, see, for example, Cattaneo (2004).

30 Finland received points under this criterion only for reporting on 2002 and 2003 activities.

31 These countries were Austria, Germany, Montenegro, Spain, Switzerland, the United Kingdom, and the United States.

32 Belgium's export control system was regionalized in August 2003. This means that each of the three Belgian regions reports separately on its arms exports. Given diverging regional reporting practices, the reports of all three regional parliaments must be provided for Belgium to be assessed in the Transparency Barometer.

33 Under the UN Register, states are requested to provide information on deliveries; however, some states provide information on licence approvals rather than actual deliveries. This is because some states do not have established data collection practices that allow them to report on actual deliveries. If, like Germany, the submitting state illustrates in its UN Register report that the data is on licences issued and not on actual deliveries, the information is scored accordingly.

34 If, under the UN Register, a country specifies that provided data concerns licences granted and not actual deliveries, then the points are scored under licences granted, but no points are attributed for deliveries. See, for example, Germany's submissions to the UN Register (UNGA, 2011a); Germany does not report on actual deliveries, but rather on licences issued.

35 The EU Report does not include quantities for which licences are granted, so countries cannot receive the 2.00 points this information can earn them. The EU Report does allow countries to earn 1.00 point for the recipient country, one half point for the values of the *licences granted*, and one half point for information on end users (that is, exports to UN-mandated or other international missions).

36 Croatia was the last South-east European country to publish a national arms export report. It was released at the end of 2010. Now all the Balkan countries publish national arms export reports.

37 Detailed information on brokering licence refusals under the EU Report can provide up to 1.25 points for this parameter. None of the EU member states under review received full points for reporting on brokering licence refusals. Only the UK received one half point for reporting on 2009 activities (CoEU, 2011a, p. 406).

BIBLIOGRAPHY

Bromley, Mark. 2010. *Regional Report on Arms Exports in 2008*. Belgrade: South Eastern and Eastern Europe Clearinghouse for the Control of Small Arms and Light Weapons. <http://www.seesac.org/uploads/armsexport/Regional_Report_on_Arms_Exports_in_2008_ENG.pdf>

—. 2011. *The Development of National and Regional Reports on Arms Exports in the EU and South Eastern Europe*. Belgrade: South Eastern and Eastern Europe Clearinghouse for the Control of Small Arms and Light Weapons. <http://www.sipri.org/research/armaments/transfers/publications/other_publ/other%20publications/the-development-of-national-and-regional-reports-on-arms-exports-in-the-eu-and-south-eastern-europe-seesac>

— and Paul Holtom. 2011. *Implementing an Arms Trade Treaty: Lessons on Reporting and Monitoring from Existing Mechanisms*. SIPRI Policy Paper No. 28. Stockholm: Stockholm International Peace Research Institute.

Cattaneo, Silvia. 2004. 'Targeting the Middlemen: Controlling Brokering Activities'. In Small Arms Survey. *Small Arms Survey 2004: Rights at Risk*. Oxford: Oxford University Press, pp. 141–71.

CoEU (Council of the European Union). 2009. *Eleventh Annual Report According to Article 8(2) of Council Common Position 2008/944/CFSP Defining Common Rules Governing Control of Exports of Military Technology and Equipment*. 2009/C 265/01. 6 November. <http://eur-lex.europa.eu/LexUriServ/LexUriServ.do?uri=OJ:C:2009:265:FULL:EN:PDF>

—. 2011a. *Twelfth Annual Report According to Article 8(2) of Council Common Position 2008/944/CFSP Defining Common Rules Governing Control of Exports of Military Technology and Equipment*. 2011/C 9/01. 13 January. <http://eur-lex.europa.eu/LexUriServ/LexUriServ.do?uri=OJ:C:2011:009:FULL:EN:PDF>

—. 2011b. *Thirteenth Annual Report According to Article 8(2) of Council Common Position 2008/944/CFSP Defining Common Rules Governing Control of Exports of Military Technology and Equipment*. 2011/C 382/01. 30 December. <http://eur-lex.europa.eu/JOHtml.do?uri=OJ:C:2011:382:SOM:EN:HTML>

Herron, Patrick, et al. 2011. 'Larger but Less Known: Authorized Light Weapons Transfers.' In Small Arms Survey. *Small Arms Survey 2011: States of Security*. Cambridge: Cambridge University Press, pp. 9–41.

Khakee, Anna. 2004. 'Back to the Sources: International Small Arms Transfers.' In Small Arms Survey. *Small Arms Survey 2004: Rights at Risk*. Oxford: Oxford University Press, pp. 99–139.

Lazarevic, Jasna. 2010. *Transparency Counts: Assessing State Reporting on Small Arms Transfers, 2001–2008*. Occasional Paper 25. Geneva: Small Arms Survey. June.

Marsh, Nicholas. 2005. *Accounting Guns: The Methodology Used in Developing Data Tables for the Small Arms Survey*. Unpublished background paper. Oslo: Peace Research Institute Oslo/Norwegian Initiative on Small Arms Transfers. 14 November.

SEESAC (South Eastern and Eastern Europe Clearinghouse for the Control of Small Arms and Light Weapons). n.d. 'Regional Reports: Online Database.' Accessed 9 March 2012. <http://www.seesac.org/new-activities/new-arms-export-controls/regional-reports/1/>

Small Arms Survey. 2012. *Small Arms Trade Transparency Barometer 2012: Sources*. Unpublished background paper. Geneva: Small Arms Survey.

—. n.d.a 'Guidelines for Scoring the Barometer for YB11.' January. <http://www.smallarmssurvey.org/fileadmin/docs/Weapons_and_Markets/Tools/Transparency_barometer/SAS-Transparency-Barometer-Guidelines-2011.pdf>

—. n.d.b 'The Transparency Barometer.' <http://www.smallarmssurvey.org/weapons-and-markets/tools/the-transparency-barometer.html>

UN (United Nations). n.d. 'The Global Reported Arms Trade.' <http://www.un-register.org/Background/Index.aspx>

UNGA (United Nations General Assembly). 1991. *Study on Ways and Means of Promoting Transparency in International Transfers of Conventional Arms: Report of the Secretary-General*. A/46/301 of 9 September.

—. 2007. *Report of the Group of Governmental Experts Established Pursuant to General Assembly Resolution 60/81 to Consider Further Steps to Enhance International Cooperation in Preventing, Combating and Eradicating Illicit Brokering in Small Arms and Light Weapons*. A/62/163 of 30 August. <http://www.un.org/ga/search/view_doc.asp?symbol=A/62/163>

—. 2009. Resolution 64/54 on Transparency in Armaments, 2 December. A/RES/64/54 of 12 January 2010. <http://www.un.org/ga/search/view_doc.asp?symbol=A/RES/64/54>

—. 2010a. *Report of the Secretary-General: United Nations Register of Conventional Arms*. A/65/133 of 15 July. <http://www.un.org/ga/search/view_doc.asp?symbol=A/65/133>

—. 2010b. *Report of the Secretary-General Addendum: United Nations Register of Conventional Arms*. A/65/133/Add.1 of 15 September. <http://www.un.org/ga/search/view_doc.asp?symbol=A/65/133/Add.1>

—. 2011a. *United Nations Register of Conventional Arms: Report of the Secretary-General*. United Nations Register of ConventionalArms. A/66/127 of 12 July. <http://www.un.org/ga/search/view_doc.asp?symbol=A/66/127>

—. 2011b. *Transparency in Armaments*. A/C.1/66/L.29 of 14 October. <http://www.un.org/ga/search/view_doc.asp?symbol=A/C.1/66/L.29>

UNODA (United Nations Office for Disarmament Affairs). 2009. *Assessing the United Nations Register of Conventional Arms*. Occasional Paper No. 16. New York: UNODA. <http://www.un.org/disarmament/HomePage/ODAPublications/OccasionalPapers/PDF/OP16.pdf>

UNSC (United Nations Security Council). 2008. *Small Arms: Report of the Secretary-General*. S/2008/258 of 17 April. <http://www.un.org/disarmament/convarms/SALW/Docs/SGReportonSmallArms2008.pdf>

Weber, Henning and Mark Bromley. 2011. *National Reports on Arms Exports*. SIPRI Fact Sheet. Stockholm: Stockholm International Peace Research Institute. March. <http://books.sipri.org/files/FS/SIPRIFS1103b.pdf>

ACKNOWLEDGEMENTS

Principal author
Jasna Lazarevic

Contributor
Thomas Jackson

A soldier with the African Union Mission in Somalia looks at a haul of weapons seized along with four suspected members of al Shabaab, March 2012. © Stuart Price/AU-UN IST/AFP Photo

Surveying the Battlefield
ILLICIT ARMS IN AFGHANISTAN, IRAQ, AND SOMALIA

10

INTRODUCTION

Shortly after midnight on 6 August 2011, a Chinook helicopter carrying NATO and Afghan security forces was preparing to land near the Tangi Valley in Afghanistan. As the helicopter was descending, a Taliban fighter fired a rocket-propelled grenade (RPG)[1] at the dangerously exposed aircraft. The helicopter burst into flames, breaking apart as it plummeted to the ground. Thirty-eight people died in the crash, including 22 US special forces troops, making it the deadliest single incident of the war for US forces (King, Dilanian, and Cloud, 2011; Riechmann, 2011). An unclassified summary of a US military investigation into the incident confirms that the helicopter was hit by a 'rocket-propelled grenade' but provides no additional details (US CENTCOM, 2011b).

Most media accounts also refer to the weapon simply as a 'rocket-propelled grenade',[2] but there are many kinds of RPGs from various sources and with very different capabilities. These differences matter—and not just in the case of the downed helicopter. Whether Taliban arsenals are filled with 30-year-old Soviet PG-7 rounds or PG-29V rounds for the deadly RPG-29 Vampir has profound implications, yet detailed data on illicit weapons in Afghanistan is scant.

These data gaps are not limited to Afghanistan. Details on the model, country of origin, age, and condition of illicit weapons are rare, and the little information that is available is usually anecdotal and incomplete. Thus, we are left to wonder what models of RPGs and other weapons are most readily available to terrorists, insurgents, and criminals. How technologically advanced are these weapons? Is it possible to keep them out of the hands of individuals and groups that are likely to misuse them? How old are they, and how many are still functional? Answers to these questions have the potential to shape efforts to stem the flow of illicit weapons and shed light on the threat they now pose. This chapter inaugurates a multi-year project to provide such answers through data-driven analysis of illicit small arms and light weapons worldwide.

The project consists of three overlapping phases. This chapter distils the findings of the first phase, focusing on 'war weapons', namely small arms and light weapons illicitly acquired and used by non-state actors in high-intensity conflict zones. Phase two will look at illicit small arms in low-intensity armed conflicts and in countries affected by high-intensity organized criminal violence. The third and final phase will examine countries affected primarily by individual criminal violence.

To date, the project has collected and analysed data on 80,000 illicit small arms and light weapons in Afghanistan, Iraq, and Somalia, making it the largest study of its kind. While the scope, specificity, and comprehensiveness of the data vary considerably from country to country, overall the data sheds new light on illicit weapons in war zones, confirming some common assumptions and challenging others. The main findings from this first phase of the project include:

- The vast majority of illicit small arms in Afghanistan, Iraq, and Somalia appear to be Kalashnikov-pattern assault rifles.[3] Other types of small arms are comparatively rare.

- Most illicit light weapons and light weapons ammunition studied appear to be versions of Soviet- and Chinese-designed weapons first fielded decades ago.
- Data compiled for this study suggests that armed groups in Afghanistan and Iraq have access to very few technologically sophisticated or latest-generation light weapons.
- Newly acquired data on weapons seized in Iraq suggests that a significant percentage of seized Iranian weapons were manufactured recently.
- Despite the large quantities of small arms and light weapons that are trafficked into Somalia, the variety of available items is limited.

The chapter begins by defining key terms and concepts. A brief overview of the data used in the study is then provided. Illicit small arms and light weapons in Afghanistan, Iraq, and Somalia are then assessed in more depth. The chapter concludes with additional observations about 'war weapons'.

USE OF TERMS

Many Iranian weapons seized in Iraq appear to have been manufactured recently.

For the purposes of this chapter, 'illicit small arms and light weapons' are weapons that are produced, transferred, held, or used in violation of national or international law. The chapter uses the term 'illicit' rather than 'illegal' to include cases of unclear or contested legality. As detailed below, the term 'small arms and light weapons' is used in accordance with established Small Arms Survey practices, with some minor modifications. In this chapter, the term 'small arms' (alternatively, 'firearms') refers to the following items:

- revolvers and self-loading pistols;
- rifles and carbines;
- shotguns;
- sub-machine guns; and
- light and heavy machine guns.

The term 'light weapons' is used to refer to the following items:

- mortar systems of calibres of 120 mm or less;
- hand-held, under-barrel, and automatic grenade launchers;
- recoilless guns;
- landmines;
- portable rocket launchers, including rockets fired from single-shot, disposable launch tubes;
- portable missiles and launchers, namely anti-tank guided weapons (ATGWs) and man-portable air defence systems (MANPADS); and
- improvised explosive devices (IEDs).

This list includes some items not typically thought of as light weapons, including improvised explosive devices and mortars of calibres of 100 mm to 120 mm.[4] Since it is not possible to separate fully data on heavy machine guns—which the Small Arms Survey categorizes as a light weapon—from other types of machine guns, heavy machine guns are assessed with small arms in this chapter. The Survey's definition of portability also shapes the choice of light weapons retained for the chapter. The weight limit for light weapons and their ammunition is 300 kg—the maximum weight that can be transported on the chassis of a typical light vehicle—but 400 kg for towed weapons (Small Arms Survey, 2008, pp. 8–11).

Ammunition and accessories for the above-mentioned weapons are also included. Data on weapons parts is not analysed because it is beyond the scope of this study. Explosives are also excluded from the study.[5]

The definitions for 'armed conflict' and 'war' are borrowed from the Armed Conflict Dataset developed by the Uppsala Conflict Data Program and the International Peace Research Institute, Oslo. The dataset defines 'armed conflict' as 'a contested incompatibility which concerns government and/or territory where the use of armed force between two parties, of which at least one is the government of a state, results in at least 25 battle-related deaths' (UCDP, 2009, pp. 1–2). Armed conflicts are further divided into 'minor' armed conflicts with 'between 25 and 999 battle-related deaths in a given year' and 'wars', which are defined as armed conflicts with 'at least 1,000 battle-related deaths in a given year' (UCDP, 2009, p. 9). It is the latter type of conflict that is the focus of this chapter.

WAR WEAPONS: UNPACKING THE DATA

As explained above, this chapter aims to gain a better understanding of illicit weapons found in war zones through the compilation and analysis of data on the type, model, origin, and—whenever possible—source, age, and serviceability of illicit small arms in conflict zones. To this end, the project assembled datasets on illicit weapons in Afghanistan, Iraq, and Somalia. Together, they contain records on more than 80,000 illicit small arms, light weapons and ammunition, and accessories, along with hundreds of thousands of rounds of small-calibre ammunition.

UN monitors produced the most comprehensive data on illicit weapons in Somalia.

The most comprehensive source of data on illicit weapons in Somalia is a series of reports published by the United Nations Monitoring Group on Somalia and Eritrea (UNSEMG). The reports contain references to thousands of weapons whose possession or transfer violated the Somali embargo after its adoption in 1992 (see Box 10.3); they also document domestic arms transfers. From October 2005 through July 2011, the UNSEMG published eight reports; each of the first six reports, covering October 2005 to December 2008, typically records around 50 incidents of transfers and seizures. As a result of changes to the UNSEMG mandate and staff, subsequent reports take a different approach, focusing on a small number of cases and providing more detailed information on associated transfers. This chapter draws primarily on information provided in the first six reports, using the later ones to assist with the overall analysis.

UNSEMG reports contain useful information on the basic types, quantities, sources, and end users of weapons in Somalia. Unlike the sources used in this study on illicit arms in Afghanistan and Iraq, the UNSEMG reports do not provide information on the age, country of manufacture, model, or condition of illicit weapons. As a result, this information is mostly absent from the database compiled for this study. To help compensate for this knowledge gap, this project included interviews with explosive ordnance disposal specialists and other experts working in Somalia as well as an examination of relevant media reports, videos, photographs, and other research articles.

There is no consolidated source of comparable information on arms trafficking to and within Afghanistan or Iraq (see Boxes 10.1 and 10.2).[6] The chapter thus relies on data on weapons recovered from seized arms caches. The Small Arms Survey requested data from 27 governments and filed 20 freedom of information requests with the US and UK governments. The governments of Australia, the UK, and the United States responded by sending multiple datasets and documents.

Data on weapons seized in Iraq comes from the following sources:

- **Documents obtained under the US Freedom of Information Act (FOIA).** Requests filed under FOIA yielded records on approximately 100 weapon caches seized in 2008 and 2009. The data, which is among the most detailed released to date, identifies the type, quantity, model, and country of origin for many weapons and the age and serviceability of a smaller percentage. The records also include dozens of photos of the seized weapons. Detailed US Army reports on small arms and light weapons that were circulating in Iraq shortly after the US invasion in 2003 were also obtained under FOIA.

- **Appendix C of *Iranian Strategy in Iraq: Politics and 'Other Means'*** (Felter and Fishman, 2008, p. 76). As part of a study on Iranian involvement in the conflict in Iraq, researchers from the Combating Terrorism Center at West Point obtained data on the contents of 74 caches that reportedly contained weapons of Iranian origin.[7] Most of the data, which was obtained directly from the US military, identifies the model and country of manufacture of the seized weapons. The data also distinguishes 'caches' from 'explosive remnants of war' and 'post-invasion' from 'pre-invasion' caches; that is, caches assembled before and after March 2003.

- **The Defense Video & Imagery Distribution System.** Funded by the US Army, this information aggregation service contains thousands of media releases summarizing the contents of seized caches in Afghanistan and Iraq. It is among the largest public sources of data on seized caches and accounts for most of the caches analysed for this study.

Most of the data on illicit weapons in Afghanistan comes from a summary of seized caches obtained from the US Army under the Freedom of Information Act. The document contains records on weapons recovered from 493 caches seized from 2006 through 2008, 162 of which are fully redacted (completely withheld). Of the remaining caches, 236 contained small arms and light weapons. The records identify the date and time of seizure and the contents of the caches. The locations and most of the contextual information about the caches are redacted. The Survey also received data on the contents of several dozen caches from the Australian Defence Force and the British Ministry of Defence; this information relates to caches seized from January 2010 through April 2011 and September 2007 through September 2008, respectively. Many of the records identify the model, country of origin, and, in some cases, condition of the seized weapons.

While it is significantly better than most other publicly available information, the collected data is not perfect. The sample of seized caches in Iraq is not random, although any selection biases are at least partially mitigated by the size of the sample and the diverse nature of the seized caches. The data also reflects the US military's apparent tendency to inspect and report on weapons of Iranian origin more thoroughly than weapons from other countries. Also problematic is the fact that a large number of records containing US data on weapons seized in Afghanistan are redacted. The US Army does not explain why the records were singled out for redaction, and it is possible that data on certain types of weapons, weapons from certain countries, or weapons seized from specific end users was systematically withheld. The Australian Defence Force only includes weapons data that has been analysed by its Weapons Intelligence Teams. Emplaced IEDs, unstable military ordnance, homemade explosives, small arms ammunition, and other weapons that are not routinely analysed by these teams are not included. Concerns about reporting biases and data gaps are allayed to some extent by the notable overlap between weapon models and countries of origin in data from different sources.

Despite the cited limitations, the data assembled for this study is significantly more detailed and complete than other open source data, most of which is aggregated by weapon type and lacks information on models and countries of origin. Thus, the data sheds new light on illicit weapons in conflict zones. It also underscores the importance of government transparency. The reluctance of governments to release detailed data on seized weapons hinders analysis by private researchers and inter-governmental bodies, such as the UN arms embargo monitoring groups.[8] The example set by the Australian, British, and US governments is worthy of emulation.

ILLICIT SMALL ARMS AND LIGHT WEAPONS IN IRAQ

This section is based on a review of the contents of more than 1,100 arms caches seized in Iraq in January 2008–September 2009. It appears to be the most comprehensive and detailed public dataset on illicit small arms and light weapons in Iraq. The data not only sheds light on the types, models, and origins of black market weapons in Iraq, but it also reveals where caches are commonly stored and how they are concealed.

Overview of seized caches

The seized caches analysed for this study range in size from a few dozen rounds of rifle ammunition to several thousand weapons of various types and calibres. In all, more than 30,000 small arms, light weapons, and rounds of light weapons ammunition were recovered from the caches, along with more than 500,000 rounds of small-calibre ammunition. Table 10.1 summarizes the data by weapon category.

Mortars and mortar rounds were found in greatest quantities, accounting for more than half of the recovered weapons. Firearms were second, representing approximately 12 per cent of seized weapons, followed by rocket-propelled grenades and launchers (10 per cent), grenades and landmines (7 per

Table 10.1 Categories of weapon recovered from caches in Iraq, 2008-09

Weapon category	Percentage of total
Mortar systems and rounds	57
Firearms	12
RPG launchers and rounds	10
Grenades*	7
Landmines	7
IEDs	5
Recoilless rifles and rounds	2
Portable missiles (MANPADS and ATGWs)	<1
Rockets in disposable launchers	<1

Note: *This category includes hand grenades, project grenades, rifle grenades, and other (unspecified) grenades.
Sources: Felter and Fishman (2008); US ARCENT (n.d.); US CENTCOM (2010; 2011a)

Table 10.2 Weapons identified by country of origin recovered from arms caches in Iraq, 2008-09

Country of origin*	Percentage of total
Iran	29
Russian Federation	16
China	13
Iraq	12
Soviet Union	9
All other countries	21

Note: *The data on countries of origin is taken directly from the source and therefore reflects any errors or omissions in the source.
Sources: Felter and Fishman (2008); US ARCENT (n.d.); US CENTCOM (2010, 2011a)

cent each), and IEDs (5 per cent). Shoulder-fired missiles, rockets in single-shot, disposable launchers, and recoilless rifles and rifle rounds were found in much smaller quantities.

The caches also contained a wide array of improvised weapons, including rocket launchers. The rockets fired from these launchers, mostly designed for use with multiple-launch rocket systems and helicopter-mounted launchers, were found in large quantities in the caches.[9]

Approximately nine per cent of the seized weapons are identified by country of origin, representing at least 19 countries, with five accounting for most of these weapons (see Table 10.2).

It should be noted that, because of the US military's emphasis on tracking weapons from Iran, the percentage of Iranian weapons identified by origin vis-à-vis weapons from other countries is almost certainly skewed. It should thus not be assumed that 29 per cent of the weapons in the dataset are of Iranian origin. It is possible that the sample more closely reflects the actual distribution of weapons from other countries, but a definitive assessment would require a larger sample size.

Data on the age of illicit weapons in Iraq remains scant. Production years are available for less than two per cent of the seized weapons analysed and involve mostly Iranian weapons. While of little value for determining the age range of seized weapons generally, the data is germane to the highly contentious issue of Iranian weapons

acquired and used by armed groups in Iraq. According to the data, at least 29 per cent of the more than 900 seized weapons reportedly of Iranian origin were manufactured since 2003, and most were less than two years old when they were seized.

Assuming that the US assessment of the weapons' provenance is accurate, the data reveals the presence of a robust supply chain feeding weapons into Iraq. Whether this chain is linked to the Iranian government is unclear. Proving the existence of such linkages is extremely difficult, and even the best open source data rarely reveals the proximate source, let alone whether the shipments had the imprimatur of the Iranian government. Yet, if the Iranian government did not approve these transfers, they certainly entailed a surprisingly rapid diversion of often newly manufactured weapons from Iranian production or storage facilities, signalling a clear need for the review and strengthening of stockpile security practices in Iran.

The records on the seized caches also contain new information on the condition, or serviceability, of nearly 4,500 of the seized weapons. The data suggests that serviceability varies significantly from cache to cache. Some of the weapons are described as 'brand new' and were found in their original packaging. Other weapons, particularly those found underground, were highly corroded. The vast majority of weapons described as 'unserviceable' were mortar rounds, most of which were found in three large caches.

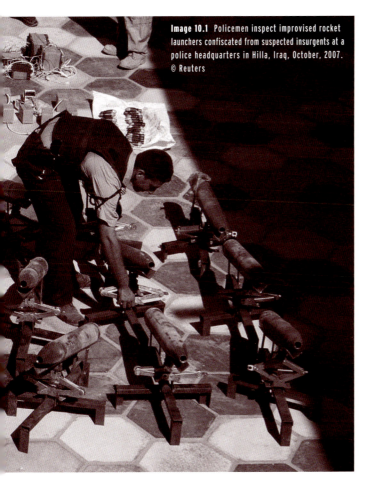

Image 10.1 Policemen inspect improvised rocket launchers confiscated from suspected insurgents at a police headquarters in Hilla, Iraq, October, 2007.
© Reuters

The contents of seized caches were not limited to small arms and light weapons. Dead bodies, stolen vehicles, insurgent platoon rosters, bottles of 'poison', tanks of nitrous oxide, and a kidnapped child were also discovered, as were currency and IED components from around the world.[10] These discoveries provide important clues about the age of the caches and the intentions of those who assembled them. The widespread presence of cellphones and other modern commercial technology suggests that many of the caches are fairly new. Al Qaeda propaganda, components for IEDs, attack plans, weapon training manuals, and kidnapping lists are indicative of the range of criminal and politically motivated violence pursued by consumers of illicit weapons in Iraq.

The caches were found in every conceivable location: bakeries, schools, mosques, tanneries, apartment buildings, fish farms, and even on the roof of the Ministry of Trade. Courtyards were common storage sites, as were houses abandoned during ethnic purges or periods of intense fighting. In some cases, displaced families returning to their homes encountered IEDs, explosives, and other hazards left behind by militia members and other armed groups.

The methods of storing and concealing caches are equally varied. Some were packed into 55-gallon drums and buried in fields or courtyards or under buildings. Others were hidden in houses and apartment buildings, sometimes under floorboards or behind walls. While searching the home of an individual accused of sectarian killings, US troops found two explosively formed penetrators in the false roof of a shower. Some of the seized caches were so well hidden or randomly placed that, without local assistance, US and Iraqi troops would not have found them. A good example is a cache of 9 mm ammunition and IED components hidden in a bag on the top of a broken elevator at the Kadamiyah hospital. Tips from local citizens were by far the most commonly cited way in which the caches reviewed for this study were found.

Small arms

Illicit small arms recovered in Iraq range from run-of-the-mill Kalashnikov-pattern assault rifles to flintlock pistols. Among the more exotic weapons found by US troops are World War II-era German rifles etched with Nazi swastikas, gold-plated AKM rifles, and a four-barrel 'duck's foot' pistol used by sea captains against a first wave of ship-boarding invaders (MilitaryNewsNetwork. 2009). Antique firearms recovered from the caches, some of which were in perfect working order, are worth as much as USD 25,000 to collectors (Minaya, 2007).

Most of the firearms recovered from arms caches are more commonplace, however. While data on firearms is less detailed than data on other weapons, it does provide some important insight into the types of illicit rifles, machine guns, and pistols available in Iraq. Interviews with government and private sector experts shed further light on these weapons.

Analysis of the seized caches is aided by two assessments of small arms and light weapons in Iraq as of 2004, both conducted by the US Army's National Ground Intelligence Center. The first is a handbook of small arms found in Iraq. The other is an assessment of small arms used by the 'anti-coalition insurgency'. The studies, which the Small Arms Survey obtained under the US Freedom of Information Act, provide a thorough overview of certain categories of small arms, light weapons, and ammunition circulating in Iraq as of mid- to late 2004.[11]

According to the US Army, most rifles in Iraq were Yugoslav and Iraqi-produced weapons of Soviet design. The handbook finds that '[t]he primary rifle to be encountered in Iraq is the Tabuk/AKM and folding stock variants' (NGIC, 2004a, p. 16). The Tabuk is a locally produced version of the Yugoslav M64/M70, which is a variant of the Russian AKM (p. 16).[12] Other rifles that were common in 2004 were the AK-47, the Belgian-designed FAL, and the German-designed G3, a version of which has been produced by several of Iraq's neighbours, including Iran, Saudi Arabia, and Turkey (NGIC, 2004a, p. 31; Jones and Ness, 2007, p. 173). Russian, Chinese, and Iraqi SKS rifles, the Russian Mosin-Nagant, the British Lee Enfield, German Mauser 98 rifles, French Lebel rifles, and French FAMAS F-1 rifles were also reportedly available, but in smaller numbers. According to the assessment

Table 10.3 **Firearms recovered from arms caches in Iraq, 2008–09**

Type	Model	Quantity
Firearm, other	Pellet gun	2
Pistol	Pistol (unspecified)	42
	Pistol (unspecified, 9 mm)	110
	Beretta	2
	Flintlock	1
	Glock/Glock 19	8
Handgun	Handgun (unspecified)	10
Revolver	Snub-nose .38 Special	1
Shotgun	Shotgun (unspecified)	9
Rifle	Rifle (unspecified)	196
	Kalashnikov-pattern rifles	2,605
	'Barno'/'Berno'/'Bruno'	6
	Enfield	4
	FAL	2
	'GC'	2
	M1	1
	M4	1
	Mauser	4
	RPK 'rifle'	2
	Seminov/Siminov/SKS	75
	Smirnov	2
	Type 85	1
Sniper rifle	Sniper rifle (unspecified)	74
	Dragunov/SVD	16
	Semenov/Siminov/Smirnov	3

of insurgent weapons, '[t]he German Mauser 98, the French Lebel, and the French FAMAS F-1 were not known to be in prewar Iraq' (NGIC, 2004b, p. 4).

Most sniper rifles and machine guns circulating in Iraq shortly after the US invasion were of Soviet design and produced in Iraq or in Eastern European countries. These include the Russian Dragunov and the Romanian FPK as well as the Iraqi Al-Kadissiya and Tabuk sniper rifles. Common light and medium machine guns include Iraqi-produced versions of Yugoslav weapons, both of which resemble the widely exported Russian RPK and PKM machine guns (NGIC, 2004a, p. 49). Other machine guns include the PKT, RPD, RPDM, and RP-46 (NGIC, 2004b, p. 4). The heavy machine gun most readily available to armed groups in Iraq was the DShK, although its use by insurgents was reportedly minimal (NGIC, 2004a, p. 55). The US Army's assessment did not look at handguns, shotguns, or sub-machine guns.

Data on firearms recovered from caches seized in 2008 and 2009 suggests that little changed following the publication of the US Army's reports in 2004. Table 10.3 provides an excerpt of the data. The majority of seized firearms were Kalashnikov-pattern assault rifles, more than 2,600 of which were recovered from the caches. The only other rifles found in significant quantities were variants of the Soviet-designed SKS. Combined, Kalashnikov-pattern and SKS rifles accounted for at least 90 per cent of all seized rifles and 99 per cent of the rifles identified by type. While little is known about the specific models of the rifles recovered from the caches, interviews with military officials and private sector analysts provide some clues as to which models are most common in Iraq. According to the Olive Group, which has had a continuous on-the-ground presence in Iraq since 2004, Soviet AKM rifles and their foreign variants are common, as are Chinese Type 56 rifles, Polish PMK, Czech VZ 58, and Iraqi Tabuk rifles.[13]

Type	Model	Quantity
Sub-machine gun	Sub-machine gun (unspecified)	5
	MP-5	4
	Port-Said	1
	SKS 'sub-machine gun'	1
	Sterling	2
	Tommy gun	1
Machine gun	Machine gun (unspecified)	54
	'BKC'	1
	Delta	1
	M60	1
	M86	4
	M240	1
	MG42	1
	PK	2
	PKC	107
	PKM	5
	RPK	39
	SGM light machine gun	1
	Anti-aircraft guns (unspecified)	2
	DShK	19
	ZPU	1
Total		3,432

Note: The figures reflect the contents of the 1,100 caches studied. Weapons are listed as reported in the source document.
Sources: Felter and Fishman (2008); US ARCENT (n.d.); US CENTCOM (2010; 2011a)

Among the most striking features of the data is the small number of assault rifles used by coalition forces from large troop-contributing countries. The four countries with the largest troop presence brought tens of thousands of M16, M4, L85, K2, and SC70/90 rifles into the country (Jones and Ness, 2007), yet the number recovered from arms caches appears to be minuscule. While a reporting bias cannot be ruled out, the authors uncovered no evidence of systematic exclusion from the cache data of weapons used by US and coalition forces. In fact, documents obtained under the Freedom of Information Act include photos and detailed descriptive information, including serial numbers, of US M4 and M16 rifles recovered from arms caches. Assuming that reporting biases are indeed minimal, the data would indicate that the stockpile security practices of US troops and the largest coalition partners are sufficiently robust to prevent widespread diversion of issued weapons.[14] However, the incomplete nature of available data precludes any firm conclusion on this issue.

> The apparent absence of Iranian-produced G3 rifles in the seized caches is noteworthy.

Also noteworthy is the apparent absence of Iranian-produced G3 rifles in the seized caches. Hundreds of Iranian mortar rounds, rockets, RPG rounds, and blocks of C4 explosives were reportedly seized from arms caches, but the only firearms identified in the data as Iranian are two sniper rifles seized in separate caches in August 2008. Caches seized before 2008 include some G3 rifles and other Iranian firearms, but their numbers appear to be comparatively small (MNC-I, 2007). It is possible that some of the unidentified rifles in the seized caches studied were G3s, although the likelihood of significant under-reporting seems small given the rigour with which the US military scrutinizes seized caches for weapons made in Iran and their tendency to highlight Iranian weapons in public summaries of seized caches.

Machine guns recovered from the caches also appear to be primarily of Soviet design, with Russian models or their foreign variants comprising at least 75 per cent of all seized machine guns and more than 98 per cent of machine guns identified by type. Only three Western-style machine guns are listed, one of which is identified as an MG42—a World War II-era German machine gun.

The data reveals little about handguns recovered from the seized caches, except that they are widespread and are mostly 9 mm pistols. This is consistent with information provided by the Olive Group, which claims that pistols are in high demand in Iraq.[15] The Olive Group identified six models that are commonly found in Iraq, five of which are 9 mm pistols.[16] The 2008–09 cache data also provides little information about the sniper rifles recovered from the caches. Dragunov rifles are mentioned most frequently, but most of the approximately 90 seized sniper rifles are not identified by type or country of origin, precluding meaningful analysis of illicit sniper rifles in Iraq.

Light weapons[17]

The following section summarizes and analyses data on nearly 27,000 light weapons and ammunition seized from arms caches in 2008 and 2009. As it is beyond the scope of this study to explore all light weapons, the section focuses on the three categories for which the most detailed data was obtained: RPGs and launchers, shoulder-fired rockets and launchers, and portable missile systems.

Rocket-propelled grenades and shoulder-fired rockets

More than 2,200 RPGs and 410 RPG launchers were recovered from the seized caches.[18] The majority of the items for which detailed information was available appear to be Soviet-designed models first fielded in the 1960s and 1970s;[19] most were in circulation in Iraq in 2004 (NGIC, 2004a). Interviews and anecdotal accounts indicate armed groups have acquired more modern—and more effective—RPGs, but in very limited numbers. Most RPGs and launchers identi-

Image 10.2 Rocket-propelled grenade launchers and jacks recovered during a raid on a weapons cache are laid out at Camp Sparrow Hawk, Iraq, September 2009.
© DVIDS

fied by model appear to be older Soviet and Chinese-designed models that are of limited utility against modern tank armour but can still be deadly when used against personnel and unprotected or lightly protected vehicles.

The US Army's assessments of weapons found in Iraq in 2004 provide a detailed overview of RPG rounds. According to the *Small Arms Handbook,* the RPG-7 was 'the primary portable rocket launcher in use by the insurgency' (NGIC, 2004a, p. 59). A version of the RPG-7, the Al-Nassira, was produced in Iraq under Saddam Hussein's regime. The RPG-16, a close cousin of the RPG-7 used by Russian airborne units, was also reportedly available, as were the RPG-18 and the RPG-22, which differ from the RPG-7 in that they are fired from single-shot, disposable launchers (NGIC, 2004a, pp. 58, 67). Smaller numbers of other rocket launchers were also reportedly available. The most notable is the Armbrust, a single-shot, disposable rocket manufactured in Singapore and Germany. The Armbrust is potentially attractive to armed groups because it is stealthy, versatile, and less likely to attract counter-fire than the RPG-7. It has no backblast, so it can be fired from enclosed spaces, has no firing signature, and is 'quieter than a pistol shot', according to the US Army (NGIC, 2004a, pp. 73–74; Jones and Ness, 2007, pp. 487–88).

RPG-7 launchers and their foreign variants fire a variety of rounds. The two main types are anti-personnel rounds and high-explosive, anti-tank (HEAT) rounds. Anti-tank rounds are further divided into models with one (unitary) warhead and those with two (tandem) warheads, the latter being much more effective against modern vehicle armour.

Most of the RPG rounds circulating in Iraq in 2004 were unitary HEAT rounds, with the most common being PG-7 and PG-7M rounds produced in Bulgaria, Iran, and the Russian Federation. Only one tandem HEAT round—a version of the Iranian NADER spotted in a news video—had been confirmed to be in Iraq as of 2004, according to the US Army (NGIC, 2004b, p. 5).

Little appears to have changed since 2004. Of the approximately 1,029 RPG rounds recovered from the seized caches that were identified by model, most were reportedly PG-7 anti-tank or OG-7 anti-personnel rounds or their Chinese or Iranian equivalents.[20] Also present in smaller numbers were the slightly improved PG-7M and the Chinese Type 69 Airburst (DZGI) rounds. Most of the anti-tank rounds identified by model have limited ranges and low armour penetration. The Iranian NADER round, several dozen of which were found in the caches, is thought to be of particularly poor quality by the US military (NGIC, 2004a, p. 85). The only higher-end anti-tank RPG rounds identified in the cache data are two PG-7L[21] rounds found in northern Iraq. The PG-7L, which was not known to be in Iraq in 2004, is described by the US Army as the 'top of the line' unitary HEAT round and is capable of penetrating up to 500 mm of armour—nearly twice the armour penetration of the basic PG-7 round (NGIC, 2004a, p. 130).[22] Given their

Table 10.4 Rocket-propelled grenades and shoulder-fired infantry rockets seized from arms caches in Iraq, 2008–09

Type	Model*	Quantity	Countries of origin*
RPG round	RPG-7 (unspecified)	88	Unspecified
	PG-7	325	Bulgaria, China, Iran, Iraq, Russian Federation, Soviet Union
	PG-7G	123	Bulgaria, China, Iraq, Russian Federation, Soviet Union
	PG-7M	86	Bulgaria, Russian Federation, Soviet Union
	PG-7L	2	Unspecified
	NADER	9	Iran
	PG7-AT-1	47	Iran
	PG-7-A2	4	Iran
	OG-7	232	Bulgaria, Iraq, Russian Federation, Soviet Union
	OG-7V	162	Bulgaria
	Type 69	7	China
	Type 69-1	1	China
	Type 69 Airburst	31	China
Single-shot rocket	RPG-18	11	Soviet Union
	RPG-22	1	Unknown
	AT-4	1	Unknown

Notes: *As identified in the source document.
The figures reflect the contents of the 1,100 caches studied.
Sources: Felter and Fishman (2008); US ARCENT (n.d.); US CENTCOM (2010; 2011a)

extremely limited numbers, however, the PG-7L rounds do not in themselves signal a clear improvement in insurgent capabilities as compared with 2004. Data collected for this study[23] and interviews with experts indicate that the number of illicit tandem HEAT and thermobaric rounds for RPG-7 launchers available in Iraq is minimal.[24] Table 10.4 summarizes the cache data for RPG rounds and shoulder-fired infantry rockets.

US troops found at least 38 optical sights for RPG launchers in the seized caches, including one night vision sight. Nearly half of the sights were found in their cases, suggesting that at least some were in working order. The use of optical sights can increase the effective range of anti-tank rounds from 200–300 m to 500 m (NGIC, 2004a, p. 58). However, RPG sights are reportedly difficult to use and require extensive training (Shea, 2006). It is unclear whether armed groups in Iraq have received such training.

Also found in the seized caches were a limited number of single-shot, disposable rockets, most of which are early model US, Soviet, and Yugoslav weapons that are comparable in range and armour penetration to most of the available RPG-7 rounds. A single 'AT-4' was recovered from a cache in eastern Baghdad in June 2008. It is unclear whether the recovered item was the modern Swedish-designed shoulder-fired rocket deployed by several NATO countries or the Soviet-designed 9K11 Fagot wire-guided anti-tank missile first fielded in 1970, since the designation 'AT-4' is used to refer to both weapons. If the item was the Swedish rocket, it is the only contemporary Western rocket identified in the cache data.

Image 10.3 A weapons cache discovered by US and Iraqi soldiers in Mosul, March 2007.
© DVIDS

Notably, there is no mention of later-model RPG series rockets in the data. Widespread proliferation of these weapons, which can penetrate modern tank armour, would pose a significant threat to armoured vehicles and other hardened targets in Iraq. These rockets include the Russian RPG-29 Vampir, a reusable rocket launcher that fires a highly effective tandem HEAT round. In the 2006 war in Lebanon, the RPG-29 was one of the few anti-tank weapons used by Hezbollah that proved capable of defeating the advanced reactive armour on Israel's Merkava 4 tanks (Ben-David, 2006).

While RPG-29s are not identified in the caches studied, armed groups in Iraq have acquired an unknown but presumably small number of them,[25] of which one was used to disable a British Challenger tank (Fox and Sheikhly, 2007; Rayment, 2007). Several videos of purported RPG-29 attacks on US tanks have also been posted on file-sharing websites (LiveLeak, n.d.). The US military claims that Iran is probably the source of the RPG-29s (AFP, 2006), which were reportedly distributed in limited quantities and 'appear to have been used in a "trial" capacity', according to the Olive Group.[26] The absence of references to RPG-29 rounds (PG-29V) in the seized caches is consistent with claims that distribution of this weapon was extremely limited.

Data on RPG attacks compiled by the US Defense Manpower Data Center indicates that they were significantly less effective than attacks with other weapons. The data reveals that RPGs were responsible for 53 deaths and 773 injuries in Iraq from 2003 through early 2012, a tiny fraction of the 35,000 deaths and injuries[27] resulting from hostile events documented during this time period. In contrast, 'explosive devices'—presumed to include IEDs—reportedly killed nearly 2,200 people and injured more than 21,000 during this period. RPG attacks also inflicted fewer casualties than firearms and indirect fire[28] (DMDC, 2012). It is unclear whether poor tactics, obsolete technology, or a combination of both accounts for the comparative ineffectiveness of RPG attacks in Iraq, but ready access to more effective anti-tank rockets almost certainly would have resulted in more casualties.

Although the percentage of seized RPG rounds identified by country of origin is relatively small, the data sheds some light on this issue. Of the roughly 2,200 RPG rounds found in the seized caches, 525 are identified by country of manufacture. Bulgaria and the Soviet Union are the most common sources, followed by Iran, China, the Russian Federation, and Iraq.

Portable missiles

Data collected for this study provides important new information on illicit man-portable air defence systems and anti-tank guided weapons in Iraq. MANPADS are lightweight, portable surface-to-air missiles that are often fired from a launch tube that rests on the shoulder of the operator. ATGWs are lightweight missiles that are used to attack armoured vehicles, bunkers, buildings, and exposed personnel. Both missile types are widely exported and have been used by armed groups worldwide.

Portable missiles comprise less than 0.1 per cent of the small arms and light weapons recovered from the caches reviewed for this study. Twenty of the 21 missiles and launchers for MANPADS identified by model were early model Soviet-designed systems, as were all of the seized ATGWs identified

Table 10.5 **Portable missiles and launchers recovered from arms caches in Iraq, 2008–09**

Model	Quantity
SA-7 missile	17
SA-7 launcher	2
SA-14 missile	1
SK-10 launcher	1
9M14 Sagger missile	9

Note: The figures reflect the contents of the 1,100 caches studied.

Sources: Felter and Fishman (2008); US ARCENT (n.d.); US CENTCOM (2010; 2011a)

by model.[29] The remaining item was an SK-10 launcher (gripstock), which is part of the Chinese QW-1 MANPADS. Table 10.5 lists the portable missiles seized in the 1,100 caches studied.

The only ATGWs identified in the cache data are early model, wire-guided Sagger missiles, which are difficult to operate. In its 2004 assessment of insurgent weapons, the US Army identifies four additional models of missiles used with ATGWs in Iraq, including later model systems that are much easier to use, namely the Russian Fagot and Konkurs and the Euromissile MILAN. While the Iraqi government stockpiled 'significant inventories' of these missiles prior to the 2003 invasion, most of the launchers were vehicle-mounted (NGIC, 2004b, p. 7). The lack of portable launchers may explain, in part, why armed groups adopted the IED as their primary weapon for use against coalition vehicles. This, in turn, may help to explain the large number of IEDs vis-à-vis ATGWs in the seized arms caches.

The seized MANPADS are also noteworthy for several reasons. First, the models of missiles—and the ratios of the different models in the seized caches—are largely consistent with other open source accounts of black market MANPADS in Iraq. US officials have identified four models of MANPADS that armed groups in Iraq have acquired: the Soviet-designed SA-7, SA-14, and SA-16 (Baker, 2007), and the Misagh-1 (Fox and Sheikhly, 2007), which is reportedly a variant of the Chinese QW-1 produced in Iran. Two of these models—the SA-7 and the SA-14—were found in the seized caches studied. While estimates of Iraq's pre-war MANPADS inventory are imprecise, available information suggests that the regime stockpiled far fewer SA-14s and SA-16s than SA-7s, which may explain why far fewer of these systems have been found in seized arms caches (SIPRI, n.d.). The Misagh-1 has also been found in much smaller quantities (Schroeder, 2008).

The US military report on the seized SK-10 launcher is among the most notable documents obtained as part of this study (US CENTCOM, 2011a). The associated QW-1 MANPADS is a second-generation system first unveiled in 1994. It is faster, has a longer range, and is able to engage targets at higher altitudes than the other MANPADS recovered from the caches; moreover, it is 'all aspect', meaning that it can engage targets from any direction (O'Halloran and Foss, 2010, pp. 9, 39). The report contains a photograph of the seized launcher—the first publicly available hard evidence of a QW series MANPADS in Iraq.[30]

The photograph is a significant contribution to the ongoing debate over Iranian-supplied weapons in Iraq. Iran is a major importer of Chinese weapons and is believed to produce a copy of the QW-1 called the Misagh-1 (SIPRI, n.d.).[31] According to the Stockholm International Peace Research Institute (SIPRI), Iran produced an estimated 1,100 Misagh-1 MANPADS in 1996–2006 (SIPRI, n.d.). The US military claims that Iran has provided Misagh-1 missiles to armed groups in Iraq but has offered little hard evidence to support this claim (Fox and Sheikhly, 2007). Information from other sources is scant. The most significant is an amateur video of what appears to be a Misagh-1 attack perpetrated by members of Kata'ib Hizballah, an armed group in Iraq with alleged ties to Iran (Schroeder, 2010). While impossible to authenticate, the video was the strongest publicly available evidence of Chinese-designed MANPADS in Iraq until the release of the report on the seized SK-10 gripstock.

There is some circumstantial evidence linking Iran to the launcher, including the cache in which it was found. In addition to the launcher, the cache reportedly contained nine blocks of Iranian C4 explosives and four recently manufactured Iranian RPG rounds (US CENTCOM, 2011a). Conclusively linking the launcher to Iran is difficult, however. There are other potential sources of QW-1 missiles, and it is unclear whether Iran imported any gripstocks directly from China or if its own gripstocks are marked with the same designations. Answers to these questions are necessary, but not sufficient, for linking the Iranian government to the gripstock, which may have made its way to Iraq through

> The only ATGWs identified in the cache data are early model, wire-guided Sagger missiles.

Box 10.1 Trafficking in small arms and light weapons in Iraq

Arms trafficking in Iraq takes many forms. Actors may circulate weapons internally or traffic them across Iraq's borders with Iran and Syria. To a lesser extent, arms dealers have smuggled arms into Iraq from the United States and Europe. Using open source information, this box describes some of this trafficking activity, including the routes and methods used by smugglers.

Of the alleged sources of weapons trafficked into Iraq, the most politically sensitive is Iran. US officials have repeatedly accused the Iranian government of facilitating this trafficking, a charge that Iranian officials deny. They claim that any Iranian weapons in Iraq are either vestiges of the Iran–Iraq war or were purchased on the open market. Data on seized weapons compiled for this study includes references to dozens of weapons reportedly manufactured in Iran. In addition, there are reports of arms traffickers caught smuggling weapons from Iran into Iraq, although they are difficult to confirm (AP, 2007).

In addition to Iran, there are several reports identifying Syria as an entry point for illicit weapons into Iraq. Arms smugglers detained by the US military have reportedly 'confessed to bringing weapons, foreign fighters, and money for [insurgent groups] across the Syrian border into Iraq' (AP, 2005). Some sources contend that Iraqi insurgents continue to obtain arms through the channels that Saddam Hussein utilized to bypass the arms embargo, some of which go through Syria and involve the provision of false end-user certificates (Intelligence Online, 2004). Nonetheless, there is little open source evidence indicating that the Syrian government is directly involved in trafficking.

In addition to cross-border trafficking, many arms transfers occur within Iraq. Former soldiers have sold some of the weapons that now circulate illicitly in Iraq. Recent reports also indicate that members of the reconstituted Iraqi army and police have sold their weapons on the black market, including those procured with US government funding. In 2006, C.J. Chivers of *The New York Times* wrote that, among the 370,000 weapons purchased by the United States for Iraq's security forces, three of the principal types could readily be found in Iraqi shops and bazaars (Chivers, 2006). In 2007, arms dealers reportedly told Reuters that many Iraqi soldiers and policemen were selling their pistols on the black market. In the same report, an Iraqi soldier explained that this decision was motivated by the need to 'feed my family until I find a safer job' (Rasheed and Colvin, 2007). Data on the pistols seized from insurgent caches in 2008–09 is too vague to assess these claims.

Open source reports indicate that international and Iraq-based traffickers utilize a variety of routes. In their paper, *Iranian Strategy in Iraq*, Joseph Felter and Brian Fishman describe several common routes for smuggling arms from Iran into Iraq, stating that traffickers often receive shipments of explosively formed penetrators from Iran through Amara, Basra, and Diwaniyah, transfer them to Sadr City, and then distribute them to Special Groups in outlying provinces (Felter and Fishman, 2008). Militants mention the marshy areas next to the Iraqi border town of Qal'at Sali as a common trafficking route for small boats carrying illicit weapons. Traffickers have also been detained in the village of Qasarin near the border with Iran (AP, 2007). Media reports mention several popular locations for selling illicit arms inside Iraq, including the Meridi market in Sadr City and the arms bazaars in Amara (Hasni, 2003; Sherwell, 2004).

When trafficking weapons, smugglers and insurgents utilize a variety of transportation and concealment methods. US intelligence reports and interviews with Iraqi gun-runners reveal that weapons are smuggled in trucks hauling cigarettes, sheep, cement, crates of fruit, or propane gas cylinders. Other shipments are carried by boat via the marshes of Qal'at Sali (Felter and Fishman, 2008). Smugglers unload the boats at various points along the shoreline and place the weapons in the beds of pickup trucks, covering them with reeds. Many smugglers use local dirt roads to evade checkpoints manned by coalition forces. A 2004 interview with an Iraqi arms trafficker illustrates the precision and planning that characterizes many smuggling efforts:

> Travelling with fake documents, he is guided by spotters who drive ahead in cars and check with street observers who advise them on which routes are guarded and where diversions are needed [. . .]. Once in Baghdad, the gun runners offload the lorries and distribute the arms in cars or minibuses, relying on the laxness of the Iraqi police to see them through (Hider, 2004).

US Marines claim that Sunni insurgents who traffic arms and fighters into Fallujah have used ambulances and aid trucks to transport weapons between neighbourhoods (Navarro, 2004). US marines have, in fact, found false bottoms in vehicles filled with assault rifles and RPG launchers as well as an anti-aircraft gun hidden in a truckload of aid (IWPR, 2005). In April 2008, Iraqi policemen identified an abandoned truck stacked with hay that had been transporting arms through the area. The shipment was reportedly 'moved up the al-Kut highway to be broken down into smaller packages for movement to Baghdad' (Sowers, 2008).

Author: Chelsea Kelly

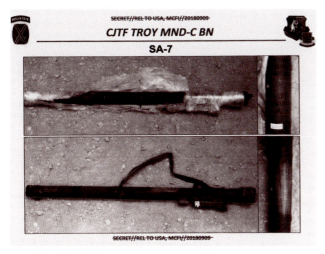

Image 10.4 SA-7b MANPADS recovered from an arms cache in Iraq.

other means. This analysis of the SK-10 and its possible origins provides yet another example of the difficulty of tracking illicit weapons back to their sources using publicly available data.

Also notable is the limited number of models of MANPADS found in the seized caches. Twenty countries have produced more than 30 different models of MANPADS (USGAO, 2004), yet only five have been identified in Iraq, and the vast majority appear to be Soviet-designed systems first fielded in the 1960s and 1970s. The data also suggests that the number of missiles seized per cache is shrinking, which may indicate that the number of MANPADS looted from government depots in 2003 and still outside of government control may be dwindling.

In the first few years after the Baathist regime fell, coalition troops regularly encountered caches containing dozens of missiles. In June 2003, US forces seized 87 SA-7s from a terrorist training camp in Anbar Province (Wright and Reese, 2008). Three months later, a cache of 23 SA-7s was found near Tikrit (First Battalion, 2003). Seizures of this size continued for years, revealing just how many looted MANPADS had entered the black market in 2003. By 2008, however, the number of missiles seized per cache had dropped significantly; none of the caches studied contained more than two missiles identified as MANPADS. It is possible that larger caches were found and not reported, but there is little evidence of a shift in reporting practices from previous years. The most likely explanation is that insurgent usage, buy-back programmes, and cache seizures have reduced the number of illicit MANPADS available in Iraq to levels more consistent with other war zones.

Finally, the data provides a rare, if limited, glimpse at the age and serviceability of illicit MANPADS in Iraq. Documents obtained from the US government contain photos of the markings on seized MANPADS, few of which are publicly available (see Image 10.4). The markings on one of the seized missiles indicates that it is a Soviet SA-7b manufactured in 1978, which suggests it is one of the thousands looted from Iraqi stockpiles in 2003. According to SIPRI, the Iraqi government imported 6,500 SA-7 missiles from 1975 to 1986, placing the manufacture of the photographed missile within the delivery timeframe (SIPRI, n.d.). The date is important in determining whether the missile and others like it are operational. Since the reported shelf life of most MANPADS is roughly 10–20 years (Schroeder, 2011; Hunter, 2001), it is possible that the missile was either no longer operational or would not have performed as intended. On the other hand, the documents also highlight the danger of assuming that MANPADS are unserviceable simply because they are first-generation systems. Documents obtained from the US military contain serviceability assessments for five of the seized MANPADS. Explosive ordnance disposal experts deemed three of the five missiles 'serviceable'. While this sample size is too small to extrapolate to the broader population of illicit MANPADS in Iraq, the documents do raise the possibility that at least some of the loose SA-7 MANPADS are operational despite their advanced age.

ILLICIT SMALL ARMS AND LIGHT WEAPONS IN AFGHANISTAN

This section is based on a review of three previously unreleased datasets. They contain:

1. records compiled by the US Army that summarize the contents of 331 caches seized in Afghanistan from 2006 through 2008;
2. information on weapons recovered from 82 caches seized in the Afghan province of Uzurgan from January 2010 through April 2011 and analysed by Weapons Intelligence Teams of the Australian Defence Force; and
3. information on 409 weapons recovered from arms caches by British forces in Helmand Province from September 2007 to September 2008.

Even though different governments compiled the datasets, and they cover different time periods, the similarities are often striking. The following section summarizes this data, supplementing it with open source accounts from journalists and government officials.

Overview of seized caches

Nearly 9,600 small arms and light weapons and approximately 200,000 rounds of small arms ammunition were recovered from the seized caches summarized in the data provided by the US Army and the British Ministry of Defence.[32] As in Iraq, mortars and mortar rounds were the most frequently encountered items, but here they only accounted for about 40 per cent of all seized weapons. Roughly 27 per cent of the weapons were grenades (hand, rifle, and spin-stabilized). RPGs and recoilless rifles and rounds were found in roughly equal numbers, each accounting for about 13 per cent of seized weapons. Notably, firearms accounted for only four per cent of all seized small arms and light weapons as compared to more than ten per cent of items seized in Iraq. Also significant is the comparatively small number of seized IEDs. Only 100 IEDs were reportedly recovered from the caches in Afghanistan, 77 of them in just three caches. In contrast, nearly 1,700 IEDs were recovered from 287 out of the 1,100 Iraqi caches studied.

Approximately 1,157 items (12 per cent) of small arms and light weapons recovered from the seized caches are identified by country of origin (see Figure 10.1). While this sample size is too small to extrapolate beyond the 1,157 items, the predominance of Chinese and Eastern Bloc weapons in the sample is noteworthy. Nearly 70 per cent of the weapons in this sub-set were identified as 'Chinese'. These items include hundreds of mortar and recoilless rifle rounds, dozens of tactical rockets, grenades, and RPG rounds, and a small number of firearms and landmines. 'Russian' and 'Soviet' weapons combined account for about 17 per cent of the seized weapons, followed by weapons from Pakistan, which comprised about seven per cent of the seized items. Given the repeated accusations of Iranian support for armed groups in Afghanistan, the data contains surprisingly few references to weapons of Iranian origin, namely one single anti-tank mine and four mortar rounds. The number of Western weapons identified in the data is also small. Combined, weapons made in Australia, Europe, and the United States account for six per cent of the 1,157 weapons. Very little data on the condi-

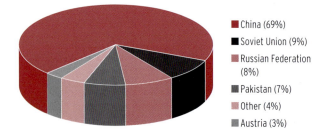

Figure 10.1 **Seized weapons by country of origin in Afghanistan**

- China (69%)
- Soviet Union (9%)
- Russian Federation (8%)
- Pakistan (7%)
- Other (4%)
- Austria (3%)

tion and date of manufacture was provided, precluding any meaningful analysis of the serviceability and age of the seized weapons.

As in Iraq, items other than weapons found in the caches provide clues regarding when the caches were assembled and the interests of those who assembled them. The items include cellphones, voter registration cards, brass knuckles with spikes, an arms sales ledger, narcotics, and IED components, including a 'remote control fob trigger device'. In a cache discovered in April 2007, US forces found 'school teaching material [including] weapons manuals [that] appears to be focused on children or young adults' (US ARCENT, 2011).

Small arms

The data on seized caches includes records of 217 firearms recovered by the US military and an additional 169 firearms seized by the Australian Defence Force and the British military (see Table 10.6). The seized weapons range from early 20th-century Lee-Enfield rifles to an AKM assault rifle 'never before fired' and still covered in packing grease. As in Iraq, several novelty firearms were also found, including a musket, a pen gun, and a World War-II-era M3 submachine gun. The caches contained few surprises; nearly all of the types and models of seized firearms have been available in Afghanistan for many years and have been documented elsewhere (Bhatia and Sedra, 2008, pp. 65–66; Chivers, 2011).

Rifles accounted for more than half of all seized firearms, with Kalashnikov-pattern assault rifles being the most numerous. A smaller but still significant number of bolt-action rifles were also found in the caches; most of them appear to be chambered for the .303 round used by various iterations of the British Lee-Enfield rifle, which was first produced more than a century ago. This mix of seized rifles documented by the United States is roughly the same as that documented by Australia and is consistent with accounts from journalists (Chivers, 2011).

The large number of .303 bolt-action rifles is a departure from caches found in Iraq and merits closer attention since they illustrate the widespread and continuous proliferation of firearms technology and the resulting difficulty of definitively identifying the sources of seized firearms. Despite their vintage, Lee-Enfield-style rifles have many potential sources for Afghans: local supplies left over from the US government's covert war against the Soviets in the 1980s; versions of the rifles produced in Pakistan and India; and 'Khyber Pass copies'—that is, local copies of the Lee-Enfield and other bolt-action rifles made by Pakistani gunsmiths near the Afghan border.[33]

Too little is known about the seized rifles to determine from which of the aforementioned sources they came, although on-the-ground reporting by Chivers provides some clues. In 2011, Chivers was given access to five Lee-Enfield rifles and other weapons seized by US troops in eastern Ghazni Province. At least two of the five Lee-Enfield rifles had anomalous markings. One of the weapons, a Short Magazine Lee-Enfield rifle, had a date stamp of 1881—35 years before that model was first fielded. The second rifle had a 'weird and wonderful' logo that was clearly different from the factory stamp usually found on that particular model. Both rifles, it turns out, were 'Khyber Pass' copies (Chivers, 2011).

Despite their apparent ubiquity, .303 bolt-action rifles are not widely used by armed groups in Afghanistan, according to Chivers. He speculates that a lack of ammunition may be a major reason (Chivers, 2011), a theory that is supported by caches studied in this analysis. Of the tens of thousands of rounds of small-calibre ammunition found in the seized caches, only one 'bag' of ammunition for .303 calibre bolt-action rifles was identified.

Most of the machine guns seized in the caches are widely exported models of Soviet design. More than half are PKMs, which are among the most widely exported machine guns in the world. Smaller numbers of RPD and RPK light

The large number of .303 bolt-action rifles is a departure from caches found in Iraq.

Table 10.6 Firearms seized in Afghanistan by Australian, British, and US troops between 2006 and 2011

Seized weapons		Quantity		
Type	Model*	US Army (January 2006–December 2008)	Australian Defence Force (January 2010–April 2011)	UK Ministry of Defence (September 2007–September 2008)
Anti-aircraft gun	DShK	6	–	5
	ZGU-1	1	–	–
	ZUK	1	–	–
Handgun	Makarov pistol	–	4	–
	Pistol (unspecified)	10	4	1
	Revolver (unspecified)	1	–	–
	Tokarov pistol	–	1	–
Machine gun	DPM machine gun	–	–	1
	Gernov	1	–	–
	M382	–	1	–
	Machine gun (unspecified)	5	1	1
	PAPASHA	4	–	–
	PK	1	–	–
	PKM	17	2	7
	RPD	2	2	–
	RPK	2	1	–
	WW2-era M3 sub-machine gun	–	1	–
Rifle	AK-47/unspecified Kalashnikov-pattern assault rifles	69	32	37
	AK-47M/AKM	–	2	–
	Bolt action (unspecified)	10	4	–
	.303 calibre (model unspecified)	–	15	–
	Dragunov sniper rifle	–	–	1
	Enfield/Enfield-style	6	4	1
	Mauser	–	1	–
	SKS	1	1	–
	Sniper (unspecified)	1	–	1
	Unspecified	15	–	–
Shotgun	Shotgun (unspecified)	7	2	35
Other	Golden gun type weapon	–	–	1
	Guns	55	–	–
	Musket	1	–	–
	Pen gun	1	–	–
Total		217	78	91

Note: *As identified in the source document.

Sources: Australian DOD (2011); UK MOD (2011); US ARCENT (2011)

machine guns were also found, as were several DShK heavy machine guns. Shotguns and handguns were also found in the caches.

While roughly 20 per cent of the seized firearms are identified by country of origin, most are from a single cache of 55 Pakistani 'guns' seized in September 2008. Of the 18 other weapons identified by country of origin, all but two—an unspecified British rifle and a Spanish pistol—are labelled as either Russian or Chinese. This is consistent with historical assessments of Afghanistan's weapons stocks (Bhatia and Sedra, 2008, p. 65) and recent on-the-ground cataloguing of seized caches (Chivers, 2009; 2010).

As in Iraq, very few US- and European-designed rifles and machine guns are listed in the contents of the seized caches. It is possible that some of the firearms not identified by type are of Western origin, but even if half of the unidentified firearms were manufactured in the United States or Europe—which is unlikely—they would still only constitute a small share of seized firearms. Furthermore, only a small percentage of the ammunition recovered from the seized caches were 5.56 mm rounds used in most US and Western European rifles. Of the 283 identifiable ammunition magazines found in the caches, only four were identified for M4 and M16 rifles.[34]

The data reveals little about the security of the Afghan National Security Forces (ANSF) inventories. Most of the firearms in ANSF arsenals are Kalashnikov-pattern assault rifles and PK series machine guns—the same types of firearms used by insurgents (*Jane's World Armies*, 2012).[35] The media regularly produce accounts of weapons that were issued to Afghan soldiers or police officers and subsequently seized from insurgents or recovered from arms caches. Among the more compelling examples is Chivers' analysis of 30 seized Kalashnikov magazines, which finds that 'at least 17 of the magazines contained ammunition identical to the cartridges issued by the United States to Afghan government forces' (Chivers, 2010).[36] Whether this account is representative of insurgent holdings more generally is difficult to determine, however, since little aggregate data on seized ANSF weapons has been made public.

Few US- and European-designed rifles and machine guns are listed in the contents of the seized caches.

Light weapons

Despite persistent concerns about the acquisition of new, deadly weapons by the Taliban (Danahar, 2007; Vanden Brook, 2008), an analysis of the seized caches and other reports suggest that most illicit light weapons in Afghanistan are primarily older models that have been available in Afghanistan for decades. While they still pose a threat, few, if any, of the Taliban's weapons appear to be technologically sophisticated game-changers comparable to the Stinger missiles acquired by their predecessors in the 1980s.

Data on light weapons recovered from the seized caches that were studied suggests that many are early generation systems of Chinese, Russian, or Soviet design. Of the more than 1,100 light weapons found in the caches that are identified by country of manufacture, almost 90 per cent are identified as Chinese, Russian, or Soviet. Many of these systems were first fielded in the 1960s or early 1970s. For example, most of the models of seized RPG launchers and rounds identified in the data were fielded at least 40 years ago,[37] and several dozen of the rounds were developed in the 1950s. As in Iraq, no tandem warhead or thermobaric rounds are listed.

Furthermore, literature on previous conflicts in Afghanistan suggests that the models of RPGs—and most of the other seized light weapons—were first introduced in the country many decades ago. The United States and its allies supplied thousands of Chinese and Soviet weapons of various models to the Mujahideen in the 1980s (Bhatia and Sedra, 2008, p. 44). Many of the same models have been recovered from the seized caches. Thousands more weapons were acquired from the Soviet-backed regime, either on the battlefield or from government depots when the regime collapsed in 1992 (p. 49). Whether the recently seized weapons themselves date back that far is not clear, but the data does suggest that the types of weapons—and their technological sophistication—has changed little over the last 20 years.

Equally striking is the apparent absence of portable missiles and other technologically sophisticated light weapons in the seized caches. Contrary to claims of large-scale trafficking of 'sophisticated weapons', there is little evidence—in the cache data or elsewhere—of widespread acquisition and use of MANPADS or ATGWs by armed groups in Afghanistan (Jalalzai, 2011). No MANPADS or ATGWs are identified in the data, and the only component of such weapons is a single battery (power unit) for a Stinger missile launcher found in November 2006. A review of other open source documentation suggests that illicit portable missiles were more commonplace during the era when the Taliban were in power. A US Defense Department summary of weapons seized by US troops during the first half of 2002 includes 319 shoulder-fired surface-to-air missiles (Rhem, 2002). The Defense Department does not identify the model of the missiles, but subsequent statements by US military officials indicate that most MANPADS seized during this period were first-generation Chinese HN-5 missiles (Murphy and Freedberg, 2002). MANPADS continue to circulate in Afghanistan, but the quantities appear to be extremely limited. In 2007, US journalist Philip Smucker received photos of HN-5 MANPADS from a Taliban 'weapons expert', who claimed that the missiles were recent acquisitions (Smucker, 2007; see Image 10.5). Markings on the launch tubes of these missiles appear to confirm them as HN-5s, which are comparable to first-generation Soviet SA-7s. Few, if any, MANPADS have been seized from Taliban arms caches in recent years (Schroeder, 2010).

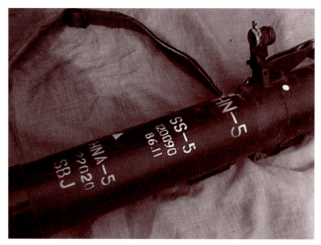

Image 10.5 HN-5 MANPADS reportedly acquired by the Taliban, 2007. Courtesy of Philip Smucker

Illicit anti-tank missiles appear to be a bit more common than MANPADS, but available data suggests that stocks are limited and most are older models. Coalition forces reportedly found 17 missiles in a single cache in August 2005 (AFPS, 2005) and additional anti-tank missiles have been recovered as recently as 2010 (US ARCENT, 2010). Little is known about many of the seized missiles, but those identified by model are all Soviet-designed Sagger missiles, which were first fielded in the early 1960s (Jones and Ness, 2007, p. 473). The Taliban have also acquired MILAN anti-tank missiles, one of which is displayed in a Taliban propaganda video released in December 2010 (see Image 10.6). The origin of the missile is unclear, but it could be one of two missiles abandoned by French soldiers

Image 10.6 MILAN anti-tank missile fired by a Taliban fighter. © Florian Flade

Box 10.2 Trafficking in small arms and light weapons in Afghanistan

Illicit small arms and light weapons in Afghanistan are legacies of decades of international and regional conflict. With thousands of small arms and light weapons left over from the Soviet invasion and the ongoing conflict, civilians and insurgent groups within Afghanistan have easy access to unregulated arms. Nonetheless, weapons left over from previous conflicts are not the only illicit arms. Weapons continue to flow across Afghanistan's porous borders and particularly across the borders with Iran, Pakistan, and Tajikistan. This box assesses the regional sources of illicit arms in Afghanistan as well as the routes and methods used by smugglers.

Reports of weapons intercepted along Afghanistan's border with Pakistan as well as interviews with arms dealers underscore the importance of the trade in the mountainous frontier region. For example, in 2005, the Afghan Defence Ministry reported 475 seizures of weapons, including more than 2,000 rockets, 4,000 land mines, and 5 million cartridges on the border with Pakistan, some of which reportedly belonged to former Taliban and al Qaeda fighters currently stationed in Pakistan (IWPR, 2005). Similarly, in January 2007, Afghan forces found 40 truckloads of machine guns, explosives, and rockets belonging to the Taliban that were hidden in mountain caves near the border with Pakistan (Khost, 2007).

Arms are also trafficked across the border with Tajikistan, to the north. There are multiple documented reports of arms caches found along the border, and arms dealers have described the region as a prime smuggling route during media interviews (Starkey, 2008). Weapons seized along the border include Kalashnikov-pattern rifles, small arms ammunition, and hand grenades reportedly abandoned by Afghan smugglers in Tajikistan's Moskovskiy district in March 2001 (ITAR-TASS, 2001). Media interviews with arms smugglers in 2008 indicate that Russian arms dealers and Taliban drug lords often utilize this border to exchange Russian arms for Afghan opium (Starkey, 2008).

Possible arms trafficking across the border with Iran receives significantly more attention than trafficking at other points along the Afghan border. Many US and Afghan officials believe that the Iranian government is at least turning a blind eye to activities of certain armed groups.[38] Part of this suspicion stems from the March 2011 seizure of 48 122 mm Iranian-produced rockets by British troops in Nimruz Province (Borger and Norton-Taylor, 2011). Similar incidents include the reported capture of Iranian arms traffickers in October 2009 and the seizure of a large shipment of anti-tank mines and mortars crossing from Iran into Afghanistan in June 2009 (AP, 2010; Parks, n.d.). Corroborating claims that the Iranian government is involved in such incidents using open source data is extremely difficult, however, and even the US military concedes that groups other than those directly affiliated with the Iranian government may be responsible for at least some of the arms trafficking. For example, the Joint Improvised Explosive Device Defeat Organization (JIEDDO) suggests that Baluchi insurgents (instead of Iranian intelligence or al-Quds) may be the ones transferring many of the Iranian weapons (Parks, n.d.).

The routes used by traffickers are many and varied. Some common routes from Afghanistan to Pakistan run through the Logar and Wardak Provinces of Afghanistan to border towns of Pakistan such as Bannu, Miranshah, and Wana (IWPR, 2005). While there is little publicly available data on the precise locations of trafficking from Tajikistan, an Afghan police commander claims that smugglers use the Darqad Pass between Tajikistan and the northern Afghan province of Takhar (Jalalzai, 2011). Furthermore, during media interviews, arms smugglers have claimed that Russian arms dealers meet Taliban drug lords to make exchanges at a bazaar near the old Afghan-Soviet border, deep in Tajikistan's desert, and then transport the arms from Tajikistan towards conflict areas in Afghanistan. According to JIEDDO, many of the arms trafficked between Afghanistan and Iran go through the Afghan border town of Islam Qala, on the road between Herat and Mashhad, Iran. The British *Sunday Times* asserts that much of the trafficking through Islam Qala is conducted by a drug trafficker in southern Afghanistan who is believed to have bought weapons from the Iranian government and sold them to the Taliban (Albone, 2007).

Smugglers and insurgents utilize a variety of methods for transporting and concealing weapons. Arms are often trafficked in cars, in trucks, and on the backs of donkeys and horses. For example, in March 2006, Afghan highway police stopped a Toyota Corolla loaded with Kalashnikov-pattern rifles reportedly destined for the Taliban (IWPR, 2006). In January 2003, Afghan authorities seized weapons being carted over the Pakistan border on donkeys and horses near the city of Jalalabad (News Wire Services, 2003). As navigating the mountainous terrain of Afghanistan's borders can be daunting, smugglers sometimes use a guide who specializes in wilderness border crossings (JIEDDO, n.d.). Traffickers often conceal their weapons with building materials, agricultural produce, livestock, flour sacks, cigarettes, and cement (Hussain and Hayadat, 2011; Parks, n.d.; Moscow Interfax, 1998). In 1998, the Kyrgyz security service detained 16 railroad cars carrying weapons from Iran to Afghanistan disguised as sacks of flour provided as humanitarian aid (Moscow Interfax, 1998). Sometimes, guns are disassembled to facilitate transport (Starkey, 2008). Smugglers have also been known to disguise themselves as cloth traders, labourers, humanitarian aid workers, and police. An official from the Afghan interior ministry stated that police vehicles are sometimes used in narcotics and weapons smuggling (Jalalzai, 2011).

Author: Chelsea Kelly

during an engagement with the Taliban in August 2008 (Flade, 2010). If so, the Taliban's supplies could be quite limited. Interviews with government officials and private analysts appear to confirm this. When queried about illicit ATGWs in Afghanistan, a NATO spokesman noted that '[e]arlier variants of MILAN could be available on the black market. Though it is not a common occurrence, insurgents have been known to hold such weapons.'[39] Similar views were expressed by private analysts.[40]

As evidenced by the data on the seized caches, the number and types of modern, technologically sophisticated light weapons accessible to armed groups in Afghanistan appear to be very limited. The constraining effect on Taliban operations is most clearly evident in the number of aircraft shot down in Afghanistan. The US has lost 17 helicopters to hostile fire in ten years (Peter, 2011) —a small number compared to the 270 Soviet aircraft shot down by Afghan armed groups in the late 1980s (Kuperman, 1999). While training and tactics partially explain this difference, the weapons themselves—and the US military's familiarity with them—undoubtedly have played a major role. The Stinger, which the Mujahideen never would have obtained without US assistance, was latest-generation technology that the Soviets were initially incapable of countering (Schroeder, Stohl, and Smith, 2007, pp. 84–85). None of the weapons supplied to—or acquired by—the Taliban are technologically comparable to the Stinger missile in the 1980s.

Whether the Taliban's light weapons are first generation or latest generation matters only in certain contexts, however. While consistently outgunned by NATO forces, the Taliban control an arsenal that is often equivalent to or better than the Afghan forces, which are the most frequent targets of its attacks. The Afghan police are particularly ill-equipped. A 2011 report by the US Defense Department's Inspector General concludes that, '[w]hen they are attacked by insurgents, the [Afghan Uniformed Police] often cannot defend themselves, or the population they are supposed to protect' (USDoD IG, 2011, p. 29). The insurgents, observes the Inspector General's office, are 'more heavily armed with better quality AK-47s, robustly supplied rocket-propelled grenades (RPGs), heavy machine guns, mortars and other weapons' (p. 30). Nonetheless, while the Taliban are apparently able to acquire sufficient quantities of certain types of small arms and light weapons, evidence suggests that national controls on portable missiles and the latest generation of other advanced weaponry are fairly robust and that the few governments that are willing to provide armed groups with such weapons are not willing to do so in Afghanistan.

ILLICIT SMALL ARMS AND SOMALIA

The information provided in this section is derived from an analysis of the reports of the United Nations Monitoring Group on Somalia and Eritrea from 2004 through 2011. As mentioned above, these reports include information on individual incidents of international and domestic transfers, as well as seizures of arms caches. They offer unique insight into the sources of illicit weapons in Somalia, patterns of movement to and within Somalia, and the types of weapons that are circulating in the country.

Background

UNSEMG reports from May 2004 through July 2011 record 445 instances of arms transfers (international or domestic) or seizures, involving almost 50,000 small arms and light weapons. Small arms were the most voluminously recorded items, appearing in approximately 72 per cent of the weapons records,[41] with light weapons accounting for roughly 28 per cent.[42] The reports reveal surprisingly little diversity in weapons. Kalashnikov-pattern assault rifles,

PKM machine guns, DShK heavy machine guns, RPG-2s and RPG-7s, and B-10 recoilless rifles are the items most frequently cited, often as part of the same transaction, suggesting their widespread use by many armed groups. The vast majority of the records concern weapons purchased at Somali arms markets (85 per cent), followed by unauthorized international transfers (10 per cent) and cache seizures and interdicted shipments (4 per cent) (SOMALILAND).

The UNSEMG data spans a period of continuous conflict and fundamental power shifts in Somalia. The formation of the Transitional Federal Government (TFG) provided Somalia with an internationally recognized government, even though, until recently, it only controlled very limited portions of the capital and countryside (Wezeman, 2010, pp. 1–2). The year 2006 witnessed the rise of the Islamic Courts Union (ICU), a political alliance that became the dominant power for a short period, briefly ending the warlordism that had reigned since the collapse of the Barre government. The ascension of hard-line factions within the ICU prompted Ethiopia to launch an offensive into Somalia in December 2006, in an effort to counter the influence of the ICU and bolster the TFG (Menkhaus, 2009). In 2007, the African Union Mission in Somalia (AMISOM) was established in the country to, among other tasks, assist in security stabilization programmes (Wezeman, 2010, p. 2).

Following the quick defeat of the ICU in early 2007—largely at the hands of Ethiopian forces—an Islamist nationalist insurgency, al Shabaab, emerged as the principal opposition force in the country, gaining control over most of Mogadishu and south and central Somalia (Menkhaus, 2010, p. S332). Ethiopian troops withdrew from Somalia in December 2008, but al Shabaab continued its insurgency against the TFG, affiliated militias, and AMISOM peacekeepers. Other armed groups—mainly clan militias—have gained importance as local proxies of the TFG, AMISOM, Ethiopia, and Kenya.[47] By late 2008, the most formidable TFG-allied militia, Ahlu Sunna wal Jama'a (ASWJ), was considered a 'legitimate local security sector institution' (UNSC, 2010, p. 13). Clashes between al Shabaab and the TFG and allied forces continued into 2011. Al Shabaab withdrew from Mogadishu in mid-2011 but accelerated its use of IEDs and suicide bombings against government targets (BBC News, 2011b). In October 2011, Kenyan forces mounted an offensive against

Box 10.3 The UN embargo on Somalia

In response to the political instability that has defined Somalia since the ousting of President Mohamed Siad Barre in 1991, the UN Security Council has maintained an arms embargo on the country (UNSC, 1992a; 1992b; 2002a; 2008b). UN Security Council Resolution 733, adopted in January 1992, decreed 'a general and complete embargo on all deliveries of weapons and military equipment to Somalia until the Council decides otherwise' (UNSC, 1992a, para 5). The Transitional Federal Government (TFG) formed in 2004 as a result of internationally sponsored peace talks (Menkhaus, 2010, p. S331). The TFG has received greater international support than previous transitional governments since the Barre era (Bryden and Brickhill, 2010, p. 259). Beginning in 2007, in an effort to bolster the TFG, the UN arms embargo was amended to include exemptions for weapons and training destined for the fledgling authority[43] and the punishment of countries seeking to destabilize or overthrow it (Wezeman, 2010, p. 2).[44]

Requests for embargo exemptions in support of the TFG are submitted to the sanctions committee and are not made public.[45] As a result, a comprehensive list of countries that have supplied the TFG armed forces is not available. In 2005 and 2006, prior to the adoption of Resolution 1744, both Ethiopia and Yemen reportedly supplied arms to the TFG armed forces.[46] The United States received committee authorization in 2009 to supply the TFG armed forces with weapons and ammunition that were largely acquired from Ugandan stocks (UNSC, 2010, p. 54).

The UNSEMG distinguishes two kinds of embargo violations: 'technical' violations, which support the TFG but are not approved in advance by the sanctions committee, and 'substantive' violations, which would not have qualified for an exemption (UNSC, 2011, p. 41). Since 2007, the UNSEMG has denounced several states for their failure to receive prior approval before supplying military equipment or training to the TFG police and military forces. They include Ethiopia, Kenya, Sudan, Uganda, the United Kingdom, the United States, and Yemen (UNSC, 2008a, pp. 23–24; 2010, pp. 55–56). While no action was taken in these cases, the Security Council imposed an arms embargo on Eritrea in 2009, largely in response to that country's support for armed opposition groups in Somalia (UNSC, 2009).

al Shabaab along the Kenya–Somalia border, while Ethiopian forces moved farther into the southern regions of the ailing state through the end of 2011.[48]

Small arms

Illicit small arms in Somalia are similar to those in Iraq and Afghanistan in that most are variants of widely exported Soviet-designed assault rifles and machine guns. Kalashnikov-pattern assault rifles are by far the most numerous types of assault rifles, while the PKM and DShK have near complete monopolies on the light and heavy machine gun markets, respectively. Combined, these three types of firearms account for 92 per cent of all firearms recorded in the UNSEMG reports (see Table 10.7).

Kalashnikov-pattern assault rifles are the most common firearms found throughout the country, with 28,695 cases of illicit transfer or seizure recorded by the UNSEMG during the period studied (October 2005–December 2008). In fact, the UNSEMG reports identify 93.2 per cent of all assault rifles as 'AK-47s', combining different generations of Kalashnikov-pattern assault rifles, including non-Soviet variants, under this heading. The reports identify other types of assault rifles in Somalia but in much smaller quantities. These are: the German-designed G3 (and G3A3), the Belgian FAL, the US M16, and the Singaporean SR-88. Together, these models represent less than 7 per cent of all illicit assault rifles documented in the UNSEMG reports. Moreover, according to one panel member, these weapons are becoming increasingly scarce as their ammunition becomes less available.[49]

As previously mentioned, the UNSEMG reports contain very little information on the country of manufacture, age, or condition of the illicit weapons that they document. However, anecdotal accounts from other sources provide some insight into these characteristics. For instance, investigations conducted by journalists into the Bakaara arms market, the most notorious arms market in Somalia, have revealed the presence of Kalashnikov-pattern rifles from Bulgaria, China, India, North Korea, the former Soviet Union, and Ukraine (Coker, 2001; Sheikh, 2009). The Chinese Type 56 is believed to be the most common version in service, probably in part due to its lower cost, which can be less than a third that of a Russian AKM (UNSC, 2010, p. 74).

Table 10.7 Firearms identified in UNSEMG reports

Type	Model/calibre	Quantity
Pistol	Unspecified	745
Rifle	Kalashnikov-pattern assault rifles	28,695
	FAL	193
	G3	1,646
	G3A3	Unspecified
	M16	65
	SAR-80	Unspecified
	SR-88	55
	Unspecified	137
Machine gun	Browning .30 Cal	Unspecified
	PKM	3,754
	RPD	20
	SG-43	Unspecified
Heavy machine gun	DShK	446
	12.7mm	Unspecified
	14.5mm	2
Total		≥35,758

Sources: UNSC (2005; 2006; 2007; 2008a; 2008c)

Ages of Kalashnikov-pattern rifles in Somalia vary greatly. Older rifles of this type, some probably remnants of Barre's stockpiles, are still used throughout the country.[50] Until 1977, the Soviet Union supplied Somalia with substantial amounts of military hardware, from fighter jets and tanks to Kalashnikov rifles, as a result of the countries' cold war alliance (Jane's SSA, 2010). As the Barre regime collapsed, so did controls over government stockpiles, leading to massive looting (Forberg and Terlinden, 1999, p. 26). Newer Kalashnikov variants are also in circulation, however. Production years for many of the Chinese Type 56 reported in Somalia, for instance, range from 1976 to 1991 (UNSC, 2011, p. 37), but newer models of this firearm, from 2007, clearly shipped to Somalia post arms embargo, are also circulating (UNSC, 2008a, p. 37).[51]

With one exception, Soviet-designed machine guns are the only versions reported by the UNSEMG. Machine guns are reported almost as frequently as Kalashnikov-pattern rifles in transfers and seizures, although overall quantities were lower. The Soviet-designed PK series is the most frequently mentioned type of machine gun in the dataset and the second-most cited weapon (3,754 recorded). Although foreign variants of the PK are manufactured, the UNSEMG's reports mention only the Russian-built PKMs. The other Soviet-designed machine guns reported by UN monitors were the RPD and SG-43. The only Western-designed machine gun was a single US Browning .30 calibre.

> In the 1970's and 1980's, the Somali government imported a number of missile systems.

Likewise, the Russian-built 12.7mm DShK was the only heavy machine gun identified. Given their size, DShKs are often mounted on vehicles known as 'technicals' (Somaiya, 2010). The term 'technical' in Somalia also applies to vehicles mounted with recoilless rifles, such as the B-10, and larger anti-aircraft cannon, such as the 23mm ZU-23. Technicals are common in Somalia; their role in both providing heavy fire and transporting troops was recorded in the 1990s (Guled, 1996). More recently, they have been utilized in many major battles, including the ICU capture of Mogadishu in 2006[52] and fighting against Kenyan troops in 2011.[53]

An analysis of the UNSEMG data reveals several patterns. Transfers and seizures typically consist of a large number of Kalashnikov-pattern rifles, a few PKMs, two or three DShK machine guns, and one or two other types of light weapons, often RPGs. This lack of diversity has logistical benefits in that the same or closely related weapons use the same type of ammunition and parts. The fact that TFG and AMISOM forces also use many of the same types of weapons has made them another potential source of arms and ammunition. As described below, there is growing evidence that weapons and ammunition supplied by AMISOM to the TFG armed forces and its affiliated militias, such as ASWJ, are finding their way into arms markets and the hands of al Shabaab.[54]

Light weapons

According to UNSEMG reports, the only large conventional weapons[55] available to Somali armed groups are anti-aircraft cannon and artillery shells. While these weapons fall outside of the scope of this study, it is worth noting that al Shabaab is converting large-calibre artillery shells—155 mm for instance—into IEDs, giving them one of their most effective weapons against heavy armour (VOA News, 2011). Recent media reports on cache seizures indicate that large-calibre artillery shells are still numerous in the country and will probably continue to threaten the TFG and AMISOM.[56]

Illicit light weapons currently circulating in Somalia appear to consist primarily of older model anti-tank weapons, grenades, mines, and mortars, along with a few surface-to-air and anti-tank missiles.

Guided light weapons systems

In the 1970s and 1980s, the Somali government imported a number of missile systems that, at the time, were relatively advanced. These included Soviet (SA-7b Strela) and US (FIM-43C Redeye) MANPADS as well as ATGWs, such as the French MILAN and Soviet AT-3 Sagger (SIPRI, n.d.). The same missiles, from the same period, have been spotted in

Somalia in recent years. In Mogadishu in 2008, AMISOM troops seized an arms cache containing French-designed MILANs as well as Russian anti-tank missiles.[57] Markings on the side of one of the MILANs indicate that it was manufactured in 1978, which suggests that the missiles were part of a 1978–79 shipment to Somalia[58] (see Image 10.7). Given the advanced age of these missiles, many may no longer be operational.

Evidence suggests that, since the UN Security Council imposed the arms embargo in 1992, a limited number of portable guided missiles have entered Somalia. During the period under review, UNSEMG reports record six incidents of shoulder-fired missiles involving SA-7b ('Strela') MANPADS,[59] unspecified 'shoulder-fired surface-to-air' missiles,[60] and disputed[61] reports of second-generation anti-tank missiles.[62] More recent UNSEMG reports also confirm the existence of three SA-18 Igla MANPADS (UNSC, 2011, p. 241).

Image 10.7 MILAN anti-tank guided missiles and rocket launchers, and an M79 rocket launcher designed in the former Yugoslavia, seized from militants by AMISOM troops in Mogadishu, 2008. Provided by Steve Priestley

Although opposition forces appear to possess some advanced weaponry, its overall impact appears to be minimal. In 2007, there were two confirmed MANPADS attacks (one successful, one not) and a third attack that was claimed but not confirmed (UNSC, 2011, p. 241). In 2010, the UNSEMG reported seeing a small number of wire-guided anti-tank weapons in the possession of armed opposition forces (UNSC, 2010, p. 6), but the only confirmed usage of an anti-tank guided missile occurred in 2010, when a Soviet designed AT-7 Saxhorn was fired at AMISOM troops (UNSC, 2010, p. 48). While the origins of the AT-7 are unclear, it seems unlikely that it was part of the Barre regime's stocks since the weapon did not enter into service until 1979, two years after Somalia stopped receiving support from the Soviet Union.[63]

Grenades

The UNSEMG reported the transfer and seizure of 6,570 hand grenades during the period under review (see Table 10.8), with the Soviet-designed F1 being the only model named in the reports. The F1, the first generation of which was designed in WWII, is an anti-personnel hand grenade that has been widely produced and deployed (Jones and Ness, 2007, p. 788). Although no other models are named in the UNSEMG reports, other sources indicate that several grenade varieties circulate in Somalia. Of the 493 hand grenades destroyed by the Mines Advisory Group in Puntland from 2008 to 2011, for example, explosive ordnance disposal experts most often observed the Italian Breda 35 type and the Czech RG4, with the Soviet F1, RGD5, British type 36/23, and US M67/M33 making up the rest. While the experts did not keep details on the serviceability or age, the grenades reportedly range from poor condition to brand new and vary in age, although many were identified as pre-1991.[64]

Grenades have become a common feature of insurgent warfare in Somalia. In 2010, the UNSEMG recorded 155 grenade attacks (UNSC, 2011, p. 18); many of these were surprise attacks, involving the use of a single grenade against specific targets or buildings. The technique is primarily associated with al Shabaab, whose attacks largely target AMISOM and TFG troops, police, and international organizations (UNSC, 2011, p. 18).

Table 10.8 Light weapons reported in UNSEMG reports

Type	Model	Calibre	Quantity*
Anti-tank weapon	Unspecified	75 mm	–
	'2nd generation (infrared)'		–
	M-72 light anti-tank weapon		175
	Shoulder-fire anti-tank		18
Under-barrel grenade launcher	For Kalashnikov-pattern assault rifle		350
Grenade launchers	M-79		1,107
Hand grenades	F1		5,000
	Unspecified		1,570
MANPADS	PZRK Strela		–
	SA-7b		18
	Unspecified		175
Mortars, launcher	Unspecified	60 mm, 120 mm, unspecified	6
Mortars, rounds	Unspecified	60 mm, 80 mm, 120 mm, unspecified	834
Mortars, unknown	Unspecified	80 mm, 82 mm, 120 mm, unspecified	294
Recoilless rifle	B-10		77
	M-40		–
RPG, launcher	RPG-2, RPG-7, and unspecified		430
RPG, round	RPG-2, RPG-7, and unspecified		2,149
RPG, unknown			1,955
Total			**14,158**

Note: *Dashes indicate that although the weapons were reported, the quantities were not specified.
Sources: UNSC (2005; 2006; 2007; 2008a; 2008c)

Anti-vehicle weapons

The UNSEMG reports indicate that RPGs, and specifically the RPG-2 and RPG-7, are the most common anti-tank weapons in circulation in Somalia. RPG launchers and ammunition present in Somalia have been attributed to Bulgarian, Chinese, and Russian manufacturers (UNSC, 2010, p. 75).

A range of RPG rounds is available in Somalia; the HEAT type PG-7V, PG-7VL, and PG-7VM are the models most frequently encountered by UN monitors (UNSC, 2010, p. 75). The UNSEMG confirms that 18 Chinese-manufactured 40 mm Type-69 grenades, seized from pirates by international naval patrols, were made in 2008 and transferred from

China to an unidentified 'East African' government. Older RPG rockets are also circulating. In 2010, UN monitors identified seized pirate-held Russian RPG expulsion charges that went out of production in 1987 (UNSC, 2011, pp. 36–37).

Firing an 82 mm cartridge (Hogg, 1989, p. 414), the Soviet-designed B-10 recoilless rifle is one of the largest anti-vehicle weapons frequently listed in the UNSEMG reports. Originally fitted on wheels for easier transport, the B-10 was designed to fire off a tripod for increased manoeuvrability (Hogg, 1989, p. 414). Videos of recoilless rifles in use in Somalia, however, suggest that other models of this weapon type are also available. Several videos show Somali fighters firing what are described as B-10s from the shoulder of a single shooter.[65] Given that the weight of a loaded B-10 is roughly 90 kg, it seems unlikely that this recoilless rifle was a B-10. A closer examination of the video suggests that it could be a SPG-9, a Soviet replacement for the B-10 that weighs significantly less.[66] UN monitors have also documented the much larger, US-designed 105 mm M-40 in a transfer originating in Yemen.

The most prevalent Western-produced light weapon cited in UNSEMG reports is the M-72 light anti-tank weapon, a single-shot, disposable system developed in the 1960s.[67] Western-produced grenade launchers are also identified, including the US-designed 40 mm M-79 as well as under-barrel grenade launchers for FAL assault rifles. Photos of the Swedish-designed Carl Gustav recoilless rifle on the shoulder of a Shabaab fighter appear to point to the presence, though not necessarily the widespread use, of other illicit light weapons in Somalia (see Image 10.8).

Image 10.8 A member of the armed militia for the Islamic Courts Union poses with a recoilless rifle near Mogadishu, December 2008. © AP Photo

Weapons circulation

Unlike the data on illicit weapons in Iraq and Afghanistan, which reveals little about how and when the weapons were acquired, the UNSEMG reports often contain detailed information on the movement of illicit weapons into and within Somalia. In some cases, information on the source country, the intermediaries involved in the transfer, the route to and within Somalia, and the eventual end user are reported.

Trafficking routes

During the period under review, covering incidents that took place from May 2004 through November 2008, UN monitors recorded 88 embargo violations. Three countries were identified as the primary sources of such transfers: Yemen (28 transfers), Eritrea (26 transfers), and Ethiopia (23 transfers). The UNSEMG also identifies the following other sources of illicit weapons: Iran (3 transfers), Italy (2 transfers), Libya (1 transfer), Saudi Arabia (2 transfers), Syria (1 transfer), and the United Arab Emirates (1 transfer). It is important to note that not all of these transfers were state-sponsored, but were, instead, reported simply as countries of origin.

Boats, or dhows, were the most frequent modes of transportation for imports from the primary source countries. Imports from Yemen mostly arrived via sea. The majority of items arriving by air typically originated in or arrived from Eritrea, though some also arrived by sea. Imports from Ethiopia, by contrast, were sent via land routes along its long shared border with Somalia.

As illustrated by data compiled from the UNSEMG reports, weapons enter Somalia throughout the entire country (see Map 10.1 and Table 10.9). The variety of entry points reveals the porous nature of Somalia's land and sea borders. Mogadishu (Banidir region) was the point of entry most frequently identified in the UNSEMG reports, with imports arriving from all the major source countries. Between May 2004 and November 2008, most arms imports originating in Eritrea arrived within a 100 km radius of Mogadishu. Later UN reports suggest that the port of Kismayo in southern Somalia has become a major entry point for weapons destined for opposition groups (UNSC, 2010, p. 48). Most imports from Yemen arrived in the Puntland region, particularly the port of Bosaso, and northern Somali coastline towns, such as Haradheere. According to the UNSEMG, the route through Puntland remains the 'primary gateway for arms and ammunition into Somalia' (UNSC, 2010, p. 49). Most shipments from Ethiopia were sent to interior provinces in central Somalia and southern Puntland.[68]

Arms shipments do not necessarily stay at their first destination in Somalia. Many make their way to Mogadishu. For example, it appears numerous weapons originally shipped to Bosaso, Galkayo, Garowe, and other towns in the Puntland region were eventually retransferred to Mogadishu or farther south.

Table 10.9 Number of arms shipments to Somalia, by mode of transit and origin, May 2004–November 2008

	Ethiopia	Eritrea	Yemen	Other	Total
Sea	0	9	18	7	34
Air	3	14	2	4	23
Land	14	1	0	0	15
Unspecified	6	2	8	0	16
Total	23	26	28	11	88

Sources: UNSC (2005; 2006; 2007; 2008a; 2008c)

Map 10.1 Trade routes into Somalia

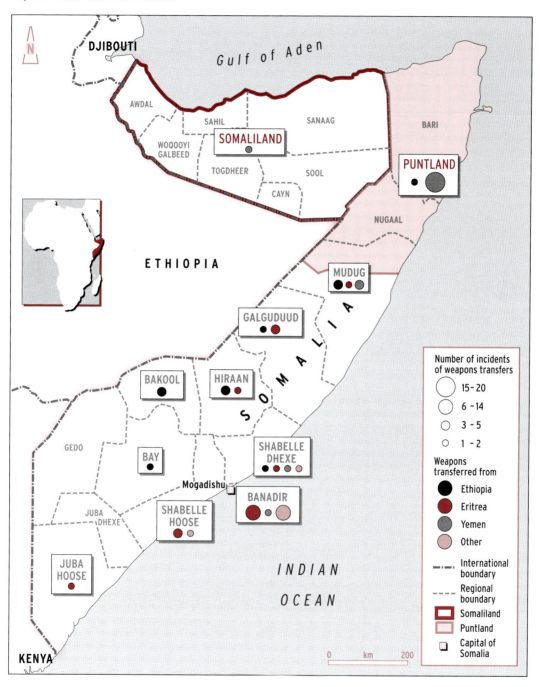

Direct support from foreign states to TFG, AMISOM, and non-state armed actors

The UNSEMG reports show that arms shipments are sent from source countries, both to specific armed actors and to undetermined end users at arms markets. According to UN monitors, direct support to specific Somali non-state actors and state forces, including the TFG, has been most evident in the case of Eritrea and Ethiopia. Since the late

1990s, Eritrea and Ethiopia have provided weapons and ammunition to the TFG armed forces, various warring factions, and clan militias in Somalia in a kind of proxy war (*Jane's Intelligence Review*, 2009). All shipments without sanctions committee authorization are formally violations of the embargo (see Box 10.3), yet the UNSEMG distinguishes between deliveries to support the TFG or AMISOM, which are eligible for exemption, and those to armed opposition groups or arms markets, which are not eligible. Arms sent in support of the TFG without prior approval are considered 'technical' violations, while others are characterized as 'substantive' violations.

Throughout the October 2005 through December 2008 reports, UN monitors document a regular supply of Eritrean-sourced weapons to Islamic opposition groups in Somalia. Of the 26 shipments the UNSEMG reports identify as originating in Eritrea, 25 (96 per cent) were reportedly intended for the ICU, al Shabaab, or other militant Islamist groups. The UNSEMG finds that, in recent years, Eritrea has dramatically reduced its support of these groups; indeed, there is no hard evidence that Eritrea has directly supplied any arms in recent years,[69] although it maintains financial support to al Shabaab (UNSC, 2010, p. 47). In November 2011, Kenyan and Somali officials accused Eritrea of sending several arms shipments to al Shabaab through the town of Baidoa (Teyie and Wabala, 2011). These reports were not verified by the UNSEMG (BBC News, 2011a) and, in fact, the UNSEMG's preliminary findings, released in January 2012, indicate that the accusations 'were incorrect' (Maasho, 2012). The UN Security Council nevertheless imposed stronger sanctions on Eritrea in December 2011 for continuing to support al Shabaab (Reuters, 2011).

The UNSEMG has alleged that the more advanced guided weapons systems documented in the UN reports have entered Somalia via Eritrea. Probably the most significant evidence of this are two SA-18 missiles seized from al Shabaab in 2007 and 2008 that UN monitors traced back to a 1995 shipment of MANPADS from the Russian Federation to Eritrea (UNSC, 2011, p. 243; see Table 10.10).

The UNSEMG reports document weapons entering from Ethiopia largely for the benefit of the state authorities in Somalia and Puntland, as well as allied armed groups, such as ASWJ (UNSC, 2010, p. 48). Although conducted in support of the TFG, the transfers documented by UN monitors were all considered 'technical violations' of the embargo as they were made either prior to Resoultion 1744 or subsequently without the consent of the sanctions committee (UNSC, 2010, p. 47).

Reports by the UNSEMG also indicate that Ethiopia sent arms into Somalia to support the operations of its own troops or the TFG armed forces and its allies. Resupplying efforts to either Ethiopian troops or the TFG armed forces accounted for 40 per cent of the total number of deliveries entering Somalia from Ethiopia. An additional nine per cent of the deliveries identified Puntland authorities as the end recipient. The UN monitors indicate that weapons

Table 10.10 Selected weapons by country of origin as reported by UNSEMG

Weapon	Eritrea	Ethiopia	Yemen
Assault rifles	7,019	6,026	10,671
Machine guns	295	549	2,782
RPGs (launchers and rounds)	272	455	2,008
Grenades	4,000	1,000	300
MANPADS	68	–	–

Sources: UNSC (2005; 2006; 2007; 2008a; 2008c)

have also supported warlords and clans firmly or loosely allied with the TFG. Each of the warlords or clan leaders identified in this regard, such as Mohammed Dheere of Jowhar, have held high positions in the TFG government at some point (Hanson and Kaplan, 2008). Despite the lack of UN approval for the weapons shipments from Ethiopia, their intended purpose, namely to support the officially recognized government of Somalia, puts them in something of a grey area as far as illicit transfers are concerned—hence their 'technical violation' status. The same is true for the Yemeni government's support of the TFG in 2005 and 2006; UNSEMG reports link 18 per cent of the deliveries from Yemen to the Somali government.

Somali arms markets

Somali arms markets are largely autonomous and, according to UNSEMG reports, supply the full range of Somali clans, warlords, al Shabaab, and the TFG police and military (UNSC, 2011, p. 41). These markets have provided a steady source of weapons for these actors, remaining open for all but brief periods during the embargo.

According to UN monitors, weapons bound for the commercial markets in Somalia most often originate in Yemen (UNSC, 2010, p. 6). Data compiled in UNSEMG reports appears to support this claim; all imports arriving to arms markets from an identified source country reportedly originated in Yemen. It must be noted that most (approximately 70 per cent) of the imports coming from Yemen are linked to local arms traders and not the government. Until 2007, 'arms were openly sold in Yemen in at least 18 arms markets'; however, arms control campaigns in 2007 and 2009 shut down most of the arms markets[70] (Small Arms Survey, 2010, p. 6).

Since the bulk of the reports in the database compiled for this study cover periods before the end of 2008, it is not possible to determine the impact this development has had on arms shipments to Somalia. Yet the 2011 UNSEMG report does note the continued predominance of Yemeni weapons in Somali markets (UNSC, 2011, p. 41). There are also allegations of the diversion of official Yemeni arms to Somalia. According to *Jane's Defence Weekly,* poor stockpile management in Yemen is probably contributing a large portion of these supplies, with Yemeni military officials suspected of selling their own inventory (Knickmeyer, 2010). The Yemeni government denies these claims (UNSC, 2011, p. 41).

Leakage from weapons stockpiles in Somalia is another important source of weapons fuelling the country's illicit market. This involves the unauthorized sale of weapons and ammunition by individual officers or corrupt officials from the TFG or allied armed groups. Opposition forces also take weapons from seized state or African Union arms depots or from dead soldiers and police. Desertion and defection of TFG police and military are also a great concern, with a 2008 report suggesting an attrition rate of 80 per cent; when these trained security and defence forces leave their jobs, they take their weapons and uniforms with them (UNSC, 2008c, p. 12). These weapons are often sold at arms markets.

A 2010 report by the UNSEMG indicates that this practice has probably increased since 2008. This is particularly true of ammunition (UNSC, 2011, p. 43). The sale of ammunition has become a means for many TFG troops and ASWJ members to supplement their low salaries, according to UN monitors, and presents a difficult dilemma for countries wishing to support the TFG with military equipment, as they may inadvertently feed the illicit market. In a survey of ammunition available in the Bakaara arms market, eight of 11 rounds had markings consistent with AMISOM holdings (UNSC, 2011, p. 44). It is estimated that government and pro-government forces typically sell between a third and a half of their ammunition (UNSC, 2011, p. 44).

> Weapons bound for the commercial markets in Somalia most often originate in Yemen.

CONCLUSION

As illustrated throughout this chapter, illicit small arms and light weapons in the three countries studied have several common characteristics. Most weapons identified by model are early designs of Eastern Bloc or Chinese weapons that have proliferated widely and are often significantly less capable than their modern counterparts. Many of the models have also been circulating in the same countries for years, or decades, revealing a remarkable continuity in the models and technological sophistication of illicit weapons.

Among the most significant findings of this chapter is the absence of latest-generation weapons in the countries studied. Most portable missiles recovered from seized caches in Iraq are Soviet-designed systems initially fielded several decades ago and stockpiled in large quantities by Saddam Hussein's regime. No latest-generation Russian or Chinese MANPADS are identified, and no MANPADS of any generation from other countries are listed. Similarly, most of the ATGWs found in Iraqi caches are first-generation Soviet-designed missiles, and no late-generation RPG systems[71] are identified in the cache data.

Evidence also suggests armed groups in Afghanistan are no better equipped. Armed groups in Somalia have acquired limited numbers of third-generation SA-18 MANPADS manufactured in the 1990s, but no latest-generation systems are identified, and most of the MANPADS appear to be first-generation SA-7s. UN monitors have not found any modern anti-tank missiles in Somalia.

The weapons acquired by armed groups in recent years contrast sharply with the arsenals of their better-armed counterparts from previous decades, such as the Mujahideen in Afghanistan in the 1980s. As explained above, the United States provided hundreds of cutting-edge Stinger MANPADS to the rebels, who used them to shoot down hundreds of Soviet and Afghan aircraft, severely affecting Soviet and Afghan air operations and possibly hastening the Soviet withdrawal from Afghanistan. Other examples of armed groups that acquired substantial numbers of latest-generation light weapons include UNITA in Angola in the 1980s (Hunter, 2001).

Without extensive government assistance, however, it is highly unlikely that the Mujahideen would have acquired any Stinger missiles at all, let alone the large quantities and training necessary to disrupt Soviet air operations. UNITA was equally dependent on the United States for its Stingers. Thus, as illustrated by these cases, assistance from producing states is an essential factor in determining the speed and extent to which latest-generation portable missiles enter the black market.

With the possible exception of Iran, there is no compelling reason for the few producers and importers of latest-generation MANPADS and ATGWs to supply them to armed groups in Afghanistan, Iraq, or Somalia, and the diplomatic consequences of doing so—deliberately or inadvertently—would be severe. Indeed, there is little substantiated evidence that *any* armed group has acquired latest-generation MANPADS—a testament to the growing awareness of the terrorist threat posed by these weapons and the effect of global efforts to address it.

It is less clear why armed groups have not acquired and used more third-generation portable missiles and advanced anti-tank rockets. These weapons are in the arsenals of many more countries than are latest-generation systems, including countries with poor stockpile and export controls and those accused of arming non-state groups. One possible explanation is that less sophisticated but readily available weapons and ammunition were effective enough that it was not worth the cost and trouble of acquiring more sophisticated weapons from abroad. In Iraq, armed groups had access to thousands of tons of weapons looted from government depots in 2003. Large-calibre ammunition from these depots has been used extensively in improvised explosive devices, which accounted for approximately 40 per cent of US casualties from 2003 to 2006 (Roane and Pound, 2004; O'Hanlon and Livingston, 2011, p. 8). In Afghanistan, IED makers also use locally available components, including ammonium nitrate fertilizer produced in Pakistan and widely used by Afghan farmers. Journalist Greg Jaffe claims that these devices cost about USD 40 to make (PRI, 2011); along with other IEDs, they account for more than 60 per cent of US casualties in Afghanistan (DMDC, 2012). In both theatres, IEDs appear to be an adequate substitute for advanced ATGWs and anti-vehicle rockets, which are almost certainly more expensive and difficult to obtain, particularly in the quantities necessary to sustain the high tempo of insurgent activity in Iraq and Afghanistan.

This explanation is less compelling with respect to MANPADS. While RPGs and firearms are sometimes used to shoot down helicopters, MANPADS are particularly well suited for use against military aircraft. MANPADS have a significantly longer effective range than RPGs and most firearms and, because they are guided, they are much more likely to hit their target. Given these attributes, and the heavy reliance of US and NATO troops on air operations during the time period studied, the apparent inability of armed groups in Afghanistan and Iraq to acquire third-generation MANPADS is more probably explained by supply-side dynamics than a lack of demand.

Another possible explanation is that the best-equipped armed groups are reliant on sympathetic governments for certain light weapons, and these governments are withholding (or limiting the number of) weapons viewed as particularly sensitive. There is evidence that state sponsors of armed groups in Iraq and Somalia have, at times, reduced the flow of weapons in response to international pressure. Whether this pressure has had a constraining effect on the quantities and types of portable missiles and other sensitive weapons provided by these governments is unclear, in part because of a lack of publicly available evidence conclusively linking individual illicit weapons to the governments accused of supplying them. More and better data on sanctions-defying arms shipments and weapons seized from arms caches would help to answer these questions.

Despite these gaps, the data has clear implications for policy-makers. The apparent absence of latest-generation MANPADS in Afghanistan, Iraq, and Somalia strongly suggests that national and international initiatives to eliminate the illicit proliferation and use of these weapons are bearing fruit. Studying these initiatives—and identifying the most effective control strategies employed as part of these efforts—could yield important insight into controlling other latest-generation small arms and light weapons. Applying these lessons to the most sensitive weapons, including portable sensor-fused weapons, programmable ('smart') airburst grenades, and guided mortars, could significantly reduce the likelihood that they will end up in the wrong hands.

At the same time, the data also underscores the need to control the availability of early generation weapons. As evidenced by data on usage of illicit weapons in all three countries studied, even the simplest weapons can be extremely destructive when deployed in large numbers and innovative or tactically savvy ways. Of particular importance is the prevention of excessive stockpiling, which led to the accumulation of millions of tons of weapons and ammunition in Afghanistan, Iraq, and Somalia. When the regimes controlling these stockpiles collapsed, the weapons quickly spread to the armed groups that filled the resulting power vacuum. These weapons were used extensively in the years following the looting, and data collected for this study suggest that at least some of these weapons remain in circulation today. Preventing similar stockpiling in countries prone to instability—and right-sizing existing (excessive) stockpiles—would help to limit the number of illicit weapons in current and future war zones.

LIST OF ABBREVIATIONS

AMISOM	Africa Union Mission in Somalia
ANSF	Afghan National Security Forces
ASWJ	Ahlu Sunna wal Jama'a
ATGW	Anti-tank guided weapon
FOIA	Freedom of Information Act
HEAT	High-explosive anti-tank
ICU	Islamic Courts Union
IED	Improvised explosive device
JIEDDO	Joint Improvised Explosive Device Defeat Organization
MANPADS	Man-portable air defence system
RPG	Rocket-propelled grenade
TFG	Transitional Federal Government
SIPRI	Stockholm International Peace Research Institute
UNSEMG	United Nations Monitoring Group on Somalia and Eritrea

ENDNOTES

1   'RPG' is a reverse acronym for the Russian term 'ruchnoy protivotankovy granatomyot,' which means 'hand-held anti-tank grenade launcher'.
2   See, for example, Riechmann (2011) and King, Dilanian, and Cloud (2011).
3   Unless otherwise specified, the chapter uses the term 'Kalashnikov-pattern assault rifles' to refer to all weapons derived from the Kalashnikov AK-47 rifle and subsequent models (such as the AKM and AK-74), including foreign variants, such as the Chinese Type 56 rifle.
4   See UNGA (1997) and Small Arms Survey (2008, pp. 8–11).
5   For the purposes of this chapter, improvised explosive devices are considered 'light weapons'.
6   Reports on the UN arms embargo on al Qaeda and the Taliban contain little detailed data on illicit small arms and light weapons. See, for example, UNSC (2002b, p. 17).
7   The data obtained by Felter and Fishman covers 166 incidents of Iranian weapons discovered in Iraq, but only 74 arms caches, which are the focus of this study.
8   UNSC (2002b, p. 17).
9   Rockets fired from rails and used with improvised launchers are not included in the dataset because of the difficulty of separating rockets used in these capacities from other types of projectiles.
10 One cache contained 5,620 US dollars, 75,000 Iraqi dinars, 245 Indian rupees, and 140 United Arab Emirates dirhams. IED components included items as diverse as Indian ball bearings, Chilean blasting caps, and Bulgarian fuses.
11 The assessment does not claim to be comprehensive, noting that 'virtually any weapon [...] in the world can be found in Iraq' (NGIC, 2004a, p. 4); that claim is consistent with the broad array of illicit arms seized since 2004.
12 In its 2004 assessment of insurgent weapons, the US Army identifies other variants of the AKM that were available in Iraq, including 'the Chinese Type 56, the Iranian KLF, the Hungarian AMD-65, the Romanian Model 63 AKM, the Bulgarian AKM, and the Polish Kbk-AKM' (NGIC, 2004b, p. 4).
13 Author correspondence with the Olive Group, 25 July 2011.
14 Since the Iraqi security forces primarily use Kalashnikov-pattern rifles, the data reveals little about their stockpile security practices.
15 Author correspondence with the Olive Group, 25 July 2011.
16 Author correspondence with the Olive Group, 25 July 2011. The one pistol identified by the Olive Group that is not 9 mm is the 7.65 mm Russian Makarov Model D.
17 Note that this section does not include heavy machine guns, which are categorized as small arms and assessed in the previous section.
18 Excluded from this figure are approximately 500 items identified as 'RPGs' in which it is unclear whether the item in question was an RPG round or an RPG launcher.
19 A smaller number of the weapons were fielded more recently, including the RPG-22 (1985) (Jones and Ness, 2007, p. 479).
20 The Iranian NADER was not designed by the Soviets but is a contemporary of its Soviet-designed counterparts.
21 While the designation 'PG-7L' technically refers only to the warhead, the source appears to use the designation as shorthand for the entire round. The designation for the complete round is 'PG-7VL'.

22 Jane's Information Group claims that the PG-7VL can penetrate up to 600 mm of armour (Jones and Ness, 2007, p. 477).
23 While one source indicates that the PG-7-AT designation is used for rounds with tandem warheads, photographs of at least some seized rounds bearing this designation are not consistent with photographs of known Iranian tandem rounds.
24 There are media reports of insurgent acquisition and usage of tandem PG-7VR rounds, but they cannot be confirmed independently (Roos, 2003).
25 As of mid-2006, US forces had only found one RPG-29 in Iraq, according to Gen. John Abizaid (AFP, 2006).
26 Author correspondence with the Olive Group, 25 July 2011.
27 Note that this total includes casualties from unknown and 'miscellaneous' causes.
28 Data on casualties from artillery, mortar, and rocket attacks is combined in a single category.
29 Note that the caches contained 11 'anti-tank missiles' that were not identified by model.
30 The US military previously released a photograph of an Iranian variant of the QW-1 (Misagh-1) (US MNF-I, 2007).
31 It is unclear whether the SK-10 launcher, which is used with the Chinese QW-1 MANPADS, is also used with the Iranian Misagh-1.
32 Since the data only includes firearms, grenades, and RPGs, it was excluded from the aggregate statistics but is included in the commodity-specific analysis.
33 Bhatia and Sedra (2008, p. 65); Chivers (2011); Crile (2003, p. 158); Pegler (2012, p. 70).
34 Since US manufacturers produce ammunition for Kalashnikov-pattern rifles and other Soviet-designed weapons, the data says little about the quantity of US-manufactured ammunition seized from the caches.
35 The US decision to supply the ANSF with M16 rifles and other US-made weapons may soon result in the decommissioning of many of the ANSF's Soviet-designed firearms.
36 The US Government Accountability Office has also reported accusations of diversion from Afghan arsenals, including the theft of 47 pistols from the Afghan National Army's central depot (USGAO, 2009, pp. 17–18).
37 The only exceptions are five rounds for the Chinese Type 69-1 RPG, which was fielded in the 1980s.
38 See ABC News (2007); AFP (2011a); Albone (2007); Channel 4 News (2010); Parks (n.d.); and Solomon (2011).
39 Author correspondence with a spokesman of the International Security Assistance Force, British Royal Navy, 1 September 2011.
40 Author correspondence with private analysts, 1 September 2011.
41 In UNSEMG reports, ammunition is often difficult to distinguish from weapons. Reports often list several different types of ammunition together, with no breakdown of quantities. Most often, ammunition types are grouped together under the generic term 'variety', meaning ammunition for various small arms, light weapons, and, in a few cases, larger conventional weapons systems that are beyond the scope of this study. Quantities, when provided, are given in individual units, bags, and tons.
42 The UNSEMG reports record anti-tank weapons (excluding anti-aircraft cannon and howitzers), explosives, grenades, landmines, MANPADS, mortars (launchers and rounds), recoilless rifles, and RPGs (launchers and rounds).
43 See UNSC (2007).
44 See UNSC (2008b; 2009).
45 Author interview with a member of the UNSEMG, Geneva, 4 January 2012.
46 For more information, see Wezeman, Wezeman, and Béraud-Sudreau (2011, pp. 30–31).
47 Author correspondence with Ken Menkhaus, professor of political science, Davidson College, 18 December 2011.
48 Author correspondence with Ken Menkhaus, professor of political science, Davidson College, 18 December 2011.
49 Author interview with a member of the UNSEMG, Geneva, 9 January 2012.
50 Author interview with a project manager, HALO Trust, Hargeisa, Somaliland, 14 August 2011.
51 See also Coker (2001).
52 See BBC News (2006).
53 See Xinhua News Agency (2011).
54 Author interview with panel member of the UNSEMG, Geneva, 9 January 2012.
55 During the cold war, Somalia was one of the most heavily armed countries in Africa. Militarily supported by the Soviet Union until the start of the Ogaden War in 1977, and subsequently by the United States, the government built large arsenals of conventional weapons, no longer featured in Somalia's weapons landscape (Jane's SSA, 2010). Today, the only verified use of large, mobile conventional weapons systems (such as tanks or aircraft) is by AMISOM and countries allied with the TFG: Ethiopia's offensive in Somalia to remove the ICU; Kenyan air incursions (AFP, 2011b); and US air attacks on al Shabaab targets in 2009 and 2011 (CBS, 2009; Walsh, 2011).
56 Demilitarization records from Puntland provided by the Mines Advisory Group show that 997 projectiles with calibres greater than 120 mm were destroyed between July 2008 and February 2011. VOA News reports that the African Union destroyed 137 shells found in 2011 (VOA News, 2011).
57 Author correspondence with Steve Priestley, explosive ordnance disposal expert, 7 August 2011.
58 The SIPRI Arms Transfers Database shows that 1,000 MILAN anti-tank missiles were delivered to Somalia from an unidentified country between 1978 and 1979 (SIPRI, n.d.). Explosive ordnance disposal expert Steve Priestley assisted in assessing the markings on the MILANs pictured.
59 See UNSC (2005, p. 13; 2006, p. 14; 2010, p. 50).

60   See UNSC (2006, pp. 13, 21).
61   Note that the accuracy of these reports has been questioned. As SIPRI analyst Siemon Wezeman points out, second-generation missiles do not use infrared guidance technologies but instead use wire, radio, or laser guidance systems. According to Wezeman, there is no evidence that missiles with infrared guidance are in Somalia (author correspondence with Siemon Wezeman, analyst, SIPRI, 14 September 2011).
62   See UNSC (2006, p. 13).
63   Author correspondence with Paul Holtom, senior researcher, SIPRI, 2 November 2011.
64   Data provided by the Mines Advisory Group. Additional information attained through author correspondence with Jack Frost, technical field manager, Mines Advisory Group, Puntland, 2 August 2011.
65   See, for example, LiveLeak (2008).
66   Author correspondence with Steve Priestley, explosive ordnance disposal expert, 17 September 2011. Priestley indicated that the weapon was probably an SPG-9 but could not confirm this because of poor image quality.
67   See FAS (n.d.).
68   Of the 14 shipments the UNSEMG identifies as entering Somalia from Ethiopia over land, ten were sent to one of these provinces: Bakool, Galguduud, Hiraan, and Mudug.
69   Author correspondence with a member of the UNSEMG, Geneva, 9 January 2012.
70   In correspondence with the Small Arms Survey, a former panelist of the UNSEMG stated that the main Yemeni arms market supplying Somalia—Mukalla—was not among the markets that have closed. Author correspondence with a panellist of the UNSEMG, Geneva, 9 January 2012.
71   The term 'late-generation RPG systems' is used here to refer to the RPG-28, RPG-29, RPG-30, and RPG-32.

BIBLIOGRAPHY

ABC (American Broadcasting Company) News. 2007. 'Document: Iran Caught Red-Handed Shipping Arms to Taliban.' 6 June. <http://abcnews.go.com/blogs/headlines/2007/06/document_iran_c/>.

AFP (Agence France-Presse). 2006. 'New Weapons from Iran Turning up on Mideast Battlefields.' 19 September.

—. 2011a. 'Iran Arms Afghan Insurgents: NATO.' 28 February.

—. 2011b. 'Kenyan Troops Enter Somalia to Attack Rebels.' 16 October. <http://www.capitalfm.co.ke/news/2011/10/kenyan-troops-enter-somalia-to-attack-rebels/>

AFPS (American Forces Press Service). 2005. 'Afghan, Coalition Forces Find Weapons, Detain Enemy Fighters.' 3 August.

Albone, Tim. 2007. 'Iran Gives Taliban Hi-Tech Weapons to Fight British.' *Sunday Times* (United Kingdom).

AP (Associated Press). 2005. 'U.S. Attacks Iraq Arms Smugglers.' 12 April.

—. 2007. 'Baghdad Raid Nets 4 Suspected Arms Smugglers.' 27 July. <http://www.armytimes.com/news/2007/07/ap_capture_070727/>

—. 2010. 'Joint Raid on Arms Smugglers in Iraq Kills 5 Villagers.' 12 February.

Australian DOD (Department of Defence). 2011. '110701 ADF WPN Seizures: AFG 2010-2011 F366962.' Received 5 July.

Baker, Fred III. 2007. 'Despite Risks, Air Still Safest Travel in Iraq.' American Forces Press Service. 13 February.

BBC (British Broadcasting Company) News. 2006. 'Somali Islamists Win City Battle.' 11 July.

—. 2011a. 'Eritrea Denies Sending Weapons to al Shabab in Somalia.' 2 November. <http://www.bbc.co.uk/news/world-africa-15559584>

—. 2011b. 'Kenya Planes Attack Somali "Militant Camp" in Hosingow.' 21 December. <http://www.bbc.co.uk/news/world-africa-16284795>

Ben-David, Alon. 2006. 'Israeli Armour Fails to Protect MBTs from ATGMs.' *Jane's Defence Weekly*. 30 August.

Bhatia, Michael and Mark Sedra. 2008. *Afghanistan, Arms and Conflict: Armed groups, Disarmament and Security in a Post-war Society*. London: Routledge.

Borger, Julian and Richard Norton-Taylor. 2011. 'British Special Forces Seize Iranian Rockets in Afghanistan.' 9 March. <http://www.guardian.co.uk/world/2011/mar/09/iranian-rockets-afghanistan-taliban-nimruz>

Bryden, Matt and Jeremy Brickhill. 2010. 'Disarming Somalia: Lessons in Stabilization from a Collapsed State.' *Conflict, Security, and Development*, Vol. 10, No. 2, pp. 239–62.

CBS. 2009. 'U.S. Strikes in Somalia Reportedly Kill 31.' 11 February. <http://www.cbsnews.com/stories/2007/01/08/world/main2335451.shtml>

Chivers, C. J. 2006. 'Black-Market Weapon Prices Surge in Iraq Chaos.' *The New York Times*. 10 December.

—. 2009. 'Arms Sent by U.S. May be Falling into Taliban Hands.' *The New York Times*. 19 May.

—. 2010. 'Reading (Rifle) Magazines.' *At War: Notes from the Front Lines. The New York Times* blog. 1 February.

—. 2011. 'Taliban Gun Lockers: The Rifles of Rural Ghazni Province.' *At War: Notes from the Front Lines. The New York Times* blog. 31 January.

Coker, Margaret. 2001. 'Somalia: Amid War, Famine, Selling Guns "Guarantees My Family Will Be Fed."' *Atlanta Constitution*. 9 July.

Crile, George. 2003. *Charlie Wilson's War*. New York: Grove Press.

Danahar, Paul. 2007. 'Taleban "Getting Chinese Arms."' BBC News. 3 September.

DMDC (United States Defense Manpower Data Center). 2012. 'Casualty Summary by Reason Code.' 3 January. Washington, DC: DMDC, United States Department of Defense. <http://siadapp.dmdc.osd.mil/personnel/CASUALTY/gwot_reason.pdf>

FAS (Federation of American Scientists). n.d. 'M-72 Light Anti-tank Weapon (LAW).' Updated 12 September 1998. <http://www.fas.org/man/dod-101/sys/land/m72.htm>

Felter, Joseph and Brian Fishman. 2008. *Iranian Strategy in Iraq: Politics and 'Other Means.'* 13 October.

First Battalion. 2003. 'Operation Iraqi Freedom.' 22nd Infantry. <http://1-22infantry.org/pics/iraqifreedompagesixteen.htm>

Flade, Florian. 2010. 'The Taliban's Latest Toy—European Anti-Tank Missiles.' *Jih@d: News of Terrorism, Jihadism & International Politics.* 13 December.

Forberg, Ekkehard and Ulf Terlinden. 1999. 'Small Arms in Somaliland: Their Role and Diffusion.' BITS Report 99.1. Berlin: Berlin Information-center for Transatlantic Security. March. <http://www.bits.de/public/pdf/rr99-1.PDF>

Fox, Mark and Tahseen Sheikhly. 2007. 'Rear Adm. Mark Fox and Dr. Tahseen Sheikhl, Sept. 23.' Press conference. <http://www.usf-iraq.com/news/press-briefings/14274>

Grau, Lester. 1998. 'The RPG-7 on the Battlefields of Today and Tomorrow.' *Infantry.* May–August.

Guled, Mohamed. 1996. 'Somalis Bury Aydiid.' Reuters. 2 August.

Hanson, Stephanie and Eben Kaplan. 2008. 'Somalia's Transitional Government.' Washington, DC: Council on Foreign Relations. 12 May. <http://www.cfr.org/somalia/somalias-transitional-government/p12475#p5>

Hasni, Mohamed. 2003. 'Kalashnikovs Best Sellers at Meridi Market.' Agence France-Presse. 21 April.

Hider, James. 2004. 'Iraqi Gun Runners "Too Professional" to be Caught Out.' Times Online. 18 February.

Hogg, Ian, ed. 1989. *Jane's Infantry Weapons 1989–1990.* Fifteenth edn. Coulsdon, Surrey: Jane's Information Group.

Hunter, Thomas. 2001. 'The Proliferation of MANPADS.' *Jane's Intelligence Review.* September.

Hussain, Tom and Amjad Hadayat. 2011. 'Poverty, Not Ideology, Fuels Afghan Gun-running Routes.' *National.* <http://www.thenational.ae/news/worldwide/south-asia/poverty-not-ideology-fuels-afghan-gun-running-routes?pageCount=0>

Intelligence Online. 2004. 'Saddam's Secret Network.' 24 September.

ITAR-TASS. 2001. 'Drugs and Weapons Cache Found on Tajik-Afghan Border.' 28 March.

IWPR (Institute for War and Peace Reporting). 2005. 'Taleban Buying Up Smuggled Guns.' <http://iwpr.net/report-news/taleban-buying-smuggled-guns>

—. 2006. 'Taleban Find Unexpected Arms Source.' <http://iwpr.net/report-news/taleban-find-unexpected-arms-source>

Jalalzai, Musa Khan. 2011. 'Analysis: Kidnapping and Arms Smuggling in Afghanistan.' *Daily Times* (Pakistan). 23 June.

Jane's Intelligence Review. 2009. 'The Enemy's Enemy—Eritrea's Involvement in Somalia.' September.

Jane's SSA (*Sentinel Security Assessment–North Africa*). 2010. 'Procurement: Somalia.' 29 October.

Jane's World Armies. 2012. 'Afghanistan.' 30 January.

Jones, Richard and Leland Ness. 2007. *Jane's Infantry Weapons 2007–2008.* Coulsdon, Surrey: Jane's Information Group.

Khost. 2007. 'Afghan Forces Find 40 Truckloads of Taliban Weapons.' 17 January.

King, Laura, Ken Dilanian, and David Cloud. 2011. 'SEAL Team 6 Members among 38 Killed in Afghanistan.' *Los Angeles Times.* 6 August.

Knickmeyer, Ellen. 2010. 'Yemen Military Turns to Arms Brokers.' *Jane's Defence Weekly.* 5 November.

Kuperman, Alan. 1999. 'The Stinger Missile and U.S. Intervention in Afghanistan.' *Political Science Quarterly*, Vol. 114, No. 2. Summer.

LiveLeak. 2008. 'Scenes from Somalia as Mujahids Enter Mogadishu.' <http://www.liveleak.com/view?i=95c_1230766376>

—. n.d. 'RPG-29 "Vampyr" vs. M1 Abrams.' Added 29 April. Accessed 4 March 2012.

Maasho, Aaron. 2012. 'Eritrea Did Not Fly Arms to al Shabaab Last Oct: UN.' Reuters. 16 January.

Menkhaus, Ken. 2009. 'Current Somalia Crisis & the Role of the Diaspora (Senate Hearings 03/11/2009).' 11 March. <http://www.youtube.com/watch?v=WQ07pBEo_bg>

—. 2010. 'Stabilisation and Humanitarian Access in a Collapsed State: The Somali Case.' *Disasters*, Vol. 34, Suppl. Iss. 3, pp. S320–41.

MilitaryNewsNetwork. 2009. 'Unusual Weapons Found in Iraq.' <http://www.youtube.com/watch?v=ndHgEBAP3so>

Minaya, Zeke. 2007. 'Soldiers Find Vintage Weapons among Caches in 2007.' *Stars and Stripes.* 25 May.

MNC-I (Multi-National Corps–Iraq). 2007. 'Golden Dragons Discover Cache Site, VBIED Production Site.' 30 March.

Moscow Interfax. 1998. 'Kyrgyzstan: Kyrgyz Seize Weapons from Iran Destined for Anti-Taleban.' 12 October.

Murphy, Chuck and Sydney Freedberg. 2002. 'Seized Afghan Arsenal Betrays Potential.' *St. Petersburg Times* (United States). 4 August.

Navarro, Lourdes. 2004. 'Rebels Smuggle Supplies into Iraqi City.' Associated Press. 12 April.

News Wire Services. 2003. 'Afghans Snag 330 Smuggled Rockets.' 3 January.

NGIC (National Ground Intelligence Center). 2004a. *Iraq: Small Arms Handbook.* NGIC-1142-7005-05. Charlottesville, VA: NGIC, United States Army Intelligence and Security Command. October.

—. 2004b. 'Iraq: UPDATE—Small Arms (Infantry Weapons) Used by the Anti-Coalition Insurgency.' 17 December.

O'Halloran, James and Christopher Foss. 2010. *Jane's Land-Based Air Defence 2010–2011.* Surrey, UK: IHS Jane's.

O'Hanlon, Michael and Ian Livingston. 2011. 'Iraq Index: Tracking Variables of Reconstruction and Security in Post-Saddam Iraq.' Washington, DC: Brookings Institution. 30 November.

Parks, Eric. n.d. 'Iranian Weapons Smuggling Activities in Afghanistan.' Washington, DC: Joint Improvised Explosive Device Defeat Organization.

Pegler, Martin. 2012. *The Lee-Enfield Rifle*. Oxford: Osprey Publishing.

Peter, Tom. 2011. 'For Many Afghans, US Helicopter Crash Confirms Taliban Momentum.' *Christian Science Monitor*. 7 August.

PRI (Public Radio International). 2011. 'IEDs in Afghanistan Costing U.S. Billions, Despite Being Dirt-Cheap to Build.' 29 November.
<http://www.pri.org/stories/world/middle-east/ieds-in-afghanistan-costing-u-s-billions-despite-being-dirt-cheap-to-build-7200.html>

Rasheed, Ahmed and Ross Colvin. 2007. 'Iraqi Police Selling Weapons on Black Market.' Reuters. 6 February.
<http://www.reuters.com/article/2007/02/05/us-iraq-weapons-idUSCOL34438320070205>

Rayment, Sean. 2007. 'MoD Kept Failure of Best Tank Quiet.' *Telegraph* (UK). 13 May.

Reuters. 2011. 'Eritrea: U.N. Increases Sanctions.' *The New York Times*. 6 December.
<http://www.nytimes.com/2011/12/06/world/africa/eritrea-un-increases-sanctions.html>

Rhem, Kathleen. 2002. 'Myers Doubts Chinese Government Helping al Qaeda.' *American Forces Press Service*. 9 August.

Riechmann, Deb. 2011. 'Afghan Witnesses: Chinook Ablaze When It Crashed.' Associated Press. 11 August.

Roane, Kit and Edward Pound. 2004. 'A Mess of Missing Ordnance.' *U.S. News & World Report*. 31 October.
<http://www.usnews.com/usnews/news/articles/041108/8weapons.htm>

Roos, Peter. 2003. 'The Threat of the Unknown: What Felled M1A1 Abrams Tank? Mystery Puzzles Army Officials.' *Navy Times*. 3 November.

Schroeder, Matt. 2008. 'Black Market Missiles Still Common in Iraq.' *Missile Watch*, No. 3. 8 December.
<http://www.fas.org/programs/ssp/asmp/issueareas/manpads/Missile_Watch_3_Black_market_missiles_still_common_in_Iraq.pdf>

—. 2010. 'Iraq: Shoulder-fired Missile in Video of Insurgent Attack Could Be Iranian.' *Missile Watch*, Vol. 3, Iss. 2. June.

—. 2011. 'Holy Grails: Libya Loses Control of Its MANPADS.' *Jane's Intelligence Review*. May.

—, Rachel Stohl, and Dan Smith. 2007. *The Small Arms Trade*. Oxford: Oneworld Publications.

Shea, Dan. 2006. 'Raffica Special: The RPG-7 System.' *Small Arms Review*. December.

Sheikh, Abdi. 2009. 'Factbox: Prices of Arms in Mogadishu Market.' Reuters. 9 June.

Sherwell, Philip. 2004. 'British Forces Turn Blind Eye as Iraqis Sell Arms to Militia.' *Telegraph* (UK). 30 May.

SIPRI (Stockholm International Peace Research Institute). n.d. Arms Transfers Database. <http://www.sipri.org/databases/armstransfers>

Small Arms Survey. 2008. *Small Arms Survey 2008: Risk and Resilience*. Cambridge: Cambridge University Press.

—. 2010. *Fault Lines: Tracking Armed Violence in Yemen*. Yemen Armed Violence Assessment Number 1. May.

Smucker, Philip. 2007. 'Taliban Uses Weapons Made in China, Iran.' *Washington Times*. 5 June.

Solomon, Jay. 2011. 'Iran Funnels New Weapons to Iraq and Afghanistan.' *Wall Street Journal*. 2 July.

Somaiya, Ravi. 2010. 'Guerrilla Trucks: Why Rebels and Insurgent Groups the World over Love the Toyota Hilux Pickup as Much as Their AK-47s.' *Daily Beast*. 14 October.

Sowers, Joe. 2008. 'National Police Seize Cache Southeast of Baghdad.' 3rd Brigade Combat Team, 3rd Industry Division Public Affairs.
<http://www.dvidshub.net/news/18420/national-police-seize-cache-southeast-baghdad>

Starkey, Jerome. 2008. 'Drugs for Guns: How the Afghan Heroin Trade Is Fuelling the Taliban Insurgency.' *Independent*.
<http://www.independent.co.uk/news/world/asia/drugs-for-guns-how-the-afghan-heroin-trade-is-fuelling-the-taliban-insurgency-817230.html>

Teyie, Andrew and Dominic Wabala. 2011. 'Kenya: Eritrea Arming al Shabaab.' AllAfrica.com. 2 November.
<http://allafrica.com/stories/201111021319.html>

UCDP (Uppsala Conflict Data Program). 2009. *UCDP/PRIO Armed Conflict Dataset Codebook*.
<http://www.pcr.uu.se/digitalAssets/63/63324_Codebook_UCDP_PRIO_Armed_Conflict_Dataset_v4_2011.pdf>

UK MOD (United Kingdom Ministry of Defence). 2011. 'Data on Weapons Seized, Discovered, or Collected from Weapons Caches in or near Helmand Province in Afghanistan from September 2007 through September 2008.' Reference number 07-06-2011-111621-004. Received 5 December.

UNGA (United Nations General Assembly). 1997. *Report of the Panel of Governmental Experts on Small Arms*. A/52/298 of 27 August.
<http://www.un.org/Depts/ddar/Firstcom/SGreport52/a52298.html>

UNSC (United Nations Security Council). 1992a. Resolution 733. S/RES/733 of 23 January.

—. 1992b. Resolution 1992. S/RES/794 of 3 December.

—. 2002a. Resolution 1407. S/RES/1407 of 3 May.

—. 2002b. *Second Report of the Monitoring Group Established Pursuant to Security Council Resolution 1363 (2001) and Extended by Resolution 1390 (2002)*. S/2002/1050 of 20 September.

—. 2005. *Report of the Monitoring Group on Somalia Pursuant to Security Council Resolution 1587 (2005)*. S/2005/625 of 4 October.

—. 2006. *Report of the Monitoring Group on Somalia Pursuant to Security Council Resolution 1676 (2006)*. S/2006/913 of 22 November.

—. 2007. *Report of the Monitoring Group on Somalia Pursuant to Security Council Resolution 1724*. S/2007/436 of 18 July.

—. 2008a. *Report of the Monitoring Group on Somalia Pursuant to Security Council Resolution 1766 (2007)*. S/2008/274 of 24 April.

—. 2008b. Resolution 1844. S/RES/1844 of 20 November.

—. 2008c. *Report of the Monitoring Group on Somalia Pursuant to Security Council Resolution 1811*. S/2008/769 of 10 December.

—. 2009. Resolution 1907. S/RES/1907 of 23 December.

—. 2010. *Report of the Monitoring Group on Somalia Pursuant to Security Council Resolution 1853.* S/2010/91 of 10 March.
—. 2011. *Report of the Monitoring Group on Somalia and Eritrea Pursuant to Security Council Resolution 1916.* S/2011/433 of 18 July.
US ARCENT (US Army Central). 2010. 'Afghans Awarded in Counter IED Program.' Defense Video and Imagery Distribution System. 1 January.
—. 2011. 'Data on Weapons Seized, Discovered, or Collected from Weapons Caches in Afghanistan from September, 2006 through September, 2008.' Received 4 April.
—. n.d. Defense Video and Imagery Distribution System. <http://www.dvidshub.net/>
US Army (United States Army). 2011. 'Data on Weapons Seized, Discovered, or Collected from Weapons Caches in Afghanistan from September, 2006 through September, 2007.' Released 4 April.
US CENTCOM (United States Central Command). 2010. 'Documents and Photographs of "…Two Shoulder-fired Missiles, Nine Iranian-manufactured 240 mm Rockets, a Mortar Sight and an Aiming Tripod" Discovered in "a Weapons Cache near the Town of Salman Pak in Iraq's Wasit Province at 7:30 am" on September 9, 2008.' Released 10 May under the US Freedom of Information Act.
—. 2011a. 'Documents and Photographs of Weapons Seized from Arms Caches in or Near Baghdad from September 2008 until September 2009.' Released 19 May under the US Freedom of Information Act.
—. 2011b. 'Memorandum for Commander, United States Central Command.' 9 September.
USDoD IG (Inspector General of the United States Department of Defense). 2011. *Assessment of U.S. Government Efforts to Train, Equip, and Mentor the Expanded Afghan National Police.* Report No. SPO-2011-003. 3 March.
USGAO (United States Government Accountability Office). 2004. *Nonproliferation: Further Improvements Needed in U.S. Efforts to Counter Threats from Man-Portable Air Defense Systems.* May. <http://fas.org/asmp/resources/govern/GAO_04_519.pdf>.
—. 2009. *Afghanistan Security: Lack of Systematic Tracking Raises Significant Accountability Concerns about Weapons Provided to Afghan National Security Forces.* GAO-09-267. January.
US MNF-I (United States Multi-national Force–Iraq). 2007. 'Iranian Support for Lethal Activity in Iraq.' PowerPoint presentation. 11 February.
Vanden Brook, Tom. 2008. 'Troops Find Stash of Advanced Rocket Grenades in Afghanistan.' *USA Today.* 18 August.
VOA (Voice of America) News. 2011. 'Somalia Declares State of Emergency in Former al-Shabab Areas.' 14 August. <http://www.voanews.com/english/news/africa/Somali-Govt-Declares-State-of-Emergency-in-Former-al-Shabab-Areas-127679913.html>
Walsh, Declan. 2011. 'US Extends Drone Attacks to Somalia.' *Guardian.* 30 June. <http://www.guardian.co.uk/world/2011/jun/30/us-drone-strikes-somalia>
Wezeman, Pieter. 2010. 'Arms Flows and the Conflict in Somalia.' Stockholm: Stockholm International Peace Research Institute. October.
—, Pieter, Siemon Wezeman, and Lucie Béraud-Sudreau. 2011. 'Arms Flows to Sub-Saharan Africa.' SIPRI Policy Paper 30. Stockholm: Stockholm International Peace Research Institute. December.
Wright, Donald and Timothy Reese. 2008. *On Point II: Transition to the New Campaign—The United States Army in Operation IRAQI FREEDOM: May 2003–January 2005.* Washington, DC: US Government Printing Office.
Xinhua News Agency. 2011. 'Ethiopian Troops Amass at Common Border with Somalia.' 21 November.

ACKNOWLEDGEMENTS

Principal authors
Matt Schroeder and Benjamin King

Contributors
Chelsea Kelly

INDEX

A

accessories
    authorized transfers 2, 5–7, 269–75
        estimating 250–1
    costs 275
        definitions 245–6
        types of 261–9
Acioli, Patricia 70
Afghan National Security Forces (ANSF) 333
Afghanistan
    illicit transfers from, to Kazakhstan 123
    illicit weapons 2–3, 7, 313–14
        country of origin 330, 333
        data sources 315–16
        light weapons 333–6
        seized caches 330–6
        small arms 331–3
        sources 348–9
        trafficking 335
African Union Mission in Somalia (AMISOM) 337, 339, 340, 345–7
Ahlu Sunna wal Jama'a (ASWJ) (Somalia) 337, 346
aiming lasers 245
airline security, US to Puerto Rico 12–13
AK-47 assault rifles
    Central America 18
    drug violence, Mexico 17, 29
    illicit transfers of, to Jamaica 31
    Iraq 320
    use in homicides, Bogotá (Colombia) 29
AKM assault rifles, Iraq 321
Al-Kadissiya sniper rifles, Iraq 321
Al-Nassira rocket launcher 323
al-Shabaab (Somalia) 154, 337–8, 346
Albania, exports, UN Register 287
Alemão favela (Brazil), drug violence 65–6, 71
AMISOM see African Union Mission in Somalia
ammunition
    authorized transfers 242, 251
    destruction programmes, Kazakhstan 134–5

non-lethal armed violence 82
ammunition depots, explosions, Kazakhstan 108, 135, 136–8
AN/PAS-13 thermal sight 264, 267, 268
AN/PSQ-23 rangefinder 267, 268
Andorra, exports, UN Register 287
Anna LLP (Kazakhstan), civilian firearms 119
ANSF see Afghan National Security Forces
anti-tank guided weapons (ATGWs)
    illicit
        Afghanistan 334–6
        Iraq 326–9
        Somalia 340
    parts 244
anti-tank rounds, Iraq 323–5
anti-tank weapons, Somalia 341–2
Antigua and Barbuda, exports, UN Register 287
AR-15 assault rifles, drug violence, Mexico 29
Argentina
    diversion of firearms 32
    exports
        Comtrade data 275
        UN Register 287
    firearm homicides 9, 13, 16, 24
    gun buy-back programme 24
    homicide rates 15
    transparency 293
Armbrust rocket launchers, Iraq 323
armed conflict see conflict
armed robbery, at sea 192
armed violence
    see also non-lethal armed violence
    definition 81
    Somaliland 3, 151–3, 159, 162
Armenia, exports, to Kazakhstan 121
arms markets, Somalia 347
arms trade see authorized transfers; transfers
Arys (Kazakhstan), ammunition depot explosion 135, 136–8
assault rifles
    drug violence, Latin America 29
    parts 243–4

Somaliland 149, 152
Assault Weapons Ban (AWB) (United States) 57
ASWJ see Ahlu Sunna wal Jama'a
AT-4 single-shot rockets, Iraq 325
AT-7 anti-tank missiles, Somalia 340
ATGWs see anti-tank guided weapons
AUC see Autodefensas Unidas de Colombia
Australia
    exports
        Comtrade data 275
        UN Register 287
    transparency 293
Austria
    exports
        Comtrade data 274
        to Kazakhstan 120, 121
        UN Register 287
    transparency 291, 292, 302
authorized transfers
    see also transfers
    accessories 2, 5–7, 269–75
    ammunition 242, 251
    estimating 247–8
    global estimates 2, 5–7, 241–2, 251
    light weapons 242, 251
    parts 2, 5–7, 251–61
    small arms 242, 251
    trends 2000–2009 2, 275–7
Autodefensas Unidas de Colombia (United Self-Defence Forces of Colombia) (AUC), demobilization 22
AWB see Assault Weapons Ban

B

B-10 recoilless rifles, Somalia 342
Bakassi Boys (Nigeria) 163
ballistics, firearm identification 25
Baltic and International Maritime Council (BIMCO) 204
Bangladesh, exports, UN Register 287
Barre, Mohamed Siad 148, 150, 337
Batalhão de Operações Policiais Especiais (Special Police Operations Battalion) (BOPE) (Brazil) 66–7, 70–1

Belarus, exports, to Kazakhstan 121
Belgium
    exports
        Comtrade data 274
        to Kazakhstan 121
        parts 255
        UN Register 287
    imports, parts 256
    transparency 292, 298, 302, 305
Beltrame, José 66, 67
Beretta ARX160 246
Beretta pistols
    transfer of parts 257
    use in homicides, Bogotá (Colombia) 29
Best Management Practices for Protection against Somalia Based Piracy (BMP) 205–6, 211
Bharat Dynamics (India) 261
BIMCO *see* Baltic and International Maritime Council
BMP *see* Best Management Practices for Protection against Somalia Based Piracy
BMS *see* United Nations, Biennial Meeting of States
Bogotá (Colombia)
    firearm homicides 22
        ballistic identification 25
        types of weapons used 29–30
Bolivia, homicide rates 15
bolt-action rifles, Afghanistan 331
BOPE *see* Batalhão de Operações Policiais Especiais
Bosnia and Herzegovina
    exports, UN Register 287
    transparency 293, 300
Boutros-Ghali, Boutros 283
Brazil
    control measures 24
    diversion of firearms 32
    drug violence 3–4, 64–72
    exports
        Comtrade data 274
        to Kazakhstan 121
    firearm homicides 9, 13, 16
        types of weapons used 30–1
    homicide rates 15, 23–4

illicit arms trade 30
    transparency 293
brokering, reporting on 300
Browning pistols, use in homicides, Bogotá (Colombia) 29
Bulgaria
    exports
        Comtrade data 275
        to Kazakhstan 121
        UN Register 287
    transparency 293
Burao (Somaliland)
    armed violence 159, 160, 162
    security perceptions 155–6
    security providers 163, 166
    weapons holdings 149–50

C

Cabral, Sérgio 65
caches
    illicit weapons
        Afghanistan 330–6
        Iraq 317–29
Calderón, Felipe, and drug violence 2, 17, 41, 46, 48, 54, 72
Cali (Colombia), firearm homicides 22
Canada
    exports
        Comtrade data 274
        to Kazakhstan 121
        parts 259–60
        UN Register 287
    imports, parts 256, 261
    transparency 292, 302
Caribbean
    see also Latin America and the Caribbean
    homicide rates, trends 14–15, 16, 19–20
Carl Gustaf recoilless rifles
    parts 260
    Somalia 342
    thermal sights 273, 275
Carmel, Stephen M. 196
celebratory fire, cause of injury 86

Central America
    gangs 63–4
    homicide rates, trends 14–15
    Northern Triangle
        drug violence 60–4
        firearm homicides 17–18
        Mexican drug cartels 3–4, 61–2, 63–4
Chihuahua (Mexico), homicide rate 16–17
children, firearm injuries, South Africa 97–8
Childsafe South Africa 97
Chile
    firearm homicides 9, 13, 16, 24
    homicide rates 15
    imports, weapon sights 269–71
China
    exports
        Comtrade data 274
        parts 255
        weapon sights 269–71
    illicit transfers from, to Iraq 317
    transparency 293
civilian firearms, Kazakhstan 107, 116–28
civilian market, weapon sights, South America 269–71
Clip-on Sniper Night Sight 266
cocaine
    price correlation with violence 48, 49, 50
    route to US 63
COCOS COmmando COntrol System 268
codes of conduct, Somali pirates 202–3
Collective Security Treaty Organization (CSTO) 122
Colombia
    drug trafficking
        state response 47
        violence 48, 49
    exports, UN Register 287
    firearm homicides 9, 13, 16, 20–1
        reduction 22
        types of weapons used 29–30
    homicide rates 15, 20–2
    illicit weapons market 31
Colt revolvers, use in homicides, Bogotá (Colombia) 29
Comando Vermelho (CV) gang (Rio de Janeiro) 42, 44, 65

communal conflicts, Somaliland 166–8
computerization, record-keeping 228
Comtrade *see* United Nations, Commodity Trade Statistics Database
confidentiality, weapons tracing 231
conflict
    see also war zones
    definitions 315
    Somalia 337–8
conflict deaths, Somaliland 153–4
Connecticut Spring & Stamping Corporation (United States) 253
Costa Rica
    firearm homicides 13, 16, 18–19
        types of weapons used 26, 28
    homicide rates 15
craft guns, use in homicides, Latin America and the Caribbean 25–6
craft production
    Kazakhstan 122
    marking 226
crime, weapons-related, Kazakhstan 126, 127
criminal violence, Somaliland 159–66
Croatia
    exports
        Comtrade data 274
        UN Register 287
    transparency 291, 292, 300
CSTO *see* Collective Security Treaty Organization
Cuba
    firearm homicides 9, 13, 16
    homicide rates 15
    homicides, proportion using firearms 13
customs data, weapon parts 248
CV *see* Comando Vermelho
Cyprus
    exports
        Comtrade data 275
        to Kazakhstan 121
        UN Register 287
    transparency 293
Czech Republic
    exports
        Comtrade data 274
        to Kazakhstan 120, 121
        parts 255
        UN Register 287
    transparency 292, 300, 303

D

Dandong Xunlei Technology Company (China) 253
Danish Demining Group (DDG), Somaliland 150
Datamyne transfers 269, 276
DDG *see* Danish Demining Group
DDR *see* disarmament, demobilization, and reintegration
Defense Video & Imagery Distribution System 316
Denmark
    exports
        to Kazakhstan 121
        UN Register 287
    transparency 289, 292, 298, 300, 305
disarmament, demobilization, and reintegration (DDR), Somaliland 149
Djibouti Maritime Security Services (DMSS) 210
DMSS *see* Djibouti Maritime Security Services
Dominican Republic
    firearm homicides 13, 16, 19
    homicide rates 15
Dragunov SVD assault rifles, Iraq 320, 321, 322
drug trafficking
    'balloon effect'
        Central America 63, 64
        Rio de Janeiro 71
    conflict 48
    gangs 42
    links with illicit arms trade
        Kazakhstan 122–3
        Latin America and the Caribbean 31–2
    producers 43
    retailers 43–4
    state policy and response 47–51
    transhipment and smugglers 43
drug violence
    actors 42–4
    Central America 17–18, 60–4
    Latin America 41–72
    links with illicit arms trade 31–2
    Mexico 2, 3–4, 17, 29, 41, 46, 48, 49, 51–60
    police tactics, Rio de Janeiro 66–9
    reasons for 48
    Rio de Janeiro (Brazil) 41, 64–72
    types of firearms used 29, 55
    typology 44–6
DShK heavy machine guns
    Iraq 321
    Somalia 339
Dzhaksybekov, Adilbek 132

E

East Timor, firearm injuries 98–101
Ecuador
    firearm homicides 13, 16
        types of weapons used 26, 28–9
    homicide rates 15
Egal, Mohamed Haji Ibrahim 149, 157
Egypt, imports, parts 256
El Salvador
    drug violence 60–4
    exports, UN Register 287
    firearm homicides 9, 13, 16, 17–18
    homicide rates 15
embargoes, Somalia 337
Eritrea
    arms embargo 21
    illicit transfers from, to Somalia 344–7
Escobar, Pablo 45, 46, 47, 48
Ethiopia, illicit transfers from, to Somalia 344–7
ethnic violence, Kazakhstan 115
Euromissile 252
European Union (EU), Annual Report, transparency 285, 300, 303, 307
exports
    Comtrade data 2009 273–5
    national reporting 289, 306, 308
    parts 255, 259–60

F

F1 grenades, Somalia 340

Faina, MV, hijacking 176, 198

FAL rifles, Iraq 320

FAMAS F-1 rifles, Iraq 320–1

FARC *see* Fuerzas Armadas Revolucionarias de Colombia

FCS *see* fire control systems

Fiji, exports, UN Register 287

Finland
> exports
>> Comtrade data 274
>> to Kazakhstan 121
>> UN Register 287
> transparency 292, 299

fire-control systems (FCS) 245, 268–9

firearm homicides
> *see also* homicides
> Central America 17–18
> compiling statistics 10, 11
> gang violence 33
> impunity 34
> Latin America and the Caribbean 1–2, 3, 9–35
>> availability of weapons 32–3
>> origins of weapons used 29–32
>> patterns 11–13
>> trends 14–15
>> types of weapons used 9, 25–9
> non-lethal injuries comparison 91–4
> organized crime 33–4
> Small Arms Survey Database (1995–2010) 11

firearm injuries
> *see also* non-lethal armed violence
> costs 94
> surveillance systems 94–101

firearm suicides, United States 84, 87

firearm violence
> lethal and non-lethal 91–4, 97
> Somalia 160–1, 162

FN Herstal (Belgium), weapons used in drug trafficking, Mexico 29

FOIA *see* United States, Freedom of Information Act

Fox, Vicente 54

FPK sniper rifles, Iraq 321

France
> exports
>> Comtrade data 274
>> to Kazakhstan 121
>> parts 259–60
>> UN Register 287
> imports, parts 256
> transparency 292, 299, 303

Freixo, Marcelo 70

Fuerzas Armadas Revolucionarias de Colombia (FARC), drug trafficking 42

G

G3 assault rifles, Iraq 320, 322

Galkayo (Somalia), armed violence 160, 162

gang violence, drug-related 17–18, 19–20

gangs
> drug trafficking 43–4
> drug violence, Central America 63–4
> firearm homicides 33
>> Caribbean 19–20
>> Central America 18
> Somaliland 159

Geneva Declaration on Armed Violence and Development (2006) 1

Georgia, exports, UN Register 287

Germany
> exports
>> Comtrade data 274
>> to Kazakhstan 120, 121, 134
>> parts 255
>> UN Register 287
> imports, parts 256
> transparency 292, 299, 300, 302, 305

Ghana, exports, UN Register 287

Glock (Austria), exports, parts 257

Glock pistols, use in homicides, Bogotá (Colombia) 29

Greece
> exports, UN Register 287
> transparency 291, 292, 302

grenades, Somalia 340

Grupo Aeromóvil de Fuerzas Especiales (Mexico) 54, 71

Guatemala
> drug violence 60–4
> firearm homicides 9, 13, 16, 17–18
>> impunity 34
>> types of weapons 25
> homicide rates 15

guided light weapons systems, Somalia 339–40

Guyana
> exports, UN Register 287
> firearm homicides 13, 16
> homicide rates 15

Guzmán, Joaquín 'El Chapo' 63

H

Haiti, exports, UN Register 287

hand grenades, Somalia 340

HDI *see* Human Development Index

HEAT *see* high-explosive, anti-tank rounds

heavy machine guns (HMGs), Iraq 321

high-explosive, anti-tank (HEAT) rounds, Iraq 323–4

HN-5 MANPADS, Afghanistan 334

holographic sights 266

homemade guns *see* craft guns

homicides
> *see also* firearm homicides
> drug-related 51
> proportion using firearms, Latin America and the Caribbean 11–13, 16
> trends, Latin America and the Caribbean 14–15

Honduras
> drug trafficking, and illicit arms trade 31–2
> drug violence 60–4
> firearm homicides 9, 13, 16, 17–18
>> types of weapons used 28
> homicide rates 15

Hong Kong, exports, weapon sights 270–1

hostages, Somali pirates 200

Human Development Index (HDI), Kazakhstan 109

Hungary
    exports
        Comtrade data 275
        UN Register 287
    transparency 292, 300
hunting weapons, Kazakhstan, imports 120

I

IBIS *see* Integrated Ballistics Identification System
ICD *see* World Health Organization, International Classification of Diseases
ICU *see* Islamic Courts Union
ICVS *see* International Crime Victims Survey
IEDs *see* improvised explosive devices
illicit transfers
    Afghanistan 335
    Brazil 30
    Iraq 328
    links with drugs trafficking
        Kazakhstan 122–3
        Latin America and the Caribbean 31–2
    parts, marking 226
illicit weapons
    Afghanistan 2–3, 7
        data sources 315–16
        light weapons 333–6
        seized caches 330–6
        small arms 331–3
    data sources 315–17
    definition 314
    Iraq 2–3, 7
        data sources 315–16
        light weapons 322–9
        seized caches 317–29
        small arms 319–22
    Kazakhstan 122–3
    Somalia 2–3, 7, 336–47
    sources 348–9
    war zones 2–3, 7, 313–49
image-intensifying sights 266
IMB *see* International Maritime Bureau
IMI *see* Israel Military Industries
IML *see* Instituto de Medicina Lega

IMO *see* International Maritime Organization
import marks, weapons tracing 225
imports
    Kazakhstan 119–22, 133–4
    parts 248–50, 256, 260–1
improvised explosive devices (IEDs)
    Afghanistan 330, 349
    Iraq 319, 327
India
    exports, Comtrade data 274
    imports, parts 260–1
    transparency 293
INDUMIL *see* Industria Militar (INDUMIL)
Industria Militar (INDUMIL) (Colombia), Llama revolver 30
injuries *see* non-lethal armed violence
Institutional Revolution Party *see* Partido Revolucionario Institucional
Instituto de Medicina Lega (Institute of Forensic Medicine) (IML) (El Salvador) 10
Integrated Ballistics Identification System (IBIS), Latin America 25
International Convention for the Safety of Life at Sea 211
International Crime Victims Survey (ICVS), non-lethal armed violence 89–90, 91
International Maritime Bureau (IMB), Somali piracy data 192, 195
International Maritime Organization (IMO), Somali piracy 191, 205–6, 208–9, 211
International Tracing Instrument (ITI) 1
    cooperation 229–31
    international assistance and capacity-building 234–6
    marking 223–7
    national frameworks 231–3
    record-keeping 227–9
    regional cooperation 233–4
    UN Meeting of Governmental Experts 5, 219, 223, 236–7
interpersonal violence, definition 81
interpolation method, transfer estimates 247–8
Iquique Free Zone, exports, weapon sights 270–1

Iran
    illicit transfers from
        to Afghanistan 335
        to Iraq 317, 318, 327, 328
    transparency 283, 289, 293, 298
Iraq
    illicit transfers, from Iran 317, 318, 327, 328
    illicit weapons 2–3, 7
        categories of weapon 317–18
        country of origin 317, 318
        data sources 315–16
        portable missiles 326–9
        rocket-propelled grenades and shoulder-fired rockets 322–6
        seized caches 317–29
        serviceability 318, 329
        small arms 319–22
        sources 348–9
        trafficking 328
    stockpiles, looting 322, 348
iron sights 264–6
Islamic Courts Union (ICU) (Somalia) 198, 337
Israel
    exports
        Comtrade data 274
        to Kazakhstan 120, 121
        parts 255
    transparency 293
Israel Military Industries (IMI), joint production with Kazakhstan 133
Italy
    exports
        Comtrade data 274
        to Kazakhstan 120, 121
        parts 255
        UN Register 287
        weapon sights 270–1
    imports, parts 256
    transparency 289, 292, 300
ITI *see* International Tracing Instrument

J

Jamaica
    exports, UN Register 287

firearm homicides 9, 13, 16, 19–20
    types of weapons 25
firearm injuries, costs of 94
homicide rates 15
illicit weapons market 31
Japan
    exports
        Comtrade data 274
        parts 255
        weapon sights 270–1
    imports, parts 260
    transparency 293
Jericho pistols, use in homicides, Bogotá (Colombia) 29
Jonglei State (Sudan), firearm injuries 99–100, 101

K

Kaibiles (Guatemala) 62, 64
Kalashnikov-pattern rifles
    Afghanistan 331, 333
    Iraq 320, 321
    production, Venezuela 23
    Somalia 338–9
    Somaliland 149, 152, 168
Kalshale (Somaliland), land dispute 152, 154, 157–8, 166, 167
Kanal operation (Kazakhstan) 122
Karaoy (Kazakhstan), ammunition depot explosion 135, 136–8
Kazakhstan
    ammunition, destruction 134–5
    ammunition depots, explosions 108, 135, 136–8
    civilian firearms 107, 116–28
        authorized sources 119–22
        control measures 123–7
        illicit sources 122–3
        impact 127–8
    control measures
        civilian firearms 123–7
        state holdings 134–5
    crime 110–11, 126, 127
    destruction programmes 108
    exports 133
    homicide rate 108, 110, 111

illicit weapons, sources 122–3
imports
    parts 256
    small arms 119–22, 133–4
organized violence 114–16
private security companies 118–19
security outlook 4, 109–16
state holdings
    control measures 134–5
    sources 131–4
    stockpiles 129–31
victimization surveys 111–14
weapons collection 124–7
Kazakhstan Engineering, civilian firearms 119
KazArsenal (Kazakhstan) 134–5, 136–7
Kazspetseksport (Kazakhstan) 133
Kenya
    firearm injuries
        costs of 94
        victimization survey 90
KNB *see* National Security Committee
Kongsberg remote turret 260
Kyrgyzstan
    exports, to Kazakhstan 121
    illicit transfers from, to Kazakhstan 123

L

land disputes, Somaliland 166–7
Las Anod (Somalia), armed violence 160, 162
laser rangefinders 245, 268
laser sights 267–8
Latin America and the Caribbean
    availability of firearms 32–3
    drugs trafficking, links with illicit firearms 31–2
    firearm homicides 1–2, 3, 9–35
        origins of weapons used 29–32
        patterns 11–13
        trends 14–15
        types of weapons used 25–9
    gang violence 33
    homicides, proportion using firearms 11–13, 16

Latvia
    exports
        to Kazakhstan 121
        UN Register 287
Lebanon, exports, UN Register 287
Lebel rifles, Iraq 320, 321
Lee-Enfield rifles
    Afghanistan 331
    Iraq 320
legislation
    Kazakhstan 123–5
    record-keeping 227
Liberia, firearm violence, data recording 98
Liechtenstein, exports, UN Register 288
light weapons
    accessories 245–6, 261–9
    authorized transfers 2, 242, 251
    definitions 5, 243, 314–15
    illicit
        Afghanistan 333–6
        Iraq 322–9
        Somalia 339–42
    parts 244
    transfers 258–61
Lithuania
    exports, UN Register 288
    transparency 293
Llama revolvers, use in firearm homicides, Colombia 29–30
Los Zetas (Mexico) 17–18, 31, 54, 55–6, 71
    activity in Central America 61, 62
Lothar Walther (Germany) 253
Luxembourg
    exports, UN Register 288
    transparency 293

M

M-72 anti-tank weapon, Somalia 342
M16 assault rifles, illicit transfers of, to Jamaica 31
M68 Close Combat Optic (CCO) 267, 268
M110 semi-automatic sniper system 275
M150/M151 fire control system 275

machine guns
    Afghanistan 331–3
    drug violence, Latin America 29
    Iraq 321, 322
    Somalia 339
McLay, Jim 222, 236
Maersk Alabama, hijacking 198
Mali, exports, UN Register 288
Malta, exports, UN Register 288
man-portable air defence systems (MANPADS)
    illicit
        Afghanistan 334, 348–9
        Iraq 326–9, 348–9
        Somalia 339–40
    illicit transfers, Kazakhstan 122
    parts 244
MANPADS *see* man-portable air defence systems
maras, Central America 18, 64
marking, UN Open-ended Meeting of Governmental Experts 223–7
Mauser 98 rifles, Iraq 320, 321
Medellín (Colombia), firearm homicides 22
medical care, non-lethal armed violence 82–3
Meeting of Governmental Experts *see* United Nations, Open-ended Meeting of Governmental Experts
Metallist Urlask Plant (Kazakhstan) 131
Mexican Mafia (MM) (United States), drug violence 42, 44
Mexico
    cocaine 49, 50
    drug trafficking 43
        cartel typology 56
        cartels in Central America 61–2, 63–4
    drug violence 2, 3–4, 41, 46, 48, 49, 51–60
        firearms 29, 55, 56–60
    exports
        Comtrade data 274
        parts 255, 257–8
        UN Register 288

    firearm homicides 13, 16–17
    homicide rates 15
    illicit transfers to, from the United States 56–60
    transparency 291, 293
    victimization survey 90
MG42 machine guns, Iraq 322
MGE *see* United Nations, Open-ended Meeting of Governmental Experts
MILAN ATGWs
    Afghanistan 334–6
    production 252, 260–1
    Somalia 339–40
milícias paramilitaries, Rio de Janeiro (Brazil) 70–1
military firearms, parts, transfers 258–61
military procurement, accessories 272–5
Misagh-1 MANPADS, Iraq 327
MM *see* Mexican Mafia
modular weapons 246
Mogadishu (Somalia), armed violence 160, 162
Moldova, exports, UN Register 288
Mongolia, exports, UN Register 288
Montenegro
    exports, UN Register 288
    transparency 292, 302, 305
Montevideo Free Zone, exports, weapon sights 270–1
mortar bombs, parts 244
mortars, parts 244
Mosin-Nagant rifles, Iraq 320
Mossberg, O. F. & Sons (United States), shotgun barrels, tranfers 257–8

N

NADER anti-tank rockets, Iraq 324
National Electronic Injury Surveillance System (NEISS) (United States) 95, 96
National Injury Mortality Surveillance System (NIMSS) (South Africa) 96, 97

National Security Committee (KNB) (Kazakhstan) attacks on 107, 116, 136
Nazarbayev, Nursultan 109, 116, 132
neighbourhood watch groups (NWGs), Somaliland 163, 166
NEISS *see* National Electronic Injury Surveillance System
Nelegal operation (Kazakhstan) 122
Netherlands
    exports
        Comtrade data 275
        UN Register 288
    non-fatal firearm injuries 96
    transparency 292, 298, 299, 300
New Zealand, exports, UN Register 288
Nicaragua
    firearm homicides 13, 16, 18–19
    homicide rates 15
Nigeria, non-lethal armed violence 82, 83
night-vision equipment 245
NIMSS *see* National Injury Mortality Surveillance System
non-lethal armed violence 2, 4, 79–101
    access to medical care 82–3
    ammunition used 82
    comparison with fatal injuries 91–4
    costs of injuries 94
    data sources 87–91
    intentional 84–7
    monitoring systems 80–1, 94–101
    severity 81–2
    stray bullets 86
    terminology 81
Norinco (North Industries Corporation) (China), weapons seized in Rio de Janeiro (Brazil) 30
North Korea, transparency 283, 289, 293, 298
Norway
    exports
        Comtrade data 274
        parts 259–60
        UN Register 288
    imports, parts 261
    transparency 289, 292, 299, 300
Nurkadilov, Zamanbek 115
NWGs *see* neighbourhood watch groups

O

Observatory of Conflict and Violence Prevention (OCVP) (Somaliland) 150, 154–5, 159–60, 168

Oceans Beyond Piracy 196

OCVP *see* Observatory of Conflict and Violence Prevention

Olive Group 321, 322

Operation Fast and Furious 59–60

Organization for Security and Co-operation in Europe (OSCE), Kazakhstan presidency 109

organized crime
- Cuba 13
- firearm homicides 33–4

organized violence, Kazakhstan 114–16

Ortaderesin (Kazakhstan), ammunition depot explosion 136–8

OSCE *see* Organization for Security and Co-operation in Europe

P

'pacification' police units *see* unidades de polícia pacificadora

Pakistan
- exports, to Kazakhstan 121
- illicit transfers from, to Afghanistan 335
- transparency 293

Panama
- exports, UN Register 288
- firearm homicides 9, 13, 16, 18–19
  - types of weapons 25
- homicide rates 15

pandillas, Central America 18

Paraguay
- firearm homicides 13, 16, 24
- homicide rates 15
- illicit transfers 30
- imports, weapon sights 269–71

Partido Revolucionario Institucional (Institutional Revolution Party) (PRI) (Mexico) 47, 52–4, 72

parts
- authorized transfers 2, 5–7, 251–61
- estimating 248–50
- definitions 243–4, 246
- repair and maintenance 254
- supply chains 252–3

Pasper (Argentina), weapons involved in crime 32

Peru
- firearm homicides 9, 13, 16
- homicide rates 15
- imports, weapon sights 269–71

PG-7 rockets 324

PG-7L ant-tank rounds 324–5

Philippines
- exports
  - Comtrade data 275
  - UN Register 288
  - weapon sights 270–1
- transparency 293

piracy
- definition 192
- private security companies 191–2, 204–12
- Somali 3, 5, 174–89, 191–213
  - high-risk area 194
  - pirate groups 192–5
  - ransoms 195, 206–7
  - rules of behaviour 202–3
  - trends 195–6
  - violence 200–2
  - weapons 197–9

pistols
- criminal use, Latin America and the Caribbean 25–32
- illicit, Iraq 320, 322
- parts, transfers 254–7
- Somaliland 152

PK machine guns, Somalia 339

PKM light machine guns, Afghanistan 331

PMK rifles, Iraq 321

PoA *see* United Nations, Programme of Action to Prevent, Combat and Eradicate the Illicit Trade in Small Arms and Light Weapons in All Its Aspects

Pointer system 263

Poland
- exports
  - Comtrade data 274
  - to Kazakhstan 121
  - UN Register 288
- transparency 289, 292, 302–3

police, anti-drug strategy, Rio de Janeiro 65–70

political violence
- Kazakhstan 114–15
- Somaliland 156–9

polymer frame firearms, marking 224

Portugal
- exports
  - Comtrade data 274
  - UN Register 288
- imports, parts 256
- transparency 292, 299

PRI *see* Partido Revolucionario Institucional

private security companies (PSCs)
- Kazakhstan 118–19
- Somali piracy 3, 5, 191–2, 204–12
  - firearm procurement 208–11
  - firearm types 207–8
  - firearm use 211–12

production, parts 252–3

Programa Nacional de Entrega Voluntaria de Armas de Fuego (Argentina) 24

PSCs *see* private security companies

public health data, firearm injuries 87

Puerto Rico
- firearm homicides 9, 12–13, 16
- homicide rates 15
- illicit transfers to, from United States 12–13

Q

qaat addiction, Somaliland 159

QinetiQ (United Kingdom) 263

Qioptiq (United Kingdom) 263

QW-1 MANPADS, Iraq 327

R

rangefinders *see* laser rangefinders

ransom payments, Somali pirates 195

Rathore, Dipendra 175
recoilless rifles, Somalia 342
record-keeping, UN Open-ended Meeting of Governmental Experts (MGE) 227–8
reflex sights 266
regional cooperation 233–4
repair and maintenance, weapon parts 254
Revolutionary Armed Forces of Colombia (FARC) see Fuerzas Armadas Revolucionarias de Colombia
revolvers
  criminal use, Latin America and the Caribbean 25–32
  parts, transfers 254–7
Rheinmetall DeTec (Germany), Vingmate fire-control system 268–9
rifles
  illicit
    Afghanistan 331, 332
    Iraq 320–1, 322
    Somalia 338–9
Rio de Janeiro (Brazil)
  drug violence 3–4, 41, 64–72
  homicide rates 70
  milícias paramilitaries 70–1
  police
    anti-drug tactics 66–70
    killings of civilians 65–6
  seizures of illicit weapons 30–1
  rocket launchers, parts 244
  rocket-propelled grenade launchers (RPGs)
    illicit
      Afghanistan 313
      Iraq 322–6
      Somalia 341–2
    optical sights, Iraq 325
    rounds, Iraq 323–5
rockets, seizures in Iraq 324–6
Romania
  exports
    Comtrade data 274
    UN Register 288
  transparency 283, 289, 292, 298, 299, 305, 306–7
Rossi (Brazil), weapons seized in Rio de Janeiro 30
RPG-7 rocket-propelled grenade launcher
  Iraq 323–4
  Somalia 341
RPG-16 rocket launchers, Iraq 323
RPG-18 rocket launchers, Iraq 323
RPG-22 rocket launchers, Iraq 323
RPG-29 Vampir rocket launchers, Iraq 313, 326
RPGs see rocket-propelled grendade launchers
Ruger (United States)
  weapons seized in Rio de Janeiro (Brazil) 30
  weapons used in Bogotá (Colombia) 29
Russian Federation
  exports
    Comtrade data 274
    to Kazakhstan 120, 121
  illicit transfers from
    to Iraq 317
    to Kazakhstan 123
  transparency 293

S

SA-7 MANPADS
  Iraq 327, 329
  Somalia 340
SA-14 MANPADS, Iraq 327
SA-16 MANPADS, Iraq 327
SA-18 Igla MANPADS, Somalia 340
SA-18 MANPADS, illicit supply 122
Sagger missiles
  Afghanistan 334
  Iraq 327
  Somalia 339–40
Saint Lucia, exports, UN Register 288
São Paulo (Brazil), homicide rates 23–4
Sarsenbayev, Altynbek 114
Saudi Arabia, transparency 293, 298
Sea Scorpion 210
Senegal, exports, UN Register 288
Serbia
  exports
    Comtrade data 274
    UN Register 288
  transparency 292, 299–300, 302, 305, 306–7
Serbia and Montenegro, exports, to Kazakhstan 121
sexual assault, Somalia 161–2
shipping
  see also piracy
  private security companies 204–12
shotguns, barrels, transfers 257–8
shoulder-fired rocket launchers, illicit, Iraq 322–6
Sig Sauer pistols, use in homicides, Bogotá (Colombia) 29
sights see weapon sights
'Silanyo', Mohamed Mohamoud 157, 167
Sinaloa (Mexico), drug cartel 43, 56
Singapore
  exports, Comtrade data 275
  transparency 293
Sirius Star, MV, hijacking 198
SK-10 launcher, Iraq 327, 329
SKS automatic rifles, Iraq 320, 321
Slovakia
  exports, UN Register 288
  transparency 292, 305
SLPF see Somaliland Police Force
small arms
  accessories 261–9
  authorized transfers 2, 242, 251
    estimation techniques 247–8
  definitions 5, 243, 314
  illicit
    Afghanistan 331–3
    Iraq 319–22
    Somalia 338–9
  parts 243–4
Small Arms Trade Transparency Barometer
  2012 update 2, 7, 283–4, 289–94
  background 284–9
  parameters 284–5, 295
  trends 2001–2010 294–308
    access and consistency 296–8
    brokering activities 300
    clarity 298–300
    comprehensiveness 300–2
    deliveries 302–3

licences granted 303–4, 305
licences refused 304–6
timeliness 295–6
Smith & Wesson (S&W) (United States)
weapons seized in Rio de Janeiro (Brazil) 30
weapons used in Colombia 29–30
SNA *see* Somali National Army
sniper rifles, Iraq 320, 321, 322
SNM *see* Somali National Movement
Somali National Army (SNA) 148, 157
Somali National Movement (SNM) 148, 150, 157
Somalia
arms embargoes 337
arms markets 347
civil war 148, 156–7
criminal violence 160
illicit transfers to 344–7
illicit weapons 2–3, 7, 313–14, 336–47
data sources 315, 336–8
light weapons 339–42
small arms 338–9
sources 348–9
trafficking 344–7
piracy 3, 5, 174–89, 191–213
high-risk area 194
pirate groups 192–5
ransoms 195, 206–7
rules of behaviour 202–3
trends 195–6
violence 200–2
weapons 197–9
stockpiles, looting 3, 339, 347, 349
Somaliland
armed violence 4, 151–3, 159, 162
clan system 166–7
communal violence 166–8
conflict 148, 153–4, 156–9
conflict resolution 167–8
criminal violence 159–66
disarmament programmes 149
homicide rates 154
local security providers 162–6
political violence 156–9
security perceptions 154–6
security situation 147–68
small arms holdings 149–51, 152
Somaliland Police Force (SLPF) 159, 164, 166

Sool, Sanaag, and Cayn militia (SSC) (Somaliland) 152, 157–9
South Africa
firearm violence, lethal and non-lethal 96, 97, 98
transparency 293, 298–9
South America
see also Latin America
homicide rates, trends 14–15
South Korea
exports
Comtrade data 274
to Kazakhstan 121
parts 255, 259–60
UN Register 288
weapon sights 270–1
imports, parts 260
transparency 293
South-east Europe, exports, regional reporting 306
Southern Cone, firearm homicides 24
Spain
exports
Comtrade data 274
to Kazakhstan 121
UN Register 288
weapon sights 270–1
imports, parts 256, 260
transparency 292, 298, 302
Special Police Operations Battalion *see* Batalhão de Operações Policiais Especiais
sporting firearms, parts, transfers 254–8
Sri Lanka, private security companies, weapons rental 210–11
SSC *see* Sool, Sanaag, and Cayn militia
Steyr firearms, Kazakhstan 119
Stinger missiles, Afghanistan 336, 348
stockpile security, Iraq 322
stockpiles
Kazakhstan
diversions from 122
state holdings 129–31
Somalia, looting of 3, 339, 347, 349
Stoeger Silah Sanayi (Turkey) 257
stray bullets, cause of injury 86
Suez Canal, firearms on ships 209

supply chains, weapons parts 252–3
Suriname
firearm homicides 9, 13, 16
homicide rates 15
Swaziland, exports, UN Register 288
Sweden
exports
Comtrade data 274
to Kazakhstan 121
parts 259–60
UN Register 288
transparency 292, 298, 299
Switzerland
exports
Comtrade data 274
to Kazakhstan 121, 134
UN Register 288
transparency 283, 289, 292, 299, 301, 303, 304
Syria, illicit transfers from, to Iraq 328

T

Tabuk rifles, Iraq 320, 321
Taiwan
exports, Comtrade data 274
transparency 293
Tajikistan
illicit transfers from
to Afghanistan 335
to Kazakhstan 123
Taliban
drug trafficking 42, 43
weapons holdings 333, 334–6
Taurus (Brazil)
weapons seized in Rio de Janeiro (Brazil) 30
weapons used in Bogotá (Colombia) 29
technology, weapon accessories 262–3
telescopic sights 266
terrorist violence, Kazakhstan 115–16
TFG *see* Transitional Federal Government
Thailand
imports, parts 261
transparency 293
thermal sights 266–7, 272–3

Timor-Leste *see* East Timor

Togo, exports, UN Register 288

Tokyrau (Kazakhstan), ammunition depot explosion 135, 136–8

tracing, UN Open-ended Meeting of Governmental Experts (MGE) 229–31

trafficking
see also drug trafficking
Afghanistan 335
Iraq 328
Somalia 344–7

transfers
see also authorized transfers
background information 286–9
Comtrade data 2009 273–5
parts and accessories 241–77
transparency 276

Transitional Federal Government (TFG) (Somalia) 337, 339, 345–7

transparency
see also Small Arms Trade Transparency Barometer
transfers 276
trends 2001–2010 2, 294–308

Trinidad and Tobago
exports, UN Register 288
firearm homicides 13, 16, 20
homicide rates 15

Turkey
exports
Comtrade data 274
to Kazakhstan 121
parts 255
UN Register 288
imports, parts 260
transparency 293

Turkmenistan, exports, to Kazakhstan 121

Tuur, Abdirahman Ahmed Ali 148–9, 157

Type 56 assault rifles, Iraq 321

U

Uganda, injury data 87

Ukraine
exports
to Kazakhstan 121

UN Register 288
transparency 293

UN Register *see* United Nations Register of Conventional Arms

unidades de polícia pacificadora ('pacification' police units) (UPPs) (Brazil) 66–9, 71

United Arab Emirates (UAE)
exports, to Kazakhstan 121
transparency 283, 289, 293

United Kingdom (UK)
exports
Comtrade data 274
to Kazakhstan 121
parts 259–60
UN Register 288
imports, parts 256, 260–1
military procurement, accessories 272–3
transparency 283, 289, 292, 298, 299, 300, 301, 302

United Nations (UN)
Biennial Meetings of States (BMS) 220–1
Commodity Trade Statistics Database (Comtrade)
2009 analysis 273–5
Kazakhstan imports 119, 120–2
Small Arms Trade Transparency Barometer 285–6
weapon parts 247, 249–50, 251, 254–5, 258–9
weapon sights 250
Convention on the Law of the Sea 212
embargoes, Somalia 337
Open-ended Meeting of Governmental Experts (MGE) 3, 5, 219–37
findings 236–7
history 220–3
implementation of International Tracing Instrument 5, 223
international assistance and capacity-building 234–6
marking of firearms 223–7
record-keeping 227–8
regional cooperation 233–4
tracing 229–31
parts and accessories 250, 251
Programme of Action to Prevent, Combat and Eradicate the Illicit Trade

in Small Arms and Light Weapons in All Its Aspects (PoA), implementation 1, 7

United Nations Monitoring Group on Somalia and Eritrea (UNSEMG), illicit weapons data 315, 336–7, 338–47

United Nations Register of Conventional Arms
exports 2003-2010 286–9
Kazakhstan 120

United Nations Security Council (UNSC), Contact Group on Piracy off the Coast of Somalia 199

United States (US)
exports
Comtrade data 274
to Kazakhstan 120, 121, 134
parts 255, 259
weapon sights 270–1
firearm injuries
costs of 94
intentional 84, 91
surveillance systems 95
Freedom of Information Act (FOIA) 316
illicit transfers from
to Mexico 56–60
to Puerto Rico 12–13
imports, parts 256, 257–8, 260–1
stray bullets 86
transfers 275–6
transparency 289, 291, 292, 299
victimization survey data 91
weapons control measures 57

UNSEMG *see* United Nations Monitoring Group on Somalia and Eritrea

UPPs *see unidades de polícia pacificadora*

Uruguay
exports, weapon sights 270–1
firearm homicides 9, 13, 16, 24
homicide rates 15
imports, weapon sights 269–71

V

Venezuela
firearm homicides 9, 13, 16, 20–3
homicide rates 15

victimization surveys

Kazakhstan 111–14
non-lethal armed violence 89–91
vigilante groups, community security 163
Vingmate fire-control system 268–9
VZ 58 rifles, Iraq 321

W

Walther pistols, use in homicides, Bogotá (Colombia) 29
war zones
see also conflict
illicit weapons 2–3, 7, 313–49
WAVE 300-Tolkyn small arms system 133
weapon sights
definition 245
transfers
estimates 250
military procurement 272–5
South America 269–71
types 264–8
weapons collection programmes, Kazakhstan 124–7
weapons registers 228
Western Kazakhstan Machine-Building Company (ZKMK) 119, 131–3
WHO *see* World Health Organization
women, violence against, Somalia 161–2
World Health Organization (WHO), International Classification of Diseases (ICD) 80–1, 96

Y

Yemen
armed violence 101
illicit transfers from, to Somalia 344–7

Z

ZKMK *see* Western Kazakhstan Machine-Building Company